STRUCTURAL DESIGN IN WOOD

STRUCTURAL DESIGN IN WOOD

Judith J. Stalnaker, Ph.D., P.E.
University of Colorado at Denver

and

Ernest C. Harris, Ph.D., P.E.
University of Colorado at Denver

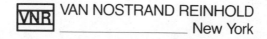

VAN NOSTRAND REINHOLD
New York

Thirteen design tables in this book (specifically Tables C-1, C-2, C-3, C-4, C-5, C-7, C-8, C-9, C-12, C-15, C-16, and C-17 from Appendix C and Table B-7 from Appendix B) were taken from the National Design Specification for Wood Construction, 1986 edition. Users are advised to obtain a complete copy of that publication for use in wood structural design. Copies are available in technical bookstores as well as from the Publications Department of the National Forest Products Association, 1250 Connecticut Avenue N.W., Washington, D.C. 20036.

Copyright © 1989 by Van Nostrand Reinhold
Library of Congress Catalog Card Number   88-20842
ISBN 0-442-23300-0

Printed in the United States of America

Van Nostrand Reinhold
115 Fifth Avenue
New York, New York 10003

Van Nostrand Reinhold International Company Limited
11 New Fetter Lane
London EC4P 4EE, England

Van Nostrand Reinhold
480 La Trobe Street
Melbourne, Victoria 3000, Australia

Nelson Canada
1120 Birchmount Road
Scarborough, Ontario M1K 5G4, Canada

16 15 14 13 12 11 10 9 8 7 6 5 4 3 2

Library of Congress Cataloging-in-Publication Data

Stalnaker, Judith J.
    Structural design in wood.

    Includes index.
    1. Building, Wooden. 2. Structural design.
I. Harris, Ernest C. II. Title.
TA666.S67   1989   624.1'84   88-20842
ISBN 0-442-23300-0

Structural Engineering Theory

STRUCTURES: FUNDAMENTAL THEORY AND BEHAVIOR by Richard Gutkowski

STRUCTURAL DYNAMICS, 2nd Ed., by Mario Paz

MICROCOMPUTER-AIDED ENGINEERING: STRUCTURAL DYNAMICS by Mario Paz

EARTHQUAKE RESISTANT BUILDING DESIGN AND CONSTRUCTION, 2nd Ed., by Norman Green

SEISMIC DESIGN FOR BUILDINGS HANDBOOK edited by Farzad Naeim

Steel Design

STEEL DESIGN FOR ENGINEERS AND ARCHITECTS by Rene Amon, Bruce Knobloch and Atanu Mazumder

Concrete Design

HANDBOOK OF CONCRETE ENGINEERING by Mark Fintel

STRUCTURAL DESIGN GUIDE TO THE ACI BUILDING CODE, 3rd Ed., by Paul F. Rice, Edward S. Hoffman, David P. Gustafson and Albert J. Gouwens

TORSION OF REINFORCED CONCRETE by T. Hsu

MODERN PRESTRESSED CONCRETE, 3rd Ed., by James R. Libby

Masonry

SIMPLIFIED MASONRY SKILLS by R. Kreh

Wood

MECHANICS OF WOOD AND WOOD COMPOSITES by Jozsef Bodig and Benjamin Jayne

STRUCTURAL USE OF WOOD IN ADVERSE ENVIRONMENTS edited by Robert W. Meyer and Robert M. Kellogg

STRUCTURAL DESIGN IN WOOD by Judith J. Stalnaker and Ernest C. Harris

Tall Buildings

DEVELOPMENTS IN TALL BUILDINGS by the Council on Tall Buildings

ADVANCES IN TALL BUILDINGS by the Council on Tall Buildings

SECOND CENTURY OF THE SKYSCRAPER by the Council on Tall Buildings

Other Structures-Related Books

ICE INTERACTION WITH OFFSHORE STRUCTURES by A.B. Cammaert and
   D.B. Muggeridge
FOUNDATION ENGINEERING HANDBOOK by Hans Winterkorn and H.Y. Fang
FOUNDATION ENGINEERING FOR DIFFICULT SUBSOIL CONDITIONS by
   Leonardo Zeevaert
ADVANCED DAM ENGINEERING edited by Robert Jansen
TUNNEL ENGINEERING HANDBOOK by John Bickel and T.R. Kuesel
SURVEYING HANDBOOK edited by Russell Brinker and Roy Minnick

Preface

Why another textbook on the design of wood structures? In many years of teaching structural design in wood, the authors have used virtually every textbook available, as well as using only a code and no textbook at all. The textbooks used have included both the old and the relatively modern; some have been fairly good, but in our opinion each has deficiencies. Some books have too few solved examples. Others omit important material or have an arrangement making them difficult to use as formal teaching tools. By writing this book, we intend to correct such deficiencies.

The prime purpose of this book is to serve as a classroom text for the engineering or architecture student. It will, however, also be useful to designers who are already familiar with design in other materials (steel, concrete, masonry) but need to strengthen, refresh, or update their capability to do structural design in wood. Design principles for various structural materials are similar, but there are significant differences. This book shows what they are.

The book has features that the authors believe set it apart from other books on wood structural design. One of these is an abundance of solved examples. Another is its treatment of loads. This book will show how actual member loads are computed. The authors have found that students, more often than not, have difficulty recognizing how load is transferred from one member to another—for example, how to proceed from a specified intensity of floor live load and type and thickness of floor material to knowing the actual load per unit length reaching the beam in question. Worked-out examples and student homework problems will illustrate the process.

Another significant feature that we believe sets this book apart is its inclusion of "structural planning." Most textbooks show only the selection of member proportions or number of connectors in a joint to satisfy a given, completely defined situation. This book, on the other hand, shows the thinking process needed to determine whether or not the member is required in the first place. Following this, the spacing and continuity of the member are decided, its loads are determined, and finally its shape and size are selected.

We believe that illustrating structural planning as well as detailed member and connection design is of considerable value in helping the student make the transition from the often simplistic classroom exercises to problems of the real world. Problems for solution by the student follow the same idea. The first problems in each subject are the usual textbook-type problems, but in most chapters these are followed by problems requiring the student to make structural planning decisions as well. The student may be required, given a load source, to find the magnitude of the applied loads and decide upon a grade of wood. Given a floor plan, the student may be required to determine a layout of structural members. The authors have used most of the problems in their classes, so the problems have been tested.

The book presents many of the design examples in the form of "computation sheets," solutions in the form of actual design office computations. This is intended to reinforce the instruction given in Chapter 1 regarding neatness and orderliness in design computations.

The book refers frequently to two codes—the National Design Specification for Wood Construction and the Uniform Building Code. Wherever possible, however, the basic principles behind code requirements are explained, and where the authors are aware of code shortcomings, that too is pointed out. We refer frequently to code requirements in the belief that theory with no exposure to real life is not good education. What is needed is balance between the theoretical and the practical (the latter is *not* a bad word).

The authors have included an example of designing for shear using "two-beam" action, so that users of the book can better understand the NDS provisions on the subject. However, the authors do not recommend using "two-beam" action for design.

In general, the book assumes that the user will be familiar with structural analysis. In fact, the authors intend that the book will reinforce the principles of structural analysis using wood as a vehicle.

The authors use the book in a three semester-hour beginning course and find that sufficient material remains that a second course could be taught from the same book. The prerequisite for the first course should be mechanics of materials as a minimum, including the subjects of shear and moment diagrams, flexural deflections, and axial forces in truss members. A more rigourous course in structural analysis, though desirable as a prerequisite, is not essential.

Chapter 5 (Connections—Nails, Screws, and Bolts) is intentionally in an unconventional position. The authors realize that most books cover connections after member design has been presented. They realize also that (1) because of its position in the course, connection design is something most students learn about almost as an afterthought (as though the subject is of lesser importance than the design of members); and (2) structural failures are far more frequent in the connections than in the members themselves. Further, design of members can be done more effectively if the designer considers how the member will be connected, before making a final selection of member size. With these thoughts in mind, the authors present the subject of connections *before* going on to the design of structural members.

Those who prefer to present the subject in the conventional sequence can merely go on to Chapter 6, delaying the study of Chapter 5 until after Chapter 8 (glulam design). It should work equally well with either sequence.

In the authors' beginning three-hour course in timber structure design, Chapters 1 through 7, 9, 10, 11, most of 8, and parts of 12, 13, and 14 are included.

The book is written to stand alone. Our many instructors may find it beneficial to ask their students to purchase the National Design Specification for Wood Construction from the Publications Department of the National Forest Products Association, 1250 Connecticut Avenue N.W., Washington, D.C. 20036.

JUDITH J. STALNAKER
ERNEST C. HARRIS
Denver, Colorado

Contents

STRUCTURAL DESIGN IN WOOD

1

Introduction

From ancient times, wood and stone have been important construction materials. Stone has diminished in importance, but wood is still our most versatile building material. It is reasonable to claim that wood and wood products are among our more important construction materials, ranking along with structural steel and reinforced concrete for building construction. By weight, more wood is used in construction each year than cement or steel (1).

1-1. EVOLUTION OF TIMBER DESIGN

Over the centuries, man learned to use wood effectively, developing rules of thumb to aid in planning and building wood structures. These rules were learned gradually, based only on experience—both successful and unsuccessful. Using these rules, skilled artisans were able to construct in wood, producing durable, long-lasting structures of utility, often of great beauty, and almost always of adequate strength and serviceability. Even today a large volume of wood construction depends, more or less, on such rules of thumb for its design.

More recently, however, engineers and architects have learned to design wood structures in ways that are based on engineering principles. Thus today's designers, using more rigorous design procedures, are able to ensure that a particular design will achieve the desired level of structural safety and stiffness, as well as economy. As an example showing the economy possible with engineered design, the same amount of wood that was needed in the past to build a 320-ft² log cabin can be used today to construct a 3500-ft² home, with those parts of the trees not suited to lumber production being adequate to manufacture a 30-year supply of paper and tissue products for an average family (2).

Much progress has been made in moving from rule-of-thumb design to present-day engineered designs. Yet our knowledge is not static, and among the design methods shown in this book are some that undoubtedly will some day be replaced by newer and better methods.

As new design methods evolve, greater reliability (with regard to safety) and greater economy (in terms of total volume of wood used) are the usual result. Generally, and this is unfortunate, the improved design methods are more difficult to apply than the earlier methods they supplant, but this is the price to be paid for greater reliability and economy. Occasionally an improved design concept proves so complex that it has to be drastically simplified for the designer to use easily. When this is done, the basic principles involved may become obscured. The "handbook engineer" is not bothered by this, but the more competent *thinking* designer may be prevented from doing the best possible design job. (An example of this situation will be seen in the section of Chapter 7 dealing with members under combined compression and bending.)

That the structural designer will play an increasingly important role in all types of wood construction is inevitable. Whether the demand for wood construction and other wood products remains constant or increases, the supply of readily available and usable trees to produce wood of good quality will almost certainly decrease. By one forecast, the demand for various forms of wood and wood products used for buildings is expected to rise by 1–4% per year (3). To meet this demand will require better use of our resources. "Engineered" structures will become more prevalent and those that use wood wastefully will become less common. Structural products manufactured from wood will replace much of the lumber we use today. We see this trend today in the introduction of wood products such as flakeboard, waferboard, plastic/wood laminates, members made from thinly laminated wood, and shop-prefabricated structural components. Engineering know-how will be involved in every step we take toward making better use of wood in construction.

1-2. MATERIAL PROPERTIES

To design effectively, a structural designer must be familiar with the properties and behavior of the material to be used. Unfortunately, the properties and behavior of wood are unlike those for other building materials and much more complex. With the possible exception of soils, wood has more idiosyncrasies than any other material the structural designer uses. Thus, it is even more important in the case of wood than for other construction materials that the designer have a good knowledge of the peculiarities of the material.

Later chapters will show that a wood designer's problems stem partly from the fact that wood is a natural material—one over which, as yet, we have little control. Through forest management, we are beginning to control certain characteristics of the lumber a forest will produce (3). However, we still cannot force a tree to produce a particular quality of material, nor can we compel different trees in a forest to produce exactly the same quality of material. Similar trees, grown in different localities, may produce wood of dissimilar properties. Consequently, we find that the engineering properties of wood are extremely variable. Changes in the environment in which the wood is used further affect its properties and behavior. For these reasons, wood properties and behavior are dealt with at great length in Chapters 2, 3, and 4.

1-3. TYPES OF CONSTRUCTION

Considering only buildings, there are several types of wood construction in use today. These include:

Light-frame construction
Post-and-beam construction
Glulam construction
Heavy-timber construction
Pole building construction
Wood shell and folded-plate construction
Prefabricated-panel construction
Log building construction

Commonly, combinations of these types with each other or with structural steel are used.

Light-Frame Construction

Light-frame construction is the type most commonly used for housing. It is used mostly for low structures, but finds occasional use for buildings up to four stories high. Frequently, light-frame construction receives little or no engineering design. Its requirements are spelled out in some detail by various building codes. Generally, knowing the code requirements and working from just a floor plan and perhaps one or two cross-section drawings, skilled carpenters often construct small light-frame buildings without engineering design assistance.

The resulting structure will normally be satisfactory as to safety and durability, but often it may be lacking in economy. With both labor costs and material costs increasing sharply, research has been directed toward designing more efficient structures. It is likely that light-frame construction of the future will require increased engineering design effort to bring about greater economy without sacrificing essential strength, stiffness, and durability.

Today, builders frequently require engineering design for certain components, such as lintels and headers. Loading conditions for these members vary from building to building, so rule-of-thumb methods may be either inadequate or wasteful. In fact, present building codes require that certain types of light-frame structures be engineered.

Fig. 1-1. Cross section through light-frame wall.

2 × 4s or 2 × 6s and are spaced at 12 in., 16 in., or 24 in. on centers. Floor joists and roof joists or rafters are also 2-in. nominal material placed with their wider dimension vertical. Where internal geometry of the building permits it, more recent light-frame structures usually have roofs supported by factory-built trusses made of 2 × 4s or 2 × 6s. Trusses and other prefabricated members may also replace solid sawn lumber for floor joists, headers, lintels, etc.

The 12-, 16-, and 24-in. spacings most commonly used in the United States are logical and economical, since most plywood sheets, wallboards, and insulation boards have a 4-ft width. Thus 4 ft, or 48 in., is the usual "module" for building construction in wood.

Post-and-Beam Construction

Post-and-beam construction, also known as timber-frame construction, had its beginnings around 200 B.C. The main, load-supporting members in this type of wood construction are larger than those of light-frame construction, and both vertical members (posts) and horizontal beams are farther apart than the studs and joists of light-frame construction. Posts are usually 8 to 12 ft apart, and horizontal members attached to the posts are spaced vertically in the range from 4 ft to 8 ft (4). More closely spaced floor joists and wall material are connected to and supported by the beams.

If needed for stability, diagonal "knee" braces are used, as shown by Fig. 1-2. Often, in older buildings, the structural members were exposed on the outside of the walls, and the spaces between posts, beams, and knee braces were filled with brick or a kind of plaster. This led to the familiar Tudor style of architecture. Except for the pleasing appearance, however, the filled-in material was troublesome. As the wood expanded or shrunk with moisture changes, wall cracks opened. Also, the wall filling had little insulating value, and its great weight necessitated heavier timber framing to carry the load.

In later forms of post-and-beam construction, the beams support wall material con-

Light-frame construction (shown by Fig. 1-1) almost always uses lumber of 2-in. nominal thickness (actually $1\frac{1}{2}$ in. thick), such as the familiar 2 × 4s or 2 × 8s. Walls use vertical members called *studs*, which are either

Fig. 1-2. Post-and-beam construction.

nected to the outside face of the beams. In old barns, for example, vertical wood planks were nailed to the beams to form the outside wall. A modern version, shown by Fig. 1-3, is called "plank-and-beam" framing and uses beams and posts at an economical spacing for 2-in. floor and roof decking.

When light-frame construction was developed, post-and-beam construction declined in popularity, mostly because light-frame buildings were much cheaper and could be built much faster. Many will argue, however, that a light-frame building is not nearly as durable as a well-built post-and-beam structure.

Post-and-beam construction is staging a comeback in popularity, and this can be attributed partly to the development of new insulating wall panels and to prefabrication techniques that help combat the high cost of an otherwise labor-intensive type of construction. Using prefabricated insulated panels for the exterior walls permits placing all main structural members inside the building, protecting those members from the deteriorating effects of weather and condensation.

Purists insist that connections in this type of construction be made using only wood parts.

Mortise-and-tenon joints of several varieties were used in the older buildings, and the joints were held together by wood pegs, inserted in holes through the members being joined. Reference 4 shows many types of such connec-

Fig. 1-3. Plank-and-beam framing. (Courtesy National Forest Products Association, Washington, D.C.)

tions. Making these connections, however, is time-consuming and requires a very high level of carpentry skill. The authors feel that substituting modern weldments (see Chapter 9) with bolts or other modern timber connectors might produce just as good a structure, permitting construction to proceed more quickly and at less total expense. But, if the appearance of modern connections is offensive, then mortise-and-tenon joints with wood pegs would still be the answer.

Glulam Construction

Because large, good-quality timbers are not readily available today, glued laminated timbers (glulams) are commonly used instead. A glulam is a stack of planks (usually $1\frac{1}{2}$ in. thick) glued together under carefully controlled conditions to make a single member of larger cross section. (See Fig. 1-4.) This substitution is good for several reasons: (1) fire resistance of glulams is at least as good as that of heavy sawn timbers; (2) glulams are less apt than sawn timbers to split after they are installed; (3) glulams can be stronger and stiffer than sawn timbers of the same size; and (4) they do not warp badly as sawn timbers frequently do.

Heavy-Timber Construction

Heavy-timber construction is defined by building codes. The Uniform Building Code (UBC) (5) defines it as construction in which timber columns are not less than 8 in. in least dimension and beams (except for roof beams) are not less than 6 in. wide by 10 in. deep. Of course, UBC defines heavy-timber construction in much greater detail than is given here. Heavy-timber construction requires engineering de-

Fig. 1-4. Light-frame construction used in conjunction with glulams. (Courtesy Overly Construction Co. Photograph by authors.)

sign. Structures meeting the requirements for heavy-timber construction are much more fire resistant than other types of wood construction. (See Chapter 16.)

In earlier building codes, heavy-timber construction was referred to as "mill building construction." Typically, it was used for manufacturing buildings and storage warehouses. Architects often find these old buildings suitable for conversion to new uses, developing attractive offices, shops, restaurants, theaters, and the like in which the old timber construction is used to aesthetic advantage. Many design engineers and architects are involved in evaluating the old construction and modifying structures to adapt them to modern use.

Heavy-timber construction is not limited to buildings, but may be found in mine structures, docks, bridges, and other structures.

Pole Building Construction

Pole-type construction goes back to prehistoric times in both Europe and North America. Examples of its use in North America are seen in ancient Indian dwelling sites in the southwestern United States. The distinguishing feature of a pole-type building is that its supporting vertical members (round poles) are embedded in the ground, having been either driven into the ground or placed into excavated holes. Horizontal beams and other framing are attached to the poles, as shown later in this book by Fig. 15-5.

Obviously, a big problem with buildings of this type is decay of the embedded poles. However, with the use of chemically treated poles, the problems of decay have been alleviated but not eliminated. As a result, beginning in the 1950s, pole buildings have enjoyed a rebirth in popularity in the United States. For certain applications, pole buildings are suitable and economically feasible.

Pole buildings are used principally in agriculture, but in some localities they are used extensively for housing. They are particularly well suited for housing where the floor must be kept well above ground level to avoid flooding. Along the U.S. gulf coast, for example, pole-supported buildings are often better able to re-

sist the force of hurricane-driven waves than structures whose supporting walls extend to ground level. The poles, of course, must have adequate bracing or stiffness to resist the wave forces.

The pole acts as a column, transferring vertical load to the soil. If the end area and perimeter of the pole are too small, the hole may be dug to larger diameter and a concrete footing cast in the bottom to reduce the bearing pressure on the soil. Hole depths usually range from 5 ft to 8 ft, and footing diameters are commonly from 16 in. to 24 in. Pole buildings are normally one-story structures, with poles spaced up to 20 ft in each direction. Heavy timber or light framing is fastened to the poles to form walls and roof, and floor if any. With trusses for the roof structure, spans of up to 60 ft are practical, and a few longer spans have been constructed. Although pole buildings date back to ancient times, today's pole buildings are usually designed by engineers rather than by old rules of thumb (6).

Pole construction can also be a partial solution to problems of freezing and thawing beneath the structure, as in permafrost areas. Being a poor conductor, a wood pole carries little heat from the building above into the frozen ground below, so it helps prevent thawing. If the structure is mounted on a platform at the top of the poles, the open cold-air space below also helps prevent thawing of the soil. Poles in this type of environment must be placed into excavated holes, not driven. Having their larger diameter at the bottom, poles provide good resistance to being lifted from the ground by alternate cycles of freezing and thawing.

Wood Shell and Folded-Plate Construction

Wood shells and folded plates may have both economic and aesthetic advantages for roof construction. Roof shells can be of various shapes, the most common being the hyperbolic paraboloid. Other shapes are limited only by the ingenuity and skill of the design engineer and by the capability of the materials used to follow complex surfaces. Finding them aesthetically pleasing, many architects incorporate shell roofs into their designs. The important thing to remember is that *either shells or folded plates require engineering design*. Rules of thumb will not suffice. To design either, an engineer with proper know-how and experience should be engaged (7).

Materials for shell roofs include both boards and plywood sheets. In either case, the materials are used in two or more (usually three) layers. The layers can be connected to each other by special nails, or glue, or both.

Prefabricated-Panel Construction

There are many varieties of prefabricated panel. The object of prefabrication is to reduce cost. Because of carefully controlled factory conditions, prefabricated panels can have nearly uniform quality. They also can speed field erection, thus saving construction cost. Where stressed-skin action is provided, they may also save in total quantity of material used, thereby reducing the total dead load that the structure must be designed to support.

Stressed-skin panels are those in which a plywood skin is glued to supporting members, so that the plywood sheet serves a double purpose: (1) It acts as a beam spanning from one supporting member to another; and (2) being rigidly connected to those supporting members, it acts together with them, the combined member having strength and stiffness exceeding the sums of the strengths or stiffnesses of the separate parts. Stressed-skin panels may have the plywood on one side only or may have both top and bottom plywood sheets attached to the supporting members. Panels are usually manufactured flat, for uses such as wall, roof, or floor panels or interior partitions. But they can also be manufactured with one-way curvature.

Stressed-skin panels always require engineering design unless they are used strictly for non-load-carrying purposes.

Log Building Construction

Log buildings, in several forms, appear to be making a comeback, especially for residential construction. Strangely, many of them are en-

gineered structures. Their popularity is enhanced, of course, by their exterior attractiveness. In many cases they are economically feasible, even though they obviously use more wood than a light-frame building of equal floor area. There are two reasons for this: (1) Less total labor is involved in preparing and installing the wood components; and (2) wood is so effective an insulating material that extra wall insulation is normally not required.

1-4. HYBRID CONSTRUCTION

The authors use the word *hybrid* to describe any construction type that is mixed, and there are numerous examples of this. For example, in construction that is essentially light frame, we may see steel floor trusses substituted for the usual sawn-lumber floor joists, or floor trusses having wood top and bottom chords but metal web members. Or we may see light-frame construction in which headers and lintels will be wood trusses, or even steel beam sections. Other frequent combinations are: (1) a complete structural steel frame, with wood members (either joists or trusses) spanning from steel beam to steel beam to support the plywood or wood composite floor, or (2) wood joists or rafters supported by masonry bearing walls.

So, even though several recognized construction types were listed and defined above, there is absolutely no need that a structure conform strictly to any one of those types. Whenever economy and serviceability are best served by a combination of types, then designers will probably use it.

1-5. TIMBER BRIDGES

Bridges built either wholly or partly of timber include pedestrian, highway, and railway bridges and special types such as pipe bridges. They include trestles, simple girder, arch, or truss bridges, and even suspension bridges. Although wood is an excellent material for many types of bridge, its value for bridges is often not clearly recognized, perhaps because of a fear of either impermanence or weakness. However, properly constructed and maintained

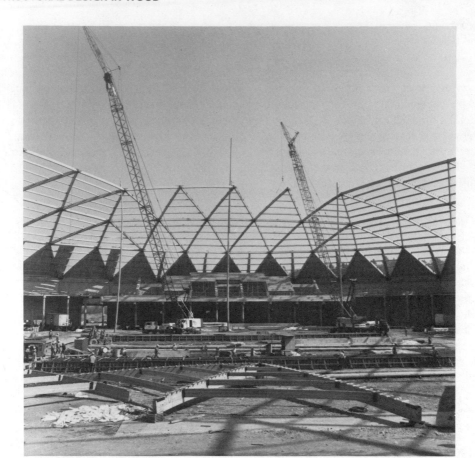

Fig. 1-5. The Tacoma Dome erection sequence. Triangular section of primary glulam beams and secondary purlins is assembled on the ground and then lifted into place, one ring at a time. The dome was erected without dense interior scaffolding. (Courtesy Western Wood Structures, Inc.)

wood bridges can be durable. The useful life of modern glulam bridges is estimated to be as high as 50 years, whereas for steel or concrete bridges (because of corrosion by deicing materials) it may be much lower. Timber bridges are not necessarily weak; they can carry very large loads, such as heavily loaded (135-ton) logging trucks (8). Examples showing durability and capability to carry heavy loads and impact are seen in the large number of existing railroad trestles and in the many covered bridges that still carry rural road traffic. The latter owe their long life, of course, to the fact that the main wood structural members and wood roadway deck are protected from the ravages of weather. According to reference 7, bridges of wood have seen useful life of over 500 years!

Timber (usually glued laminated) is a popular material for pedestrian bridges. In these, the entire superstructure—main girders and deck—are wood. The selection of wood over other materials is often made on the basis of aesthetics, but economy, durability, and ease of maintenance also play important roles in determining the choice.

In railway trestles, treated wood piles are frequently used for the foundation, and treated timbers are used for all other structural parts, including the cross ties. Highway bridges spanning up to about 30 ft can be built using sawn timbers for the main longitudinal members (girders). When glued laminated (glulam) girders are used, the practical span limit is higher, spans as long as 150 ft having been constructed. Wood is used frequently for the towers, stiffening trusses, and floor systems of suspension bridges.

Fig. 1-6. The 530-ft-diameter Tacoma Dome completely framed, with 2-inch timber decking being installed. (Courtesy Western Wood Structures, Inc.)

Arch bridges have become practical with the development of glulams. This type of timber structure has been used for pedestrian bridges, highway bridges, and pipe bridges. The decision to use wood is often based on appearance of the bridge and the blending of its appearance with the surroundings. Both two-hinged and three-hinged arches are practical using glulams, but fixed-end arches generally are not. For longer spans, wood trusses are practical. These are more commonly used in localities where timber is plentiful but steel is expensive and concrete construction difficult.

Bridge decks are built also using composite wood and concrete. In this type of deck the concrete forms the wearing (top) surface. Wood members below form the tension part of the composite flexural section and the concrete above resists the compression. The wood members also serve as supporting formwork to carry the weight of the wet concrete as the wearing surface is placed.

1-6. NOTES TO STUDENTS

Design Codes

For some of you, this will be the first design course. By *design* we mean the process of choosing the type and quality of material to use, choosing location and spacing of the members, and selecting shape and dimensions for each member so that it will (1) carry the load safely, (2) have the desired stiffness and durability, and yet (3) be as economical as possible. Principles that you learned in previous courses (mechanics of materials, for example) still apply, but now you will have to supplement those principles with design rules agreed upon by committees of experienced engineers and researchers to ensure that timber construction will be safe. These rules make up a standard of practice, sometimes referred to (loosely) as a "design code."

The standard of practice having most meaning for this course is the National Design Specification for Wood Construction (NDS) (9). The NDS may be one of the required texts for your course; if not, at least it is an available reference. It would be an excellent idea, especially if this is your first design course, to read through the NDS now, not studying its requirements in great detail but just becoming familiar with the types of instruction it gives and where to find things in it.

A student's first exposure to using a design standard is often a confusing experience. Try to realize now, at the outset of this course, that what the NDS does is merely to *define in some detail (for a particular material) how to apply the design principles to which you were introduced in previous courses*. Realizing this, you will appreciate the true purpose of each requirement and will find its application more meaningful.

Accuracy

Your instructor has probably covered the next point before, but it will bear repeating. Just because your calculator shows answers to seven or eight places does not indicate such a high level of accuracy for our purpose in design. The answers your calculator shows are no more accurate than the values you enter into it. Loads and allowable stresses, for example, are rarely known to more than three significant figures, so you should not expect the answers to be significant to more than three figures. In the process of solving a design problem (if it is more convenient than rounding off at each step), you might copy figures as the calculator shows them, that is, to more than three figures. But to retain more than three figures in the *answer* to the design problem would be pure foolishness. By all means, round off the answer to the number of significant figures that is appropriate for the data with which you started.

Occasionally, however, a student learns the above so well that rounding off goes too far, and in this case the answer may suffer. Do not arbitrarily round off to one- or two-figure accuracy early in the problem and then expect to retain three-figure accuracy in the final answer.

Order and Neatness

A designer's computations are usually kept as part of the complete, permanent record of a design project. This is done for the designer's protection, of course. But, equally important, the computations may find additional future use. The structure to which they apply may need to be revised or extended; or it may need to be adapted to a new use in which the loading differs from that of the original design. Consequently, it may be necessary for you (or someone else) to read and understand your computations clearly and easily at some distant future date. If your original computations are neat and orderly, and fully labeled to show exactly what you had in mind, they will be easy to interpret.

Engineers have long been noted for the neatness and orderliness of their work, whether it be longhand computations or working drawings. However, being neat and orderly does not come naturally to most of us; we have to *work* at it. So why wait until professional practice begins to develop this necessary habit of producing neat and orderly work? Wouldn't it make more sense to develop it during student years? To demonstrate what the authors consider proper form, several of the examples in later chapters are shown as typical designer's computation sheets.

A word of caution, however: Neatly recopying a previously made sloppy and disorderly computation is wasting time. Furthermore, the copy may suffer from inaccuracies of the sloppy original. Rather than making a preliminary version of your work, try to *think first, then write*. Orderly thinking is even more important than orderly writing.

REFERENCES

1. Smith, W. F., *Principles of Materials Science and Engineering*, McGraw-Hill, New York, 1986.
2. "Tree Improvement," Georgia Pacific, Portland, OR, 1981.
3. Stone, R. N., and J. F. Saeman, "World Demand and Supply of Timber Products to the Year 2000," *Forest Products Journal*, **27** (10), Oct. 1977.
4. Benson, T., *Building the Timber Frame House*, Scribner's, New York, 1980.
5. *Uniform Building Code*, International Conference of Building Officials, Whittier, CA, 1985.
6. Patterson, D., *Pole Building Design*, American Wood Preservers Institute, McLean, VA, 1969.
7. *Wood Structures, a Design Guide and Commentary*, American Society of Civil Engineers, New York, 1975.
8. Gibas, D., "The Timber Bridge, an Old and New Concept," *Forest Products Journal*, **33** (11/12), Nov./Dec. 1983.
9. *National Design Specification for Wood Construction*, National Forest Products Association, Washington, DC, 1986.

2

Wood Structure and Properties

2-1. WOOD AS A STRUCTURAL MATERIAL

For structural applications, wood is most commonly found as either sawn timbers, lumber, or glued laminated members (glulams). In the interest of economy and to permit using wood more efficiently, increasing amounts of wood today find their way into manufactured structural materials or members such as (1) plywood, hardboard, chipboard, flakeboard, waferboard, and plastic/wood laminates (these are known collectively as wood composites); and (2) manufactured members such as plywood-lumber beams or wood trusses. Many of these will be discussed later in this book.

2-2. PROBLEMS IN USE OF WOOD FOR STRUCTURES

For most structural applications, fortunately, wood's advantages far outweigh its disadvantages. However, structural designers must learn to cope with (1) wood's variability and (2) its response to environmental conditions.

Variability is probably the more serious of these. Wood properties vary from species to species, from one position to another in the tree, from one tree to another grown in the same locality, and between trees grown in one locality and those grown in another. Generally, humans have little control over the quality of wood a tree produces, although strides are being made in that direction by means of selective tree farming. It is hoped that by such means straighter and faster-growing trees may be developed, with more nearly uniform properties than are found in trees from natural forests (1).

The moisture content of wood installed in a structure may change with time, eventually reaching an equilibrium moisture content that depends on the average relative humidity of the surroundings. However, as the relative humidity within a building may not be constant, the equilibrium moisture content may vary with time. With any change of moisture content, wood will either shrink or swell—and warp. Collectively these size and shape changes are known as *dimensional instability*. Dimension and shape changes due to moisture change can be reduced or avoided by proper seasoning (drying) and by proper attention to details of design.

Duration of loading causes strength changes: The longer a load remains on a wood member, the weaker the wood member becomes. Luckily, the structural designer can easily consider this problem and compensate for it in the design procedure. Duration of loading has negligible effect on modulus of elasticity but, because of creep, deflections are time-dependent.

Finally, under some conditions wood's durability will be limited. Weathering, decay, insects, or fire can obviously limit the useful life of a wood structure. Yet this problem, too, can be overcome by selecting the proper kind of wood and by proper design, treatment, construction, and maintenance.

2-3. ADVANTAGES OF WOOD AS A STRUCTURAL MATERIAL

Wood's principal advantages over the other common structural materials are (1) its economy, (2) its appearance, and (3) its ease of working and reworking. Other advantages are better durability (for some applications), a high strength/weight ratio, and excellent thermal insulating properties.

A wood structure is frequently less costly to construct than a similar structure in either structural steel, reinforced concrete, precast concrete, or masonry. Of course, a completely valid cost comparison would have to include such costs as operation, maintenance, insurance, and the like, over the entire life of the structure. However, it should be obvious from observing structures all around us that, even when all these factors are considered, wood is frequently the most economical choice for buildings up to three or four stories high.

A wood structure can be either aesthetically pleasing or unattractive, depending entirely on the care that goes into its design. Early U.S. colonial structures provide excellent examples of attractive design and of construction details that ensure good durability despite the fact that wood can decay. Modern glued laminated timber makes possible very attractive buildings. We see numerous examples of this in commercial and institutional buildings, in high-quality residences, and even in interior treatment in larger buildings whose main structural members are of other materials.

Wood can easily be cut, shaped, and finished in the field. Prefabrication at another location is generally not needed, although prefabrication may be employed as a time-saving or money-saving measure. Existing wood structures can be revised or added to more easily than similar structures of any other material.

If proper care is taken in details of design, in maintenance, and in selecting suitable preservative and fire retardant treatments, wood will be durable. If untreated wood is allowed to remain moist in service, however, it will decay. The secret to ensuring durability is fourfold: (1) Use proper species and grades, (2) avoid undesirable environmental conditions, (3) use proper design details, and (4) use whatever treatment is necessary and economically feasible.

When wood is used for all components of a structure, the total weight is often less than for a structure of other materials. This is particularly true in comparison to reinforced concrete. The strength/weight ratio of wood is advanta-

Fig. 2-1. Proper detailing of joints in log building leads to long life. Upper photograph shows old-time construction. Lower photograph (Courtesy CMB Log Homes, Golden, Colorado) shows turned logs in a modern log building.

geous wherever dead load is an appreciable part of the total load.

2-4. CLASSIFICATION OF WOOD

Wood is classified as either softwood or hardwood, according to the species of tree from which it is cut. These are not necessarily good names, but they have been used for so long that they are accepted as part of the language by people who work with wood. Softwoods actually include some varieties that are quite hard, while hardwoods include species whose wood is very soft. Softwoods come from the plant group known as *gymnosperms*. Softwood species include the needle-leaved trees. Most of these are cone bearing (conifers), and most are evergreen, which means that they do not drop their leaves for the winter season. Softwoods also include the scale-leaved evergreens, most of which do not bear cones.

Trees producing hardwoods are of the plant group known as *angiosperms*. All hardwood trees have broad leaves. Most nontropical hardwood trees are deciduous; that is, they shed their leaves for the winter.

Without getting into biologic structural differences (except for one that is explained later) the above definitions are about as precise as we can get. Perhaps a better way for our purpose is merely to list common North American species of each class. Table 2-1 shows typical softwood species and Table 2-2 typical hardwoods.

The Latin names in the second column of these tables are scientific names. The first part of the scientific name is the genus and the second part the species. Ordinarily, scientific names find little use by structural designers. However, scientific names are the only reliable ones for defining precisely a particular species. Names in the first column are called *common names*. These are the names used by most people when identifying trees, but they are usually less accurate than the scientific names. For example, a species may have different common names in different regions; or, equally confusing, the same common name may be applied to more than one species.

The third column of each table gives still another version—names applied by the lumber industry. These are the names we will use in this book. Under lumber industry names we find

Table 2-1. Important North American Softwoods.

Common Name	Scientific Name	Lumber Name
Ponderosa pine	*Pinus ponderosa*	Ponderosa pine
Lodgepole pine	*Pinus contorta*	Lodgepole pine
Longleaf pine	*Pinus palustris*	Longleaf yellow pine
Slash pine	*Pinus elliottii*	Longleaf yellow pine
Virginia pine	*Pinus virginiana*	Southern yellow pine
Pitch pine	*Pinus rigida*	Southern yellow pine
Shortleaf pine	*Pinus echinata*	Southern yellow pine
Slash pine	*Pinus elliotti*	Southern yellow pine
Longleaf pine	*Pinus palustris*	Southern yellow pine
Loblolly pine	*Pinus taeda*	Southern yellow pine
Red (or Norway) pine	*Pinus resinosa*	Norway (red) pine
Black spruce	*Picea mariana*	Eastern spruce
Red spruce	*Picea rubens*	Eastern spruce
White spruce	*Picea glauca*	Eastern spruce
Sitka spruce	*Picea sitchensis*	Sitka spruce
Englemann spruce	*Picea engelmannii*	Engelmann spruce
Blue spruce	*Picea pungens*	Engelmann spruce
Douglas fir[a]	*Pseudotsuga menziesii*	Douglas fir
Pacific silver fir	*Abies amabilis*	White fir
White fir	*Abies concolor*	White fir
Eastern hemlock	*Tsuga canadensis*	Eastern hemlock
Western hemlock	*Tsuga heterophylla*	West coast hemlock
Western red cedar[b]	*Thuja plicata*	Western red cedar
Redwood	*Sequoia sempervirens*	Redwood
Bald cypress	*Taxodium distichum*	Cypress

[a]Not a true fir. (Some botanists call it the Douglas tree.)
[b]Not a true cedar. (True cedars—*Cedrus* species—are found only in Europe and Asia.)
Note: The list above is by no means complete.

Table 2-2.   Typical North American Hardwoods.

Common Name	Scientific Name	Lumber Name
Northern red oak	Quercus rubra	Red Oak
Southern red oak	Q. falcata	Red Oak
Black oak	Q. velutina	Red Oak
Pin oak	Q. palustris	Red Oak
Scarlet oak	Q. coccinea	Red Oak
White oak	Q. alba	White Oak
Chestnut oak	Q. prinus	White Oak
Sugar maple	Acer saccharum	Hard maple
Black maple	A. nigrum	Hard maple
Red maple	A. rubrum	Soft maple
Yellow birch	Betula alleghaniensis	Birch
River birch	B. nigra	Birch
White ash	Fraxinus americana	White ash
Green ash	F. pennsylvanica	White ash

Note: The list above is by no means complete.

several different species listed together under a single name (species group). The justification for this is that the species included under a single lumber industry name grow in the same locality, have similar appearances, and have similar properties. The important thing to notice is that while only the scientific nomenclature is sufficiently accurate for the botanist, lumber industry names are quite suitable for the structural designer's use.

The lists in Tables 2-1 and 2-2 are far from complete. For example, the lumber name *red oak*, for which Table 2-2 shows five different species, actually includes 14 different North American species. Similarly, the table shows only two species under *white oak*, but 13 species are actually included under that name. Other important hardwoods (lumber names), not shown in Table 2-2, are hickory, beech, pecan, poplar, cottonwood, aspen, basswood, elm, walnut, and cherry.

To emphasize that hardwoods are not necessarily hard, consider the familiar tropical wood known as *balsa*. Balsa is actually a hardwood!

2-5. WOOD STRUCTURE

Features of wood structure include those of the wood-forming substance and those of the "whole wood." Features of the wood substance are all microscopic and are of interest to industries that use wood substance (rather than whole wood) as a raw material—the paper industry, for example. On the other hand, except for individual wood cells, most whole-wood features are visible to the naked eye or by aid of merely a hand lens. These are the features that affect wood's suitability for construction, and are the ones that will be explained here.

Wood Cell

Wood's basic structural element is the cell. The hollow, tubular wood cells are many times longer than they are wide. For softwoods, the cell length is about 100 times the width. Cells are approximately rectangular in cross section and have unsymmetrical tapered ends that overlap, in staggered position, with the cells above and below. The cell walls consist principally of cellulose and lignin. Cellulose accounts for 45–50% of the weight of completely dry wood, and lignin for about 20–30%. Another ingredient, hemicellulose, amounts to 20–25% (2). Minor components, which vary from species to species, include natural resins, oils, tannin, and alkaloids. Often, it is these minor components that make a species suitable or unsuitable for some particular application.

Cellulose gives the wood its strength; it is the load-carrying material. Lignin is the "glue" that cements cellulose fibers together, filling the spaces between fibers and stiffening the fibers. The cellulose-lignin combination is actually heavier than water; it has a specific gravity of 1.54. However, wood cells are hollow, so that the specific gravity of *whole wood* ranges from 0.31 to 0.75 for most woods used in construction. If the cells were not hollow, wood would sink rather than float. A few tropical species do actually sink. Ebony, for example, weighs 70–80 lb/ft³ (versus 62.4 for water).

If cellulose forms the cell walls in all wood species and is the constituent that gives the wood structural strength, why, then, do wood properties (strength, for example) vary so much from one species to another? To understand why, we must consider further the structure of the wood itself, in particular the arrangement and shape of its cells. Figure 2-2 is a drawing

Fig. 2-3. Strength of tubular cells.

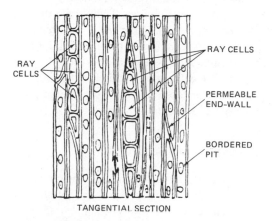

Fig. 2-2. Softwood cross sections. (Courtesy American Society of Civil Engineers, R. J. Hoyle, "Introduction," *Wood Structures, a Design Guide and Commentary,* 1975.)

showing typical cross sections through a piece of wood. Notice that most of the hollow, tubular cells are in a longitudinal position; that is, their long dimension is vertical, parallel to the trunk (stem) of the tree. A few cells, in localized bundles, run radially, that is, parallel to a radial line from the center of the tree trunk to the outside.

Because the wood cell is a hollow tube, it is very efficient for resisting a compressive force parallel to its length, as shown by Fig. 2-3a. For this reason, wood under longitudinal compression has a rather high ratio of strength to weight. However, compression forces normal to the length of the cells easily crush the cells, bending their walls as shown in Fig. 2-3b. Thus, the strength of wood is poor under compression perpendicular to the cell length (called *compression perpendicular to the grain*).

Longitudinal shear strength of wood is limited by the strength of the lignin that binds adjacent cells together, or by the shear strength of the cell wall, whichever is lower. Longitudinal shear strength of wood is lower than its compression strength in any direction and also lower than its longitudinal tensile strength. The only strength that is lower than longitudinal shear strength is the transverse tensile strength.

Interestingly, the hollow-cell arrangement has a *good* effect on both longitudinal compressive strength and bending strength. Given a specified weight of wood in the form of either a beam or a column, its strength is higher than it would be if all void spaces were eliminated and the wood substance moved together to form a solid member of lesser cross-sectional area.

Cell walls are permeable, and the end walls of wood cells are particularly so. In living parts of the tree, fluids necessary to growth can move vertically from one longitudinal cell to another. By capillary attraction and surface tension, fluids are raised by way of these cells all the way from the roots of the tree to the uppermost leaves.

Cambium

Figure 2-4 shows the cross section of an entire tree trunk. New wood cells, formed by cell division, are produced by the cambium layer. As new cells are added on the inner side of the cambium, the diameter of the tree increases. At

Fig. 2-4. Tree cross section.

the same time, the cambium adds new cells to the inner bark, just outside of the cambium layer. Immediately after a cambium cell divides, the newly created cell begins to grow in both length and diameter. During this growth the cell wall is thin and pliable, but after growth ceases a secondary layer is added to the cell wall, thickening it. At this stage the cell wall is primarily cellulose. After wall thickening ceases, lignin is deposited, cementing the fibers together in a process called lignification. The cell now has structural strength that enables it to help carry the weight which later is added to the tree.

Earlywood and Latewood

During springtime, growth is rapid and the newly created cells have relatively large cell openings (cavities) and thin walls. Portions of new wood created during this season are known as *earlywood* (or springwood). New cells created later in the year, when less moisture is available, have smaller cavities and thicker walls. This portion of the wood is called *latewood* (or summerwood). Latewood contains more cellulose per unit of cross-sectional area than earlywood; consequently, latewood is stronger than earlywood.

Annual Rings

Each year a band of earlywood and then a band of latewood are added to the tree. Together, these two bands comprise an annual ring. Annual rings arc easily visible on all North American hardwoods and softwoods. Generally, the latewood portion of the annual ring is darker

than the earlywood. In some tropical woods, grown where differences between seasons are not marked, annual rings may be either obscure or absent.

Medullary Rays

Medullary rays are ribbonlike bundles of cells arranged in a radial direction in the tree. That is, they are perpendicular to the annual rings, running from the center (pith) toward the bark. Medullary rays are prominent in some species but nearly invisible in others. The rays serve a useful structural function: They brace the longitudinal cells so that their buckling strength is higher. Tensile strength perpendicular to the annual rings may be benefited a little, but tangentially (in a direction parallel to the rings and perpendicular to the rays) tensile strength may actually be *reduced* by the rays.

Heartwood and Sapwood

Wood in the annual rings nearest the outside of the tree—the newer portion—is called *sapwood*, since it still transports the tree's life fluids. Portions nearer the center no longer carry these fluids nor store food, and they are called *heartwood*. As the tree grows, new annual rings of sapwood are added to the outside, and the innermost (oldest) sapwood rings convert to heartwood. Usually, heartwood is darker than sapwood, the darker color being due to by-products formed as the cells' function for food storage terminates. However, in some species it is hard to distinguish heartwood from sapwood by color alone.

The dark products that color the heartwood are often toxic to insects and fungi; therefore, heartwood of most species resists decay better than sapwood. An exception is sweet gum, whose heartwood decays readily. An exception to the usual color difference is white cedar, whose heartwood is light in color but is very resistant to decay. Unless juvenile wood is involved, the strength properties of lumber sawn from sapwood or from heartwood (each dried to the same moisture content) are the same, since each contains the same amount of cellulose per unit cross-sectional area. However,

Fig. 2-5. Cross section of strongly ring-porous oak. Note distinct medullary rays. (Photograph by authors.)

because of its ability to transport fluids, sapwood may behave differently from heartwood in such properties as permeability and ability to absorb stain or hold paint.

Pores

A structural difference between hardwoods and softwoods is that hardwoods have vessels or pores, while softwoods do not. These are structures whose only function is to carry water. Pores are large-diameter structures having thin walls made up of individual cells. In many hardwoods the pores are large enough to be seen easily by the naked eye. Hardwood is classified as either diffuse-porous or ring-porous, according to whether the diameters of the pores are similar throughout the annual ring or larger in the earlywood. Ring-porous wood has

the appearance of having pores in the earlywood only. The importance of pores is mainly in the appearance of the wood. Figure 2-5 shows a cross section of ring-porous wood.

2-6. JUVENILE WOOD

A tree grows faster in its early years of life than in later years, and wood produced during those early years has a larger-than-normal percentage of earlywood. This portion of the tree is called *juvenile wood*. The specific gravity and strength of juvenile wood are less and the longitudinal shrinkage more than for wood grown by the mature tree. Just how much of the tree's interior is juvenile wood depends on the species and other factors, but usually ranges between 5 and 20 annual rings. Genetically improved trees grown on intensively managed tree farms grow fairly rapidly; they can be harvested earlier and, therefore, have a higher percentage of juvenile wood than trees from either natural or regrowth forests (3). Today most trees grown on tree farms are used by the paper industry, so the structural designer is not affected. In the future, however, more of such trees will undoubtedly be used for construction lumber. When this happens, the code-writing committees that establish allowable stresses for the designer's use will have to account for the increased percentage of juvenile wood in the lumber.

2-7. WOOD AXES

Because of its internal structure, wood is orthotropic, which means that its properties (physical and mechanical) differ in the three main, mutually perpendicular directions—longitudinal, radial, and tangential. Figure 2-6 shows the axes in these directions and the three reference planes associated with the axes.

Strength and modulus of elasticity vary in the three directions, and there are six values of Poisson's ratio! Shrinkage (or swelling) occurring as wood's moisture content changes also differs in the three directions; this is what may cause wood to warp as it either dries out or takes on additional moisture.

The nonisotropic nature is probably the most serious characteristic with which the designer of wood structures must cope.

2-8. PROPERTIES OF INTEREST TO THE STRUCTURAL DESIGNER

Mechanical properties of wood, determined from small, nearly perfect wood specimens (small clear specimens) become the bases for the allowable stresses given by design codes and specifications. Of these properties, those concerning strength and stiffness interest the structural designer most. The designer is interested in other properties, of course, such as creep, durability, hardness, dimensional stability, and thermal insulating qualities.

The strength and stiffness properties of most interest in structural design are:

1. compressive strength parallel to the grain
2. modulus of elasticity parallel to the grain
3. tensile strength parallel to the grain
4. compressive strength perpendicular to the grain

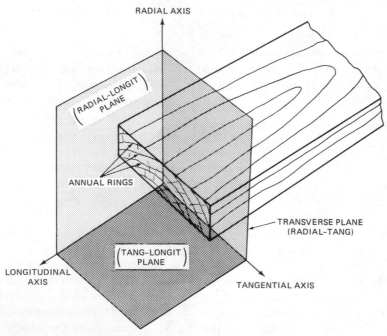

Fig. 2-6. Orthotropic axes and reference planes in wood.

5. modulus of rupture (bending strength)
6. longitudinal shear strength (horizontal shear)
7. shear modulus

These strength and stiffness properties are determined by a series of standardized tests defined by the American Society for Testing and Materials (ASTM) (4). A few of these tests are described briefly below.

Test for Compression Parallel to the Grain

In this ASTM test a clear (defect-free) specimen measuring a full $2 \times 2 \times 8$ in. is used, with the longitudinal axis of the wood parallel to the 8-in. dimension. Longitudinal load is applied, increasing until the compressive stress is well beyond a defined elastic limit or until the specimen fails. From the loads applied and longitudinal strains measured during the test, a stress-strain curve can be plotted, as shown by Fig. 2-7. This curve shows several points of interest to the designer.

Ultimate strength is the highest value of stress reached, shown by the ordinate at point *A*.

Modulus of elasticity, also called Young's modulus, is the slope, *E*, of the initial straight-line part of the stress-strain curve. In the U.S. system, it is expressed in pounds per square inch (psi). (Modulus of elasticity can also be found from a bending test, but its value will differ slightly from that found from the compression test. The reason is that the mea-

sured beam deflections are affected by shear strains as well as by flexural strains.)

Proportional limit is the value of stress at the upper end of the straight-line portion of the curve, that is, the stress level above which stress is no longer proportional to strain. The proportional limit is sometimes loosely referred to as the elastic limit, but the two are not precisely the same. Point *B* of the curve in Fig. 2-7 is at the proportional limit.

Notice that at the beginning of the test, the curve of Fig. 2-7 shows a slight reverse curvature. This is due to local crushing of unintentional high spots as the ends of the specimen become seated on the platen of the testing machine. To correct for this, the straight-line portion of the curve may be extrapolated downward to locate an adjusted zero mark. Various types of compression failure seen in this type of test are shown by Fig. 2-8.

Test for Tension Parallel to the Grain

The ASTM test for tensile strength parallel to the grain uses a 1-in.-square clear specimen 18 in. long, carefully tapered to measure only $\frac{3}{8}$ in.

CRUSHING   BROOMING   SHEARING

SPLITTING   WEDGE SPLIT   COMBINED SHEARING AND SPLITTING

Fig. 2-8. Types of failure under compression parallel to grain.

Fig. 2-7. Stress-strain curve, compression parallel to grain.

by $\frac{3}{16}$ in. at its narrowest portion. A stress-strain curve for this test shows elastic action almost to the breaking point. For defect-free wood, tensile strength parallel to the grain (longitudinally) is appreciably higher than compressive strength in the same direction. However, designers can seldom utilize this high tensile strength, since wood in sizes practical for construction always has strength-reducing defects, such as cross-grain and knots. Tensile strengths in tangential and radial directions are very low compared to the longitudinal strength, about 3 percent and 5 percent, respectively (5).

Test for Compression Perpendicular to the Grain

The ASTM test for compression perpendicular to the grain illustrates clearly how much weaker wood is in compression perpendicular to the grain than in compression parallel to the grain. Figure 2-9 shows schematically how this test is performed, and also a typical stress-strain curve. Notice that there is no true ultimate strength value. As the wood nearest to the surface of the specimen is crushed, it merely becomes consolidated and is harder to crush further. As the load increases, additional wood cells farther from the surface also collapse and the strength increases again. The test could proceed until the load capacity of the testing machine was reached. The wood specimen would still be there, and could accept more load, but of course it would be severely flattened. For this type of loading (in design we call it *bearing*), the failure criterion is not reaching an ultimate load, but rather reaching a limit beyond which the degree of distortion is considered unacceptable.

Test for Flexure

The ASTM static bending test defined by reference 4 uses a full 2-in.-square clear specimen 30 in. long, tested with a midspan concentrated load and a span of 28 in. The load-deflection curve for this test shows a definite ultimate load. Values that can be computed from this curve are given below.

Modulus of rupture: The modulus of rupture (MR) is the value of Mc/I one would compute using the bending moment caused by the ultimate load. Modulus of rupture is *not* the value of extreme-fiber bending stress at failure. Bending stresses under elastic conditions vary as a straight line, as shown by Fig. 2-10a, whereas at ultimate load they vary as shown by Fig. 2-10b. Nevertheless, the MR is a useful quantity for comparing bending strengths of different specimens, and it is used to establish allowable stress values for the designer's use.

Modulus of elasticity: By equating the theoretical expression for midspan deflection to an observed deflection on the straight-line portion of the load-deflection curve, the modulus of elasticity, E, can be computed. This value of E is the more reliable one to use for computing deflection of other beams, although it may not agree exactly with the values obtained from tension or compression tests.

Test for Shear Strength Parallel to the Grain

Shear strength parallel to the grain is less than in any other direction. In wood beams, the grain is normally parallel to the span, and shear failure occurs along a plane parallel to the span, as in Fig. 2-11. (Wood beams will not fail by

Fig. 2-9. Compression perpendicular to grain.

(a) ELASTIC   (b) AT ULTIMATE

Fig. 2-10. Bending stress variation.

shearing at right angles to the span.) Since beams are usually horizontal, the measured ultimate shear stress is often called *horizontal shear strength*. In the ASTM test for shear strength there is no gradual yielding; failure is sudden and only the ultimate load is observed.

ASTM D 143 defines other tests also, including tension perpendicular to the grain, impact bending, and hardness. With each test, moisture content of the wood is also determined and recorded.

2-9. FACTORS AFFECTING STRENGTH

Major factors affecting strength and stiffness properties of small clear wood test specimens are:

1. species
2. moisture content
3. specific gravity (indicator of cellulose amount)
4. duration of loading

Strength and stiffness properties of construction-sized lumber are affected by all of the above factors, and also by:

5. size and shape of the wood member
6. nature, size, and location of defects (strength-reducing characteristics)

2-10. MOISTURE CONTENT

Moisture content (MC) is expressed as a percentage of the weight of the completely dry wood. To determine MC by ASTM's standard test (see reference 4), a sample is first weighed, then oven-dried until its weight is constant, cooled, and weighed again. The weight lost during drying is the total weight of water present in the original sample. Expressed as a percent,

$$MC = 100(\text{orig. wt.} - \text{dry wt.})/\text{dry wt.} \tag{2-1}$$

Moisture content of seasoned (dried) wood for most construction purposes is equal to or less than 19%; it usually averages about 15%. Green wood, however, can have MC from as low as 40% to as high as 240%! That is, the weight of water in the green wood may be as high as 2.4 times the weight of dry wood (6).

Moisture present in wood (1) may be adsorbed by the cell walls, or (2) after the walls are saturated may occupy the void space within the cell cavities. The first type is called *bound water* and the second *free water*. Adsorbed water softens the cellulose/lignin material of the cell wall. Therefore, it weakens the wood, reducing all strength and stiffness properties except perhaps impact strength. Free water within the cell cavity, on the other hand, has no effect whatever, except to increase the weight of the wood member and possibly to decrease its impact strength when the cell cavities are nearly full.

For practical purposes, we may say that all strength properties decrease as MC increases, but only up to the moisture percent known as the *fiber saturation point* (FSP). This is the moisture content at which the cell wall material

Fig. 2-11. Horizontal shear failure.

is completely saturated but the cell cavities are essentially empty. Any further moisture taken up by the wood is not adsorbed by the cell walls, but merely occupies the cell cavities. The fiber saturation point varies among species, as illustrated by Table 2-3.

Wood is hygroscopic, which means that its moisture content adjusts to reach equilibrium with the temperature and relative humidity of the atmosphere in which it is used. If wood dried at the mill to the usual MC is used in a very dry location, such as the interior of a heated building, it may give up moisture to the atmosphere, shrinking and perhaps warping as it does so. On the other hand, if the wood is used in a very moist location, such as in a laundry or over a swimming pool, it will absorb moisture from the atmosphere, expanding and possibly warping. Such taking on or giving up of moisture is a slow process, so that if atmospheric conditions fluctuate, the MC adjusts to a condition near the average.

The moisture content that remains constant under a given atmospheric condition is called the *equilibrium moisture content*, sometimes abbreviated EMC. Table 2-4 shows average EMC values for various combinations of temperature and relative humidity.

Moisture affects the size of a piece of wood, and perhaps its shape. With increasing MC (again, only up to the fiber saturation point) the cell walls swell and all dimensions of the piece become larger. With reduction of MC, all dimensions become less; that is, the wood shrinks. Swelling and shrinkage are very small in the longitudinal direction, varying from 0.1% to 0.2% as green wood is completely dried (7). However, they are considerably more in the radial direction and highest of all tangentially. Table 2-5 compares, for a few species commonly used for construction, the percentages of shrinkage in radial and tangential directions, as the green wood is completely dried. It is the *difference* of shrinkage and swelling in these two directions that causes wood to warp as its MC changes.

Amount of shrinkage to expect with a given change of MC may be estimated using Tables 2-3 and 2-5. The position of the annual rings is usually not known in advance. Thus, in estimating the dimension change, the designer should assume the rings to be nearly parallel to the dimension in question; that is, the larger dimension change (tangential) should be anticipated. Note that the percentages in Table 2-5 are based on dimensions of the green wood (MC = FSP). (An approximate method often used is based on the fact that the average FSP for various species is about 30%, and the average shrinkage for various species as they dry from EMC to oven-dry is about 6%. Thus, the lateral dimensions of wood change, on the average, by about 1% for each 5% change of MC.) However, it should be recognized that the shrinkage in different pieces of wood is quite variable—from 50% to 150% of that estimated using Tables 2-3 and 2-5.

Table 2-3.   Fiber Saturation Point of Different Woods at Room Temperature.

Species	FSP (%)	Species	FSP (%)
Ash, white	24.0	Pine, loblolly	21.0
Basswood	32.0	Pine, longleaf	25.5
Birch, yellow	27.0	Pine, red	24.0
Cedar, Alaska	28.5	Pine, slash	29.0
Cedar, western red	22.0	Pine, shortleaf	30.0
Douglas fir	26.0	Poplar, yellow	31.5
Fir, red	30.0	Redwood	22.5
Hemlock, western	28.0	Spruce, Sitka	28.5
Larch, western	28.0	Spruce, red	27.0
Oak, white	32.5	Tamarack	24.0
Oak, swamp	31.0	Teak	22.0

Condensed from reference 9, courtesy Van Nostrand Reinhold Company.

Example 2-1

A coast region Douglas fir plank measures 10.25 in. wide when its MC is 10%. What will its width be when MC is raised to 35%?

By Table 2-3, the fiber saturation point for Douglas fir is 26%, so raising the water content above 26% will have no effect. The change from 10% to 26% will cause the wood to swell. The final dimension will be the same as the dimension, D, when green. From Table 2-5, tangential shrinkage is 7.6% as MC reduces from FSP to dry.

$$10.25 = D[1 - 0.076(16/26)]$$
$$D = 10.75 \text{ in.}$$

Table 2-4. Moisture Content of Wood in Equilibrium with Stated Dry-bulb Temperature and Relative Humidity.[a]

Temperature (dry-bulb) (°F)	Relative Humidity (%)																			
	5	10	15	20	25	30	35	40	45	50	55	60	65	70	75	80	85	90	95	98
30	1.4	2.6	3.7	4.6	5.5	6.3	7.1	7.9	8.7	9.5	10.4	11.3	12.4	13.5	14.9	16.5	18.5	21.0	24.3	26.9
40	1.4	2.6	3.7	4.6	5.5	6.3	7.1	7.9	8.7	9.5	10.4	11.3	12.3	13.5	14.9	16.5	18.5	21.0	24.3	26.9
50	1.4	2.6	3.6	4.6	5.5	6.3	7.1	7.9	8.7	9.5	10.3	11.2	12.3	13.4	14.8	16.4	18.4	20.9	24.3	26.9
60	1.3	2.5	3.6	4.6	5.4	6.2	7.0	7.8	8.6	9.4	10.2	11.1	12.1	13.3	14.6	16.2	18.2	20.7	24.1	26.8
70	1.3	2.5	3.5	4.5	5.4	6.2	6.9	7.7	8.5	9.2	10.1	11.0	12.0	13.1	14.4	16.0	17.9	20.5	23.9	26.6
80	1.3	2.4	3.5	4.4	5.3	6.1	6.8	7.6	8.3	9.1	9.9	10.8	11.7	12.9	14.2	15.7	17.7	20.2	23.6	26.3
90	1.2	2.3	3.4	4.3	5.1	5.9	6.7	7.4	8.1	8.9	9.7	10.5	11.5	12.6	13.9	15.4	17.3	19.8	23.3	26.0
100	1.2	2.3	3.3	4.2	5.0	5.8	6.5	7.2	7.9	8.7	9.5	10.3	11.2	12.3	13.6	15.1	17.0	19.5	22.9	25.6
110	1.1	2.2	3.2	4.0	4.9	5.6	6.3	7.0	7.7	8.4	9.2	10.0	11.0	12.0	13.2	14.7	16.6	19.1	22.4	25.2
120	1.1	2.1	3.0	3.9	4.7	5.4	6.1	6.8	7.5	8.2	8.9	9.7	10.6	11.7	12.9	14.4	16.2	18.6	22.0	24.7
130	1.0	2.0	2.9	3.7	4.5	5.2	5.9	6.6	7.2	7.9	8.7	9.4	10.3	11.3	12.5	14.0	15.8	18.2	21.5	24.2
140	0.9	1.9	2.8	3.6	4.3	5.0	5.7	6.3	7.0	7.7	8.4	9.1	10.0	11.0	12.1	13.6	15.3	17.7	21.0	23.7
150	0.9	1.8	2.6	3.4	4.1	4.8	5.5	6.1	6.7	7.4	8.1	8.8	9.7	10.6	11.8	13.1	14.9	17.2	20.4	23.1
160	0.8	1.6	2.4	3.2	3.9	4.6	5.2	5.8	6.4	7.1	7.8	8.5	9.3	10.3	11.4	12.7	14.4	16.7	19.9	22.5
170	0.7	1.5	2.3	3.0	3.7	4.3	4.9	5.6	6.2	6.8	7.4	8.2	9.0	9.9	11.0	12.3	14.0	16.2	19.3	21.9
180	0.7	1.4	2.1	2.8	3.5	4.1	4.7	5.3	5.9	6.5	7.1	7.8	8.6	9.5	10.5	11.8	13.5	15.7	18.7	21.3
190	0.6	1.3	1.9	2.6	3.2	3.8	4.4	5.0	5.5	6.1	6.8	7.5	8.2	9.1	10.1	11.4	13.0	15.1	18.1	20.7
200	0.5	1.1	1.7	2.4	3.0	3.5	4.1	4.6	5.2	5.8	6.4	7.1	7.8	8.7	9.7	10.9	12.5	14.6	17.5	20.0
210	0.5	1.0	1.6	2.1	2.7	3.2	3.8	4.3	4.9	5.4	6.0	6.7	7.4	8.3	9.2	10.4	12.0	14.0	16.9	19.3
220	0.4	0.9	1.4	1.9	2.4	2.9	3.4	3.9	4.5	5.0	5.6	6.3	7.0	7.8	8.8	9.9	*	*	*	*
230	0.3	0.8	1.2	1.6	2.1	2.6	3.1	3.6	4.2	4.7	5.3	6.0	6.7	*	*	*	*	*	*	*
240	0.3	0.6	0.9	1.3	1.7	2.1	2.6	3.1	3.5	4.1	4.6	*	*	*	*	*	*	*	*	*
250	0.2	0.4	0.7	1.0	1.3	1.7	2.1	2.5	2.9	*	*	*	*	*	*	*	*	*	*	*
260	0.2	0.3	0.5	0.7	0.9	1.1	1.4	*	*	*	*	*	*	*	*	*	*	*	*	*
270	0.1	0.1	0.2	0.3	0.4	0.4	*	*	*	*	*	*	*	*	*	*	*	*	*	*

[a]Asterisks indicate conditions not possible at atmospheric pressure.
Source: Courtesy USDA Forest Service, Forest Products Laboratory.

Table 2-5. Typical Radial and Tangential Shrinkage in Drying from FSP (Green) to Oven-Dry.

Species	Radial[a]	Tangential[a]
Douglas fir		
Coast	4.8	7.6
Interior, west	4.8	7.5
Fir		
California red	4.5	7.9
Hemlock		
Eastern	3.0	6.8
Western	4.2	7.8
Larch		
Western	4.5	9.1
Pine		
Eastern white	2.1	6.1
Ponderosa	3.9	6.2
Longleaf southern	5.1	7.5
Spruce		
Red	3.8	7.8

[a]Percent, based on dimensions when green.
Note: For a more nearly complete list, see reference 7.

Example 2-2

What is the width of the same plank when its MC is raised from 10% to 19%?

The width at 19% would be the width at FSP minus the shrinkage that would occur as the MC was lowered from FSP to 19%.

$$D = 10.75[1 - 0.076(26 - 19)/26]$$
$$= 10.53 \text{ in.}$$

Note that, compared to the 10.25 in. width when the plank was installed, it would now be over $\frac{1}{4}$ in. wider. To install the planks without any space between might cause trouble if the moisture content did actually rise to 19%.

2-11. SPECIFIC GRAVITY

All strength and stiffness properties depend on how much cellulose is present, so we should

Table 2-6. Typical Relationships of Strength of Clear, Straight-Grained Wood to Specific Gravity (Approximate).

Strength Property	Strength Property in Terms of Specific Gravity	
	For Green Wood	For Wood with 12% Moisture
Static bending:		
Bending stress at proportional limit (psi)	$10,200 SG^{1.25}$	$16,700 SG^{1.25}$
Modulus of rupture (psi)	$17,600 SG^{1.25}$	$25,700 SG^{1.25}$
Modulus of elasticity (million psi)	$2.36 SG$	$2.80 SG$
Compression parallel to grain:		
Proportional limit (psi)	$5250 SG$	$8750 SG$
Ultimate (crushing) strength (psi)	$6730 SG$	$12,200 SG$
Modulus of elasticity (million psi)	$2.91 SG$	$3.38 SG$
Compression perpendicular to grain:		
Proportional limit (psi)	$3000 SG^{2.25}$	$4630 SG^{2.25}$

expect heavier woods to be both stronger and stiffer than lighter woods. Extensive tests on many species have shown this to be the case. Results of those tests are in the form of equations relating various strength properties to specific gravity (8). These equations are all of the form

$$F = K(SG)^n \qquad (2\text{-}2)$$

Table 2-6 shows typical equations for strength of clear, straight-grained wood.

Equations showing these relationships are interesting, but have limited value in structural design. The reason is that so many other factors (besides specific gravity) affect wood strength, and it is the combination of *all* these factors that controls the actual strength of any member.

When the term *density* is used in connection with certain lumber grading rules, it does not refer to density as defined for use in physics or chemistry; rather, it is a qualitative term referring to the percentage of summerwood present. Dense lumber weighs more and has higher strength than nondense lumber.

2-12. TIME-DEPENDENT BEHAVIOR OF WOOD

Two time-dependent characteristics of wood are (1) creep under sustained load and (2) a relationship between ultimate strength and load du-

ration. These phenomena are related, but much yet has to be learned about each of them.

Creep

When a wood structural member is loaded, it deflects immediately. If the load remains on the member, deflection increases with time. This increase above and beyond the initial deflection is called *creep* deflection. Creep can occur in any type of member (beam, column, or tension member) but is of greatest concern for beams. Under the allowable stress magnitudes normally used in structural design, the *rate* of creep deflection becomes less as time passes, so that after a few years no further increase is seen. Figure 2-12 illustrates the variation of deflection with time, the solid portion of the curve becoming practically horizontal after a few years. If the load were kept on the member for a long enough time, the curve would eventually continue as shown by the dotted line in Fig. 2-12. Similar curves could be plotted for any type of loading, the exact shape depending on the type of loading and level of stress applied.

The complete curve of Fig. 2-12 may be used to define three stages of creep—primary, secondary, and tertiary. During primary creep (segment *AB* of the curve) deflection continues to increase after the load has been applied. Secondary creep is represented by segment *BC*, the essentially horizontal part of the curve. If the load is maintained long enough, tertiary creep

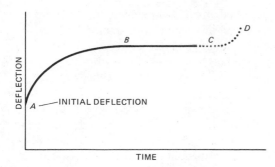

Fig. 2-12. Creep in deflection of a wood beam.

occurs as deflection begins to increase again as shown by segment *CD*. During this stage, the *rate* of creep deflection increases and failure follows (9). For members with very low stress, the overall life of the structure is not long enough for us to observe tertiary creep. For members of very high stress level, however, tertiary creep may be reached quickly. In this case, collapse will follow unless the increasing deflection alerts the user to either unload the member or reinforce it.

Figure 2-13 shows a set of creep curves for four test beams, each with a different level of applied stress. In the beam represented by curve 1, the applied stress was fairly low, and because the test was not continued long enough, the beam never entered tertiary creep. Curve 2 is for a similar beam, but stressed to a higher level. It just started to enter tertiary creep at the end of the test. Curve 3, for a beam with still higher stress level, entered tertiary creep and then failed. Curve 4 is for a beam having a very

high initial flexural stress, just slightly below the ultimate stress determined from a conventional short-term flexural test. This beam had a negligible period of secondary creep, then quickly entered tertiary creep and failed.

The beam tested for curve 1 of Fig. 2-13 did not enter tertiary creep. Would tertiary creep eventually occur if the test were continued? Actually, we don't know, for the reason that such a test would take many years—perhaps centuries. It *may* be that tertiary creep will occur eventually, or it *may* be that under such a low stress level tertiary creep will never occur. What is important is this: We know that if the stress level is kept low enough, failure will not occur *during the expected life of the structure.*

Load-Duration Effect

Since the length of time that a load is present on a wood member affects the ultimate or breaking strength of that member, the designer of wood structures must consider that effect. Otherwise, to ensure a desired level of safety, the designer would have to be overly conservative, designing as though all loads were to be in place forever. To consider the load-duration effect, the structural designer uses a method based on the relationship shown by Fig. 2-14. This graph is the result of extensive research by the Forest Products Laboratory (10). Chapter 4 shows how the designer takes into account this load-duration effect.

Fig. 2-13. Creep at various stress intensities.

Fig. 2-14. Effect of load duration on strength. (Courtesy Board of Regents, University of Colorado.)

2-13. STRENGTH-REDUCING CHARACTERISTICS

Wood strength is determined first by laboratory tests of the nearly perfect, small clear specimens, but these measured strengths cannot be used directly for design. Structural members, being of larger size, will not be so nearly perfect as the small clear specimens. The larger pieces will almost certainly have some strength-reducing defects, such as the following: (1) knots, (2) cross grain, (3) checks, (4) shakes, (5) compression wood, (6) wane, and (7) decay.

Knots

Knots are formed by the change of wood structure that occurs where limbs grow from the main stem of the tree. The limb, extending approximately radially in the main trunk, has its own annual rings and rays, and this local arrangement of cells interrupts the normal pattern for the main portion of the tree. The knot has a weakening effect that is illustrated by Fig. 2-15. The grain direction changes severely as the wood fibers pass around the knot (cross grain). In this region, the applied load causes tensile stress components normal to the grain of the knot. Since tensile stength normal to the grain is very low, the knot weakens the member significantly.

Tensile strength is affected most by the knot, but compressive strength is reduced also. Bending strength is reduced too, the amount of reduction depending on where the knot is located on the beam cross section. In drying, wood shrinks more radially than longitudinally, so knots frequently become loose and may even fall out. Then, obviously, they are as damaging to bending, compressive, or shear strength as they are to tensile strength.

The manner in which the tree is cut to produce lumber influences the degree of effect a knot will have on strength. This will be illustrated in Chapter 3.

Cross Grain

When the longitudinal axis of the cells is not parallel to the edge of a piece of wood, the piece is said to have cross grain. Cross grain

Fig. 2-15. Effect of knot on stresses.

involves both spiral grain and diagonal grain. Cross grain occurs at knots and other locations. Nearly all lumber has this strength-reducing characteristic to some degree. Figure 2-16 illustrates several cases of cross grain, differing from each other according to the orientation of the annual rings with the faces of the piece of lumber.

Many trees grow in a spiral fashion, so that the direction and degree of cross grain varies along the length of a piece of lumber. This special case of cross grain is called spiral grain. Spiral grain makes wood very difficult to work. Further, as its moisture content changes, the wood warps—all plane surfaces become twisted, warped surfaces. Spiral grain is common and is a hereditary trait in some trees.

Diagonal grain is caused by the way in which a piece of lumber is cut from the log, rather than being due to an inherent defect of the tree itself. However, there will always be variations in grain direction in the tree (cells are not everywhere perfectly parallel), so it is not possible in the mill to cut lumber so as to completely eliminate diagonal grain.

Figure 2-17 shows how cross grain reduces the strength of a piece of wood. Almost everyone has tried to bend a stick with cross grain, only to have it break in the manner shown by Fig. 2-17a. This break occurs because the flexural stress (which is parallel to the surface) has a component perpendicular to the grain, and tensile strength in that direction is very low. The Mohr's circle in Fig. 2-17a shows how appreciable perpendicular-to-grain tension can occur with small cross-grain angles.

Similarly, Fig. 2-17b shows how a compression member with cross grain is weakened. Here, it is either the low strength in compression perpendicular to the grain or the low longitudinal shear strength that precipitates failure.

Strength at an angle to the grain may be estimated by a simple equation called *Hankinson's formula*. This equation, shown in later chapters, is used in designing members and connections where the forces to be transferred are neither parallel to nor perpendicular to the grain.

Fig. 2-16. Types of cross grain. (Courtesy of USDA–Forest Service, Forest Products Laboratory.)

CROSS GRAIN
BENDING FAILURE

$f_b = Mc/I$

MOHR'S CIRCLE

2α

f_t|

(a) BENDING

(b) COMPRESSION

Fig. 2-17. Effect of cross grain on strength.

Fig. 2-18. Checks and shakes in a cottonwood log. (Photograph by authors.)

Shakes and Checks

Sometimes the generic term *splits* is used to cover both shakes and checks. Shakes and checks are longitudinal planes of separation (cracks) in the wood (see Fig. 2-18). Shakes originate in the tree. They may be in the longitudinal/radial plane, in which case they are called *heart shakes* or *rift cracks*. When they are in the longitudinal/tangential plane they are called *ring shakes*, the plane of the break being parallel to the annual rings. Heart shakes are thought to be caused by tensile stresses in the tangential direction, the stresses resulting from growth pressures within the tree. Ring shakes are thought to be caused by shear due to wind, perhaps combined with weakness due to cell damage by frost. Figure 2-19 shows types of shake.

Checks occur as the wood seasons after the tree is felled. They may occur in the log itself or in pieces of lumber cut from the log. Checks are longitudinal cracks, usually in the longitudinal/radial plane but occasionally in the longitudinal/tangential plane. They occur because of stresses resulting from differential shrinkage in the tangential and radial directions during drying, or from uneven drying in different portions of the lumber. Too-rapid drying can cause severe checking. Thus, checking can be reduced by careful attention to drying procedures during lumber production.

Figure 2-20 shows the position and names of types of check. Surface checks may sometimes be removed by planing; however, if drying continues, they may reappear. The major structural effect of checks is to reduce longitudinal shear strength, often the deciding factor in designing timber beams.

Compression Wood

Compression wood is a form of "reaction wood," in which an unsymmetrical growth pattern has been caused by long-term bending stresses in the live tree. For example, a tree that leans badly during its growth will have to resist a high bending moment caused by the tree's own weight. Compressive stresses due to

SURFACE
SHAKE

THROUGH
SHAKE

RING
SHAKE

PITH
SHAKE

Fig. 2-19. Types of shake.

SURFACE
CHECK   THROUGH CHECKS

Fig. 2-20. Types of check.

this bending occur on the lower side, and tensile stresses due to bending occur on the other side. The compressive stresses, being combined with the direct vertical compressive stress, P/A, are the highest stresses for the entire cross section. These high stresses cause the tree to grow thicker annual rings on that side, as shown by Fig. 2-21.

Cell walls in compression wood are thicker, and specific gravity is higher than for normal wood. Also, in the internal microstructure of a compression wood fiber, the angle of inclination of its component fibrils is altered in a way that affects physical properties (11, 12). Compression wood is undesirable mainly because, with moisture change, its *longitudinal* shrinkage and swelling are as much as ten times greater than for normal wood.

Wane

Wane is the lumberman's term for the absence of wood. Often wane consists of bark remaining on the surface of the finished lumber. Obviously, the strength is reduced by wane, so wane is taken into account in the allowable design stresses assigned to a given grade of lumber. Inclusions of bark, called bark pockets,

occasionally occur within the tree. These result from tree injury that is covered by new wood as the tree grows. The injury giving rise to bark pockets is usually something minor, such as holes made by bird pecks or boring insects.

Decay

Decayed spots represent portions of the wood that should be considered to have no strength, since the wood fibers are interrupted by the decay. Further, decay can easily spread to adjacent areas. Wood with decay should not be used in new construction, but a designer may be called upon to evaluate the strength of an existing structure in which decay is already present. If the decay can be removed or its growth arrested, the structure can conceivably continue to be used.

All of the above defects affect the structural value of the wood, and they are considered in "grading," the process by which wood is evaluated and classified to indicate suitable allowable design stresses. This will be shown in Chapters 3 and 4. Other defects, not mentioned above, affect only appearance and durability.

2-14. THERMAL PROPERTIES OF WOOD

Effect of Temperature on Mechanical Properties

Temperature affects both dimensional stability and strength of wood. Wood expands as its temperature increases, as do other construction materials. Its coefficients of expansion vary with direction, being largest radially and tangentially, and least longitudinally. Expressed as length change per unit of original length per degree Fahrenheit, wood's radial and tangential coefficients of expansion range from $0.000025 \times SG$ (specific gravity) for dense hardwoods to $0.000045 \times SG$ for softwoods such as Douglas fir or redwood (see reference 8). Considering that the specific gravity of these woods is from 0.42 for redwood to 0.68 for common dense hardwoods, the radial and tangential coefficients of expansion are seen to be about 0.000017 to 0.000019 in./in./°F, or

HIGHER DENSITY
BUT
LOWER TENSILE AND
IMPACT STRENGTH

Fig. 2-21. Compression wood.

about three times as large as for structural steel or concrete.

Wood's longitudinal expansion coefficient is much less. It is independent of *SG* but varies with species from about 0.0000017 to 0.0000025 (13), which is only about 26–42% as much as for other common construction materials.

Exposure to temperatures of 150°F or more for extended periods will permanently weaken wood (13). For temperatures between zero and 150°F, the 70° strength of dry wood changes by $\frac{1}{3}-\frac{1}{2}$% per degree, being weaker at high temperatures and stronger at low. If the exposure to abnormal temperatures in this range is not prolonged, wood's strength properties can be expected to return to approximately their original values when conditions return to normal.

Thermal Conductivity

Wood's thermal conductivity is low compared to that of other common structural materials. In other words, wood can do a better job of insulating against heat loss. Often, therefore, less insulation is needed with wood construction than with other types. The total thermal transmission of a building wall, of course, depends on the properties and thickness of each of the various layers of material used. Engineering information on computing the total thermal transmission (or resistance) of a wall may be found in sources such as the *ASHRAE Handbook* (14).

Thermal conductivity depends on several factors, including direction of heat movement with respect to the grain, specific gravity, moisture content, and type and location of defects. Thermal conductivity is about equal in radial and tangential directions, but is about 2.5 times as great along the grain. Table 2-7 shows average values of thermal conductivity for wood and other commonly used construction materials and compares their insulating efficiencies. Notice that wood's thermal efficiency far surpasses that of all other listed materials that might be used for structural members.

REFERENCES

1. "Tree Improvement," Georgia Pacific, Portland, OR, 1981.
2. Thomas, R. J., "Wood Anatomy and Ultrastructure," in *Wood: Its Structure and Properties*, by F. F. Wangaard, Pennsylvania State University, University Park, PA, 1981.
3. Pearson, R. G., and R. C. Gilmore, "Effect of Fast Growth Rate on the Mechanical Properties of Loblolly Pine," *Forest Products Journal*, **30**(5), May 1980, pp. 47–54.
4. *Standard Methods of Testing Small Clear Specimens of Timber*, American Society for Testing and Materials, ASTM Designation D 143, Philadelphia, PA, 1965.
5. Schniewind, A. P., "Mechanical Behavior and Properties of Wood," in *Wood: Its Structure and Properties*, by F. F. Wangaard, Pennsylvania State University, University Park, PA, 1981.
6. Tarkow, H., "Wood and Moisture," in *Wood: Its Structure and Properties*, by F. F. Wangaard, Pennsylvania State University, University Park, PA, 1981.
7. *Wood Handbook*, U.S. Forest Products Laboratory, 1974.
8. *Wood Structures, A Design Guide and Commentary*, American Society of Civil Engineers, New York, 1975.
9. Bodig, J., and B. A. Jayne, *Mechanics of Wood and Wood Composites*, Van Nostrand Reinhold, New York, 1982.
10. Wood, L. W., "Behavior of Wood Under Continued Loading," *Engineering News Record*, Dec. 11, 1947.
11. Stamm, A. J., *Wood and Cellulose Science*, Ronald Press, New York, 1964.
12. Mark, R. E., "Molecular and Cell Wall Structure of Wood," in *Wood: Its Structure and Properties*, by F. F. Wangaard, Pennsylvania State University, University Park, PA, 1981.
13. American Institute of Timber Construction, *Timber Construction Manual*, 3rd ed., Wiley, New York, 1986.
14. *ASHRAE Handbook of Fundamentals*, American So-

Table 2-7. Thermal Conductivity and Relative Insulating Efficiency of Building Materials.

Material	Conductivity (BTU/hr/ft²/in. thickness/°F)	Insulating Efficiency Compared to Wood (%)
Average softwood	0.80	100
Clay brick	4.8	16.5
Glass	5.5	14.7
Limestone	6.5	12.3
Sandstone	12.0	6.6
Concrete	12.6	6.3
Steel	312.0	0.25
Aluminum	1416.0	0.06

Fig. 2-22. Two nailing methods for Problem 2-7.

ciety of Heating, Refrigerating and Air-Conditioning Engineers, Inc., New York, 1972.

PROBLEMS

2-1. A wood sample weighs 67.2 grams (g) when removed from a compression test specimen. After drying in an oven until its weight is constant, it weighs only 56.4 g. What was the moisture content (percent) at the time of the compression test? Was the MC above or below the fiber saturation point?

2-2. The average specific gravity of Douglas fir south may be found from Appendix Table C-1. What is the average unit weight for that species (pounds per cubic foot, or pcf)?

2-3. A 2 × 2 clear specimen is tested as a beam on a 28-inch simple span. The beam fails under a midspan vertical concentrated load of 900 lb. Measurements of actual dimensions showed the specimen to be 2.02 in. wide and 1.96 in. deep (vertically). What is the modulus of rupture?

2-4. Assuming that the bending test of Problem 2-3 took five minutes to perform, what would be the probable value of modulus of rupture for a similar specimen under a loading of ten years' duration?

2-5. When the load at the center of the beam of Problem 2-3 was 400 lb, the midpoint of the beam had deflected 0.12 in. Considering that the effect of shear deformation is negligible, what is the modulus of elasticity, E?

2-6. For a seasoned wood whose specific gravity is 0.47, what would be the approximate small-clear values for: (a) ultimate compressive strength parallel to the grain; (b) compressive strength (proportional limit) perpendicular to the grain; and (c) modulus of elasticity in compression parallel to the grain?

2-7. Figure 2-22 shows two possible ways of nailing of board-and-batten siding for exterior use. Which method would be less troublesome in service, and why?

2-8. Green western larch 2 × 10s are installed as floor joists in a heated building where the average temperature will be 65°F and the relative humidity 25%. The green 2 × 10s measured $9\frac{1}{2}$ in. deep when installed. What will be their probable depth when the EMC is reached?

2-9. Planks of 2 × 12 coast D. fir will be used flatwise to form a cover over a water tank. The planks will dried to about 15% MC and will measure $11\frac{1}{4}$ in. wide when installed. In use they will be exposed for long periods to 90% relative humidity and 80°F temperature. Why should they not be placed in side-by-side contact with each other? How much space should be allowed between the planks?

2-10. Figure 2-23 shows two arrangements for bolting a wood 2 × 10 to a steel plate. The plank is installed green (unseasoned) and will dry out after it is connected. Which arrangement is more likely to result in the member splitting, and why?

2-11. Figure 1-1 showed the cross section of a two-story light-frame wall. Assume that the sill is a 2 × 6, floor joists are 2 × 12s, and the length of the studs in each story is 7 ft 10 in. (See Appendix Table B-1 for actual dimensions.)

(a) Recognizing that shrinkage in thickness of the plywood floor sheathing is negligible, how much will the distance from

Fig. 2-23. Connections for Problem 2-10.

the top of the concrete foundation wall to the underside of the second-floor ceiling joists decrease, assuming that

(1) all wood is ponderosa pine;
(2) MC was 19% when the wood was installed;
(3) average relative humidity in the building and surroundings will be 30%;
(4) average temperature of the wood will be 70°F; and
(5) annual rings are in the worst position for vertical shrinkage.

(b) What sort of trouble could this cause?

3

Production and Grading of Sawn Lumber

3-1. LUMBER PRODUCTION

Lumber production has grown from the old axe and two-person saw process to a sophisticated operation involving specialized machinery, such as multiple band saws or laser-beam cutters with digital control, and even optimization techniques to increase the yield of useful material from the log. To the engineer or wood scientist who wishes to specialize in it, wood production presents many interesting technical challenges. The structural designer, however, needs to know little about lumber production, except as it might affect either selection of material or the performance of that material in the completed structure.

Where trees are grown on tree farms, lumber quality and yield are influenced by the care given to the trees—thinning and pruning, for example. Lumber production begins with logging, the felling and trimming of the trees. Following logging the steps in lumber production may include (but not necessarily in this order) debarking, air seasoning, rough sawing and finish sawing (by circular saws, multiple band saws, or laser), as shown in Fig. 3-1, surfacing (by planing or sanding), chemical treatment (preservative or fire retardant), and kiln drying. For the structural designer, the important production features are those dealing with standard sizes, type of surfacing, types of "cuts," and moisture condition.

3-2. STANDARD SIZES OF LUMBER

Softwood lumber is normally available in even-foot lengths, from 6 ft to 24 ft or 26 ft. In some instances, the structural designer can save money for a client by avoiding designs that call for lengths that are just slightly longer than these standard lengths. For example, if a floor system required a total of 48 joists, each 18 ft 3 in. long, the builder would have to purchase 48 pieces each 20 ft long. A total of 84 lin ft of joist material would be wasted, unless a use could be found for the 1 ft 9 in. pieces cut off. Being made aware of the possible saving, the client or the architect might consider the wisdom of reducing the width of the structure by three inches to permit using joists 18 ft long. Hardwoods are also available in 2-ft length increments, and occasionally it is possible to obtain odd-foot lengths.

Quoted lengths of lumber are actual dimensions. However, widths and thicknesses of lumber for structural use are expressed in *nominal* dimensions. These are whole-inch dimensions corresponding approximately to the size of the unseasoned, rough-cut pieces from which surfaced lumber is produced. For example, a piece of wood measuring almost 2 in. thick by 4 in. wide in the rough condition may be surfaced (smoothed) to produce a piece of dressed lumber measuring 1.5 in. by 3.5 in., but this smaller piece is still called a 2 × 4.

The thickness of wood removed to smooth the surfaces of rough lumber varies with the rough dimension and with the moisture condition of the wood. Larger rough material will probably not be as straight as smaller-sized rough material, and surface defects may be larger. For either reason, a greater wood thickness may have to be removed from large sizes of rough lumber than from smaller sizes. The

Fig. 3-1. Band saws reducing log to slabs. (Courtesy Western Wood Products Association.)

American Softwood Lumber Standard (1) defines the final dimensions required for the finished lumber.

Table 3-1 shows the minimum standard sizes for dressed structural lumber in the United States. Notice that there are two sets of standard dimensions—one for lumber that is surfaced after it is dry and one for lumber that is surfaced while green. The reason for this is that all lumber eventually reaches an equilibrium moisture content (EMC) for the average temperature and relative humidity where it is installed. So lumber surfaced green is made thicker (and wider) than similar lumber that is surfaced when it is dry (19% or less MC). The object of having these two standards is to cause

Table 3-1.   Minimum Standard Sizes for Dressed Lumber for Structural Use.

	Thickness (in.)			Width (in.)		
		Minimum			Minimum	
	Nominal	Surfaced Dry	Surfaced Green	Nominal	Surfaced Dry	Surfaced Green
Dimension lumber	2^a	$1\frac{1}{2}$	$1\frac{9}{16}$	2^a	$1\frac{1}{2}$	$1\frac{9}{16}$
	3	$2\frac{1}{2}$	$2\frac{9}{16}$	3	$2\frac{1}{2}$	$2\frac{9}{16}$
	4	$3\frac{1}{2}$	$3\frac{9}{16}$	4	$3\frac{1}{2}$	$3\frac{9}{16}$
				6	$5\frac{1}{2}$	$5\frac{5}{8}$
				8	$7\frac{1}{4}$	$7\frac{1}{2}$
				10	$9\frac{1}{4}$	$9\frac{1}{2}$
				12	$11\frac{1}{4}$	$11\frac{1}{2}$
				14	$13\frac{1}{4}$	$13\frac{1}{2}$
				16	$15\frac{1}{4}$	$15\frac{1}{2}$
Timbers	5 or more	—	$\frac{1}{2}$ less than nom.	5 or more	—	$\frac{1}{2}$ less than nom.

^{a}In 2-in. nominal thickness, widths greater than 14-in. nominal are usually not available.

pieces of each variety to be of the same size when they reach a common EMC, the surfaced-green lumber shrinking as it dries and the surfaced-dry material either swelling or shrinking, depending on whether the EMC is more than or less than 19%.

With few exceptions, *nominal* dimensions of lumber normally available in the United States for structural purposes vary by 2-in. increments. Thus (except for nominal 3-in. and 5-in. material that is sometimes available) nominal thicknesses and widths, in inches, are even numbers. Table B-1 in the Appendix of this book shows the standard nominal sizes of sawn lumber and the actual dimensions expected at 19% MC.

Older Size Standards

The dimensions shown in Table 3-1 and Appendix Table B-1 were accepted as the standard for the United States in about 1970. A designer doing remodeling, repair, or preservation work on older buildings should be aware that prior to 1970 the standard dimensions were different. For lumber manufactured before 1970, thickness and width (for lumber of up to 12 in. nominal width) were generally $\frac{1}{8}$ in. to $\frac{1}{4}$ in. more than shown by Tables 3-1 and B-1. For example, a 2 × 4 measured $1\frac{5}{8}$ in. by $3\frac{5}{8}$ in., as opposed to today's $1\frac{1}{2}$ in. by $3\frac{1}{2}$ in.

However, in buildings dating back to the early part of this century or before, lumber dimensions may not conform even to that pre-1970 standard. For example, in working on restoration and remodeling of an historic building, the authors encountered existing first-floor joists that were actually a full 8 in. deep by $2\frac{1}{16}$–$2\frac{1}{4}$ in. wide. Second-floor joists were only slightly different. Each were probably referred to as 2 × 8s. When adding modern 2 × 8s to reinforce the floor, it was necessary to install shims under the ends of the new joists to make their tops flush with the tops of the original joists. Designers working with old structures should always *measure* to determine the actual size of existing wood members.

The problem raised by combining lumber of older and present size standards is compounded by the effects of shrinkage. Structural members

in place in old structures probably have a very low moisture content, whereas new lumber being added probably has higher MC. Thus, some shrinkage of the newer lumber may be expected, and the amount of shrinkage may add to the difference between old and new dimensions.

3-3. FINISH DESIGNATIONS

Most lumber used in building design is surfaced on both faces and on both edges and is referred to as *S4S* (surfaced on four sides). If a designer specifies a nominal size of lumber but does not define which faces are to be finished, the material supplied will most likely be S4S. If some other combination of surfaced and rough faces is desired, however, the designer may specify it. Other combinations are defined by symbols *S* for sides and *E* for edges. Examples of these combinations include:

S2S1E (surfaced on two sides and one edge)
S2S (surfaced on two sides only)
S1S2E (surfaced on one side and two edges)

If lumber other than S4S is desired, the designer will have to specify it. If the item is not stocked, a special mill order may be required, and the builder may have to wait for the material to be produced. A probable exception is rough material (not surfaced), which is frequently available in some of the larger standard sizes.

3-4. CUTTING PATTERNS

Figure 3-2 shows three patterns in which lumber may be cut from the log. The difference between patterns is only the position of the annual rings relative to the wide face of the piece. The importance of this difference is that shrinkage, warping, abrasion resistance, ease of painting, appearance, and other properties all differ according to the position of the annual rings on the cross section.

Figure 3-2a shows lumber cut so that the wide face is nearly parallel to the annual rings. In softwoods, this is called *slash cut* lumber, and in hardwoods *plain sawed*. Figure 3-2b

(a) SLASH CUT
(PLAIN SAWED)

(b) RIFT CUT
(QUARTER SAWED)

(c) BASTARD CUT
(BASTARD CUT)

Fig. 3-2. Types of cut (hardwood names in parentheses).

SLASH CUT

RIFT CUT

BASTARD CUT

Fig. 3-3. Shape change with loss of MC (exaggerated).

shows lumber in which the rings are more nearly perpendicular to the wide face. This is called *rift cut* for softwoods and *quarter sawed* for hardwoods. Of course, there will be many pieces intermediate between these two. Pieces with the rings crossing at about 45 degrees, as shown by Fig. 3-2c, are known as *bastard cut*. Rift-cut lumber is also known as *vertical grain* or *edge grain*, and slash cut as *flat grain*.

In most respects, rift-cut lumber is superior to slash cut. Figure 3-3 shows one reason. As the wood dries, it shrinks, and shrinkage is larger in a transverse direction parallel to the annual rings than in any other direction. Consequently the length along the trace of a ring shortens, and the piece bends (warps) as shown. The slash-cut piece, with its wide face nearly parallel to the rings, distorts appreciably. The rift-cut piece, on the other hand, shows such distortion only on its narrow edges; over the wider dimension it remains flat, although that dimension reduces a little because of shrinkage in a radial direction. The bastard-cut piece deforms to become diamond-shaped, rather than rectangular.

Slash-cut lumber used as floor material is very apt to wear by having whole annual rings pull away from the rest of the piece, leaving the floor with an irregular, often hazardous, surface. Rift-cut material, subject to the same severe use, merely develops small furrows where the exposed edges of the softer springwood erode more easily than the harder summerwood.

If rift-cut lumber is generally superior to slash cut, why not cut the log so that the majority of pieces are rift cut? The answer is that sawing to produce a higher percentage of rift-cut lumber would require turning the log more frequently and would be more expensive. Two common methods used in cutting logs are shown by Fig. 3-4. A third commonly used method is a combination of the two shown. For any of these methods, the majority of the pieces will be either slash cut or bastard cut.

One disadvantage of rift-cut lumber is that it can be severely weakened by knots. As shown by Fig. 3-5, a knot in slash-cut lumber might affect only a small part of the member width, whereas in rift-cut lumber it could affect the entire width.

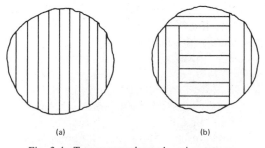

(a)   (b)

Fig. 3-4. Two commonly used sawing patterns.

SLASH CUT   RIFT CUT

Fig. 3-5. Effect of knots on slash-cut and rift-cut lumber.

If the type of cut is not specified, the lumber received will be a mixture of all three cuts, but usually with a majority of slash cut. However, if the better properties of rift-cut lumber are required (as for floors subject to abrasion), the designer can specify that rift-cut lumber be provided, but may expect to pay a higher price than if rift cut were not specified.

3-5. DRYING

Lumber may be either air dried or kiln dried, the object being to reduce the MC to a level close to the equilibrium MC for the conditions under which the lumber will be used. For drying by either method, the lumber is placed in piles, with the layers of wood separated by narrow, intermittent, transverse pieces of wood called *stickers*. The stickers separate the moist pieces of lumber so that air can circulate through the pile, allowing for easy evaporation of adsorbed moisture.

Air drying is an economical method when carried out in well-designed drying yards or sheds, using properly arranged piles, and with favorable drying weather. Air drying may be slow natural seasoning, or it can be accelerated by using fans to circulate the air through the piles of lumber and by using a small amount of heat to raise the air temperature and reduce its relative humidity.

With kiln drying, higher temperatures are used and the rate of air circulation is much faster, and the wood may be dried to any desired MC. Another advantage of kiln drying is that the temperatures used are sufficient to kill any decay-causing organisms present in the wood. Still another advantage is that pitch in resinous woods is set by the heat. However, kiln drying is not without its hazards. The temperature in the kiln must be carefully controlled to avoid damaging the lumber; drying too rapidly can result in shrinkage-caused defects, such as checks, end splits, honeycombing, collapse, warping, and loosening of knots (2).

To What MC Should Lumber Be Dried?

Table 3-2 shows recommended levels of moisture for sawn lumber for various uses and various areas of the United States. Obviously, such a set of "lumped" recommendations cannot provide the optimum MC for each application. The designer should be alert to local conditions

Table 3-2.   Recommended Moisture Content at Time of Installation.

	Recommended Percent MC					
	Dry Southwestern States[a]		Damp Southern Coastal States[b]		Remainder of Continental U.S.	
Lumber Use	Avg.	Range[c]	Avg.	Range[c]	Avg.	Range[c]
Interior finish woodwork and softwood flooring	6	4–9	11	8–13	8	5–10
Hardwood flooring	6	5–8	10	9–12	7	6–9
Siding, framing,[d] exterior trim, and sheathing	9	7–12	12	9–14	12	9–14

[a]Eastern California, Nevada, southeastern Oregon, southwestern Idaho, Utah, and western Arizona.
[b]Coastal California (south of San Francisco), coastal N. and S. Carolina, southern Georgia and Alabama, Florida, Louisiana, southern Arkansas, southeastern Texas (Laredo, San Antonio, Dallas).
[c]Range is more important than average for applications where shrinkage effects are critical.
[d]Framing lumber ordinarily not available manufactured to these recommended percentages.
Source: Courtesy American Society of Civil Engineers, R. J. Hoyle, "Introduction," *Wood Structures, A Design Guide and Commentary*, 1975.

or uses that might require using some other MC level in wood for a particular project. Normally, as it leaves the mill, seasoned or dry lumber has a maximum of 19% MC. The average is about 15%.

When Is Lumber Surfaced?

Lumber may be surfaced while the wood is green (undried) or after drying (seasoning). It is important to know which type of material (surfaced green or surfaced dry) is being used, since the allowable unit stresses for the two types differ (3). (As an example of this, see Table B-3, in Appendix B, under southern pine.) One can identify lumber of either type by the grading stamp applied to each piece. Lumber that is seasoned before surfacing is required to be stamped "S-DRY." If lumber does not bear this stamp, the user should assume that it was surfaced while unseasoned (green). Another mark that may be used is "MC-15," which indicates seasoning to 12% average MC and 15% maximum. Allowable stresses for lumber of this type are higher for some species.

Though seasoning or kiln drying of lumber up to 4-in. nominal thickness is practical, it is difficult and time-consuming to do so with larger material. Consequently, pieces of larger thickness (such as the Beam-and-Stringer and the Post-and-Timber size groups explained later) are usually furnished unseasoned. This means that such pieces are quite prone to splitting as they dry to their EMC, and it is a good reason for turning to glued laminated members (Chapter 8) rather than using large sawn timbers.

3-6. LUMBER GRADING

Grading is the process of classifying lumber according to *quality for a particular use*. This section will introduce the reader to the principles of lumber grading and to its value to the structural designer. (This section is *not* intended, however, to teach the reader to do lumber grading.) Grading is accomplished through rules that define both desirable characteristics and limits to undesirable characteristics of the lumber, and provide methods of quantifying the

undesirable characteristics. Many of the rules governing lumber grading are developed by persons with structural-design expertise, but the actual task of implementing those rules is best left to professionals for whom lumber grading is the principal, full-time job.

The purpose of grading rules is to maintain a standard among lumber mills manufacturing from the same or similar species, so that the quality of the lumber for some particular use can be defined and controlled, regardless of the overall character of the logs from which the lumber is produced. Thus, lumber of a particular grade from different mills will have essentially the same strength and stiffness properties. The structural designer can specify lumber of a particular species group and grade, and can be reasonably assured that the lumber received will be suitable for the intended structural use, regardless of which mill produced it. The grade stamp placed on each piece of lumber will define the quality of material to the user.

History

Lumber grading has been done on some scale for many centuries. Carpentry skill was known in Biblical times. Then, even without formal rules for defining quality, the effect of certain defects was well recognized, and good craftsmen selected lumber carefully to obtain the desired qualities. The first known formalized grading system for selecting lumber was devised in 1754 by a Scandinavian named Swan Alverdson. In the United States, the first grading rules were used in Maine in 1833. As the focus of the lumber industry shifted to the central and western United States, grading rules were refined to adapt them to the new species encountered. By the end of the nineteenth century, grading rules were in use even for west coast species.

Grading rules were developed by various groups, representing manufacturers, distributors, and users of lumber. Naturally, the standards developed by one group often diverged from those of another. In 1924, through the American Lumber Standards Committee, voluntary product standards came into being (see reference 1). Today, even though various regional groups write their own grading rules, those rules fall within the framework of the voluntary standard. As a result, uniform size standards for lumber under 5 in. thick are observed by all U.S. lumber producers. Also, uniform grade names and grading provisions assure designers and users of reliable performance for lumber from most U.S. commercial species.

3-7. TYPES OF GRADING

Grading rules and methods differ according to the use to which the lumber will be put. Properties desired in wood for making furniture are not necessarily the same as those sought in wood for structural members or for interior wall paneling. Thus, different grading rules have been developed for various lumber uses. Even among those for a specific use, grading rules may differ according to the size of the piece being graded. So grading is related to both size and use. This will be shown later by a partial outline of grading rule classifications of the Western Wood Products Association (WWPA).

Structural designers, of course, are interested in strength and stiffness, so modern grading rules provide for what is sometimes called *stress grading*. In stress grading, qualities that affect various strength properties and stiffness are evaluated, the assigned grade name permitting the structural designer to specify a material having a desired level of strength. Two methods used for stress grading are visual grading and machine grading. Today, visual grading is probably the more common. However, machine grading (machine stress rating) is finding increasing use and may soon be the more common method for grading of structural lumber up to 2 in. thick.

Visual grading can be used to determine quality for any given use. In some cases the grader must evaluate appearance, or suitability for use in furniture, and so on. In visual grading of lumber for structural uses, the grader examines each piece to determine the type, location, and size of various defects that might affect its structural strength. Then, according to rules that quantify the effect of each defect, the piece is assigned to a "grade." The grade name or symbol is stamped on the piece, giving the user a means of determining the probable strength and elastic modulus of the material. (Chapter 4 will show how the allowable stresses for design are related to the lumber grade. It will also illustrate how allowable stresses for various lumber species and grades are determined.)

Machine grading is used to determine lumber properties by nondestructive testing. It is based on the principle that all strength properties bear some relationship to the modulus of elasticity, E (see Chapter 2). Figure 3-6 shows, schematically, how a grading machine works. Machine grading now is done only for thinner material, of 2–4 in. nominal thickness. In machine grading, each piece is first subject to visual grading. Then it passes through the grading machine, which bends the piece to a predetermined curvature and measures the force required to bend (deflect) it. Deflection and force required are then used to determine the modulus of elasticity, E. Allowable stress levels are determined from the E value.

For either type—visual or machine—grading cannot be considered an exact science. In one case, it depends on visual capability and personal judgment. In the other, it depends on the accuracy of the relationships that we assume to exist between flexural modulus of elasticity and

$$E = f(\Delta, F)$$

Fig. 3-6. Exaggerated schematic of lumber-grading machine.

strength properties. Nevertheless, the grading process is sufficiently well defined that pieces graded should not fall more than 5% below grade (4).

3-8. DEFINITIONS

Grading rules involve many terms which the structural designer and other users of graded lumber must understand. Knowing the meaning of these terms may also be helpful to understanding subsequent chapters of this book.

Thickness—the smaller lateral dimension of the piece of lumber, shown in Fig. 3-7. (To call that dimension the "width" might *seem* logical when the member in question is a beam, but that is inconsistent with the terminology for grading. When speaking of the piece of lumber, call the smaller dimension the "thickness.")

Width—the larger lateral dimension, shown in Fig. 3-7. (When the piece is used as a beam, with width vertical, this dimension can also be referred to as "depth.")

Rough lumber—lumber that has been sawed but has not had its four longitudinal surfaces finished. Saw marks show on each surface.

Dressed lumber—lumber whose surfaces have been finished (by planing or other means) so that the saw marks on the surfaces are removed. Dressed lumber may be S1S, S2S, S1E, or S4S, according to which faces have been finished (surfaced).

Worked lumber—lumber that has been dressed and has also been matched, shiplapped, or patterned. Matched lumber has a groove on one longitudinal edge and a corresponding tongue on the other. The two pieces join together, side by side, as the projecting tongue of one piece is fitted into the groove of the other. It is sometimes called *tongued and grooved* or *dressed and matched*. Shiplapped lumber has a rectangular notch called a rabbet along each edge, the notch being on one face at one edge and on the other face at the other edge. Adjacent pieces overlap slightly as they are installed side by side. Patterned lumber fastens together side by side with something fancier than a plain tongue and groove or even plainer rectangular rabbets.

Dimension Lumber—Surfaced lumber of 2 in. through 4 in. nominal thickness designed for use as framing members such as joists, planks, rafters, studs, and small timbers.

Joists and Planks—lumber 2–4 in. thick (nominal) and 5 in. or more (nominal) in width.

Beams and Stringers—abbreviated B&S, pieces 5 in. or more in thickness, with width *more* than 2 in. greater than the thickness. (A 6 × 10 could qualify as B&S, but a 6 × 8 could not.) B&S lumber is graded assuming that its principal use will be in bending, not axial load, although it may actually be used for either purpose.

Posts and Timbers—abbreviated P&T, pieces measuring 5 in. × 5 in. or more, with the width *not* more than 2 in. greater than the thickness. (A 6 × 6 or a 6 × 8 would qualify as P&T, but a 6 × 10 would not.) P&T lumber is graded assuming that its principal use will be in axial load rather than bending, although it may actually be used for either purpose.

Light Framing—Pieces with both thickness and width in the range from 2 in. nominal to 4 in. nominal, graded for use in light-frame structures (2 × 2s, 2 × 4s, and 4 × 4s all qualify).

Boards—boards ranging from $\frac{3}{4}$ in. to $\frac{16}{4}$ in. (called "16-quarters") in nominal thickness. They are graded for nonstructural uses such as shelving, wall facing, and cabinets. Appearance is an important criterion in the grading of boards.

Factory Lumber—Lumber is for remanufacturing into items such as doors, window frames, moldings, furniture, pencils, and the like. Just what criteria are important in grading Factory Lumber depends on the type of use. Conse-

Fig. 3-7. Thickness and width of lumber.

WIDTH

THICKNESS

quently there are many sets of grading rules for various types of Factory Lumber.

3-9. MODERN GRADING RULES

Associations that have produced grading rules for structural lumber include the following:

Northeastern Lumber Manufacturers Association (NELMA)

Northern Hardwood and Pine Manufacturers Association (NHPMA)

Redwood Inspection Service (RIS)

Southern Pine Inspection Bureau (SPIB)

West Coast Lumber Inspection Bureau (WCLIB)

Western Wood Products Association (WWPA)

National Lumber Grades Authority (for Canada) (NLGA)

Through the voluntary Product Standard (1) mentioned earlier, the grading rules of the agencies listed above are sufficiently similar that a structural designer can be reasonably assured of obtaining the quality lumber desired.

As an example, consider the grading rules of one of the above agencies, the Western Wood Products Association (given in reference 4). The condensed list below shows only those lumber classifications, size/use groups, and grades that might be of interest to the structural designer. Following this list is a rather detailed illustration of the grading rules for just one of these size/use groups. It should be pointed out that some of the grades listed may be hard to obtain. For example, select structural lumber is difficult to obtain in large quantities.

Size/Use Group	Grades
Light Framing and Studs	
Light Framing	Construction
	Standard
	Utility
	Economy
Studs	Stud
	Economy Stud
Structural Light Framing	Select Structural
	No.1
	No. 2
	No. 3
	Economy

Size/Use Group	Grades
Special Dimension	
Appearance Framing	Appearance
Machine Stress Rated	
Foundation Lumber	
Decking	Selected
	Commercial
Structural Laminations (see Chapter 8, on glulams) (1″ and 2″ nominal thickness)	
	Laminating One (L1)
	Laminating One-C (L1-C)
	Laminating Two (L2)
	Laminating Two-Dense (L2-D)
	Laminating Three (L3)
Joists and Planks	
Structural Joists and Planks	Select Structural
	No. 1
	No. 2
	No. 3
	Economy
Timbers	
Beams and Stringers	Select Structural
	No. 1
	No. 2 (No. 1 Mining)
	No. 3 (No. 2 Mining)
Posts and Timbers	Select Structural
	No. 1
	No. 2 (No. 1 Mining)
	No. 3 (No. 2 Mining)

If we were to expand the above list to include Factory Lumber, Boards, paneling, and other items that normally do not interest the structural designer, it would be several times as long.

3-10. EXAMPLE OF VISUAL GRADING OF BEAMS AND STRINGERS

Four grades of Beams and Stringers (B&S) are defined by the WWPA rules—Select Structural, No. 1, No. 2, and No. 3, the first being the one of highest quality and highest allowable stresses. WWPA rules specify allowable stresses for only the highest three grades; no allowables are quoted for grade No. 3.

In WWPA's rules for each grade defined for Beams and Stringers, there is first a general statement showing the purpose of the grade. This is followed by a list of characteristics and limiting provisions. For example, the list of characteristics and limits for the Select Structural grade is as follows (4):

Grain—medium.

Stained sapwood.

Firm heart stain—10% of width, or equivalent.

Splits equal in length to half the width of the piece or equivalent of end checks.

Seasoning checks—single or opposite each other with a sum total equal to one-fourth the thickness.

Pinholes limited.

Heavy torn grain.

Pitch streaks.

Slope of grain—1:15 in middle third of length, balance of piece may be 1:12.

Occasional skips—$\frac{1}{16}$ in. deep, 2 ft in length.

Medium pitch pockets.

Wane—one-eighth of any face, or equivalent slightly more for a short distance.

Shake—one-sixth the distance on end.

Knots, sound, tight, and well-spaced, are limited as follows:

Width or Thickness (in.)	Narrow Face and Edge of Wide Face, Knots in Middle Third of Length (in.)	Knots at Ends and Along Centerline of Wide Faces (in.)
5	$1\frac{1}{4}$	
6	$1\frac{1}{2}$	
8	$1\frac{7}{8}$	2
10	2	$2\frac{5}{8}$
12	$2\frac{1}{8}$	$3\frac{1}{8}$
14	$2\frac{3}{8}$	$3\frac{3}{8}$
.	.	.
.	.	.
.	.	.

Knot sizes on narrow faces and at the edge of wide faces may increase proportionally from the size permitted in the middle one-third of the length to twice that size at the ends of the piece, except that the size of no knot shall exceed the size permitted at the center of the wide face.

The above excerpt from the grading rules may appear lengthy, but actually it is somewhat condensed. Obviously then, lumber grading is a job for an expert—one who is thoroughly familiar with the subject. It is a task the structural designer is usually happy to leave to that expert. Nevertheless, knowing something about the basis of grading is helpful to the designer. That knowledge may be useful in deciding bor-derline cases and may help to avoid construction and design errors.

3-11. GRADING STAMPS

Each piece of graded lumber receives a distinctive stamp telling the assigned grade, species (or species group), moisture condition at time of surfacing, identity of the mill that produced the lumber, and perhaps other information. By the grade stamp, the builder can be assured of getting the specified quality of lumber, and the owner or the building inspector can be assured that the material actually being furnished conforms to the construction specifications.

Figure 3-8 shows one example of a grade stamp for visually graded lumber. In this example stamp, the numeral 12 at the upper left identifies the mill that produced the lumber. (Sometimes the mill is identified by name rather than number.) The logo in the circle at the lower left certifies that the grading was done under WWPA rules and supervision. The species group shows in the square at the lower right. In the upper right, the label "S-DRY" tells that the wood was dry (seasoned) when it was surfaced. The large numeral in the center shows that the piece bearing this stamp is of grade No. 1. It is not necessary for all parts of the grade stamp to be in the same position or form shown by this illustration. Also, the stamp may contain a value for allowable fiber stress. If it does so, the value shown is the "single-member" allowable stress for bending (explained in Chapter 4) about the major axis.

Figure 3-9 shows an example grade stamp for machine-stress-rated lumber. This stamp, too, tells which mill produced the lumber and graded it, under what agency rules and supervision it was graded, its moisture condition when it was surfaced, and the species or species group, and finally, an allowable bending

Fig. 3-8. Example of stamp for visually graded lumber. (Courtesy of Western Wood Products Association.)

MACHINE RATED

(WWP)® 12 S-DRY [HEM FIR]

1650f 1.5E

Fig. 3-9. Example of stamp for machine-stress-rated lumber. (Courtesy of Western Wood Products Association.)

stress and modulus of elasticity. In this example, E is 1.5 million psi, and the allowable stress for major-axis bending is 1650 psi.

3-12. CAUTION TO DESIGNER AND BUILDER

The grade stamp on a piece of lumber applies only to the entire piece that was graded. A piece cut off from the original graded piece can be of either the same grade, a higher grade, or a lower grade than the stamp implies. Why? The reason is that in grading, the *location* of knots and other defects is considered.

For example, consider the beam-and-stringer rules that were outlined above. Assume that you have ordered a 20-ft length of Select Structural 6 × 12. For some reason, you now decide to have that length cut to produce two pieces 10 ft long. Do you now have two 10-ft pieces of Select Structural? Possibly so. But if the original member had a 1 : 12 grain slope in the end one-thirds of its 20-ft length, wood having that grain slope now appears in the central one-third of the new 10-ft lengths, a location where the limit allowed is 1 : 15. That limit for the central one-third of length is now exceeded; consequently, the grade of the shorter pieces is *lower* than the grade of the original 20-ft piece. (Further, it may be that only one of the two pieces has the necessary stamp to verify to the building inspector that it is graded lumber. The other 10-ft piece could be rejected as ungraded.)

As another example, assume that you have excess 2 × 12s on the job site, but you need 2 × 10s badly. Can you have the carpenters rip (saw) off a 2-in. strip from the full length of each 2 × 12 to produce 2 × 10s of equal grade? For some of the pieces you probably could not, since knots near the interior of the original width would now become edge knots. In bending members, the most damaging knots are edge knots, that is, knots near the top or bottom extreme fibers, where flexural stresses are the highest. The new 2 × 10s produced by this method would have to be regraded to be certain that they still met the requirements of the grade shown by the grade stamp.

The authors know of a case in which a large number of 2 × 12s were ripped to produce 2 × 10s. The ripped members were installed before the inspector could regrade them. The members having been installed, their upper surfaces were not visible and it could not be confirmed that no edge knots were present. Needless to say, the solution was very expensive for the builder.

As a designer, could you cause trouble for yourself because of the way Beams and Stringers are graded differently in the middle third of their length than in the end portions? Possibly, if you ordered long pieces of sawn lumber with the idea of using them for multiple-span beams. Continuous beams usually have the highest bending moments at their interior supports. In such cases it would be well to check the grading rules to be certain that you have the required quality at all locations along the length of the piece.

3-13. BOARD MEASURE

The quantity of wood in a timber of given length and cross section could be measured in cubic inches or cubic feet. However, the lumber industry in the United States has preferred to use the board foot. A board foot, abbreviated "fbm" for foot-board-measure, is the amount of wood present in a piece one foot long and having an end area of 12 square inches. For example, a 1 × 12 one foot long contains one board foot. So does a 2-ft length of 2 × 3 or a 3-ft length of 2 × 2.

The number of board feet of lumber in a given project is a convenient quantity to use in cost estimating, since lumber prices are often quoted in terms of dollars per Mfbm—cost per 1000 board feet.

Nominal sizes are used in determining the number of board feet. Thus, the fbm figure indicates not the amount of wood present in the finished product, but rather the amount of wood consumed in producing the finished product. Board measure is used routinely for sawn lumber. For glulams, the board measure is that of the pieces from which the glulam is made.

Example 3-1

What is the probable cost of a truckload of lumber that includes the items listed below? Assume that the average price for these items will be $500 per Mfbm. (Today's unit price can be estimated from reports in publications such as the *Engineering News Record*. These prices, however, are usually for carload lots of any one size, so there will be considerable price markup to obtain retail prices for smaller loads.)

170	2 × 4s,	8′ long
38	2 × 4s,	22′ long
85	2 × 10s,	20′ long
16	4 × 4s,	12′ long
85	2 × 12s,	20′ long
1	6 × 14,	22′ long

Consider each item separately. Compute the fraction of a board foot per foot of length, and multiply this by the total length per piece and by the number of pieces. For example, a 2×4 has 8 in.^{2} of nominal end area. Each foot of length, therefore, contains $\frac{8}{12}$ of a board foot. A piece 8 ft long contains 8 times $\frac{8}{12}$, and 170 such pieces contain $170\ (8)(\frac{8}{12})$ board feet. Tabulating this computation for each item:

$$170\ (2 \times \tfrac{4}{12})(8) = 907 \text{ fbm}$$

$$38\ (2 \times \tfrac{4}{12})(22) = 557 \text{ fbm}$$

$$85\ (2 \times \tfrac{10}{12})(20) = 2833 \text{ fbm}$$

$$16\ (4 \times \tfrac{4}{12})(12) = 256 \text{ fbm}$$

$$85\ (2 \times \tfrac{12}{12})(20) = 3400 \text{ fbm}$$

$$1\ (6 \times \tfrac{14}{12})(22) = \underline{154} \text{ fbm}$$

$$\text{Total} = 8107 \text{ fbm} = 8.107 \text{ Mfbm}$$

$$\text{Probable cost} = 8.107\ (\$500) = \$4053.50$$

REFERENCES

1. *American Softwood Lumber Standard, Product Standard PS-20-70*, U.S. Department of Commerce, U.S. Government Printing Office, Washington, DC, 1970.
2. *The Encyclopedia of Wood, Wood as an Engineering Material*, Drake Publishers, New York, 1977.
3. *National Design Specification for Wood Construction*, National Forest Products Association, Washington, DC, 1986.
4. *Western Lumber Grading Rules 81*, Western Wood Products Association, Portland, OR, 1981, and 1983 Supplement.

PROBLEM

3-1. How many fbm are used to produce each of the following?
  (a) one 10 × 16, 24 ft long
  (b) one 8 × 12, 16 ft long
  (c) 70 3 × 4s, 18 ft long

4

Loads and Allowable Stresses

This chapter is in two parts. Part I deals with loads—the forces to which timber structures may be subjected. It concerns the sources of these loads and shows how to compute their magnitudes. Part II covers the method code writers use to establish allowable stresses and shows how the structural designer selects and modifies those allowables to design by the service load method.

Part I. Loads

4-1. GENERAL

Loads and stresses are presented here using the current U.S. system, although at some future date the international system (SI) may be used routinely in the United States. Lumber sizes are given in inch units, with lengths in feet and inches. Appendix Table B-1 shows standard lumber sizes and cross-section properties.

Service loads are the maximum loads one may reasonably expect to occur. Under the service load design method, they are the loads a structure must be designed to resist. Except for loads due to the dead weight of the structure, service loads are usually spelled out by a local building code and tend to be somewhat higher than the *normally expected* loads. For example, the design service load for a classroom is often specified as high as 50 pounds per square foot (psf). However, in a normal classroom situation, with desks occupying about 10 ft² each, the weight of the movable desk and the person occupying it actually averages much lower than 50 psf, perhaps as low as 25 psf. The full specified service load would not occur normally, but it might be approached under abnormal conditions. For example, the load intensity near the classroom exit during a fire drill could become even higher than the specified design service load.

Thus the specified service load is conservative for normally expected situations, but does not necessarily cover the extreme case. The margin of safety provided in the allowable stress used for design is usually adequate to prevent catastrophic collapse for overload conditions such as the one illustrated above. The designer should not be too complacent, however; serious failures, often well publicized, do occasionally occur.

Under service load design (also known as either working-stress or allowable-stress design) safety is provided by using an allowable stress that is low enough to protect against (1) variation in material properties, (2) errors in design theory, and (3) uncertainties as to the exact load. The National Design Specification (1) is based on the working-stress design method.

The ratio of the load that would cause failure to the load for which the structure is designed is called the factor of safety. Because stress redistribution and inelastic action occur as failure is approached, the factor of safety is *not* the same as the ratio of ultimate stress to allowable stress.

Another design approach, probabilistic design, is discussed briefly in Section 4-11. The main difference between the two design methods, service load and probabilistic, is simply that the former method ensures safety by using a relatively low allowable stress, while the latter method provides a more nearly uniform factor of safety for all load conditions and types of member.

Load Types

Loads are classified according to their source and duration. Types to be considered in timber design are dead load, live load, snow load, ponding load, wind load, seismic load, and occasionally impact. For timber bridges, water pressure and ice loads may also be important.

Design dead loads are computed by the structural designer. Values of other loads, however, are usually specified by various codes. Of these, the Uniform Building Code (UBC) (2) is probably the one most widely used in the United States. Local building codes are often patterned after the UBC, differing only with respect to unusual local conditions or more conservative local preferences.

4-2. DEAD LOADS

Of all the loads a designer must consider, dead load is the one that can be determined most precisely, and it generally remains constant for the life of the structure. Dead load is the force due to gravity acting on the mass of the structure itself. Ordinarily, it is not a specified load, but one that must be determined by the designer. For the design of any particular member, it includes the self weight of that member, plus all dead-load forces brought to it by other members or components that it supports. In computing the amount of dead load, a designer should remember to include the weights of items such as ceilings, floor covering, ventilating equipment, light fixtures, fixed partitions, wall panels, and windows. The weights of some of these may have to be estimated, but others may be computed precisely. Also, for office buildings and other buildings in which interior partitions may be added or relocated, the location and amount of dead load due to partition weight is uncertain. To provide for this, UBC requires the designer to consider a uniform dead load of 20 psf in addition to other known floor

dead loads. Tables of building material weights are helpful. One such table is Table A-1 in Appendix A.

Our concern is designing wood members, so it will be necessary to estimate the weight of various sizes of wood member. Two Appendix tables are particularly helpful. Table B-1 gives the weights per lineal foot of standard U.S. lumber sizes for various unit weights (pounds per cubic foot) of wood. However, unit weight may not be known in advance. In that case, one can assume an average value. A reasonable approach might be to use Table C-1 showing average specific gravities for various species. From the average specific gravity, the average weight per cubic foot for *oven-dry wood* may be computed. Wood in the structure will ordinarily contain 11–19% moisture, so the computed weight per cubic foot will have to be adjusted upward to account for moisture content.

At the start of design for a particular member, the member's final size is not known, so self-weight can only be estimated. If this estimate is sufficiently wrong as to affect the design (i.e., correcting it will change the choice of member size), the design procedure should be repeated using a corrected estimate. In any case, a faulty estimate of member weight should be corrected before going on to the design of other members whose loads may depend on this estimate.

Example 4-1

What is a reasonable unit weight to use in designing a member of lodgepole pine?

From Table C-1, average specific gravity = 0.44 (oven dry).
Average unit weight, oven dry = $62.4 \times 0.44 = 27.5$ pcf.
Unit weight, corrected for 19% moisture = $1.19 \times 27.5 = 32.7$ pcf.

Table C-1 does not mention for what *grade* of lumber the value of specific gravity is given. Stronger grades may weigh more than weaker ones. So the designer must still use judgment in rounding off the 32.7 figure. (The American Society for Testing and Materials suggests a more accurate method of determining unit weight which corrects for volume change due to change in moisture content. However, the precision implied by that method is seldom justified for the design of structural members, since self-weight of a member is usually only a small percentage of the total load the member carries.) With experience, a designer can quickly estimate the unit weight.

Example 4-2

Figure 4-1 shows the plan and a cross section of a building floor. Compute the dead load carried by beam A.

This is a second-floor member, so it must support the weights of ceiling, light fixtures, and ventilation ducts for the story below. Assume that D. fir–larch

will be used for the joists and beams. Finish floor will be $\frac{7}{8}$-in.-thick oak. (This same building will be used several times in this chapter for computing various types of load.)

The floor plywood acts as a continuous beam, spanning several 16-in. distances between floor joists and carrying the floor live and dead loads, including the weight of the plywood itself. The joists support this continuous beam, and reactions for the continuous beam of floor plywood become loads on the joists. Next, each joist acts as a beam, carrying all load it receives from the floor plywood, plus the weight of the joist itself.

(a) 2ND FLOOR PLAN

(b) VIEW AT SECTION 1-1

(c) BEAM A

Fig. 4-1. Floor for Examples 4-2, 4-3, and 4-6.

The joists in turn are supported by the beams, and the joist reactions become loads for the beams, such as beam A. Beam A carries the loads (joist reactions) transferred to it by all joists it supports, plus the weight of beam A itself.

Each joist supports load from one-half the floor plywood span on one side of the joist plus that from one-half the plywood span on the other side (shaded in Fig. 4-1b). This is equivalent to saying that each joist supports the load from a *tributary area* extending from the center of the span on one side of the joist to the center of the space on the other side. Per foot of joist length, this tributary area is

$$1(16/12) = 1.33 \text{ ft}^2$$

The floor dead load per square foot consists of the weight of the plywood plus the weight of the finish flooring material. Assuming the plywood to weigh 40 pcf, the weight of $\frac{5}{8}$-in.-thick plywood is approximately

$$40(0.625/12) \qquad\qquad = 2.1 \text{ psf}$$

$\frac{7}{8}$-in. finish floor (SG of oak = 0.67)

$$0.67(62.4)(0.875/12)(1.19) \qquad = \underline{3.6}$$

$$\text{Total} \qquad\qquad\qquad = 5.7 \text{ psf}$$

The total dead load per square foot of tributary area includes the weight from below—ceiling, light fixtures, and ventilation ducts.

$$\frac{5}{8}\text{-in. gypsum board (Table A-1)} = 2.5 \text{ psf}$$
$$\text{Lighting and ducts, estimated} = \underline{4}$$
$$\text{Total approx.} = 6.5 \text{ psf}$$

The weight of the joist itself must be estimated. Using the approximate method, Douglas fir–larch with 19% moisture weighs about $0.51(62.4)(1.19) = 38$ pcf, and the joist weight is

$$38(1.5)(11.25)/144 = 4.45 \text{ lb/ft}$$

The total dead load per linear foot of joist is

$$4.45 + 1.33(12.2) = 20.7 \text{ lb/ft}$$

(Most designers would round this to 21 lb/ft. Had the more accurate method been used, a slightly lower weight would have resulted.)

Since the joists span 18 feet, each joist has a reaction of 9×20.7, or 186 lb. Beam A supports joists from *each* side, so it receives a concentrated dead load of 2×186, or 372 lb at each joist location.

Beam A is not yet designed, so we must estimate its weight. Let's assume that the beam will be a 4 × 14. If so, and if it, too, is D. fir–larch, it will weigh about 13 lb/ft. The loads on beam A are its own uniformly distributed weight plus a series of 372-lb concentrated loads as shown in Fig. 4-1c. When solving for the bending moment due to such a series of closely spaced loads it is sufficiently accurate to replace them by a uniform load. Thus, the series of 372-lb loads at 16-in. centers along the beam could be replaced by a uniform load of $(12/16) \times 372$, or 279 lb/ft. Adding for the beam weight, the total design dead load for beam A is $279 + 13 = 292$ lb/ft.

The solution above went into considerable detail so that beginning designers would not lose sight of how load is transferred from one member to the next: from floor to joists, and from joists to main beams. However, the end purpose is to determine a uniform design dead load for beam A, so a simpler approach, demonstrated next, can be used.

Example 4-3

Repeat the above example, but by a simpler and more direct method.

The total tributary area bringing floor dead load to beam A extends from the center of the joist span on one side of beam A to the center of the joist span on the other side. The entire tributary area measures 20 ft by 18 ft. Each square foot of floor area has the following average dead load:

Floor plywood	= 2.1 psf
Finish floor	= 3.6
Ceiling below (including ducts and lighting)	= 6.5
Floor joists $(12/16) \times 4.45$	= $\underline{3.3}$
Total	= 15.5 psf

Total load on beam A = $(18 \times 20 \times 15.5)$ + beam weight. Estimate the beam weight to be 13 lb/ft. Total dead load per foot of beam is

$$w = (18 \times 20 \times 15.5/20) + 13 = 292 \text{ lb/ft}$$

the same as found by the lengthy procedure.

4-3. VERTICAL LIVE LOADS

For buildings, vertical live loads include the weight of anything supported by, but not a permanent part of, the structure. Thus, carpeting, furnishings, movable equipment, personnel, and stored materials would all be called *live load*. For bridges, live load is the weight of the traffic the bridge carries—vehicles for a highway or railway bridge, or people for pedestrian bridges.

The weight of snow meets the above definition, but in the United States snow load is usually considered separately from other live loads. This is a reasonable practice, and a convenient one to use in designing timber structures, since allowable stresses for wood vary according to the load duration.

Roof Loads

Roof Live Loads. Even in areas where snow is not expected, some live load must be considered in designing members that support roofs. This live load is based on the realization that the weights of construction supplies and personnel must be carried by the roof members. Appendix Table A-3 shows minimum roof live loads specified by the Uniform Building Code (UBC). Note that the improbability of full live load occurring simultaneously over large areas is considered by specifying smaller load intensities for larger tributary areas. ("Method 2" shown by the three columns at the right refers to a reduction method that will be shown here for floor live loads, but which applies to roof live loads also.) Wherever local specified snow loads cause larger stresses, the snow loads are used instead of the minimum specified live loads. Use whichever is worse, not both.

The designer should consider *unbalanced* live load wherever it causes a more severe condition than full loading for either members or connections. Examples would be alternate-span loadings for continuous members or loading of one slope only for a pitched roof.

Snow Loads. Design snow load intensity for various localities is usually specified by local building codes for those areas—the Denver

Building Code, for example. Occasionally, though, terrain and climatic conditions are such that a code-specified snow load is unrealistically low. One such case reaching the authors' attention was for a structure in the Colorado mountains. The specified snow load was 60 psf, but weather records for that particular spot showed 135 psf to be a more reasonable design load. The designer should be alert to such unusual conditions.

Snow load is specified in pounds per square foot *of horizontal projection* rather than pounds per square foot of roof. Also, steeply sloping roofs can hold a lesser depth of snow than can flat roofs or those with a mild slope. Where the specified snow load exceeds 20 psf, UBC allows the following reduction of snow load intensity for each degree of roof slope over 20°:

$$R_s = (S/40) - 1/2 \qquad (4\text{-}1)$$

In this equation, R_s is the snow load reduction in pounds per square foot per degree of slope above 20°, and S is the specified design snow load in pounds per square foot of horizontal projection for a flat roof.

Snow load is normally considered to be uniform over the entire span of members supporting a roof. However, *unbalanced* snow load due to drifting must also be considered. Failures have occurred because of considering only balanced loading. The UBC requires designers to consider the "potential accumulation of snow at valleys, parapets, roof structures, and offsets in roofs of uneven configuration." (The main body of UBC does not suggest how the designer might do this, but a UBC appendix gives details on recommended drift depths for various types of structure irregularity.)

Other Roof Loads. In addition to snow, live loads that may affect roof support members include rain, ice, building materials or equipment stored on the roof during construction or repair, and the weight of personnel working on the roof. The last two loads *may* be adequately covered by the minimum specified roof live loads (Table A-3), but the first two are not.

Loads from rain are important in the case of flat, or nearly flat, roofs whose supporting members have fairly long spans and are fairly flexible. *Ponding* on these roofs may cause complete collapse. In ponding, the weight of rainwater causes the member to deflect slightly, forming a small depression in the roof surface. Water filling the depression causes further deflection, allowing more water to accumulate, and so on. Under certain combinations of load and stiffness, the process stops, a point of stability being reached. However, with roof systems of inadequate stiffness, the effect is that of a diverging series. The structure becomes unstable, filling and deflecting continuing until the system collapses. Ponding failure can occur in roof systems of any structural material, particularly in regions where roofs are not designed to carry snow loads. Fortunately, proper attention by the designer can prevent ponding failures. (See Chapter 6.)

Rain (or snow melt water) may also cause significant load when roof drains become blocked, especially if the roofs have parapet walls around their perimeter. If the drains cannot remove water from the roof, the water may accumulate, applying even greater load than the snow load for which the structure was designed. In one case known to the authors, several wood trusses supporting a nearly flat roof had failed. Investigators determined that a soft-drink can found on the roof had blocked the drain.

Floor Loads

Building floor live loads are specified by building codes and vary according to building use. Table A-2 (Appendix) shows the UBC minimum floor live loads. A distributed-load intensity is shown for all building types; for some a single concentrated load (located to cause worse effect) also must be considered. As pointed out earlier, the specified design load (service load) is usually conservative. However, the designer should be alert to unusual circumstances in which the actual load might be larger than the specified design load.

Full specified live load is not normally present. It is even less frequent that *large* floor areas receive full design live load simultaneously on each unit of area. Building codes consider this improbability by allowing the designer to use a reduced value of live load in designing members that receive live load from large areas. For example, UBC allows reducing the unit floor live load by 0.08% for each square foot of area in excess of 150 supported by the member, but with limits, as follows:

1. The total reduction may not exceed 40% for members receiving load from one level only, nor 60% for other members.
2. The total reduction may not exceed R (percent), where

$$R = 23.1(1 + D/L) \qquad (4\text{-}2)$$

(In this equation, D and L are the dead and live loads per unit area, respectively.)
3. Reductions are not allowed for places of public assembly or for areas having design live loads larger than 100 psf.

(Roof live loads may be reduced in the same manner as floor live loads, except the percentage reduction and maximum reduction are given by Method 2 in Table A-3.)

Example 4-4

Figure 4-2a shows a building cross section. The governing code specifies snow load of 35 psf of horizontal projection and allows the UBC reduction for roofs sloping more than 20°. In designing the roof beams, what load per lineal foot should be used? (Consider both snow and dead loads.)

The roof slope is 26.6°, so design snow load may be reduced by

$$R_s = (35/40) - 1/2 = 0.375 \text{ psf}$$

for each degree over 20°. Thus, for design, we may use snow load equal to $35 - 0.375(26.6 - 20) = 32.5$ psf *of horizontal projection.*

Figure 4-2b shows a short section of roof decking and a mass of snow resting on it. The thickness (normal to the sketch) is one foot. The volume of snow shown weighs 32.5 lb, and the sloping length of roof on which it acts is $1.0/\cos 26.6 = 1.12$ ft.

Roof dead loads *per square foot of sloping roof surface* are

Fig. 4-2. Roof for Examples 4-4 and 4-5.

Roofing (Table A-1)   = 6.0 psf

Decking (2.5/12) ×

(0.42 × 62.4)(1.19) = 6.5

Insulation (est.)   = 2.0

　　　Total   = 14.5 psf of roof

Snow loads are per square foot of horizontal projection, but dead loads are per square foot of roof area. One or the other of these loads must now be resolved to permit adding them together. In this case, let's express both in pounds per square foot of sloping roof area.

　　　The 32.5-lb snow load acts vertically on a sloping roof area measuring 1.12 ft by 1.0 ft. The snow load per square foot of sloping roof is $32.5/1.12 = 29.0$ lb, and the total vertical load per square foot of sloping roof area is $29.0 + 14.5 = 43.5$ lb.

The supporting beams (rafters) are at 10-ft centers, so each one-foot length of beam receives 435 lb of load from the roof plus the self-weight per foot of beam. Assuming the beam to weigh 17 lb/ft, the total vertical load on a beam is 452 lb per lineal foot. For computing the beam bending moment, it is sometimes easier if we resolve this vertical load into two components, one transverse to the beam and one parallel, as shown by Fig. 4-2c. The transverse load (404 lb/ft of beam) can be used to determine the beam shear and bending moment. Designers should not forget that the other component, 202 lb/ft, causes axial force in the beam.

Example 4-5

What is the vertical load applied to the column (vertical member) at point A of Fig. 4-2a, assuming that the column spacing is 10 ft?

We have already determined the vertical load per square foot of roof and, for snow, per square foot of horizontal projection. Each rafter is supported at one end by the ridge beam and at the other end by a column. From Example 4-4, the total vertical load on the rafter is 452 lb per sloping foot, which amounts to $1.12(452) = 506$ lb per horizontal foot. Figure 4-2d shows the rafter and its vertical load. The reaction at point A is the load in the column:

$$R_A = 24(506)(12/20) = 7290 \text{ lb}$$

For the design of the column, the self-weight of the column and the weight of any wall material supported by the column would have to be added to this reaction.

Example 4-6

Consider again the floor plan of Fig. 4-1. The specified floor live load is 50 psf, as given by Table A-2 for offices. What is the design live load per foot of length for beam A?

The tributary area for beam A is 20 ft by 18 ft = 360 ft^2 (one area 20 ft by 9 ft north of beam A and a similar area to the south). The tributary area exceeds 150 ft^2, and the specified live load does not exceed 100 psf. Therefore, the design live load may be reduced. A reduction of $0.08(360 - 150)$, or 16.8% is allowed based on the area supported by the member. From Example 4-3, we know that the dead load is 15.5 psf, ignoring the weight of beam A. Equation 4-2 gives a limit of

$$R = 23.1(1 + 15.5/50) = 30.3\%$$

The reduction based on tributary area, 16.8%, exceeds neither R nor the 40% ceiling for members receiving their load from one level, so the 16.8% reduction controls. The design live load for beam A is

$$50 \text{ psf}(1 - 0.168)(18 \text{ ft}) = 749 \text{ lb/ft}$$

and the total load for design is 749 (live) + 292 (dead), or 1041 lb/ft.

Example 4-7

For the same structure, assume that two such floors (but no roof) deliver live load to column B-2 (at the intersection of grid lines B and 2). What is the design live load for column B-2?

The tributary area per floor for column B-2 is 360 ft^2. For the two supported floors it is 720 ft^2. The allowable live-load reduction for designing the column is $0.08(720 - 150)$, or 45.6%. The limits are 60% (since the column receives load from more than one level), and $R = 30.3\%$ by Eq. 4-2. The latter value controls, and the design live load for the interior column is

$$50 \text{ psf}(1 - 0.303)(720 \text{ ft}^2) = 25,090 \text{ lb}$$

Example 4-8

What is the total live load for design of column B-2 if it also supports a nearly flat roof and the specified snow load is 30 psf?

No reduction for area is allowed for snow load (although it is allowed for other types of roof live load), so the total design live load per interior column is

$$25,090 + (360 \times 30) = 35,890 \text{ lb}$$

4-4. WIND LOADS

Wind forces are normal to the surface against which the wind acts. Thus, building walls receive horizontal forces, while horizontal roofs are subject to vertical forces from wind. On sloping roofs the wind forces are normal to the roof and have both horizontal and vertical components. In each case, the force may be either toward the surface (pressure) or away from it (negative pressure, or suction). The importance of suction is shown by failures where roofs are lifted from buildings rather than being blown down into the buildings. The chance of such failure is increased where wall openings allow the wind to increase air pressure inside the building, the internal force from this pressure adding to the outward-directed force from outside suction.

Wind forces for structural design are usually specified by local building codes. Some codes specify the intensity of horizontal wind pressure to be applied to a vertical surface. These codes also include rules for determining pressure or suction intensities on sloping roof surfaces or on the underside of open, overhanging surfaces. Many codes, however, have changed to the method shown by the Uniform Building Code (2). UBC's wind force requirements are based on methods first shown by the American National Standards Institute (ANSI) (3). The main points of the UBC method are summarized below.

Figure 4-3 is UBC's map giving basic wind speeds for the United States. The map shows *minimum* wind speeds for design; where local records show higher 50-year wind speeds at a standard distance above the ground, the higher speeds should be used.

Exposure affects the wind pressure a structure will receive. Defined exposure C occurs where the terrain is flat and generally open for one-half mile or more from the site. Exposure B occurs where buildings, forest, or surface irregularities 20 ft or more in height cover 20% or more of the area extending one mile or more from the site. The height above the ground and defined exposure are used with Appendix Table A-4 to determine a combined height, exposure and gust factor, C_e. Another UBC table, shown here as Table A-5, gives pressure coefficients, C_q, for various parts of the structure. These C_q coefficients allow for pressure computation by either of two methods defined by UBC—the normal-force method or the projected-area method. Next, a wind stagnation pressure, q_s, is found from Appendix Table A-6.

Fig. 4-3. Basic wind speeds. USA. Reproduced from the 1985 edition of the Uniform Building Code, copyright © 1985, with permission of the publishers, the International Conference of Building Officials.

Finally an importance factor, I, is selected. I is specified as 1.15 for essential facilities—ones which, after a windstorm, must be safe and usable for emergency purposes to preserve the safety and health of the public, and assembly buildings holding more than 300 persons in one room. For all other structures, I of 1.0 minimum may be used. The above factors are then entered into the following equation to determine the design pressure:

$$p = C_e C_q q_s I \qquad (4\text{-}3)$$

Some codes permit (or even require) design wind loads for certain types of structure to be determined experimentally (wind tunnel tests).

For structural design, all directions of horizontal wind must be considered. For buildings whose plan is symmetrical about two mutually perpendicular axes, usually the worst wind direction for design is parallel to one or the other of these axes. For buildings of irregular shape, diagonal winds may control. Further, wind may cause such buildings to twist, in which case the designer will have to consider stresses due to torsion.

4-5. STABILITY UNDER WIND LOADS

The entire structure must be connected to its foundation so that it will be stable under wind load; it must neither slide nor overturn. Overturning moment of the wind forces on the entire structure must not exceed two-thirds of the resisting moment from dead load alone (i.e., the factor of safety against overturning must be at least 1.5). Also, each element (part) of a structure must be strong enough to withstand the wind force and must be adequately connected to the rest of the structure.

The UBC allows using (with some restrictions) either of two methods for determining the applied wind forces. Using these applied loads, the designer can then analyze for stability or for wind-induced stresses.

Method 1, known as the Normal-Force Method, is a theoretically correct method with no simplifying approximations. It may be used for any structure and is the only method allowed for stress analysis of gabled rigid frames.

In Method 1, wind forces normal to the surfaces are applied simultaneously to all external surfaces. Figure 4-4a shows these wind pressures and forces for a typical building. These consist of positive pressures on the windward side and negative pressures (suctions) on the leeward side. The lee-side suction is computed using C_e at the mean height of the roof. Stability against overturning is computed by taking moments about point B for all of the wind forces and for the weight of the building.

Method 2, called the Projected-Area Method, is allowed for any building under 200 ft high, but *not* for buildings with gabled rigid frames. Method 2 approximates the true wind forces by using horizontal pressures on the vertical projected area of the entire structure and vertical pressures on the entire horizontal projected area of the building. Figure 4-4b shows these resultant wind forces. Notice that positive pressure and suction are replaced by positive

Fig. 4-4. Wind loading methods.

pressure applied to the windward wall only. Appendix Table A-5 provides different pressure coefficients for use with the two methods.

4-6. LOAD COMBINATIONS

What load combinations should the designer consider? For building design, UBC requires the following:

1. Dead load and floor live load, plus either roof live load or snow.
2. Dead load and floor live load, plus either wind or earthquake.
3. Dead load and floor live load, plus wind, plus one-half snow load.
4. Dead load and floor live load, plus snow, plus one-half wind load.
5. Dead load and floor live load, plus snow, plus earthquake. (With approval by local building officials, snow loads in this combination may be ignored if not over 30 psf and may be reduced if over 30 psf.)

Dead load is present in every combination above, and designers often overestimate dead load. This is a safe practice only for combinations in which the dead-load effect is of the same sign as for other loads. If the effect of a load combination is reduced by dead load, then a conservatively *low* amount of dead load must be used.

Example 4-9

Compute wind forces and check stability of the condominium building shown in Fig. 4-5. The building has two stories, is 60 ft long, and is built without roof overhangs at its ends. The roof is supported by light-frame trusses. Designers estimate the weights of various parts of the building to be as shown in parentheses in Fig. 4-5. Use UBC wind requirements, assuming that the location is an open area where the basic wind speed is 80 miles per hour. Use the Normal-Force Method (UBC's Method 1).

Exposure is type C. From Table A-4, the combined height, exposure and gust factors, C_e, are:

1.2 for the lowest 20 ft of building height
1.3 for heights from 20 to 40 ft

Fig. 4-5. Building for Examples 4-9, 4-10, and 4-11.

For the leeward wall, C_e of 1.3 at the 28.17-ft mean height of *roof* is used for the entire height of wall.

From Table A-6, the stagnation pressure, q_s, is 17 psf. The importance factor is 1.0. Table A-5 gives pressure coefficients, C_q, for various parts of the structure. These coefficients and the design pressures obtained by Eq. 4-3 are:

Structure Part	Factor C_e	Coefficient C_q	Pressure = $(C_e)(C_q)(q_s)(I)$ psf
Windward wall			
Lower 20 ft	1.2	0.8 inward	16.3
20–40 ft	1.3	0.8 inward	17.7
Leeward wall	1.3	0.5 outward	−11.0
Windward roof	1.3	0.7 outward	−15.5
Leeward roof	1.3	0.7 outward	−15.5

Computed pressures are shown in Fig. 4-5, positive pressure on the left wall and suction (negative pressure) on the right. The windward and leeward walls are assumed to act as simple beams, transferring the wall wind pressures to the second-floor level and second-floor ceiling level as forces $F(1)$ and $F(3)$ on the windward side, and as $F(2)$ and $F(4)$ on the leeward side. Forces were calculated at these levels to be useful in diaphragm design (Chapter 12). At the left, Fig. 4-5 shows the simple beams for computing these forces on the windward side. For example, the force at the second-floor level on the windward side is

$$F(1) = 60\big[(16.3 \times 11.0/2) + (16.3 \times 9.92/2)$$
$$+ (1.4 \times 0.92 \times 0.46/9.92)\big]$$
$$= 10{,}230 \text{ lb}$$

Similarly, at the second-floor ceiling level

$$F(3) = 60\big[(16.3 \times 9.92/2)$$
$$+ (1.4 \times 0.92 \times 9.46/9.92)\big]$$
$$= 4920 \text{ lb}$$

Amounts of all other resultant forces shown in Fig. 4-5 are computed and entered into the table below. To check stability against overturning, we compute and tabulate the overturning moment (moment about point B) of each of these wind forces. Note that the inclined forces on the sloping roof are replaced by their horizontal and vertical components. This is simpler than using the diagonal total wind forces on the sloping roof. However, the horizontal components, H, cancel each other, so they are omitted from the table.

Base shear = 27.0 kips + bottom reactions to wind on first-story walls (not included above).

The overturning moment must now be compared to the moment of the dead load alone, a moment which tends to *prevent* overturning. The latter moment should exceed the overturning moment by whatever factor of safety is desired (UBC requires at least 1.5). If the dead-load moment does not provide adequate protection against overturning, then special provisions are needed to anchor the structure.

The resultant of all the dead loads shown in Fig. 4-5 is 77.4 kips. By symmetry, it acts at the center of the 24-ft building width, so the moment of the dead load about point B is

$$12 \times 77.4 = 928.8 \text{ ft-k}$$

and the factor of safety against overturning is only

Force	Horiz. (k)	Vert. (k)	Arm to B (ft)	Moment about Point B (ft-k)
$F(1)$	10.23		11.00	+112.5
$F(2)$	6.90		11.00	+ 75.9
$F(3)$	4.92		20.92	+102.9
$F(4)$	3.28		20.92	+ 68.6
$F(5)$ 17.7 × 1 × 60	1.06		21.42	+ 22.7
$F(6)$ 10.2 × 1 × 60	0.66		21.42	+ 14.1
$F(7)$ 15.5 × 15 × 60	—	13.95	19.50	+272.0
$F(8)$ 15.5 × 15 × 60	—	13.95	4.50	+ 62.8
			Total wind moment (ft-k) about point B =	+731.5

928.8/731.5 = 1.27. This is less than the required factor of 1.5, so additional anchorage must be provided.

Example 4-10

What total vertical anchor force is required at the two building corners at point A to provide the necessary stability against overturning?

To provide a 1.5 factor of safety against overturning requires a "righting" moment of 1.5 × (731.5), or 1097 ft-k. Dead load provides only 928.8 of this, leaving 168 ft-k to be furnished by anchor bolts connecting the ends of the end shear walls to the foundation. Assume the lever arm from the center of the wall at *B* to the bolts at point *A* to be 23.7 ft. (This is the approximate distance between centers of the opposite walls.) The vertical force required in the anchor bolts at the two corners at *A* is

$$168/23.7 = 7.09 \text{ k}$$

Additional anchor bolts at intervals along the end walls would act to resist the horizontal shearing force of 27.0 k. Chapter 12 discusses anchorage further in the sections covering shear walls.

4-7. SEISMIC LOADS

Seismic loads are forces due to the inertia of a structure as it resists being moved by an earthquake. Although earthquakes can occur anywhere, some localities are more prone than others to damaging ones. In the United States, seismic design is usually done following UBC recommendations. The UBC indicates the probable severity of earthquakes by a map dividing the U.S. into five seismic-risk zones, as shown by Fig. 4-6. These zones are defined as follows:

Zone 0: No damage expected.
Zone 1: Minor damage; distant earthquakes may damage structures with a fundamental period longer than 1.0 second. Corresponds to intensities V and VI on the Modified Mercalli (MM) scale.
Zone 2: Moderate damage. Corresponds to MM intensity VII.
Zone 3: Major damage. Corresponds to MM intensities VII and higher.

Zone 4: Areas in zone 3 that are near certain major fault systems.

The UBC permits the designer to analyze for seismic forces by dynamic structural analysis, but also shows an approximate method that is adequate for most structures. In this approximate method, the total *base shear* for the structure is given as

$$V = ZIKCSW \qquad (4\text{-}4)$$

Factors on the right side of the equation are those that affect seismic inertial loads. The factor *W* is the total weight of the structure, including the weight of partitions. Live load is not included, except for storage buildings, in which case 25% of the stored material weight must be added to the dead load. Where snow loads greater than 30 psf will occur, a portion of the snow weight must also be included in *W*. The other factors are as follows:

Z considers the seismic-risk zone in which the structure is located (3/16 for zone 1; 3/8 for zone 2; 3/4 for zone 3; and 1.0 for zone 4).
I is an "importance factor" for the type of structure. It is 1.50 for essential facilities—structures that must necessarily continue to be usable after an earthquake occurs. Examples of essential facilities are hospitals, medical centers for emergency treatment, and fire and police stations. For buildings used primarily for assembly of more than 300 persons in one room, *I* is 1.25. For all other structures it is 1.0. (Notice that the importance factors for earthquake are different from those for wind.)
K is a numerical coefficient accounting for a structure's ability to absorb energy as it is distorted by seismic forces. *K* values for various structure types are given by a UBC table, shown here as Appendix Table A-7. *K* is smaller for structures that are ductile under severe distortion.
C is a factor that depends on *T*, natural period of vibration of the structure. The UBC

SEISMIC RISK MAP OF THE UNITED STATES

ZONE 0 - No damage.
ZONE 1 - Minor damage; distant earthquakes may cause
damage to structures with fundamental periods
greater than 1.0 second; corresponds to
intensities V and VI of the M.M.* Scale.

ZONE 2 - Moderate damage; corresponds to intensity VII of
the M.M.* Scale.

ZONE 3 - Major damage; corresponds to intensity VII and
higher of the M.M.* Scale.

ZONE 4 - Those areas within Zone No. 3 determined by the
proximity to certain major fault systems.

*Modified Mercalli Intensity Scale of 1931

Fig. 4-6. Seismic risk areas of United States. (Reproduced from the 1985 edition of the Uniform Building Code, copyright © 1985, with permission of the publishers, the International Conference of Building Officials.)

allows computing T by accurate means or an approximate equation as follows:

$$T = 0.05h_n/\sqrt{D} \qquad (4\text{-}5)$$

in which D is the length of the structure (in feet) parallel to the seismic force. (An alternate equation for T does not apply to timber construction.)

Factor C is

$$C = 1/(15\sqrt{T}) \qquad (4\text{-}6)$$

but not higher than 0.12.

S is a coefficient for site-structure resonance. The UBC gives two methods for determining a value of S, one of which is described here. Factor S depends upon the characteristic period T_s of the soil at the site. If T_s is not known, factor S is taken to be 1.5. If T_s is known from properly substantiated geotechnical data, then S can be computed by the following equations. However, site period T_s used in the equations must be in the range between 0.5 and 2.5 seconds. Also, the structure's fundamental period, T, used in the equations must be not less than 0.3 second. If the ratio T/T_s is not more than 1.0,

$$S = 1.0 + (T/T_s) - 0.5(T/T_s)^2$$
$$(4\text{-}7)$$

If ratio T/T_s exceeds 1.0,

$$S = 1.2 + 0.6(T/T_s) - 0.3(T/T_s)^2$$
$$(4\text{-}8)$$

Regardless of the values determined for factors C and S, their product, CS, does not need to be larger than 0.14.

Having values for all of the above factors, and knowing the total structure weight, W, we may then use Eq. 4-4 to compute the total base shear, V. Knowing the base shear force, V, our next step is to allocate that force to various height levels in the structure. This is done by first assigning a portion F_t to the uppermost level, then distributing the remainder of force V to various height levels (including the top level) in proportion to the product of height times weight at each level. Usually this results in maximum forces at the top and lesser forces toward the bottom. The concentrated lateral force assigned to the top level is

$$F_t = 0.07TV \qquad (4\text{-}9)$$

F_t need not exceed one-quarter of the base shear, V, and if T is 0.7 second or less, F_t may be assumed to be zero.

The remainder of the base shear, $V - F_t$, is distributed as concentrations at each level of the building—roof level and each floor level. The UBC equation for the force at each level is

$$F_x = \frac{(V - F_t)w_x h_x}{\sum\limits_{i=1}^{n} w_i h_i} \qquad (4\text{-}10)$$

In this equation h_x is the height of level x above the building base; w_x is the portion of the total load w assigned to level x; and the denominator is the sum of products, $w_x h_x$, for each level of the building. The top level receives the sum of F_t and F_x computed at height h_n, where n is the number of the top level. Forces computed by this procedure must be assumed to act in any horizontal direction. For buildings with two perpendicular horizontal axes of symmetry, the usual relevant directions are parallel to these axes. For buildings of irregular plan, other directions may control and it may be necessary to consider torsion also. Under the UBC, forces due to vertical earthquake acceleration are not considered.

Both stability and strength under seismic loads must be considered. To check stability, the designer must consider both the likelihood of overturning and that of sliding horizontally. The UBC does not specify a required factor of safety against either overturning or sliding, but designers frequently use factors of 1.25 to 1.5. Stability must be provided *at all levels in the building, not merely at the base.*

The above is a very condensed summary of the UBC's requirements for the entire building.

Other UBC sections deal with seismic forces on individual elements and nonstructural parts of the building. The force on any part of the structure or on a nonstructural component is specified as

$$F_p = ZIC_p W_p \qquad (4\text{-}11)$$

In this equation, factors Z and I are the same as for the whole building; W_p is the weight of the portion or component in question; and factor C_p is given by a UBC table shown here as Appendix Table A-8. As with other code tables, it is important to read the entire table, *footnotes included*, before selecting a value.

Example 4-11

Consider again the building of Fig. 4-5, to be built in seismic-risk zone 3. Preliminary design shows that the dead-load distribution will be approximately as shown by the numbers in parentheses in Fig. 4-5 and that the building will have both horizontal-plane and vertical-plane plywood diaphragms. Will the dead-load righting moment be sufficient to resist the overturning moment due to seismic action?

Figure 4-7 shows the solution on a designer's computation sheet. From this solution we conclude

that overturning at the base due to seismic action is obviously no problem, whereas for wind load it was. (This could have been seen early in the solution, but the entire solution is shown here to illustrate the procedure.) Do not assume that this conclusion is always true, as earthquake frequently does control the design. Figure 4-7 shows analysis for east-west quake motion only. North-south motion should also be considered, but in this case it will prove to be even less serious than the east-west motion.

In the solution, notice that the force $F(3)$ is located at the center of mass for the roof system (trusses, sheathing, and roofing), estimated to be at 3.0 ft above the ceiling. Many designers, interpreting the UBC requirements, would locate force $F(3)$ and force $F(2)$ at the same height, at the center of the ceiling rafters or at the bottom-chord level of the roof trusses. Since so large a portion of the roof mass is in the heavy roofing material and roof sheathing, it would seem more proper to locate $F(3)$ where the mass is. If it is "lumped" with $F(2)$ then, ideally, a downward force at the right and an upward one at the left should be added to compensate.

Notice also that the walls are assumed to act as vertical simple beams spanning from floor to floor. Thus, the 32.6-kip weight of the four walls is distributed as follows: 7.73 kips is added to $w(2)$, 16.30 kips is included in $w(1)$, and 8.57 kips is assumed to go directly into the foundation.

Part II. Allowable Stresses

Structural design in wood today is usually by the working-stress method, in which computed "actual" unit stresses are compared to specified allowable stresses. This part of the chapter tells how standard writers determine acceptable values for these allowable stresses. It also shows how the designer modifies the specified allowable stresses to correct them for anticipated conditions of use.

4-8. ALLOWABLE STRESSES

A designer of wood structures needs to know the modulus of elasticity parallel to the grain and allowable values for stresses as follows:

   Flexure
   Compression parallel to the grain
   Compression perpendicular to the grain

Horizontal shear (i.e., parallel to the grain)
Tension parallel to the grain

Values of these allowable stresses and elastic modulus are influenced by many factors. For manufactured materials such as steel, quality control is good and little variation occurs among samples meeting the same specification. With wood, this is not so. Both the variation and the *quality* of the wood (measured by the type, size, and location of defects and by specific gravity) affect the strength.

Therefore, to design in wood with reliable safety it is necessary either to use an unreasonably high factor of safety or to choose reasonable allowable stresses based on statistical analysis of the variables involved. The latter is the method used by writers of the National Design Specification (see reference 1).

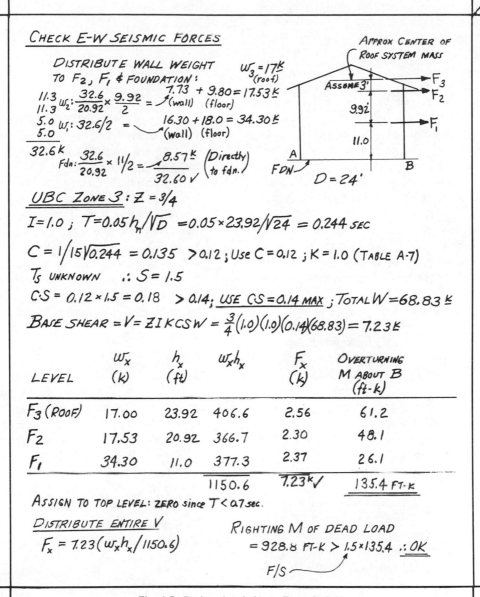

CHECK E-W SEISMIC FORCES

DISTRIBUTE WALL WEIGHT
TO F_2, F_1 & FOUNDATION:

$W_3 = 17^K$ (roof)

11.3 $W_2: \dfrac{32.6}{20.92} \times \dfrac{9.92}{2} = \nearrow$ (wall) $\dfrac{7.73 + 9.80 = 17.53^K}{\text{(floor)}}$

5.0 $W_1: 32.6/2 = \searrow$ (wall) $\dfrac{16.30 + 18.0 = 34.30^K}{\text{(floor)}}$

$\dfrac{32.6^K}{}$ Fdn: $\dfrac{32.6}{20.92} \times 11/2 = \dfrac{8.57^K}{32.60\ \checkmark}$ $\binom{Directly}{to\ fdn.}$

Approx Center of Roof System Mass

ASSUME 3' $\rightarrow F_3$ $\rightarrow F_2$

9.92 $\rightarrow F_1$

11.0

A FDN D = 24' B

UBC ZONE 3: $Z = 3/4$

$I = 1.0$; $T = 0.05 h_n/\sqrt{D} = 0.05 \times 23.92/\sqrt{24} = 0.244$ SEC

$C = 1/15\sqrt{0.244} = 0.135 > 0.12$; Use $C = 0.12$; $K = 1.0$ (TABLE A-7)

T_S UNKNOWN ∴ $S = 1.5$

$C \cdot S = 0.12 \times 1.5 = 0.18 > 0.14$; USE $C \cdot S = 0.14$ MAX; TOTAL $W = 68.83^K$

BASE SHEAR $= V = Z I K C S W = \frac{3}{4}(1.0)(1.0)(0.14)(68.83) = 7.23^K$

LEVEL	W_x (k)	h_x (ft)	$W_x h_x$	F_x (k)	OVERTURNING M ABOUT B (ft-k)
F_3 (ROOF)	17.00	23.92	406.6	2.56	61.2
F_2	17.53	20.92	366.7	2.30	48.1
F_1	34.30	11.0	377.3	2.37	26.1
			1150.6	7.23$^K\checkmark$	135.4 FT-K

ASSIGN TO TOP LEVEL: ZERO since $T < 0.7$ sec.

DISTRIBUTE ENTIRE V

$F_x = 7.23(W_x h_x/1150.6)$

RIGHTING M OF DEAD LOAD
$= 928.8$ FT-K $> 1.5 \times 135.4$ ∴ OK

F/S

Fig. 4-7. Designer's solution to Example 4-11.

The procedure followed to establish allowable stresses is defined by a series of Standards of the American Society for Testing and Materials (ASTM). This series includes the following titles:

D143. *Standard Methods of Testing Small Clear Specimens of Timber* (4). This standard defines test procedures, but equally important, it tells how to obtain representative samples for the species or species group in question, and for various localities from which the species is harvested. D143 covers tests for compression parallel to and perpendicular to the grain, tension parallel to and perpendicular to the grain, shear parallel to the grain, static and impact bending, toughness, hardness, cleavage, and nail withdrawal. It also provides for determining E (modulus of elasticity parallel to the grain).

D2555. *Standard Methods for Establishing Clear Wood Strength Values* (5). This standard tells how to analyze the results of D143 tests and of *density surveys*. A density survey provides values of specific gravity for the species being considered. Relationships of specific gravity to various strength properties are determined. (D2555 gives an alternate method for use if density survey results are not available.) D2555 specifies statistical analysis methods and requires using the "5% exclusion value" as a basis for allowable stresses. Tables in D2555 show strengths for various species and species groups and give ratios between the properties of green wood (as tested under D143) and of wood that has been dried to 12% moisture content.

D245. *Standard Methods for Establishing Structural Grades and Related Allowable Properties for Visually Graded Lumber* (6). This standard defines how the strengths of small clear specimens are modified to determine allowable design stresses. Tables in the standard give strength ratios showing the effects of various sizes, types, and positions of defect.

The series of standards above is for visually graded lumber. There are similar ASTM standards for determining allowable stresses for machine-stress-rated lumber, for round timbers, and for glued laminated lumber.

4-9. ESTABLISHING BASIC ALLOWABLE STRESSES

By *basic* the authors mean the allowable values listed by the design codes, before modification by the designer for such things as load duration, fire-retardant treatment, or size effect. For these basic allowable stresses, the industry often uses the terms *design values* or *tabular design values*. The authors prefer the term *basic* because, in their experience, it is more readily understood by the student.

Summarized briefly, the procedure defined by the ASTM and used by the NDS code writers is as follows:

1. Test a sufficiently large population of small clear specimens of the species being considered.
2. Determine the 5% exclusion value for each property to be considered. (The term *exclusion value* is explained below.)
3. Modify the 5% exclusion value as follows:
   a. By a strength ratio to account for the type, location, and size of defects in the grade of lumber considered.
   b. By a decreasing factor to account for the size effect (for bending stress only).
   c. By an increasing factor to account for the difference of moisture content between green small clear specimens and wood dried to 12% moisture.
   d. By a reducing "general adjustment" factor.

The final result of modifying the 5% exclusion value as outlined above is the code basic allowable stress. The average designer of timber structures will seldom have to go through such a procedure personally. However, if new species or obviously different-from-normal lumber were to be used, then it might be necessary to apply the procedure to establish suitable design stresses. Individual steps in the procedure are discussed further in the following paragraphs.

Small Clear Specimen Test Results

Figure 4-8 shows a hypothetical distribution curve for one particular strength property for a particular species. This is a graph of such data as might come from a large number of small clear specimen tests for that species. The abscissa of a bar represents a value of strength, and the ordinate at that same point shows the total number of test specimens having that particular strength. For example, point A, by the height of the bar at that point, shows that 38 of the 1000 specimens tested failed at between 5650 and 5750 psi. Point B, located at an abscissa of 4350 psi, shows by the sum of all bar heights to the left of that point that only 50

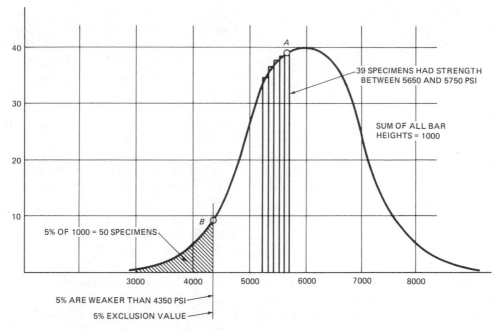

Fig. 4-8. Typical distribution of strength for 1000 specimens.

specimens were weaker than 4350 psi. The 4350-psi stress represented by point *B* is called the *5% exclusion value* since only 5% of the 1000 specimens failed to have at least that strength. ASTM D2555 specifies that the 5% exclusion value be used to establish allowable stresses for timber, and the NDS allowables are based on these values.

Seasoning Increase

Wood in service normally has lower moisture content (MC) than the green wood used for the small-clear tests. As wood dries, two significant changes take place: a volume decrease and a strength increase. ASTM D2555 shows for various commercial species the percent of strength increase as wood is dried from its fiber saturation point to 12%, a reasonable MC to expect under normal use. The factor shown for this increase is applied to the 5% exclusion value.

Size-Effect Decrease

Wood's apparent bending strength is affected by size of the cross section. Small clear test

specimens have a 2.00-in. × 2.00-in. cross section, whereas the average depth used for wood bending members is 12 in. By its size alone, the deeper member is expected to fail at a lower modulus of rupture than the small clear test specimen. The size effect applies *only* to the allowable stresses in bending. To correct the results of small clear specimen tests for use with 12-in.-deep members, the 5% exclusion value is multiplied by 0.82.

Modification for Defects

Lumber grading rules define the defects anticipated in a particular grade of lumber, and all of these defects can decrease strength. Specimens of normal size (not small clear) have been tested to determine the degree of weakening expected from various types of defect. The strength-ratio tables of ASTM D245 were developed from these tests. The tables give ratios for:

1. Knots on the narrow face of bending members
2. Centerline knots on the wide face of bending members

3. Edge knots in the wide face of bending members
4. Shakes, checks, and splits in bending members
5. Knots in compression members

General Adjustment Factor

This catch-all factor includes adjustments for stress concentrations, load duration, position of the annual rings, span-to-depth ratio, manufacturing errors, and various types of strength-reducing damage that might occur during use of the lumber. The last includes end splits that occur during or after installation, holes drilled by plumbers or electricians, and nail holes.

To test a small clear specimen takes only a few minutes, but in service wood is ordinarily under load for a much longer time. Allowable stress tables are based on a "normal duration" of load (loading that lasts for ten years, either in one continuous period or through the accumulation of shorter load periods). Failure stresses depend on load duration, as shown by Fig. 2-14. The high factor on that curve for a loading of five- to ten-minute duration shows how much higher the failure stress will be for small clear specimens than for members under normal-duration load. The 5% exclusion value for the small clear specimen tests must be reduced accordingly to establish allowables for normal-duration loading.

An adjustment for stress concentration applies only to the allowable for horizontal shear. A correction for end position is included since the compressive strength perpendicular to the grain differs for stress parallel to the annual rings and for stress normal to the rings. Span-to-depth ratio is considered since it affects the apparent value of E, modulus of elasticity (7).

The cumulative effect is called the *general adjustment factor*. It differs for different-strength properties, but in every case it is a factor by which the 5% exclusion value is reduced.

4-10. DESIGN ALLOWABLE STRESSES

Design allowable stresses are given by publications such as the National Design Specifica-

tion (1) and the Uniform Building Code (2). Appendix Tables B-3 and B-4 of allowable stresses are from the latter. Using these tables is not completely simple, however, as the designer may still have to adjust their tabulated values.

To obtain an allowable stress or elastic modulus for visually graded lumber using Table B-3, the designer must:

1. Select the species or species group.
2. Determine the size/use classification.
3. Select a grade.
4. Read the listed basic allowable stress or elastic modulus.
5. Observe which footnotes are referred to.
6. Modify the basic allowable as instructed by any of the listed footnotes that apply to the design being considered.
7. Further modify the allowable values, as required, for
   a. load duration (modulus of elasticity is *not* modified for duration)
   b. fire-retardant treatment
   c. temperature
   d. slenderness (for bending or axial compressive stresses only)
   e. size effect (for bending stresses only)

Species or Species Group

Table B-3 shows allowable stresses and E for visually graded lumber of four species. This table is first divided according to species or species group, using commercial names (Chapter 2).

Size/Use Classification

The second column of Table B-3 lists *nominal* sizes for lumber ranging from 2 in. up to 4 in. thick. For larger sizes of some species, categories according to *use* are shown. The use categories are: Beams and Stringers (B&S), Posts and Timbers (P&T), and Decking. (See Chapter 3 for definitions.)

Grade

In the left column again are listed various grades that may be available. Note that the same grade names may appear for wood of various size/use classifications, but the allowable stresses probably differ. Also, some grade names may appear in only one or two of the size/use classifications.

Allowable Stress—Basic Value

Table B-3 shows basic allowable values for several different types of stress and for modulus of elasticity. The values chosen *must be* for the correct species or species group, the correct size/use classification, and correct grade. These basic allowable stresses may now have to be modified by the structural designer, as outlined above.

Footnotes

The extreme right-hand column of Table B-3 lists the initials of the grading agency along with the corresponding UBC Standard under which the timber is graded. Also listed are the numbers of those footnotes that *may* apply to the basic stresses given for the species and size/use group in question. A complete list of the footnotes follows the end of Table B-3. Usually, very few of the footnotes referred to actually apply to the case in hand, but the designer still must look at all of them carefully to find out whether or not the tabulated allowable stress value must be modified. The student would be wise to examine the entire footnote list now, so as to become acquainted with it.

Modification for Load Duration

Figure 2-14 shows how the strength of wood varies with duration of load. If the load duration is known precisely, this curve can be used to determine how to modify the allowable stress for anything other than normal duration. For ordinary design purposes, however, the following modification factors are used:

For impact loading, 2.0
For wind or seismic load, 1.33*

*Since cumulative duration of wind load may be shorter than one day, some designers believe a higher duration factor should be used for wind.

For snow load, 1.15
For usual floor loads (normal duration), 1.0
    (i.e., no modification)
For permanent load (dead load, for example), 0.9

These load duration factors are *not* cumulative with each other. Rather, when a *combination* of load types occurs, the type of *least* duration determines the duration factor to be used for the entire combined loading. Factors are *not* prorated according to the amounts of various load types. For example, if a load combination includes dead load, live load and wind, the duration factor for wind (1.33) is used to determine the allowable stress under the *combination* of loads. In this case, however, it is also necessary to consider the other possible combinations: dead plus live load alone, using a factor of 1.0 (normal load duration), and dead load alone using a factor of 0.9. Examples that follow show how to decide which is the critical load combination and how to select the proper load duration factor.

The load duration modification is cumulative with all other types of modification required. However, it applies to allowable stresses only, not to modulus of elasticity.

Fire-Retardant Treatment

For sawn lumber that is pressure treated with a fire-retardant chemical, the NDS requires that all allowable stresses and elastic modulus be multiplied by the following factors:

Extreme fiber stress in bending	0.85
Tension parallel to grain	0.80
Horizontal shear	0.90
Compression perpendicular to grain	0.90
Compression parallel to grain	0.90
Modulus of elasticity	0.90
Fastener design loads	0.90

Size Factor

The size factor applies only to allowable bending stress. It does *not* apply to shear stress, nor to axial tension or compressive stresses. The size factor is used to reduce the allowable bending stress for all sawn lumber members that are *deeper than 12 in. and thicker than 4 in.* (Also, it applies to all glulam members that are deeper than 12 in., regardless of width.*) The size factor is given by the equation

$$C_F = \left(\frac{12}{d}\right)^{1/9} \qquad (4\text{-}12)$$

in which d is the actual member depth (distance from the compression face to the tension face) in inches. For glulams, the size factor is the value given by Eq. 4-12, but with further modification for load position and span-to-depth ratio. (See Chapter 8.)

Repetitive Members

The NDS defines repetitive members as "members in bending, such as joists, trusses, rafters, studs, planks, decking, or similar members that are in contact or not more than 24 inches on centers, are not less than 3 in number and are joined by floor, roof, or other load-distributing elements adequate to support the design load." This means that repetitive members are those that are close enough alongside each other that, if one member is weaker than normal, the adjacent parallel members can help out. Load that cannot be carried by the weaker member is transferred laterally to the stronger and stiffer members on each side. The allowable flexural stress for a repetitive member is higher than for a *single member*, one which must carry all load applied to it without depending upon assistance from adjacent members. Note that even though studs are listed in the definition above, the stress increase is only for flexural stress.

Allowable Stresses for Machine-Stress-Rated (MSR) Lumber

Table B-4 is much easier to use, because species is not a factor and the number of lumber sizes that can be machine rated is limited. Ta-

---

*New data shows that in glulams there is a volume effect which is a function of depth, thickness, and length. The designer should check with current AITC publications so that safe designs are produced for larger glulams.

ble B-4 has footnotes of its own, and (by reference) involves some of the footnotes from Table B-3 also. The first column of Table B-4 gives the grade designation, which includes both the allowable bending stress for normal-duration single-member loading and the flexural modulus of elasticity, E.

The next seven examples illustrate the use of Appendix Tables B-3 and B-4 and the modifications to allowable stress.

Example 4-12

Determine the allowable flexural stress for a 2 × 12 roof rafter of No. 2 Douglas fir–larch. The roof structure is over a swimming pool. Since the moisture content of the wood will be fairly high, the wood is preservatively treated to prevent decay. Bending moments for the rafter are:

From dead load: 1000 ft-lb
From snow load: 1800 ft-lb

Refer to Appendix Table B-3. The size/use classification described as "2 to 4 inches thick, 5 inches or wider" includes the 2 × 12 to be considered. For that classification, find the desired grade (No. 2) in the left-hand column. Now move to the right and in column 4 read 1450 psi, the allowable extreme-fiber bending stress for a repetitive member. (The rafters are spaced at not over 24-in. centers, there are at least three rafters, and they are joined by structural roof material that is strong enough to support the design load. The NDS recognizes such members as "repetitive" members.) This basic allowable is for dry conditions of use and normal load duration.

The extreme right-hand column of the table shows that several footnotes must be considered for all grades of D. fir–larch in this size group. At the end of Table B-3, find the footnotes. Reading these footnotes carefully, we find that only footnote 6 applies to this particular member. Being subjected to high humidity over the swimming pool, the rafter will undoubtedly have MC > 19%, so the allowable flexural stress F_b must be reduced using a factor of 0.86. Considering this modification, the allowable flexural stress for normal-duration loading becomes 1450 × 0.86 = 1247 psi.

Further modification must now be made for load duration. There will be no modification for size, since the member is not greater than 12 in. deep or thicker than 4 in. Under full loading, which includes snow, the allowable F_b becomes

$$1247 \times 1.15 = 1434 \text{ psi}$$

However, when dead load is the only load present, the allowable F_b is only

$$1247 \times 0.9 = 1122 \text{ psi}$$

Which value should be used? Actually both of them. Unless we use a convenient shortcut (demonstrated later) we must compute the actual stress for each load combination and compare it to the allowable stress for that combination. The section modulus for the 2 × 12 is 31.64 in.³, and the computed actual stresses are:

For dead + snow: f_b = 12 × 2800/31.64
$$= 1062 \text{ psi}$$
This is less than the allowable of 1434 psi, so the member is satisfactory for dead plus snow.

For dead load alone: f_b = 12 × 1000/31.64
$$= 379 \text{ psi}$$
This is less than 1122 psi and the member is OK for dead load alone.

Example 4-13

Repeat the above example, but use the convenient shortcut mentioned above to determine the controlling load combination.

Merely dividing the total bending moment for each load combination by the load duration factor that applies to that combination, we have

For dead + snow: 2800/1.15 = 2435
For dead alone: 1000/0.9 = 1111

The quotient for dead plus snow is the larger; therefore that load combination controls the design. The allowable flexural stress is next determined and compared to the computed actual stress for that load combination only.

$$F_b = 1434 \text{ psi} > f_b = 1062 \text{ psi} \quad OK$$

Example 4-14

No. 3 hem-fir 2 × 6s are to be used in the flat position as floor planks to carry a total dead load of 60 psf and a live load of 70 psf. Load will bear directly on the planks, without any means of lateral distribution from one plank to another. Use conditions will be very dry, so the wood will have less than 15% moisture at all times during use. Assuming that it will be possible to obtain planks that are also manufactured at less than 15% MC, what will be the allowable flexural stress? Which load condition controls?

Table B-3 shows an allowable flexural stress F_b of 575 psi for single members. Again some footnotes must be considered, but only nos. 4 and 5 apply. Footnote 4 shows that the tabulated allowable must be multiplied by 1.22 to account for the greater modulus of rupture for a 2 × 6 that is bent in the weak direction. (Footnote 4 does not apply to decking; decking is *always* used in the flat position and the allowable stresses shown take this into account.) Manufacture and use of these 2 × 6s are both under very dry conditions, so footnote 5 permits the basic allowable to be multiplied by 1.08. Thus, the modified allowable flexural stresses are:

For dead plus live load: 575 × 1.22 × 1.08 × 1.0 = 758 psi
For dead load alone: 575 × 1.22 × 1.08 × 0.9 = 682 psi

Alternatively, if we use the shortcut to find out first which load combination controls,

For dead load alone: 60/0.9 = 66.7
For dead + live:
(60 + 70)/1.0 = 130 *Controls*

Allowable stress for dead + live = 758 psi (as above).

Example 4-15

The 3 × 8 tension chord of a truss is made from select structural southern pine, surfaced dry, and with fire-retardant treatment. The member will be dry in use. Forces in the chord are

From dead load: 8 kips
From snow load: 9 kips
From wind load: 5 kips

If the code-specified load combinations are: (1) dead alone, (2) dead plus snow, and (3) dead plus wind, which load combination controls and what is the allowable tensile stress?

Totals and quotients for these combinations are

For dead alone: 8/0.9 = 8.89
For dead plus snow: (8 + 9)/1.15 = 14.78 *Controls*
For dead plus wind: (8 + 5)/1.33 = 9.77

The basic allowable tensile stress from Table B-3 is 1150 psi. Modifications to this basic allowable are

For load duration: 1.15
From footnote 15: 0.90
For fire-retardant treatment: 0.80 for tension

The modified allowable tensile stress on the *net* section is

$$1150(1.15)(0.9)(0.8) = 952 \text{ psi}$$

To consider the improbability of load combinations that include either full wind or earthquake, a recent NDS revision permits the combined loads to be reduced. UBC does the equivalent by allowing an additional increase of allowable stress. However, in the interest of simplicity, this refinement is omitted from the examples in this book.

Example 4-16

A dense No. 1 Douglas fir–larch 8 × 10 timber, graded under Post-and-Timber rules (definition in Chapter 3) and used dry as a short compression member in a storage warehouse, is subject to the following axial loads:

Dead load	25 kips
Floor live load	40 kips
Snow load	15 kips
Wind load	29 kips

What is the allowable stress?

Since the building is used for storage, it is reasonable to assume that the live-load duration will be longer than ten years. Therefore, live load will be treated as permanent load. Considering all the load combinations specified by the UBC, the combinations and quotients are as follows:

Permanent (dead plus live): $(25 + 40)/0.9 = 72.2$
$D + L + S$: $(25 + 40 + 15)/1.15 = 69.6$
$D + L + W$: $(25 + 40 + 29)/1.33 = 70.7$
$D + L + S/2 + W$: $(25 + 40 + 7.5 + 29)/1.33 = 76.3$
$D + L + S + W/2$: $(25 + 40 + 15 + 14.5)/1.33 = 71.1$

The fourth combination, $D + L + S/2 + W$, controls. Table B-3 gives 1200 psi for the basic allowable stress in compression parallel to the grain. Several footnotes are referred to, but examination shows that none of them apply to this particular case. Thus, the only modification needed is for load duration and the modified allowable is $1200 \times 1.33 = 1600$ psi. (If the member were either an intermediate-length or a long column, further modification would be needed to account for possible buckling.)

Example 4-17

It is not always possible to apply the above shortcut method directly to the amounts of the loads. Rather, bending moments, shears, or axial forces may have to be determined first so we can identify the controlling load combination. This is the case for the beam of Fig. 4-9, for use in a wet-process manufacturing plant. Loads shown include the weight of a permanently installed machine. During operation, the machine occasionally imposes an impact load.

Fig. 4-9. Beam for Example 4-17.

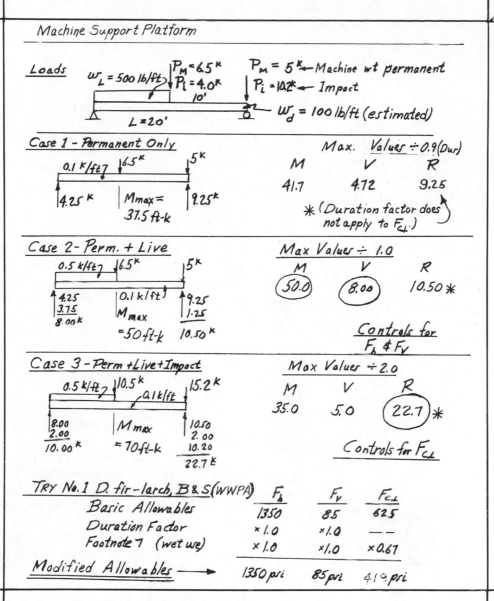

Fig. 4-10. Solution to Example 4-17.

Which load combination controls for bending and what is the allowable stress for that combination? For end shear? For end bearing (compression perpendicular to the grain)?

Figure 4-10 shows the solution in the form of a designer's computation sheet. Notice that detailed explanations are not given, but sufficient information is shown that others can check the designer's work. Equally important, the designer can easily determine at any future date the exact basis for the con-

clusions shown. For operations so simple that they can be carried out easily with a hand calculator (simple-beam reactions, for example) only the answers are given.

Example 4-18

Machine-stress-rated southern pine 2 × 10s are used as floor joists over a wet manufacturing process. Joists are 12 in. c/c and the attached floor is fairly rigid. The MSR grade is 1800 f-1.6E. Dead load totals 80 psf of floor area and live load 30 psf. Which load combination controls, and what are the values of E and allowable stress for bending and for shear?

Table B-4 shows basic values of $F_b = 2050$ (repetitive use) and $E = 1,600,000$ psi. Use $F_v = 90$ psi.

$$80/0.9 \text{ is less than } (80 + 30)/1.0,$$
$$\text{so } D + L \text{ controls}$$

The modification required for MC > 19% is the same as for visually graded lumber (footnote 6 of Table B-3). Modified values are:

$$E = 0.97(1,600,000) = 1,552,000 \text{ psi}$$
$$F_b = 0.86(2050) = 1760 \text{ psi}$$
$$F_v = 0.97(90) = 87 \text{ psi}$$

4-11. RELIABILITY-BASED DESIGN

The procedures presented above and the resulting allowable stresses are based on two assumptions: (1) that the maximum loads for which the structure is to be designed are known and (2) that using the 5% exclusion limit to establish allowable stresses ensures a reliably safe design. It does not guarantee that the design will always be safe, but more often than not it produces a conservative design.

For reinforced concrete (8) the working stress design method has almost completely been replaced by the more reliable strength design method (ultimate strength design). Load and resistance factor design (LRFD) is now accepted for structural-steel design (9). In timber, an ASCE committee and others are working toward a similar reliability-based design code.

Several names have been given to describe this newer design method. Among them are *probability-based design*, *probabilistic design*, and *load and resistance factor design*. Regardless of name, they all are based on recognizing that for a structure to be reliably safe its resistance to loads must be equal to or greater than the factored loads (10). Expressed in equation form,

$$\phi R_n \geq \sum \gamma_i Q_i \qquad (4\text{-}13)$$

In Eq. 4-13, uncertainties of each load are considered using code-recommended load factors

(γ_i). Also, variability of the material is considered by using a code-specified strength reduction factor, ϕ. Reliability concepts are used to establish values for these factors for a given level of reliability. Studies for determining strength reduction factors for timber are well under way (11, 12).

Neither the loading to be applied nor the resistance (theoretical strength times the strength reduction factor) to that loading can be known precisely for all cases. The load may be defined, for example, as the weight of a certain depth of snow, such as 25 psf. The probability is great that the actual snow load will never exceed that amount, but there is always the *possibility* that it may. That is, we cannot say, with 100% assurance, that the snow load will never exceed the quoted amount. Yet, because of the extreme probability that it will not exceed that amount, we regard the quoted snow load as a reliable figure to use in designing ordinary structures. For structures of short life or limited value, or for structures whose failure would not endanger life, we might be willing to use a lesser design load. In other words, we might be willing to accept a greater probability that the design load might be exceeded. To do so would be to use one of the elements of probabilistic design.

Both sides of Eq. 4-13 are subject to variation, so the probability of failure depends on the values used on each side of the equation. This is illustrated graphically by the two distribution curves shown in Fig. 4-11. The curve at the left is for loading, showing a high probability that the actual loading will be at or very near to the design loading. The frequency of much larger or much smaller loadings is less, as shown by the small ordinates at each end of the curve. Plotted on the same graph is a curve for the resistance of the structural member to those loads. The object is to have the resistance, R, larger than the load. The resistance of the structural member cannot be predicted precisely, but it is highly probable that the predicted resistance will be fairly correct. There is a lesser probability that the actual strength will be much less than predicted, and this is shown by the "tail" of low ordinate near the left end of the distribution curve for resistance.

Fig. 4-11. Overlap of load and resistance distribution curves.

Failure will occur when resistance is less than load. The curves overlap, and the degree of their overlap indicates the probability of failure. Expressed qualitatively only, for greater reliability, the intersection of the two curves should be at a point of very low ordinate.

REFERENCES

1. *National Design Specification for Wood Construction*, National Forest Products Association, Washington, DC, 1986.
2. *Uniform Building Code*, International Conference of Building Officials, Whittier, CA, 1985.
3. *Minimum Design Loads for Buildings and Other Structures*, ANSI 58.1-82, American National Standards Institute, Inc., New York, 1982.
4. *Standard Methods of Testing Small Clear Specimens of Timber*, D143-52 (1978), American Society for Testing and Materials, Philadelphia, PA.
5. *Standard Methods for Establishing Clear Wood Strength Values*, D2555-81, American Society for Testing and Materials, Philadelphia, PA.
6. *Standard Methods for Establishing Structural Grades and Related Allowable Properties for Visually Graded Lumber*, D245, American Society for Testing and Materials, Philadelphia, PA.
7. Biblis, E. J., "Shear Deflection of Wood Beams," *Forest Products Journal*, **15**(11), 1965, pp. 492–498.
8. *Building Code Requirements for Reinforced Concrete*, ACI 318-83, American Concrete Institute, Detroit, MI, 1983.
9. *Load and Resistance Factor Design, Specification for Structural Steel for Buildings*, American Institute of Steel Construction, Inc., Chicago, IL, 1986.
10. Bodig, J., and B. A. Jayne, *Mechanics of Wood and Wood Compostites*, Van Nostrand Reinhold, New York, 1982.
11. Bulleit, W. M., and J. G. Ernst, "Resistance Factor for Wood in Bending or Tension," *Journal of Structural Engineering, American Society of Civil Engineers*, **113**(5), May 1987, pp. 1079–1091.
12. Hendrickson, E. M., and B. Ellingwood, "Limit State Probabilities for Wood Structural Members," *Journal of Structural Engineering, American Society of Civil Engineers*, **113**(1), Jan. 1987, pp. 88–106.

PROBLEMS

4-1. Calculate the approximate dead load for a one- or two-story wood structure (your home, for example). If the species is unknown, assume one.

4-2. Compute design wind loads for the building of Problem 4-1. If you can't measure actual dimensions, estimate them as closely as possible.

4-3. For your seismic-risk zone, calculate the total seismic base shear and overturning moment for the building of Problems 4-1 and 2.

4-4. Does dead load provide adequate safety against overturning for the building of Problems 4-1, 4-2, and 4-3?

4-5. For the structure of Fig. 4-12, what is the total bending moment in one inclined rafter due to snow load, where the specified snow load on a flat roof is 25 psf? Assuming dead load to be 20 psf of roof plus the weight of the rafter, what is the total bending moment in one rafter?

4-6. Compute the bending moment per roof beam (rafter) for the building of Fig. 4-13. Snow load specified is 25 psf of horizontal projection, the maximum wind speed is 70 mph, and the terrain is open. Give answers for all reasonable combinations. Assume that the rafter weighs 17 lb/ft.

4-7. A three-story building with 16- by 20-ft bays is designed for a 70-psf floor live load. Floor joists are 2 × 12 D. fir-larch at 16 in. c/c with $\frac{5}{8}$-in. plywood sheathing above. Finish floor weighs about 3 psf. Use the UBC recommendation for weight of partitions. The ceiling below is $\frac{1}{2}$-in. drywall. Mechanical and electrical systems add about 10 psf. Beams

Fig. 4-12. Structure for Problem 4-5.

spanning the 20-ft direction between columns weigh an average of 35 lb per linear ft. The roof system weighs the same as the floor system. Considering allowable reductions, what is the design load $(D + L)$ per foot for interior beams supporting the 2×12 floor joists?

4-8. For the same building, and for the UBC minimum live load (not snow) on the flat roof, what is the design load $(D + L)$ for a bottom-level interior column? Assume each column tier to be a 10×10 D. fir, 10 ft long.

4-9. Same as Problem 4-8, but with roof snow load of 30 psf instead of roof live load. What is the bottom-tier column design load $(D + L + S)$? Assume a flat roof.

4-10. If the code specifies a 75 psf floor live load for your classroom and if the floor is supported by beams at 12-ft centers (spanning the short direction), what is the design load $(D + L)$ per foot for those beams? Make a reasonable assumption for the self-weight of the beams.

4-11. Estimated dead loads were shown on Fig. 4-5 for use in checking stability and in calculating required anchor-bolt forces in Examples 4-9 and 4-10. Ordinarily the designer tries to be safe by using a high estimate of dead loads. Is it a safe practice here? What would be the required anchor-bolt force if the true dead loads were 10% less than estimated?

4-12. Check the stability versus overturning for the office building of Fig. 4-14. Maximum wind speed is 70 mph and exposure is type B. The building length is 50 ft. Estimated dead loads (walls included) at various levels are shown. Use the Normal-Force Method.

Fig. 4-13. Structure for Problem 4-6.

Fig. 4-14. Building for Problems 4-12, 4-13, and 4-14.

4-13. Repeat Problem 4-12 using the Projected-Area Method.

4-14. Check the stability of the building of Problems 4-12 and 4-13 when subject to earthquake. Use seismic-risk zone 2. Assume a box system.

4-15. Find the basic allowable bending and shear stresses for No. 1 southern pine 2×4s used as light framing. The maximum service moisture is 19%. Spacing is 16 in.; use is repetitive.

4-16. Find the basic allowable bending and shear stresses for select structural D. fir–larch 6×6s, used singly as columns.

4-17. D. fir–larch 2×8s are to be used repetitively. They are machine stress rated as 1800f-1.6E. What are the allowable bending and shear stresses?

4-18. Southern pine commercial decking (repetitive) will have a 19% maximum service MC. What is the allowable bending stress?

4-19. No. 2 D. fir–larch 2×4s are used repetitively. What are the basic allowable bending and shear stresses? If these 2×4s are manufactured and used where MC will be 13%, how are these allowables modified?

4-20. If the 2×4s of Problem 4-19 have no splits or shakes, how are these allowables modified?

4-21. No. 3 D. fir–larch 2×4s are used repetitively. What are the basic bending and shear allowables? If they are used in a locale where the humidity is extremely high, what are the modified allowables?

4-22. Select structural California redwood 4×14 beams are used at 6 ft apart to support the roof over a laundry room. What are the allowable stresses for bending and longitudinal shear for dead load acting alone? For dead load plus snow, acting together? For dead plus wind?

4-23. If the bending moments for Problem 4-22 are: dead, 6,000 ft-lb; snow, 9,000 ft-lb; and wind, 10,000 ft-lb, what is the controlling load combination? Consider all applicable UBC combinations. Solve by both methods illustrated.

4-24. A building floor consists of $\frac{3}{4}$-in. plywood resting on (but not connected to) No. 3 southern pine 2×12 planks, used flat at 12 in. c/c. Live load is 40 psf. Assume the plywood to weigh 40 pcf. The planks are treated with fire-retardant chemicals. What are the allowable bending and shear stresses?

4-25. Dense no. 2 D. fir–larch 2×8s are used as tension chords of flat-topped roof trusses at 24-in. centers (repetitive) to support 25 psf dead load and either 20 psf live load or 35 psf snow load under moist-use conditions. Which load combination controls, and what are the allowable stresses in bending, tension, and shear?

4-26. Same as Problem 4-25, but for continuously dry use.

4-27. What are the allowable stresses in compression parallel to the grain and in bending, and the modulus of elasticity, E, for 2×6 wall studs of 1500f-1.4E grade D. fir–larch spaced 16 in. c/c? Give values for relevant UBC combinations of D, L, S, and W.

4-28. An existing structure is being investigated for possible increased loading. Beams are 8×18, No. 1 D. fir–larch, spanning 16 ft. Some beams have end splits 8 in. on their wide face. Use conditions were, and will continue to be, very dry. Uniform loads are 400 lb/ft live load (reduction already considered) and 200 lb/ft dead load. A midspan concentration is 1 kip for dead load and 1 kip for snow. Which load combinations control and what are the controlling allowables for: bending, horizontal shear, and compression perpendicular to the grain?

5

Connections—Nails, Screws, and Bolts

5-1. CONNECTION DESIGN

It is important that structural members—beams, columns, and tension members—be designed with due regard to safety and economy. But, no chain being stronger than its weakest link, it is equally important that the connections joining these members to each other be carefully designed. A connection must be able to transfer load from member to member without the connection material itself failing and also without damaging the wood of the members it joins. Design of structural members, addressed in later chapters, may be easier if the designer knows how the member will be connected and knows how the member in question may be weakened by the connection.

There is a wide variety of mechanical connectors (fasteners), ranging from the old wooden pegs to modern custom-made welded assemblies for joining together large glued laminated members. This chapter will cover the more common types—nails, staples, lag screws, wood screws, and bolts. Chapter 9 will present additional information on bolts, and also will cover modern types of timber connectors and specially designed weldments.

5-2. GENERAL PRINCIPLES

Certain principles apply to connection design in general, regardless of whether the fasteners are nails, screws, lag screws, or bolts. The capacities of each of these fasteners is affected by specific gravity and moisture content of the wood; by dimensions of the connector and the wood members; by the type, number, and arrangement of the fasteners; by the position of the fastener relative to the grain; and (in most cases) by the direction of the fastener force relative to the grain.

Specific Gravity

Wood with high specific gravity is stronger than wood with low specific gravity. This may affect the allowable load for a fastener. The National Design Specification (1) considers this effect by dividing commercial wood species into four groups according to average specific gravity. Table C-1 in the Appendix of this book shows these groups. Notice that the table has two lists. The list using Roman numerals applies to the fasteners covered by this chapter, whereas the other list applies only to the timber connectors covered by Chapter 9. Notice also that lumber grade is not involved in the classification.

For bolts, the effect of specific gravity is taken into account by the fact that tables of allowable bolt loads (see Appendix Table C-12) are arranged according to wood species.

Moisture Content

Moisture conditions of the wood also affect fastener strength. Appendix Tables C-2 and C-3 show the modifications for this effect. Notice that for many types of fastener, it is not the actual moisture content at any given time, but rather the *change* of moisture content between the time of installation and service that affects fastener strength markedly. This is because any MC change is accompanied by dimensional changes (shrinkage or swelling) that may tend to withdraw or loosen the fasteners, or in some cases split the wood.

For example, timber structures often are assembled using green lumber. This green lumber (with a high MC) then seasons after the connection is made. If the arrangement of the fasteners does not allow the wood to shrink freely, a significant force develops in each fastener as it acts to prevent the wood from shrinking. This force is present in the wood also, and may cause the wood to split. Whether the wood splits or not, the strength of the connection is severely reduced.

Dimensions

Regardless of the type of fastener, dimensions of the fastener and of the wood members it connects affect the deformed shape of the fastener and, therefore, its strength. Also, ratios of one dimension to another (bolt diameter to member thickness, for example) may be important. Other important dimensions are center-to-center spacing of fasteners, edge distance, and end distance.

Center-to-center distance between fasteners (spacing) and distance from the fastener to the end of the wood member or to its edge (end distance or edge distance) will reduce the connection strength if the distances are too small. This occurs as the wood between fasteners fails in either longitudinal shear or tension perpendicular to the grain. Required spacings are defined for lag screws and bolts, but for nails the rule is merely that the nails shall be far enough apart that the wood does not split.

Number and Arrangement of Fasteners

The number of fasteners has an obvious effect: the more fasteners, the stronger the connection. However, the arrangement of these fasteners also may affect the strength and allowable load for a connection.

Position Relative to the Grain

A fastener may be placed so that its length is transverse, that is, in a radial/tangential plane of the wood. It may also be placed so that its length is in the longitudinal direction of the wood. The position of the fastener relative to the grain affects both the lateral resistance and the pullout (withdrawal) strength of the fastener.

Direction of Fastener Force Relative to the Grain

Angle of load to grain (Fig. 5-1) affects the strength of fasteners. As load is transferred from one member to another by shear in a screw or bolt, the side of the metal shank presses against the wood. Where this bearing pressure is in the direction of the grain, the resistance will be fairly high; but where it is perpendicular to the grain, resistance will be much lower. This is because the compressive strengths of the wood differ in the two directions. Often, the fastener causes pressure at some other angle, neither parallel to the grain nor perpendicular. In this case, an allowable value for the fastener is approximated by the Hankinson equation. (Nails, spikes, staples, and wood screws are excluded from this requirement, since joint slip, rather than ultimate strength, is considered the more important criterion for nailed joint failure.) The Hankinson formula is

$$N = \frac{PQ}{P \sin^2 \theta + Q \cos^2 \theta} \qquad (5\text{-}1)$$

in which

P = the allowable load for the fastener if loaded parallel to the grain

Fig. 5-1. Angle of load to grain.

Q = the allowable if loaded perpendicular to the grain

θ = the acute angle between the load direction and the grain (longitudinal) direction

Net Section

If a hole is drilled in the member (for a bolt or lag screw, for example), the connection reduces the cross-sectional area of the member. In this case, the *net* area of the member may need to be considered when computing the member strength. The cross-sectional area is not considered to be reduced when the fastener merely pushes aside the wood fibers, as does a driven nail.

Other Factors

Load duration modifications apply also to connections. The allowable fastener load is also reduced by 10% for fasteners in wood that is fire-retardant treated.

The strength of any connection is improved when the outer pieces joined are metal plates. The added strength results from the stiffness of the plates, which reduces joint slip and tilting of the nail, screw, or bolt in the wood as the connection is loaded.

Nails, staples, wood screws, or bolts may be used in combination with glue. However, the glued connection is very stiff, whereas nails, for example, must deform appreciably before they resist significant shear. Thus, in a combination joint, the glue tends to resist the total load, the mechanical fasteners not receiving load until the glued connection is broken.

5-3. NAILS AND SPIKES

The variety of available nails and spikes is almost unbelievable—over 10,000 different varieties are produced in the United States during one year (2). An excellent summary of the available types and sizes is given by reference 3. For our purpose, it is sufficient to recognize the three major types and various sizes shown by Appendix Table C-4. Nail types are depicted in Fig. 5-2, along with some other fasteners covered in this chapter. Common nails

BOX NAIL COMMON NAIL ANNULARLY THREADED NAIL HELICALLY THREADED NAIL

WOOD SCREWS

LAG SCREW

Fig. 5-2. Nail and screw types.

have a larger diameter for a given length than either box or threaded nails. Threaded nails have been hardened by heat treating and have greater strength than common nails or box nails of equal diameter.

For lengths up to 9 in., the size of a nail or spike is specified by *pennyweight*, an antiquated term indicating the purchase price, in English pennies, per 100 nails. Today the term is still used, but may soon be replaced by naming nail sizes by diameter and length.

Nails are usually driven without first drilling a hole, but for either nails or spikes a pilot hole may be drilled. This will not reduce the allowable load for the fastener provided the hole diameter is not over 0.9 fastener diameters for Group I species (Table C-1) or 0.75 fastener diameters for other species.

Allowable loads for nails or spikes depend on several things:

Fig. 5-3. Nailing and load directions. (Courtesy Board of Regents, University of Colorado.)

1. Diameter, surface shape, and surface treatment of the shank
2. Specific gravity and moisture content of the wood
3. Depth to which the point penetrates into the wood of the member that contains the nail point
4. Direction of the nail shank relative to the grain
5. Direction of the load relative to the shank length

Since they are determined by limits of deformation rather than strength, allowable lateral (shear) loads for nails and spikes are the same regardless of whether the lateral load is parallel to or perpendicular to the grain.

Nails and spikes can transfer either lateral load or load parallel to their length (withdrawal loading), or a combination of the two. They are stronger when driven into side grain than when driven into end grain (Fig. 5-3a). They are stronger in shear (lateral load) than in withdrawal (Fig. 5-3b). Withdrawal strength from side grain is much less than for shear, and the designer should avoid this condition if at all possible. Withdrawal strength from end grain is so small and so unreliable that the NDS prohibits using connections that depend on withdrawal from end grain.

Shear Resistance in Side Grain

The allowable design lateral load *in any direction* for common or box nails or spikes driven into side grain of seasoned wood is

$$P = KD^{1.5} \qquad (5-2)$$

Factor K in this equation varies by species; K is 2040 for Group I, 1650 for Group II, 1350 for Group III, and 1080 for Group IV species; D is the shank diameter in inches. This equation is based on joint slip of 0.015 in., which occurs at approximately proportional limit loads. For softwoods, the ultimate loads are up to 3.5 times the value given by the equation; for hardwoods, up to 7 times. At ultimate load, however, joint slip is over 20 times as great as the 0.015-in. limit used to define the allowable. Table C-4 shows values given by this equation for various sizes and types of nail. Most designers prefer to use this table rather than calculating the allowable lateral load from Eq. 5-2.

Both Eq. 5-2 and Table C-4 are limited to nails or spikes having *standard penetration* into the wood member that contains the point of the nail. Penetration is the length of nail in the piece of wood that contains the nail point (Fig. 5-3). Standard penetration for each species group is defined as not less than a certain number of nail diameters: $10D$ for Group I species; $11D$ for Group II; $13D$ for Group III; and $14D$ for Group IV. Table C-4 and Eq. 5-2 are for seasoned wood and normal load duration. If the connection joins woods of two different species groups, the weaker one (higher group number) should be used to compute the design allowable for the nail or spike. Equation 5-2 and Table C-4 are based on tests in which the wood member nearest the head of the nail is 1 in. nominal or thicker.

Design allowables given by either Eq. 5-2 or Table C-4 must be modified as follows, and the modifications are cumulative:

1. For load duration, using the factors shown in Chapter 4.

(a) SINGLE SHEAR   (b) DOUBLE SHEAR

Fig. 5-4. Single- and double-shear connections.

2. For less-than-standard penetration. Use a fraction of the value shown, interpolating in straight-line manner between standard and zero penetration. In no case, however, should the nail or spike be used to carry load when its penetration is less than one-third of standard.
3. For metal side plates, if used, by increasing the allowable value by 25%.
4. For diaphragm construction, an increase of 30%.
5. For moisture content, according to Appendix Table C-2.
6. For use in double-shear connections (Fig. 5-4). The nail value may be increased by one-third if the nails fully penetrate all three members and the wood side pieces are at least one-third the thickness of the main member. If the side pieces are each as thick or thicker than the main member, the allowable increase is two-thirds (interpolate for intermediate thickness). For nails up to $12d$ size, the allowable nail load may be doubled, provided: (1) the side pieces are at least $\frac{3}{8}$ in. thick, (2) the nails project by at least three diameters, and (3) the pointed ends are clinched (bent over and pointed flat against the side piece).

Nailing of Plywood to Lumber

Plywood, being thinner than the lumber it is nailed to, is often the weak part of the connection. Using a large nail size to join plywood and lumber will not increase the allowable load of the connection. Provided the plywood and the lumber receiving the point are of similar densities, and provided the nail has full standard penetration (or more), the following plywood thicknesses have been proven by test to fully develop, for common nails, the lateral nail values shown in Table C-4.

Nail Size	APA Rated Sheathing (in.)	Structural I (in.)
$6d$	$\frac{3}{8}$	$\frac{5}{16}$
$8d$	$\frac{15}{32}$	$\frac{3}{8}$
$10d$	$\frac{19}{32}$	$\frac{15}{32}$

Shear Resistance for Nails in End Grain

The NDS allows lateral loads for nails or spikes driven into end grain to be two-thirds of the allowable in side grain. The same modifications apply.

Withdrawal Resistance of Nails or Spikes in Side Grain

Withdrawal resistance is less than lateral resistance. It depends on the nail diameter, the length of penetration, and the specific gravity of the wood. The withdrawal allowable load *per inch* of penetration into side grain of the member holding the nail point may be computed by

$$P = 1380DG^{2.5} \qquad (5\text{-}3)$$

in which D is the nail diameter (inches) and G is the specific gravity. The allowable value given by Eq. 5-3 is only for common nails and spikes, and only for those driven into side grain, either of wood that is seasoned and will remain seasoned, or of wood that is unseasoned and will remain that way in service. This equation is based on one-fifth of the ultimate load from tests. Average values of specific gravity are found in Appendix Table C-1. Appendix Table C-5 shows withdrawal design values per inch of penetration given by Eq. 5-3 for common nails, box nails, and spikes.

Nails that will be loaded in withdrawal should be spaced so as to prevent splitting.

Fig. 5-5. Toe nailing.

Woods with lower specific gravity generally do not split as easily as heavier ones, so in the lighter woods, the nail spacing can be less and the apparent advantage of denser woods, suggested by Eq. 5-3, may be offset.

Nails or spikes are frequently "cement coated." The coating is not cement, but usually a resin that increases the withdrawal resistance. In harder woods, the coating may be damaged by friction during driving. Unfortunately, the effectiveness of the cement coating diminishes with time, so that after only one or two months, withdrawal resistance may be no better than for an uncoated nail. Nails or spikes having grooved or threaded shanks also have greater withdrawal resistance than plain nails, and this added resistance tends to be of long-lasting value (see reference 2).

The structural designer should preferably avoid using nails under withdrawal loading. Furthermore, the NDS prohibits using nails driven into end grain for withdrawal loading. Table C-2 shows how severely the allowable withdrawal loads must be reduced if moisture content changes occur during service—to only 25% of the normal design allowable.

Toe Nailing

Figure 5-5 shows what is meant by *toe nailing*. This method of nailing is commonly used to connect vertical studs to the horizontal plates or sills in light-frame construction. Tests show that the highest strength is obtained if one uses the largest nail that can be driven without splitting the vertical member, having a distance from the end of the member to where the point first enters the wood equal to about one-third the nail length, driving the nail at about 30° with the vertical member, and "burying" the full length of the nail but avoiding excessive damage to the wood. Toe nailing is also more effective when the nails are driven from both sides of the vertical member, that is, cross-slant driving. The allowable design load for toe nails in withdrawal is two-thirds of the withdrawal design load for nails driven into side grain. The allowable design load for laterally loaded toe nails is five-sixths of the lateral loading design value for nails driven into side grain. Modifications for load duration, fire-retardant treatment, and less-than-standard penetration apply, but the modification for change of moisture content does not.

Example 5-1

A 3-in. (nominal) D. fir–larch tension member is spliced using metal side plates, as shown in Fig. 5-6. The wood was dry when the splice was made, but will be subject to wetting and drying in use. The member load is 400 lb and the controlling load combination includes wind. How many 8d box nails are required through each end of each splice plate?

The wood species is in Group II, and Table C-4 gives a value of 63 lb per nail. Standard penetration of 11D for Group II is 1.24 in. Actual penetration is the nail length (2.5 in.) minus the plate thickness, or 2.438 in. This exceeds 1.24 in., so no reduction is made for penetration. For wind load duration, the design value may be increased by a factor of 1.33. Because we are using metal side plates, it may also be increased by 25%. However, for moisture (Table C-2), it must be reduced multiplying by a factor of 0.75. The final modified allowable, then, is

3″ D. FIR-LARCH   1/16″ STEEL ℞

Fig. 5-6. Connection for Example 5-1.

$$(63)(1.33)(1.25)(0.75) = 78.6 \text{ lb per nail}$$

The number required to transfer 400 lb is $400/78.6$ = 5.1 nails. This, of course, is rounded up to 6 per side of the splice. There are two splice plates, so 3 nails per end per plate will be needed, for a total of 12.

Example 5-2

Figure 5-7 shows a single shear connection of a 2-in. nominal redwood member to a 4 in. nominal D. fir–larch member. How many $10d$ common wire nails are needed to transfer the controlling 400-lb reaction, which includes snow load? The wood is dry when installed and will remain dry in use.

Two species groups are involved, Group II for the column and Group III for the horizontal member. Group III must be used, and the value from Table C-4 is 77 lb per nail. Standard penetration is $13D$, or 1.92 in. However, the length of nail in the member receiving the point is only $3.00 - 1.50 = 1.50$ in. Therefore, the allowable must be reduced to $77(1.50/1.92) = 60.2$ lb. No modification is needed for moisture, but for snow load a factor of 1.15 may be used. The modified allowable is $(60.2)(1.15) = 69.2$ lb per nail. $400/69.2 = 5.78$. Use 6 nails.

Example 5-3

Analyze a double-lap connection similar to Fig. 5-4b except that each nail penetrates only one side piece and the main piece. Assume that all three members are 2 in. nominal and Group II species. If the con-

Fig. 5-7. Connection for Example 5-2.

4" D. FIR-LARCH

10d COMMON NAILS

2" REDWOOD

nection is made with a total of 24 $8d$ common nails (12 for each side piece), what is the allowable permanent load, P?

Equation 5-2 or Table C-4 gives 78 lb allowable lateral load per nail. The penetration into the main member is $2.50 - 1.50$, or 1.00 in. The nails are in single shear, since any one nail penetrates only two of the three members. Standard penetration is $11D$, or 1.44 in. The modified allowable lateral load per nail is

$$(78)(0.9)(1.00/1.44) = 48.8 \text{ lb}$$

and the allowable load P is 24×48.8, or 1170 lb.

Example 5-4

For the same connection as Example 5-3, what is the allowable load, P, if the connection is made using 24 $40d$ common nails, with their projecting points not crimped?

This is a double-shear connection. The nails are 5 in. long, so they fully penetrate all three pieces, and the points project $\frac{1}{2}$ in. beyond the surface of the side piece. Also, each side piece thickness is equal to that of the center member. Thus the load per nail may be increased by two-thirds the basic allowable. Table C-4 shows 176 lb for the basic allowable load per nail. Standard penetration $11D$ is 2.48 in. The center piece thickness of 1.50 in. should be used as the actual penetration. The modified allowable is

$$(176)(0.9)(1.50/2.48)(5/3) = 160 \text{ lb}$$

and load, P, allowed is $(24)(160)$, or 3840 lb.

Example 5-5

Same as Example 5-4, but the points of the $40d$ nails are crimped. What is the allowable load, P?

If the nail size did not exceed $12d$, the basic allowable could be doubled for the double-shear connection with nails crimped. The nails used are larger than $12d$, so the allowable P is, again, 3840 lb.

Example 5-6

What is the allowable load per nail for the withdrawal connection of Fig. 5-3b? The side piece is 2 in. nominal and the piece containing the point 6 in. nominal. The nails are $12d$ box nails, driven into side grain of No. 2 California redwood, seasoned when installed, but subject to long periods of exposure to wet use.

For California redwood, Table C-1 shows a specific gravity of 0.42 average. (The grade has no effect unless the grade causes the average specific gravity given by Table C-1 to be far out of line.) Table C-5 or Eq. 5-3 may be used to determine an unmodified (basic) allowable withdrawal load of 20 lb per inch of penetration in the piece containing the nail point. Modified for moisture change after installation (see Table C-2) the allowable load is

$$(20)(0.25) = 5.0 \text{ lb/in.}$$

The penetration is the nail length (3.25 in.) less the side member thickness (1.5 in.), so the penetration is only 1.75 in. and the allowable withdrawal load per 12d nail is (1.75)(5.0) = 8.75 lb per nail. Obviously, this connection is practically worthless for load-carrying members.

5-4. STAPLES

Wire staples are seeing increasing structural applications, such as attaching plywood diaphragms to framing members (Chapter 12). Staples are frequently used for factory-prefabricated items, such as wood pallets. They are also frequently used in combination with gluing. The reason for the growing popularity of staples is that they can be power-driven, saving hours over the time required for installing hand-driven fasteners.

The NDS does not give design values for staples, but information can be obtained from manufacturers of power staple driving equipment. The American Plywood Association has published information on the strength of staples in joints connecting plywood to lumber, as in a diaphragm (4). They recommend an allowable lateral load of 75 lb for a 14-gauge staple. For diaphragms, the duration factor of 1.33 for wind is applied to the 75-lb allowable (but the 30% increase allowed for nails and spikes in diaphragm construction is not allowed). Under wind loading, the modified allowable is, then, 1.33(75) = 100 lb per 14-gauge staple. This allowable load for 14-gauge staples was based on tests using Douglas fir framing members and 2-in. staple penetration. If the penetration of the staple into the main piece is only one inch, use 75% of the allowable value with 2-in. penetration, and interpolate for intermediate values of penetration.

Since the diameter of the staple leg is quite small, staples can be placed quite close together without splitting the wood. When plywood diaphragms are attached to framing members, the crown of the staple should be placed parallel to the long dimension of the framing member so that both legs go into the member.

Staples can have good withdrawal resistance as well as lateral load resistance. As for nails, withdrawal resistance can be improved by using coated staples.

5-5. LAG SCREWS

Lag screws, also known as *lag bolts*, are made with the unthreaded portion of their shanks in the same diameters as machine bolts. Dimensional information is shown by Appendix Table C-6. (This table also shows dimensions for small-sized lag screws that are not recognized in NDS tables of allowable loads.) Lag screws are especially useful for connections where bolts are not possible, either because the material is too thick or the far side of the connected parts is not accessible. Their behavior is much like that of bolts in single shear.

Lag screws are installed in predrilled holes. For the piece receiving the threaded part, the pilot hole should be a little less than the shank diameter: NDS specifies different hole diameters for different species groups—65–85% for Group I, 60–75% for Group II, and 40–70% for Groups III and IV. The predrilled hole in parts receiving the shank should be the same diameter as the shank. A washer under the head of the lag screw will protect the wood from damage as the screw is tightened. Installation is by turning the head with a wrench, the threads cutting their way into the wood as the screw is tightened.

Lateral Resistance

Lateral resistance of a lag screw varies with species group and with angle of load to grain. Table C-7 shows allowable loads parallel to the grain and perpendicular to the grain for lag screws in joints with wood side pieces (the part nearest the head of the lag screw), and Table

A ROW IS IN DIRECTION
OF LOAD

Fig. 5-8. Two-row connection.

C-8 shows similar values for joints with metal side pieces. When the load is neither parallel nor perpendicular to the grain, and the lag screw is installed in side grain of the main member, Hankinson's formula (Eq. 5-1) is used to determine the allowable load for the lag screw.

Other modifications that may be required are:

Reduction by 10% for fire-retardant treated lumber
Modification for load duration
Modification for number of lag screws in a row parallel to the direction of applied load
Modification for lag screw material
Modification for MC changes (Appendix Tables C-2 and C-3)
Modification for use in end grain

Modification for number of fasteners in a row applies to both lag screws and bolts. It is important to know the definition of a *row*, and this is shown clearly by Fig. 5-8. When the row contains more than two connectors, it is inaccurate to say that the fasteners in that row share the load equally, and this is true for connections in structural steel as well as in wood. When conditions are elastic, the assumption leads to impossible and incompatible conditions in the material being connected. In structural steel, yielding occurs, and for connections

of reasonable length, the assumption becomes acceptable for design. In wood, this is not the case.

In 1975, a paper by C. O. Cramer (5) presented a method by which actual loads on each connector in a row parallel to the load might be computed. His method considered the deformation of each fastener (see Fig. 5-9) and of the wood members between fasteners. Using an equation relating the elongation of the side pieces and main piece, he derived means for calculating the fastener loads. This method led to reduction factors that allow the designer to make the simplifying assumption that all fasteners in the row share load equally (even though we know that the fasteners at the ends of the row take considerably more than their equal share of the load transferred). Tables C-10 and C-11 show this modification for bolts and for lag screws. Footnotes to the tables show how the ratio of areas is computed.

Note that, according to the number in a row and ratio A_1/A_2, the effectiveness of a fastener can range from 100% to as little as 15%. Obviously, a penalty is paid for using a large number of fasteners in a row. One alternative is to use more rows but fewer lag screws per row. For connecting tension members this alternative has a disadvantage also—the net area of the member is reduced as additional rows are added. This effect can be reduced, though, by

Fig. 5-9. Deformation in a row of bolts.

Fig. 5-10. Spacing and stagger.

staggering the lag screws of one row relative to those of the next, as shown by Fig. 5-10. If the stagger is at least four bolt diameters, the net section reduction by the second row does not have to be added to that by the first row.

Lag Screw Material

Ordinarily, lag screws are made from a low-carbon steel conforming to ASTM Standard A307 and having a yield point of about 45,000 psi. If lag screws are made from a material having a different yield strength, the NDS requires modifying the design allowable lateral load in proportion to the square root of the ratio of yield-point stresses. When steels with yield strength less than 45,000 psi are used, this means a reduction of allowable lateral resistance. (If steel with a yield strength higher than 45,000 is used, this NDS requirement implies an increase in the allowable load for the lag screw. The authors believe, however, that an increase should *not* be allowed for this effect, since at higher loads, deformation of the wood becomes a more important factor than shear failure of the lag screw shank.)

Withdrawal Resistance

In resistance to withdrawal, lag screws are much more effective than nails. For shorter lengths of penetration of the threaded part into the wood, it is the wood that fails, allowing the screw to withdraw. For longer embedment of the threaded part, the wood does not fail but the screw itself fails in tension. Thus, the tensile strength of the screw is a maximum beyond which the withdrawal strength cannot go, no matter how much the effective embedded threaded length (see dimension T in Table C-6).

For embedment lengths in which the wood fails, withdrawal resistance is affected by specific gravity of the wood. This is reflected by Appendix Table C-9, which gives allowable withdrawal loads per inch of embedded effective threaded length. (Embedment of the unthreaded shank does not add to withdrawal strength.) In addition to determining an allowable load based on strength of the wood, the designer must check to make sure that the withdrawal load does not exceed an allowable based on tensile strength of the threaded shank.

Since the yield strength of material used for lag screws is usually 45,000 psi, the allowable withdrawal load based on tensile strength is

$$45,000 \, (\text{thread root area} / \text{factor of safety})$$

The NDS does not specify values for this allowable, but the designer can easily calculate it for any desired safety factor. However, the NDS does point out that the approximate tensile value of the lag screw will be developed when penetration of the threaded portion into the piece containing the point is 7 diameters in Group I species, 8 diameters in Group II, 10 in Group III, and 11 in Group IV.

Withdrawal of lag screws from end grain is not prohibited, but is inadvisible. If end grain withdrawal cannot be avoided, then the allowable value per inch should be reduced to not over three-quarters of the value given by Table C-9.

Lag Screw Spacing, Edge Distance, and End Distance

Edge distances, end distance, and spacing must be sufficient that tearout does not control the connection strength. If these distances are less, then the allowable load must be reduced. (These same principles and rules apply also for bolts.) Figure 5-11 summarizes the NDS rules for bolts and lag screws loaded either parallel to or perpendicular to the grain. The NDS avoids specifying spacing requirements for lag screws with loads at some other angle to the grain. (A guide might be found, however, in the practice shown by Chapter 9 for shear plates and for split-ring connectors.)

(a) LOAD PARALLEL TO GRAIN

(b) LOAD PERPENDICULAR TO GRAIN (OF HORIZONTAL PIECE)

Fig. 5-11. Spacing, end distance, and edge distances for lag screws and bolts. (D = shank diameter, ℓ = length in main member.)

End distances in Fig. 5-11 are for ends cut at 90° to the length. What should the end distance be when the end is cut diagonally? The NDS does not say for lag screws (or bolts); however, it does present a method of measuring such end distances for timber connectors. The authors feel that this is a conservative approach, suitable for bolts or lag screws also. It is illustrated in Example 5-8.

Example 5-7

A California redwood 2 × 6 is connected to a California redwood 4 × 6 in a single-lap connection as shown by Fig. 5-12. The permanent axial load to be transferred is 1.5 kips. Choose a size, length, and arrangement of lag screws to make the connection.

California redwood is in species Group III. Table C-7 gives lateral-load values for lag screw connections with wood side pieces. The maximum length of lag screw that can be used for this connection is 5 in., the total thickness of the two members. Many combinations of diameter, length, and number of screws will be satisfactory. By Table C-6, we see that either 4- or 5-in. lengths could be used.

Try a 4-in. length. Next, assume a diameter. For $\frac{1}{2}$-in.-diameter lag screws, Table C-7 shows a lateral load design value of 210 lb per screw in Group III species. This is for a single-lap connection, with a 1.5-in. side piece, parallel-to-grain loading, and normal duration. Modified for load duration, the allowable is $(210)(0.9) = 189$ lb per lag screw. If $\frac{1}{2}$- by 4-in. lag screws are used, the approximate number required will be

(a) SINGLE-LAP JOINT AND LOAD

(b) SOLUTION

Fig. 5-12. Connection for Example 5-7.

1500/189 = 7.9, say 8

The arrangement of the screws must now be considered. The two possibilities are to use (1) a single row of eight screws or (2) two rows of four screws each. Use Table C-10 to determine the modification factors for number in a row. Area A_1 is defined as the area of the main piece, which is the member receiving the point of the screw. Area A_2 is the area of the side piece.

$$A_1 = (3.5)(5.5) = 19.25 \text{ in.}^2$$
$$A_2 = (1.5)(5.5) = 8.25 \text{ in.}^2$$

Ratio $A_1/A_2 > 1$, so use $A_2/A_1 = 0.43$. Extrapolating from Table C-10, the ratios N_A/N are 4.29 for 8 in a row and 3.31 for 4 per row. Modification factors are $4.29/8 = 0.54$ and $3.31/4 = 0.83$, respectively. The modified allowable loads per lag screw are

$(0.54)(189) = 102$ lb if there are eight per row,
$(0.83)(189) = 157$ lb if used in two rows of four each

Assuming that we still prefer a single row, the approximate number required now becomes $1500/102 = 15$. But for 15 in a row, the modification factor is much less than 0.54, which makes the required number of lag screws much larger than 15. Obviously, there is too great a penalty to be paid for insisting on a single row in this case.

Try two rows of five each. The extrapolated ratio is 3.74 and the allowable per lag screw is

$(3.74/5)(189) = 141$ lb, and the total number required is $1500/141 = 10.6$. Recalculating with six per row shows that two rows of six $\frac{1}{2}$-in. by 4-in. lag screws each will be satisfactory, providing all spacing and edge distance requirements can be met.

Use Fig. 5-11 to determine spacing and edge distances. The spacing between rows must be at least 1.5 diameters, which is $\frac{3}{4}$ in. center-to-center. Try $2\frac{1}{2}$ in. for a practical spacing between rows. Ratio L/D for the $\frac{1}{2}$-in. by 4-in. screws exceeds 6, so the required edge distance is not less than 1.5 diameters ($\frac{3}{4}$ in.) nor less than one-half the actual spacing between rows ($1\frac{1}{4}$ in.). The latter controls. The sum of two edge minimum distances plus the $2\frac{1}{2}$-in. spacing between rows is only 5 in.; but the member is $5\frac{1}{2}$ in. wide, so the edge distances may be increased. A practical solution would be to use $1\frac{1}{2}$-in. edge distances with the two rows $2\frac{1}{2}$ in. on centers.

Spacing in each row, parallel to the load direction and the grain, must be at least four diameters, or 2 in. if the lag screws are fully loaded. (This can be reduced for lag screws that are not fully loaded, but to not less than three diameters.)

Redwood is a softwood, and the member is in tension, so the required end distance is seven times the screw diameter, or $3\frac{1}{2}$ in. Figure 5-12b shows the final answer to the example.

Example 5-8

Figure 5-13 shows a connection of a 2 × 8 bracing member to a beam. Load in the brace is 2.1 kips, including wind. Design a lag screw connection.

(a) MEMBERS AND LOAD

(b) MEASUREMENT OF END DISTANCE

Fig. 5-13. Connection for Example 5-8.

The unthreaded shank length must be at least 1.5 in., the thickness of the side piece. Try $\frac{1}{2}$-in. × 5-in. lag screws. Table C-7 gives allowables for Group II species of 440 lb and 280 lb for load parallel to and perpendicular to the grain, respectively. The load in the 4 × 16 beam is at 50° to the grain, so an intermediate allowable is computed by the Hankinson formula, as follows:

$$N = \frac{440(280)}{440 \sin^2 50° + 280 \cos^2 50°}$$
$$= 330 \text{ lb/screw}$$

Modifying for wind, the allowable is 440 lb per screw. The number required is $2100/440 = 4.77$ screws. Use five screws. A possible arrangement is as shown by Fig. 5-13. Spacing, edge distance, and end distance requirements are the same as for bolts. Following the rules shown by Fig. 5-11, the dimensions on Fig. 5-13 should be

b and c = not less than 4 diameters, or 2 in.

d = not less than four diameters, or 2 in. (as for a load perpendicular to the edge of the horizontal member)

In addition, dimension $2c$ should not exceed 5 in. unless vertical shrinkage of the 4 × 16 is checked and found to be negligible. Dimension a is calculated as shown by Fig. 5-13b. A reasonable dimension (instead of the 2.76 in.) would be $a = 3$ in. Actual dimensions would probably be much greater than these minimums.

Example 5-9

A horizontal 2 × 6 (laid flatwise) is connected crosswise to the bottom edge of D. fir–larch 2 × 10 ceiling joists at 16 in. c/c using lag screws. The 2 × 6 is to support a vertical load of 400 lb, placed anywhere along the length of the 2 × 6. Assuming one screw at each location where the 2 × 6 passes beneath a joist, what size and length of lag screw should be used?

Since the load can be placed anywhere, the maximum withdrawal load for one screw will be 400 lb plus the weight of a 16-in. length of the 2 × 6, or about 403 lb. The problem statement did not mention load duration, so the only safe assumption is that it is permanent. The required normal-duration allowable load is, then, $403/0.9 = 448$ lb.

Many different diameters could be used, but the larger ones are impractical since the lead hole for installing them would remove too much wood from the tension side of the joist cross section. Try a $\frac{1}{4}$-in.-diameter lag screw. The average specific gravity of the joists (Table C-1) is 0.51. The screws will be in side grain of the joists. Table C-9 shows an allowable load of 232 lb per inch of embedment of the threaded part into the piece containing the point (the joist).

Embedment required = $448/232 = 1.93$ in.
Add for thickness of 2 × 6   = 1.50 in.
Length required   = 3.43 in.

Try $\frac{1}{4}$-in. × 4-in. lag screws. Table C-6 shows that these lag screws will have a $1\frac{1}{2}$-in. unthreaded shank (equal to the thickness of the 2 × 6) and an effective threaded length of $T = 2\frac{5}{16}$ in., all of which will be in the joists. The tensile strength of the screw is developed (approximately) by embedment of 8D, or 2 in. The actual embedment in this case is $2\frac{5}{16}$ in. However, the required embedment is only 1.93 in., which is less than 8D, so the tensile strength of the lag screw does not control.

5-6. WOOD SCREWS

Wood screws are used mainly for connecting finish materials, cabinetry, and the like. They are often used to hold together pieces being glued. They have occasional structural uses, however, and can carry either transverse or withdrawal loads. Screw sizes are expressed in gauge numbers that indicate the size of wire from which they are manufactured. Screw heads may be either round or countersunk, as shown in Fig. 5-2.

Table C-13 shows allowable withdrawal loads in pounds per inch of embedment of the threaded part into the member containing the point. These allowables depend on the specific gravity of that member. They are for screws installed in side grain only. (Withdrawal loads from end grain are not allowed.)

Allowable transverse loads, shown by Table C-14, are for a standard penetration of seven diameters into side grain of the piece containing the point. Penetrations of less than $4D$ should not be used. For penetrations between $7D$ and $4D$, the allowable lateral load is reduced in proportion to penetration. Allowable lateral loads for wood screws in end grain are two-thirds of the values shown by Table C-14.

The values in Table C-13 and C-14 are subject to the following modifications:

1. Load duration
2. Moisture, according to Table C-2
3. Fire retardant, if used
4. Embedment length (explained above)
5. For lateral load only, a 25% increase of allowable when a metal side plate is used

To have structural value, wood screws must be installed in predrilled holes. NDS requires:

For screws loaded in withdrawal—A lead hole diameter of about 90% of the root diameter in Group I species, and about 70% of root diameter in Group II species. For Group III or IV species, no lead hole is required.

For wood screws loaded laterally—The lead hole diameter in Group I species must be the same as the shank diameter for the part receiving only the shank and the same as the root diameter for the part receiving the threaded part.

Field construction supervision will have to be especially good to ensure that these required diameters are actually provided. The designer using wood screws for structural purposes should question whether he or she can really exercise control over what diameters will be used.

Screws should be installed by rotating using a screwdriver; they should not be hammer-driven. A lubricant is sometimes used to make installation easier. Soap is often used, but this may be undesirable in that it holds moisture and can promote decay. Wax would be better and equally effective.

5-7. BOLTED CONNECTIONS

Bolt Types

Three types of bolts are commonly used in wood structures, but only one of them is suitable for major structural purposes. The word *bolt*, used without a modifier, indicates the type shown by Fig. 5-14a. This type, which may have either a square head or a hexagonal head, is suitable for structural purposes. If made of low-carbon steel, such bolts are sometimes called *common bolts*. Bolts made of high-strength steel may also be used.

Figures 5-14b and c show two other types of bolt that are frequently used in wood construction, although they are of minimal structural value. The carriage bolt is useful where the head must be smooth or where one side of the

Fig. 5-14. Bolt types.

connection cannot easily be reached to hold the bolt head while the nut is tightened. A short square section of the bolt shank cuts into the edge of the hole so that the bolt cannot rotate freely. The disadvantage to the carriage bolt is that no washer can be used under the head, so the head is drawn into the wood as the nut is tightened. Overtightening can easily damage the wood member and allow the bolt to rotate. Used carefully and without overtightening, the carriage bolt is useful for nonstructural and minor structural applications.

The stove bolt, shown in Fig. 5-14c, is threaded for the full length of shank. The threaded shank does not provide good resistance in bearing against the side of the hole. Stove bolts should not be used for structural purposes where one of the members connected tends to slide relative to the other, that is, in shear connections.

Everything that follows in this section will apply only to common bolts, the type shown by Fig. 5-14a.

Types of Bolted Connection

Bolts may be used alone to transfer load, or they may be used merely to hold the parts in firm contact while other devices (keys, timber connectors, or plugs) transfer the load. Bolts may be used in shear-type connections or tension-type connections. They may also be used where the load is in a diagonal direction, so that the shank receives both tension and shear.

In shear connections, load transfer is by bearing (compression) of the bolt shank against the wood at the side of the hole and by shear in the bolt shank. Friction between the connected parts may be sufficient to transfer some load immediately after the bolts are tightened. But this capability should not be counted on, since shrinkage (and possibly creep) may reduce the thickness of the members being joined. When this happens, the joint is loosened and compression between the parts is reduced, so that friction cannot be developed. Bolted connections that are subject to rotation (combined vertical load and moment, for example) are covered in Chapter 9.

Design of Bolted Shear Connections

Two kinds of shear connection are shown by Fig. 5-15. In each type, the load causes shear stress in the bolt shanks, and also causes compression (bearing) stress between the bolt shank and the wood at the side of the hole. If the bolts were infinitely stiff, the bearing stresses in the double-shear connection of Fig. 5-15a would be uniform along the length of the bolt in each wood member. The bolts are not infinitely stiff, however; consequently they bend, and the actual bearing stresses are as shown rather than uniform. In the single-shear connection of Fig. 5-15b, the bolts tilt as well as bend, so the bearing stresses are even less uniform than for the double-shear connection.

Average values of bearing stress are computed by dividing the force to be transferred by the projected bearing area (bolt diameter times thickness of material). From Fig. 5-15, however, it is obvious that the maximum bearing stress is higher than the average. Tests of double-shear connections show that the intensity of the maximum bearing stress depends upon the stiffness of the bolt and that this can be measured by the ratio L/D, in which L is the length of bolt in the center (main) member and D is the diameter of the bolt. The allowable load,

(a) DOUBLE SHEAR

(b) SINGLE SHEAR

Fig. 5-15. Bearing-stress variation.

P, parallel to the grain for one bolt in double shear, in a "balanced" connection with wood side pieces is given by

$$P = 0.8SrLD \qquad (5\text{-}4)$$

In this equation,

$S =$ allowable stress in compression parallel to the grain
$r =$ an experimentally determined factor to account for the nonuniformity of bearing stress (it depends on L/D for the main member)
$L =$ length of bolt in the main member
$D =$ diameter of the bolt

Product LD is the projected bearing area of the bolt shank against the wood.

A similar equation for allowable load perpendicular to the grain is

$$Q = S_1 r_1 LDc \qquad (5\text{-}5)$$

in which the experimentally determined factor c adjusts for the fact that wood fibers give higher values of perpendicular-to-the-grain compressive resistance for narrow bearing widths than for wide. Thus c is higher for small bolt diameters than for large ones (6).

These two equations give loads corresponding to the proportional limit load divided by a suitable factor of safety. For any given species and ratio L/D, the allowable shear loads, P and Q, will be constants. Thus, for design use, a simple table of allowable bolt values, such as Appendix Table C-12, can be used. Table C-12 shows allowable loads both parallel to and perpendicular to the grain for four commonly used species.

Design values in Table C-12 are for a balanced connection. A *balanced connection* (Fig. 5-16a) is a double-shear connection, having load in the same direction (either parallel to or perpendicular to the grain) in each of the three pieces connected, with all three pieces of the

Fig. 5-16. Bolted connections—balanced standard and variations.

same species, and with the side pieces each one-half as thick as the main piece (actual thicknesses, not nominal). The bolt values shown are for normal-duration loads only and only for common bolts installed in seasoned wood that will remain dry in service.

If the joint in question does not meet this definition of a balanced connection, then the allowable load per bolt must be modified. For the simpler cases, the modification can be determined by comparing the actual joint to an imaginary equivalent joint that is balanced. Examples follow.

1. Side piece thickness more than one-half that of main piece (Fig. 5-16b): No change; the allowable load is based on the thickness of the main (center) piece, and the extra thickness of the side pieces does not increase the joint strength.

2. Side piece thickness less than one-half that of main piece (Fig. 5-16c): Use the value given for an equivalent balanced joint whose main piece is twice as thick as the actual side pieces being considered.

3. Single-shear connection, as in Fig. 5-16d or e: Imagine that another member is added to make this a double-shear connection. If both members are the same species, consider the thickness of the main member for this imaginary double-shear connection to be whichever is smaller, the thickness of the thicker actual member or twice the thickness of the thinner actual member. Use one-half of the tabulated value for whichever imaginary balanced joint controls.

For other cases, use of the imaginary balanced joint is unclear. In these cases, the following NDS "cookbook" rules may be used.

4. Connections with four or more members of equal thickness, as in Fig. 5-16f: Several shear planes are present, and the design value for *each shear plane* is equal to one-half the value tabulated for a balanced connection of the same main-member thickness.

5. Connections with four or more members, not all of equal thickness, as in Fig. 5-16f: Consider the actual joint as divided into a series of adjacent, overlapping, three-member joints. Each of these joints has two shear planes. Determine the allowable load for each three-member joint, assuming that joint to act independently of the others. Divide the load for that three-member joint equally between the two shear planes it contains. Do this for all adjacent three-member joints. Then, for shear planes to which two different allowable loads have been assigned, use the lower value.

For assemblies of members in which the load is shared equally by all members, or for assemblies in which the division of load among the members is indeterminate, the design allowable for the entire joint is limited to the number of shear planes times the smallest value computed for any of the shear planes.

For assemblies of members in which the load on each individual member is known, the bolt design value for any member is the sum of the two individual bolt shear values computed for the two shear planes adjacent to that member.

6. Connections with steel side pieces: Load parallel to grain—the allowable value may be increased by 75% for bolts up to $\frac{1}{2}$-in. diameter, or 25% for $1\frac{1}{2}$-in.-diameter bolts. For bolt sizes between these limits, the percent increase is found by interpolation. Load perpendicular to grain—no increase is allowed.

7. Connections with a steel plate for the main member: Load parallel to grain of side pieces: Use the allowable load for a balanced joint in which the thickness of the main member is twice the actual thickness of the side pieces. Increase this allowable by 75% for bolts up to $\frac{1}{2}$-in. diameter, by 25% for $1\frac{1}{2}$-in.-diameter bolts, and interpolate to determine the percent increase for intermediate bolt sizes. (No increase is allowed for load perpendicular to the grain.)

8. Connections with different species of wood: The allowable bolt load is the lesser of (1) the allowable for a joint in which the actual side member species is used for all three members or (2) the allowable for a joint in which all three pieces are the same species as the main member.

9. Double-shear joints in which side members and main member are loaded in different directions: Use whichever allowable is smaller,

Fig. 5-17. Poor detail of bolted connections caused 145 of the 150 such connections in this building to split the wood. (Photograph by authors.)

the tabulated value for the main member or the tabulated value for a piece twice the thickness of the side members and loaded in the same direction as the side members.

10. Single-shear joints in which one member is loaded parallel to the grain and the other is loaded at an angle to the grain: Allowable load per bolt is the lesser of (1) one-half the tabulated value for the thickness of the member in which the bolt load is parallel to the grain or (2) one-half the value given by Hankinson's formula using tabulated P and Q values for a balanced connection with main-member thickness equal to that of the piece in which the bolt load is at an angle to the grain.

Other Modifications

In addition to the modifications mentioned above for joint configuration, the following modifications may be required:

1. Load duration.
2. Reduction for number of bolts in a row. This is the same as for lag screws (see Tables C-10 and 11).
3. Fire-retardant treatment.
4. Moisture conditions at installation and during service, using Tables C-2 and C-3. (See Figs. 5-17 and 5-18.)

Hole Diameter

To realize the full allowable design value of the bolt, it is important that the hole be prepared properly. Sharp tools must be used and the rate of drilling (feed) must be properly controlled to ensure a smooth-sided hole.

Fig. 5-18 Note how wood shrinkage caused the two-bolt connection to split the wood. (Photograph by authors.)

For proper bearing of the bolt shank against the wood of a shear-type connection, the hole diameter should be only slightly larger than that of the bolt. The NDS specifies that bolt hole diameters be $\frac{1}{32}$–$\frac{1}{16}$ in. larger than the bolt diameter. However, others recommend that hole diameters vary with the bolt size and with expected service moisture conditions. Generally, for bolts used in wood that will be at 6% MC or less, the hole diameter is equal to the bolt diameter plus $\frac{1}{16}$ in. for bolts up to $\frac{3}{4}$ in. diameter and plus $\frac{3}{32}$ in. for larger bolts. For bolts in wood at higher service moisture content, the difference should be less. At 12% MC, holes $\frac{1}{32}$ in. larger than the bolt diameter are recommended for bolts up to $\frac{3}{4}$ in., and $\frac{1}{16}$ in. for bolts larger than $\frac{3}{4}$ in. For wet service, only $\frac{1}{32}$ in. should be added for all sizes of bolt.

Washers

Washers are necessary under both the head and the nut. They serve two purposes. The force caused by tightening is spread over an area larger than the area of contact between the head or nut with the wood, thus preventing the wood from being crushed. Washers also permit the nut to be rotated easily as it is tightened. Where the head or nut bears against a steel plate of sufficient thickness to spread the load, a washer is not needed.

If the bolt is loaded in tension (or for a threaded steel rod used as a tension member) it may be necessary to use a washer that is larger than standard size and thickness. The necessary diameter can be calculated easily, as follows:

$$\text{Req. net bearing area} = \text{tensile load}/F_{c\perp}$$

$$(5\text{-}6)$$

The net area provided is the area enclosed by the outer perimeter of the washer minus the area of the hole in the wood or in the washer, whichever is larger.

Example 5-10

A double-shear connection as shown in Fig. 5-15a joins a 3 × 6 member to two 2 × 6 side pieces with a single row of five $\frac{5}{8}$-in. bolts. All wood is seasoned D. fir-larch, but frequently will be wet in service.

What is the allowable permanent load for the connection? What is the controlling net area for tension in the wood members?

For one $\frac{5}{8}$-in.-diameter bolt having a length of 2.5 in. in the main member, Table C-12 shows a basic allowable load of 1820 lb parallel to the grain. This value must be modified:

For load duration, factor = 0.9
For wet service (Table C-2), factor = 0.67
For 5 in a row, use Table C-10
  A_1 for the main member (3 × 6) = 13.75 in.^{2}
  $A_1/A_2 = 2.5/(2 \times 1.5) = 0.833$

Interpolating between the values shown for A_1/A_2 of 0.5 and of 1.0, the ratio for five bolts in a row is 4.33 and the modification factor is $4.33/5 = 0.87$. The modified allowable load per bolt is

$$(1820)(0.9)(0.67)(0.87) = 955 \text{ lb}$$

The allowable load for the connection = 5(955) = 4780 lb.

For the net area of main member, assume a $\frac{21}{32}$-in. hole diameter (bolt diameter plus $\frac{1}{32}$ in. for wet service):

$$A(\text{net}) = 2.5(5.5 - \tfrac{21}{32}) = 12.11 \text{ in.}^2$$

If the tensile stress in the wood based on this net area is greater than the allowable tensile stress, then the allowable load for the connection will be reduced below 4780 lb (see Chapter 7).

Example 5-11

A horizontal 2 × 12 joist is connected to a 4 × 4 vertical post by three $\frac{1}{2}$-in. bolts in a single vertical row. Both joist and post are D. fir–larch that is seasoned when installed and will remain dry in use. Based on the connection strength only, what is the allowable vertical reaction for a load combination that includes snow?

The basic allowable load for one bolt is whichever is lower:

One-half of P, the allowable parallel to the grain, for a balanced joint whose main member is $3\frac{1}{2}$ in. thick (1270/2 = 635 lb)
One-half the value of Q (load allowable perpendicular to the grain) for a joint whose main member is as thick as the 2-in. nominal joist (430/2 = 215 lb)

The latter value, 215 lb per bolt, controls.

There are three bolts in a row, and Table C-10 must be used to determine a modification factor. Area $A_1 = 3.5(3.5) = 12.25$ in.^{2} The side member (joist) has bolt loads perpendicular to the grain, so area A_2 is the product of member thickness times the "overall width of the fastener group." There being only one row of fasteners, the overall width is taken as the minimum parallel to grain spacing of the fasteners. Minimum parallel to grain spacing is four diameters, or $4(1/2) = 2.0$ in. Thus, area $A_2 = 1.5(2.0) = 3.0$ in.^{2} The ratio $A_2/A_1 = 0.245$, and by extrapolating on Table C-10, the modification factor (for three in a row) is

$$0.92 - 0.05(0.255/0.5) = 0.89.$$

Other modifications:

For duration, factor = 1.15
For service conditions, factor = 1.00
Modified allowable load per bolt = $215(1.15)(0.89) = 220$ lb
Allowable reaction (based on connection only) = $3(220) = 660$ lb.

(This must be compared to the allowable reaction based on shear in the joist. See Chapter 6.)

Example 5-12

Figure 5-19 shows an inclined tension member connected to a beam. The inclined member is a lodgepole pine 2 × 8, and the beam is a D. fir–larch 4 × 16. Tensile force F is 3 kips, normal duration. Assuming two rows of $\frac{3}{4}$-in.-diameter bolts, how many bolts per row are needed? In the space available, can they be positioned so as to satisfy all spacing, edge distance, and end distance requirements?

This problem is complicated in several ways: The connection is not balanced, two different species are used, and the load is parallel to the grain of one member but at an angle to the grain of the other. The problem must be solved assuming both members to be lodgepole pine, and again assuming both members to be D. fir–larch.

First, consider the diagonal. Using a main-member thickness equal to the thickness of the diagonal ($1\frac{1}{2}$ in.) $P = 1020$ lb for lodgepole pine, or 1420 lb for D. fir. The allowable for single shear, therefore, is $\frac{1}{2}$ of 1020, or 510 lb.

Second, consider the member in which the load is at an angle to the grain. Using a main-member thickness of 3.5 in., for lodgepole pine $P = 2310$ lb and $Q = 820$ lb for a bolt in double shear. For a bolt in double shear and loaded at 60° to the grain, Hankinson's formula gives

$$\frac{(2310)(820)}{2310 \sin^2 60° + 820 \cos^2 60°} = 978 \text{ lb}$$

Since the bolts are actually in single shear, the allowable is one-half of 978, or 489 lb per bolt, based on members of lodgepole pine.

For a double-shear joint in which both members are D. fir–larch, $P = 2800$ lb and $Q = 1260$ lb. It is obvious that Hankinson's formula will give a value greater than 978 lb. So the allowable, considering only the horizontal member, is 489 lb per bolt.

The final allowable for design is the lesser of the allowable computed for the diagonal or the allowable computed for the member with bolts loaded at an angle to the grain. Therefore, the design allowable is 489 lb/bolt.

Approximate number req.

$$= 3000 \text{ lb}/489 \text{ lb per bolt} = 6.13$$

Try 8 bolts (4 per row)

The modification for four bolts per row is computed as follows: Assume spacing between rows = $3\frac{1}{2}$ in. Also assume that the load is perpendicular to the grain in the angle-to-grain member (a conservative assumption for calculating A_2). Then the "overall width of the fastener group" is $3\frac{1}{2}$ in.

$$A_2 = 3.5(3.5) = 12.25 \text{ in.}^2$$
$$A_1 = 1.5(7.25) = 10.875 \text{ in.}^2$$
$$A_1/A_2 = 0.89$$

Factor for four per row (interpolated from Table C-10) is $3.61/4 = 0.90$.

$$N \text{ req.} = 3000/(0.90 \times 489) = 6.82$$
Use 8 bolts (2 rows of 4)

Arrangement of bolts: Spacing requirements must be examined for each member and the larger requirement will control. See Fig. 5-19b and c.

For the 4 × 16, ratio L/D for the $\frac{3}{4}$-in. bolts is $3.5/0.75 = 4.67$, which is less than 6. The loading in the horizontal member is nearly perpendicular to the grain, and the required distance (measured parallel to the grain) between rows is $4.17D$ (interpolated) or 3.13 in. In the 2 × 8, the load is parallel to the grain, and the spacing of the two rows must be at least $1.5D$, or 1.125 in. (which would be 1.3 in. if measured parallel to the grain of the 4 × 16). Obviously, the 3.13-in. dimension measured parallel to the 4 × 16 is the more severe requirement.

Spacing between bolts in each row in the 2 × 8 must be at least $4D$, or 3 in. Edge distance to the top

Fig. 5-19. Connection for Example 5-12.

of the horizontal member needs to be only $1.5D$ or 1.125 in. The load is directed toward the bottom edge of the horizontal member, however, so the edge distance required there is $4D$, or at least 3 in.

The 2×8 is a tension member, so its required end distance is $7D = 5.25$ in. The end of the 2×8 is not cut perpendicular to the load direction. It is suggested that this end distance be measured as shown on Fig. 5-19b: along a line $D/4$ away from the centerline of the row of bolts and toward the side for which the angle at the member end is larger than

90°. When the required end distance of 5.25 in. is measured in this manner, the top bolt in the row will be 4.64 in. away (dimension b) from the end of the member. Rounding this up to 4.75 in., suitable dimensions are as shown on Fig. 5-19c.

If space were not sufficient to place the bolts in this pattern, we could use the NDS provision allowing both smaller spacing in the row and smaller end distances, when the bolts are not stressed to their full allowable load. Each could be reduced in proportion to the load actually carried: end distance to as little as one-half the distance required for fully loaded bolts, and spacing to an absolute minimum of $3D$ rather than $4D$.

Example 5-13

A hem-fir 2×6 tension member carries 8 kips from a load combination that includes wind. Design a tension splice using $\frac{1}{2}$-in. bolts and metal splice plates. The wood is seasoned when installed, but is exposed to the weather.

From Table C-12, the basic value parallel to the grain for a $\frac{1}{2}$-in. bolt for a balanced connection is 800 lb.

Modifications required:

For wind, factor = 1.33
For steel side plates, factor = 1.75
For service conditions (Table C-2), factor = 0.75
Modified value = $800(1.33)(1.75)(0.75)$
   = 1396 lb/bolt

Approx. N req. = 8000/1396 = 5.73 bolts
Try 8 bolts (2 rows of 4 each)

Modification for four per row:

$A_1 = 1.5(5.5) = 8.25$ in.^{2}
Assume 2 side plates, $\frac{1}{8}$ in. $\times$ 5 in.;
$A_2 = 2(5)(1/8) = 1.25$ in.^{2}
$A_1/A_2 = 8.25/1.25 = 6.6$
From Table C-11, factor is $2.92/4 = 0.73$

Req. $N = 8000/(0.73 \times 1396) = 7.85$
OK Use 8 bolts (2 rows of 4 on *each* side of the joint in the 2×6)

Check net areas: For steel plates,

A (net) $= 2(5 - 2 \times 9/16) \times 1/8 = 0.969$ in.^{2}
Tensile stress = 8000/0.969 = 8260 psi
Allowable for A36 steel = 29,000 psi *OK*

For wood member,

A (net) $= 1.5(5.5 - 2 \times 9/16) = 6.56$ in.^{2}
Tensile stress = 8000/6.56 = 1220 psi

Example 5-14

Same as Ex. 5-13, but using unseasoned wood that will dry to less than 19% MC in service.

The allowable load per bolt will be affected by

seasoning of the wood after installation (see Table C-3). If a single plate (on each face of the splice) is used to contain two rows of bolts, as in Example 5-13, the allowable load per bolt has a reduction factor of 0.4. The reason is that high tensile stresses perpendicular to the grain are set up as the steel plates restrain the wood from shrinking. If each plate contains the two rows of bolts, the allowable load per bolt will be reduced to

$$800(1.33)(1.75)(0.4) = 745 \text{ lb}$$

On the other hand, if two narrow plates are used on each face of the splice (one narrow plate per row of bolts) to replace each wider plate, the factor is 1.0, so that the allowable load is

$$800(1.33)(1.75)(1.0) = 1860 \text{ lb per bolt}$$

Obviously, to use the split plates is the better solution, insofar as number of bolts is concerned.
    Approximate number required = $N = 8000/1860$
      = 4.3
Try six bolts (two rows of three each):

Assuming two side plates $\frac{1}{8}$ in. $\times$ 2 in. on each face of the 2×6
Factor for 3 per row = $2.55/3 = 0.85$ (Table C-11)

N req. = $8000/(0.85 \times 1860) = 5.1$
OK Use 6 bolts per end (2 rows of 3; i.e., 6 bolts on each side of the joint in the 2×6)

Example 5-15

Figure 5-20a shows a two-piece member spliced to a three-piece member. All wood members are D. fir–south. The loads shown are normal duration. How many $\frac{5}{8}$-in. bolts are required, assuming that they can be placed in two rows?

Each bolt has four shear planes. To solve, the connection is divided into the three three-member connections shown by Fig. 5-20b, c, and d. Basic allowable loads per bolt are computed for each three-member connection as follows:

For connection of Fig. 5-20b, 1780 lb/bolt
For connection of Fig. 5-20c, 1090 lb/bolt
For connection of Fig. 5-20d, 1780 lb/bolt

Next, divide these allowables to find the allowable shear forces per bolt *for each shear plane*, as follows:

Fig. 5-20. Connection for Example 5-15.

Shear Plane	Fig. 5-20b	Fig. 5-20c	Fig. 5-20d
1	890 lb		
2	890 lb	545 lb	
3		545 lb	890 lb
4			890 lb

Where the same shear plane has two different allowables, use the lower value. Shear allowables for each plane are the underlined values, above.

Number of bolts required at each shear plane:

Plane	Shear on Plane	Allowable	N Required
1	4600 lb	890 lb	5.17
2	−6000 + 4600 = −1400	545	2.57
3	−1400 + 2800 = 1400	545	2.57
4	4600	890	5.17 *Controls*

Try two rows of three bolts each. The smaller modification factor computed for either Fig. 5-20b or 5-20c, is 0.90, and the corrected number required is

$$N = 4600/(0.90 \times 890) = 5.7$$
OK Use 6 bolts (2 rows of 3)

5-8. BOLTS LOADED AT AN ANGLE TO THE BOLT AXIS

Figure 5-21 shows a bolted connection with load applied at an angle to the axis of the bolt. Table C-12 does not apply directly to this situation. What is necessary is first to resolve the diagonal load into two components, one normal to the bolt axis and one parallel. The component parallel to the bolt causes either tension in the bolt or compression between the two wood members, depending on whether the sloping member is in tension or compression. The component normal to the bolt axis is then handled with Table C-12, using whichever is less,

1. One-half the allowable load for a balanced double-shear connection whose main-member thickness is twice the smaller dimension, *a* or *b*, in Fig. 5-21, or

Fig. 5-21. Bolt loaded at angle to axis.

2. One-half the allowable load for a balanced joint whose main-member thickness is the larger of dimensions *a* and *b*.

Example 5-16

Assume that both members in Fig. 5-21 are No. 1 hem-fir 6 × 6s (P&T), intersecting at 45°. Force F is 4.3 kips, either tension or compression, depending on wind direction. Is the connection satisfactory with one 1-in.-diameter bolt, if dimension *b* is 3.75 in.?

The 4.3-kip force is resolved into two components, each 3.04 kips. Consider first the component that causes shear in the bolt. Dimension *b* = 3.75 in., which is smaller than dimension *a*, so the allowable bolt shear is whichever is smaller: one-half the value for a balanced, three-member connection whose main-member thickness is 5.5 in., or one-half the value for a balanced three-member connection whose main-member thickness is 2(3.75 in.) = 7.5 in. From Table C-12, we find the smaller of these two values to be

$$4820/2 = 2410 \text{ lb}$$

Modifying for wind load duration, the allowable is

$$1.33(2410) \quad \text{or} \quad 3210 \text{ lb}$$

This exceeds 3.04 kips, so the bolt is satisfactory for shear load.

Consider next the component parallel to the bolt. If the component is tensile, the unit tensile stress in the bolt, computed according to the AISC Code (7), is 3.04/0.785 = 3.87 ksi, which is well below the 20 ksi that AISC allows for A-307 bolts. (Note that AISC's allowable stress is based on using the *gross* area of the unthreaded shank of the bolt.)

If the diagonal-member force is compressive, the vertical member has compression perpendicular to the grain and the diagonal has compression at a 45° angle to the grain. For this species and grade, Table B-3 shows 850 and 405 psi for allowable stresses parallel and perpendicular to the grain, respectively. For wind, the 850 allowable is raised to 1130 psi. The length of the contact between the two members is about 6 in., so the contact area is about 6(5.5) = 33 in.^{2}. The compressive stress between the two members is 3040/33 = 92 psi, which is less than 405 psi. *OK*.

Two washers or bearing plates are needed, one at the left of the vertical member and one at the surface of the sloping member. For the washer at the left, the required bearing area is 3040/405 = 7.51 in.^{2}

At the right end of the bolt, we will need a wedge-shaped piece of material for the bolt head or nut to bear on. This can be made of either wood or metal. However, to prevent it from sliding, it would have to be keyed to the sloping member in some manner. As an alternative, the sloping member could be notched to provide a flat bearing surface, using a washer or bearing plate on this surface *under either the bolt head or nut*. If this is done, the bearing stress under the washer or bearing plate would be at 45° to the grain of the sloping member. It would be compared to an allowable value for bearing stress computed by Hankinson's formula (see Section 6-6).

End distance must also be considered. This is done by making the shear area (shaded in Fig. 5-21) equal to the shear area required for a connection in which the bolt axis is normal to the member surface. In this case, set the area shown shaded on the diagonal member of Fig. 5-21 equal to (or more than) the specified end distance times the member dimension, *h*. If the diagonal in this example is a tension member, the required shaded area would then be

$$(7D)(5.5) = 7(1)(5.5) = 38.5 \text{ in.}^2$$

if the bolt is fully loaded. The bolt load is only 3040 lb and the allowable is 3210 lb, so the required shaded area is

$$38.5(3040/3210) = 36.5 \text{ in.}^2$$

The actual shaded area is only about 9.2 in.^{2}, so the connection would be unsatisfactory for end distance if the load in the diagonal member were tensile.

REFERENCES

1. *National Design Specification for Wood Construction*, National Forest Products Association, Washington, DC, 1986.
2. *Wood Structures, A Design Guide and Commentary*, American Society of Civil Engineers, New York, 1975.
3. Hornbostel, C., *Construction Materials—Types, Uses, and Applications*, Wiley, New York, 1978.
4. *Plywood Diaphragms*, Research Report 138, American Plywood Association, Tacoma, WA, 1983.
5. Cramer, C. O., "Load Distribution in Multiple-Bolt Tension Joints," *Journal of the Structural Division, American Society of Civil Engineers*, **94**(ST5), Proc. paper 5939, New York, 1975.
6. *Western Woods Use Book*, 2nd ed., Western Wood Products Association, Portland, OR, 1979.
7. *Specification for the Design, Fabrication, and Erection of Structural Steel for Buildings*, American Institute of Steel Construction, Chicago, IL, 1978.

PROBLEMS

Unless otherwise noted, all lumber sizes are nominal and S4S, and loads are normal duration.

5-1. Two 2-in. nominal D. fir–larch planks are to be nailed together, with the nails driven into side grain. The planks are seasoned but will be wet in use. Snow loading controls. What is the allowable lateral load per 10d common nail?

5-2. Two 2 × 10 hem-fir planks are to be spliced in a double-shear connection using two 1 × 10 side pieces of the same material. Using the smallest common nail that will fully penetrate all three pieces, what is the allowable lateral load per nail?

5-3. A $\frac{1}{8}$- by 3-in. perforated steel plate is nailed to the top and bottom edges of a No. 2 grade 4 × 12 D. fir–larch, to give the wood member better bending strength. The plates are nailed to the wood using two rows of 10d common nails. What is the maximum satisfactory nail spacing per row where the total shear V is 1500 lb? Use VQ/I (in which I = moment of inertia of the transformed gross section) to compute the longitudinal shear force per linear inch between the steel plate and the wood beam.

5-4. Figure 5-22a shows the end of a 2 × 8 joist supported by a 2 × 12 header. The species is D. fir–larch. What is the allowable reaction for the joist, based on the strength of three 20d common nails in end grain, as shown?

5-5. Figure 5-22b shows an alternate detail in which the end of the joist bears on a 2 × 4 that is nailed into side grain of the 2 × 12 header. If the 2 × 4 is nailed to the header using six 10d common nails, what is the allowable reaction for the joist?

5-6. Assume that the joist of Problem 5-4 is connected to the header with only toe nailing, using six 10d common nails. What is the allowable joist reaction? Of the three details considered in Problems 5-4, 5-5, and 5-6, which is best, and why?

5-7. How many 14-gauge staples should be used to connect the short edge of an 8-ft by 4-ft plywood sheet to the roof rafters if the shear force along that edge due to wind is 400 lb/ft? The penetration of the staples into the Douglas fir rafters is $1\frac{1}{2}$ in.

(a)

(b)

Fig. 5-22. Details for Problems 5-4, 5-5, and 5-6.

5-8. Wind pressure of 20 psf acts on a light-frame wall. The wall studs are 2 × 6 hem-fir, spaced 16 in. c/c and spanning 7 ft 6 in. between top and bottom plates, which also are 2 × 6 hem-fir. The ends of each stud are connected by toe nailing. To resist the shear due to wind load, what size and number of box nails are required, if placed in a cross-slant pattern? (See Fig. 5-5.)

5-9. Two 4 × 12s of select structural D. fir-larch are to be spliced using lag screws and two $\frac{1}{2}$-in.-thick steel side plates. The tensile loads on the member are 3 kips from dead load, 4 kips from snow, and 3 from wind. Use two rows of lag screws. How many and what size are required?

5-10. A 2 × 10 California redwood beam is to be connected to the side of a California redwood 4 × 4 vertical post using lag screws. The end reaction of the beam is: 200 lb due to live load, 350 lb due to snow, and 45 lb due to dead load. The member will be seasoned when installed, but the connection will be exposed to weather. Choose size, length, and number of lag screws for the connection.

5-11. Architects revising the interior of an existing building want the appearance of a heavier type

Fig. 5-23. Connection for Problem 5-11.

of construction. One feature of this treatment is to add wood beams that really are not supporting members. Actually, the only way to hold these beams is to hang them from the bottom chord of existing wood trusses above, as shown by Fig. 5-23. What size lag screws would be suitable, and at what spacing along the bottom chord of the truss? (Assume that the net section of the truss chord is adequate.)

5-12. Figure 5-24 shows a wood sign supported by two timber poles. The total weight of the sign is 1400 lb. The sign will be in an open area where the recommended design wind speed is 70 mph. The builder wishes to connect the sign to the poles using lag screws, one verti-

cal row in each pole. Assuming that the lag screws pass through $\frac{3}{4}$-in. plywood and a 2-in. nominal member before entering the poles, what size and length of lag screws would be suitable and at what spacing? The poles are D. fir–larch, seasoned when installed, but exposed to weather.

5-13. Same as Problem 5-9, but using $\frac{1}{2}$-in. bolts.

5-14. A vertical 3 × 10 of dense No. 1 D. fir–larch, and a diagonal member of 3 × 10, No. 1 hem-fir are connected using two $\frac{3}{4}$-in. bolts in a row parallel to the diagonal. If the wood is installed and used dry, what is the allowable load, F, in the diagonal member, based on the strength of the bolted connection? Also, check the net area to be sure that the diagonal member is not overstressed in tension. The angle between the two members is 30°.

5-15. Compute the spacing, edge distance, and end distances required for the bolts of Problems 5-14. Show the bolt arrangement and these dimensions on a scale sketch of the connection.

5-16. A wood 2 × 8 D. fir ledger is bolted to a concrete wall as shown by Fig. 5-25. Plywood floor sheathing nailed to the top of the ledger applies vertical load ($D + L$) of 170 lb/ft of wall and horizontal load (parallel to the wall, from diaphragm action) of 220 lb/ft due to wind. What spacing, s, is required for the $\frac{1}{2}$-in. bolts joining the ledger to the wall? (The bolt value in the wood is based on one-

Fig. 5-24. Sign and support for Problem 5-12.

Fig. 5-25. Ledger bolts for Problem 5-16.

half of its value in a joint with main-member thickness equal to twice the ledger thickness.)

5-17. Two 3 × 12s of select structural D. fir–larch are to be spliced in a double-shear connection using 2 × 12 side members and $\frac{1}{2}$-in. bolts. The normal-duration load in the members is 13 kips tension. Design the splice using two rows of bolts. Remember to check the net section of the members.

5-18. How many $\frac{1}{2}$-in. bolts are needed for the connection of Fig. 5-20, assuming that each member on the right is known to carry one-third of the total load and that the two members on the left also share the load equally?

5-19. Design a bolted connection for the joint of Fig. 5-13, assuming that the controlling load, which includes snow, is 4.5 kips. (Do not necessarily use five bolts, unless that is what you consider the best choice.) Make a layout sketch showing all necessary dimensions.

5-20. Where would the bolt of Example 5-16 (Fig. 5-21) have to be located so that the diagonal member would be satisfactory as to end distance when the load is tensile? Draw the joint to scale to aid your computations. You will probably agree that it is a poor connection. How could it be improved?

5-21. Try repeating Example 5-16, but using two $\frac{3}{4}$-in. bolts instead of one 1-in. bolt. Position the bolts side by side, so that the side view of the connection would appear exactly the same as in Fig. 5-21. Does this solve the end distance problem?

5-22. As a challenge problem at this stage, design a bolted connection for the intersecting members shown by Fig. 5-26. All truss members shown are D. fir–larch. Loads shown are normal duration and are *maximums*, not necessarily simultaneous.

Fig. 5-26. Connection for Problem 5-22.

6

Selecting Sawn-Timber Beams

6-1. INTRODUCTION

Structural members stressed primarily in bending are called *beams*. Horizontal beams of dimension lumber, such as 2 × 8s, used repetitively in bending are usually referred to as *joists*, whereas large beams that support other beams are called *girders*. Beam sections are generally chosen on the basis of bending and then checked for other possible failure modes. Beam failure could be a result of bending, shear, lateral buckling, bearing, deflection under service loads, or deflection from creep or ponding. Beam design consists of selecting a species, grade, and cross-sectional dimensions to prevent any of these types of failure from occurring.

Some of these failures might be catastrophic, while others might affect only aesthetics or serviceability. This chapter will discuss each of these failure modes in the sequence in which they would normally be considered.

Beam sections are specified by nominal dimensions, but cross-sectional properties such as area and section modulus are always based on actual dimensions. Unless bending moment due to self-weight is trivial, one must include beam weight as a load on the beam. In this case, the specific gravities given in Appendix Table C-1 will be helpful.

6-2. DESIGN FOR FLEXURE— LATERALLY SUPPORTED BEAMS

If not braced to prevent their compression surface from moving sidewise, wood beams may buckle laterally due to instability. Fortunately, most wood beams and joists are braced at close

enough intervals along their compression face that lateral buckling is prevented. If this is the case, then the full modified allowable bending stress may be used.

Lateral support may be provided by blocking (short pieces of dimension lumber cut to fit between joists), or cross bracing called *bridging*, as shown in Fig. 6-1. The diagonal bridging members are sometimes light-gauge metal, but are more often cut from wood 1 × 3s or 1 × 4s for material economy. Bridging not only keeps the joists upright (prevents rotation), but also helps distribute loads to adjacent joists so that the repetitive-member value of F_b may be used. Lateral support can be provided by well-nailed plywood sheathing or by nailing one-inch nominal boards (subfloor) diagonally over the joists. Commercial metal hangers, besides providing a bearing seat, can also prevent the ends of beams from rotating.

The NDS (1) lists minimum requirements for lateral support to permit using the full modified bending allowable stress. These requirements, developed from past experience, are:

1. If the ratio of depth to thickness (based on nominal dimensions) of the beam is not more than 2 to 1, no lateral support is required.
2. If the ratio is 3 or 4 to 1, the ends must be held in position to prevent rotation.
3. If the ratio is 5 to 1, one edge must be held in line for its entire length.
4. If the ratio is 6 to 1, full-depth bridging or blocking must be installed at not over 8-ft centers. There are two alternatives to this: (a) Both edges must be held in line, or (b) the ends must be held in position to prevent

Fig. 6-1. Means of lateral support of beams.

rotation and also the full length of the compression edge of the member must be laterally supported by subflooring or sheathing.

Design Procedure

Sawn-timber beams are designed by working (allowable) stress design. The computed extreme fiber stress ($f_b = M/S = Mc/I$) must not exceed the modified allowable bending stress:

$$f_b \leq F_b \qquad (6\text{-}1)$$

In this equation and other equations to follow, the symbol for allowable stress (F_b in this case) indicates the particular allowable stress *including all applicable modifications*. Some of these modifications (discussed in Chapter 4) are size factor, load-duration modification, moisture content modification, and fire-retardant modification. Others are spelled out by footnotes to Tables B-3 and B-4. In bending-moment calculations for simple beams, the span is to be taken as the distance center-to-center of *required* bearing areas.

Should there be a notch in a bending member, the section modulus at that location should be reduced accordingly for calculating flexural stress. The notch depth must not exceed one-sixth the depth of the member. Notches should be avoided, especially on the tension side of the member (except at the ends), and must not be located in the middle one-third of the span where bending moment is the greatest.

The steps in the design procedure are as follows:

1. Estimate the size factor, C_F (discussed in Section 4-10); recall that it applies only to F_b.
2. Estimate beam weight per foot.
3. Solve for actual maximum bending moment.
4. Determine which load combination controls the design, and solve for section modulus required.
5. Choose a trial section.
6. Verify weight per foot, C_F, moment, and section modulus required.
7. Recycle as needed.
8. After selecting a section for bending, check for shear, deflection, and bearing (if required).

Form Factor

For circular and diamond-shaped beams, there is a modification to the allowable bending stress called the form factor, C_f. Early experiments showed that a circular log can support the same load as a square beam of the same cross-sectional area, even though the circular log has a section modulus only $1/1.18$ times as large. A square beam loaded diagonally to its face can support the same allowable load as if it were loaded perpendicular to a face. Consequently, a circular beam has $C_f = 1.18$ and a diagonally loaded square beam has $C_f = 1.414$. The allowable bending stress is multiplied by this factor. Use the section properties of the actual circle or diamond cross section.

Decking

Tongued-and-grooved (T & G) decking, used for floors and roofs, is another type of wood bending member. It is popular today in conjunction with glulam beams and arches or heavy-timber beams because the underside of the decking forms an attractive exposed ceiling for the room below at no additional cost. Tongues and grooves distinguish decking from dimension lumber used as floor planks. A projection (tongue) on the edge of one board fits into a depression (groove) in the adjacent board as seen in Fig. 6-2. The tongues and grooves constrain each board to deflect when adjacent boards deflect; that is, tongued-and-grooved members are repetitive.

Common sizes of sawn-lumber decking are 2×6, 3×6, or 4×6 nominal. Requirements for manufacture and design of decking are given in AITC Standard 112 (2). The 2-in. nominal has a single tongue and groove as in Fig. 6-2a, whereas the heavy-timber decking has double tongues and grooves as in Fig. 6-2b.

Table B-3 gives allowable bending stresses for select and commercial grade 4-in. nominal decking. Except for California redwood, this allowable bending stress should be multiplied by a modification factor of 1.10 for 2-in. nominal decking and by a factor of 1.04 for 3-in. decking. Decking is designed based on bending and deflection only. Shear stress is low and is usually disregarded. Because decking is bent in the plane of least flexural rigidity, it has no tendency to buckle laterally.

Decking may be laid in several different arrangements distinguished by location of end joints. The lay-up arrangements (see Fig. 6-3) and basis for design are:

1. Simple span: If all pieces bear on only

Fig. 6-2. Sawn lumber decking.

(a) SIMPLE SPAN

(b) TWO-SPAN CONTINUOUS

(c) COMBINATION SIMPLE AND TWO-SPAN CONTINUOUS

(d) CONTROLLED RANDOM LAY-UP

(e) CANTILEVERED PIECES INTERMIXED

Fig. 6-3. Layup patterns of decking. Courtesy of American Institute of Timber Construction.

two supports and end joints occur atop every joist for every piece of decking, then the arrangement is merely a simple span. The design for flexure and deflection would proceed exactly as for a simple span beam; that is, max $M = w\ell^2/8$ and max $\Delta = 5w\ell^4/(384EI)$.

2. Two-span continuous: If all end joints occur in line at alternate supports, as shown by Fig. 6-3b, the arrangement is two-span continuous. It is designed as a two-span continuous beam with both spans fully loaded; that is, max $M = w\ell^2/8$ and max $\Delta = w\ell^4/(185EI)$.

3. Combination simple and two-span continuous: In this lay-up, shown by Fig. 6-3c, every other piece in the *end* span is a simple span. All other pieces are continuous over two spans. The net effect is that all end joints occur over supports, but for adjacent pieces they are in alternate courses. Design for bending is based on a two-span continuous beam with both spans loaded. Design for deflection is based on the average of the maximum deflections of a simple span and a two-span continuous beam; that is, max $M = w\ell^2/8$ and max $\Delta = w\ell^4/(109EI)$.

4. Controlled random lay-up, shown by Fig. 6-3d (see reference 3 for more details): Use a maximum moment of $w\ell^2/10$ for flexural design, but use only two-thirds (tables in reference 3 use 69%) of the actual moment of inertia in bending and deflection calculations for 2-in. nominal and 80% of the moment of inertia for 3- and 4-in. nominal. These reductions are based on previous experience with random joints. Maximum deflection is considered to be as for a uniformly loaded three-span beam (max $\Delta = w\ell^4/(145EI)$). With the reduction in moment of inertia, this amounts to $w\ell^4/(100EI)$ for 2-in. nominal and $w\ell^4/(116EI)$ for thicker decking.

5. Cantilevered pieces intermixed, as in Fig. 6-3e: Flexural design is the same as for controlled random lay-up, including the reduction in moment of inertia. The deflection design formula is $\Delta = w\ell^4/(105EI)$ with no reduction of moment of inertia.

Example 6-1

A simple beam (D. fir) has a span of 16 ft and supports a uniform load of 480 lb/ft. The basic allowable bending stress (single-member use) from Table B-3 is 1300 psi. Choose a section based on flexure only.

Assume the load is of normal duration since it was not stated otherwise in the problem. Estimate $C_F = 0.95$ (an estimate between 0.95 and 0.98 is a good first trial).

$$F_b = (0.95)(1300) = 1235 \text{ psi}$$

Estimate that cross-sectional area will be about one square foot, so the beam weight is approximately

$$38 \text{ lb/ft } (0.51 \times 62.4 \times 1.19 = 38)$$
$$M = w\ell^2/8 = 518(16^2)/8 = 16,600 \text{ ft-lb}$$
$$S \text{ req} = M/F_b$$
$$= (16,600)(12)/1235 = 161 \text{ in.}^3$$

A 6×14 beam has a section modulus of 167 in.^{3} (see listing of section moduli in Table B-1) and weighs

$$(38)(74.25)/144 = 20 \text{ lb/ft}$$

For a 6×14, $C_F = (12/13.5)^{1/9} = 0.987$. Estimates for C_F and beam weight were both conservative, so the 6×14 is a safe choice. If the computations are repeated using these new values,

$$M = 500(16)^2/8 = 16,000 \text{ ft-lb}$$
$$S \text{ req} = 12(16,000)/(0.987 \times 1300) = 150 \text{ in.}^3$$

There is no smaller beam with a section modulus of at least 150, so the 6×14 is the final selection. Other sections, wider and shallower than the 6×14, could be found that would have the necessary bending strength. However, the chosen section has the smallest area and, therefore, the smallest cost.

Example 6-2

In an apartment building, the living room floor beams (joists) extend 5 ft past the exterior wall, forming a balcony that is 5 ft wide and 16 ft long. The loads on the balcony are:

5 psf dead load, including beam weight
60 psf live load due to people, furniture, and stacked firewood
40 psf snow load

Using machine-stress-rated lumber of grade 1200f-1.2E and assuming these joists will be spaced at 16 in. c/c, what size joist should be used considering flexure only? Assume that the negative bending moment at the exterior wall controls the design.

Permanent loads total 5 psf (just the dead load).
Ten-year duration loads = 5 + 60 = 65 psf
Two-month duration loads = 5 + 60 + 40 = 105 psf

Divide each of these by the load-duration factor for that type of load (the largest quotient will be the loading case that controls):

$$5/0.9 = 5.6$$
$$65/1 = 65$$
$$105/1.15 = 91.3 \quad \textit{Controls}$$

The tributary width to each joist is 16 in. (i.e., each joist carries load equal to that occurring on an 8-in. width on each side of the joist).

$$w = (105 \text{ psf})(16/12 \text{ ft}) = 140 \text{ lb/ft}$$
$$M = w\ell^2/2 = (140)(5)^2/2 = 1750 \text{ ft-lb}$$

The basic (unmodified) value of F_b is 1400 psi for repetitive members (Table B-4).

$$S \text{ req} = (1750)(12)/[(1.15)(1400)] = 13.0 \text{ in.}^3$$

Use 2 × 8s ($S = 13.1$ in.^{3}) at 16 in. center-to-center. Note that C_F does not apply to this size lumber.

Example 6-3

Exposed floor beams for a ski lodge will be made from 16-in.-diameter logs whose basic $F_b = 1250$ psi. If the beams span 30 ft, what is the allowable permanent uniform load per foot?

Cross-sectional area = 201 in.^{2}, so the equivalent square has a side of 14.2 in. Using the equivalent square, the size factor is

$$C_F = (12/14.2)^{1/9} = 0.98$$
$$F_b = 0.9(0.98)(1250) = 1102 \text{ psi}$$
$$S = 14.2^3/6 = 477 \text{ in.}^3$$
Allowable $M = F_b S = 1102(477)/12 = 43{,}800$
ft-lb = $w\ell^2/8$
Allowable $w = 389$ lb/ft (including beam weight)

An alternate to using the equivalent square is to use section properties of the actual circular section, but with F_b modified also by the form factor $C_f = 1.18$. The answer is identical.

Example 6-4

A wood house built in the early 1900s is to be converted to use as a private club. The building code requires increased load capacity because of change in use. A two-span continuous beam (each span = 13 ft) supporting the first floor consists of four 2 × 12s nailed together. Because of random joints where pieces of lumber abut, only three of the four 2 × 12s are effective at any one cross section. The lumber has been identified as Douglas fir corresponding to No. 2 grade. Measurements show the 2 × 12s to be actually $1\frac{3}{4}$ in. by $11\frac{5}{8}$ in. Find the allowable uniform load on the continuous beam.

Assuming that the duration of loads till now has exceeded ten year cumulative,

Factor for load duration = 0.9
$F_b = 0.9(1250) = 1125$ psi
$S = (3)(1.75)(11.625)^2/6 = 118$ in.^{3}

Maximum moment in a two-span beam is the same for a single-span beam so

$$w(13^2)/8 = F_b S = 1125(118)/12$$
Allowable $w = 524$ lb/ft

Example 6-5

Douglas fir T & G decking for a floor will be laid up in a controlled random pattern (Fig. 6-3d). This Commercial grade 2 × 6 decking supports a uniform live load (normal duration) of 50 psf. Self weight is the only dead load. Moisture content will not exceed 19%. What is the farthest distance apart the supporting joists may be placed (based on bending stress)?

$E = 1{,}700{,}000$ psi (Table B-3)
Basic F_b (for 4-in. decking) = 1650 psi (Table B-3)
Modification (footnote 4, Table B-3) for 2 in. nominal = 1.10
$F_b = (1.10)(1650) = 1815$ psi

Using a specific weight of 38 pcf and an actual thickness of 1.5 in.,

$$w_D = (1.5)(38)/12 = 5 \text{ psf}$$

Considering a one-foot width (approximately two planks),

$$w_D = (5 \text{ psf})(1 \text{ ft}) = 5 \text{ lb/ft per ft width}$$
$$w_L = (50 \text{ psf})(1 \text{ ft}) = 50 \text{ lb/ft per ft width}$$
$$I = (12)(1.5)^3/12 = 3.375 \text{ in.}^4 \text{ per ft width}$$
$$c = 0.75''$$

Expressing permissible span, ℓ, in inches:

$$f_b = (0.1w\ell^2)(c)/(2I/3)$$
$$= 0.1(55/12)(\ell^2)(0.75)/2.25 = 0.153\ell^2$$

Let $f_b = F_b = 1815$ psi:

$$\ell = 109 \text{ in. (based on bending stress)}$$

A maximum permissible span based on deflection will be discussed in Example 6-15.

6-3. DESIGN FOR FLEXURE— LATERALLY UNSUPPORTED BEAMS

Most beams are laterally supported by such things as attached plywood or decking, or by blocking, bridging, or diagonal bracing. If the distance between points of lateral support is too great, as in Fig. 6-4, the beam can buckle (twist and move sidewise). In this case we reduce the allowable bending stress to F_b', a value less than the full modified allowable, F_b, used for laterally braced beams. Based on their laterally unsupported length and their breadth and depth, beams are divided into three classes: short, intermediate, and long.

One must take care to distinguish between the span of a beam and its unbraced (laterally unsupported) length. The span of a beam is the distance between supports—the walls, columns, or girders that the beam bears on. The unbraced length is the distance between *lateral*

(sidewise) restraints. Lateral supports prevent movement perpendicular to the plane of the loads.

In the following, the design procedure specified by the NDS is explained first. Then the theoretical basis for the design equations is given for those who wish to understand the fundamentals of beam buckling.

To analyze a beam with only intermittent points of lateral support, first calculate its slenderness factor, C_s, from the following equation.

$$C_s = \sqrt{\ell_e d/b^2} \qquad (6-2)$$

The slenderness factor for beams is similar to the ℓ/r ratio for columns. In Eq. 6-2, b = thickness (width normal to plane of loads), d = depth, and ℓ_e = effective unbraced length. Effective unbraced length ℓ_e depends on the shape of the bending moment diagram; for example, the moment may or may not remain large over a considerable length of the beam. Effective unbraced length ℓ_e, defined in Table 6-1, is also a function of the actual unbraced length ℓ_u (actual distance between points of adequate lateral support).

If a bending member is used in a flatwise position, slenderness factor will be of no concern since the compression face will have no tendency to buckle. Also, for circular beams such as logs, conditions of lateral support need not

Fig. 6-4. Lateral buckling of a beam.

Table 6-1.   Effective Length of Beams.

Type of Span	Loading	Effective Length, ℓ_e
Single	Concentrated load at center	$1.37\ell_u + 3d$
	Equal end moments	$1.84\ell_u$
	Uniformly distributed load	$1.63\ell_u + 3d$
	Any load (conservative value)	$1.84\ell_u$ for $\ell_u/d > 14.3$
		$1.63\ell_u + 3d$ for $\ell_u/d < 14.3$
Cantilever	Uniformly distributed load	$0.90\ell_u + 3d$
	Concentrated load at free end	$1.44\ell_u + 3d$
	Any load (conservative value)	$1.84\ell_u$ for $\ell_u/d > 14.3$
		$1.63\ell_u + 3d$ for $\ell_u/d < 14.3$

be considered because a circular member cannot buckle due to bending; it is stable.

If $C_s < 10$, then the beam is a "short" beam for which lateral buckling is not possible. It is designed exactly as if it had continuous lateral support, using the full allowable bending stress with all applicable modifications; that is,

$$F_b' = F_b \qquad (6\text{-}3)$$

A value called C_k serves as the upper limit for classification as an "intermediate" beam, where

$$C_k = 0.811 \sqrt{E/F_b} \qquad (6\text{-}4)$$

If C_s is between 10 and C_k, then the beam is "intermediate" and

$$F_b' = F_b \left[1 - \frac{1}{3} \left(\frac{C_s}{C_k} \right)^4 \right] \qquad (6\text{-}5)$$

but not more than the value from Eq. 6-3. Size factor C_F may *not* be used to modify F_b in Eqs. 6-4 and 6-5, but all other modifications apply. Also, the elastic modulus, E, is never modified for load duration.

If C_s is more than C_k (but never more than a limit of 50), then the beam is "long" and

$$F_b' = 0.438E/C_s^2 \qquad (6\text{-}6)$$

but not more than the value from Eq. 6-3. It is possible for Eq. 6-3 to control the design of an intermediate or long beam since F_b in Eq. 6-3 is modified by C_F whereas F_b in Eqs. 6-4 and 6-5 is not. If the size factor is of greater influence than stability, then Eq. 6-3 might control.

Theory of Elastic Stability of Beams

For a laterally unsupported beam, if the flexural rigidity in the plane of bending is large in comparison with the lateral flexural rigidity, the beam will buckle laterally at a certain critical load in the elastic range. The critical bending moment is an ultimate load at which sudden instability occurs. To determine the critical load for the case of bending moment that is uniform along the length of a beam, consider the beam of Fig. 6-5a with applied end moments, M_o, about axes parallel to the Z-axis. The beam ends are free to rotate (about the Z-axis), but are fixed against twisting (about the X-axis).

In the deflected position, the compression edge of the beam has moved sidewise, as the cross section in Fig. 6-5b indicates. The displacement in the Z-direction is denoted as w, and $dw/dy = \beta$ is angle of twist (4). From equilibrium of the buckled beam, the bending and twisting couples at any cross section are

$$M_x \simeq M_o \frac{dw}{dx}$$

$$M_y \simeq M_o \frac{dw}{dy} = M_o\beta$$

$$M_z \simeq -M_o$$

But by the theory of elastic stability,

$$M_y = -EI_y \frac{d^2w}{dx^2} = M_o\beta$$

$$M_x = GK \frac{d\beta}{dx} = M_o \frac{dw}{dx}$$

where K is the torsional stiffness factor for a rectangular cross section (K equals about

(a)

(b) SECTION A-A

(c)

Fig. 6-5. Laterally unsupported beam.

$0.26db^3$ for wood beams of practical size). Differentiating the latter with respect to x,

$$GKd^2\beta/dx^2 = M_o(d^2w/dx^2)$$

Substituting,

$$GKd^2\beta/dx^2 = M_o(-M_o\beta/(EI_y))$$
$$d^2\beta/dx^2 + M_o^2\beta/(EI_yGK) = 0$$

By inspection, the solution to the differential equation will require that β be the sum of a sine function and a cosine function. Solving the equation gives

Critical $M = \pi\sqrt{EI_yGK}/\ell$

Critical $f_b = M_{cr}d/(2I_x) = \pi d\sqrt{EI_yGK}/2I_x\ell$

$$= \frac{\pi d\sqrt{E(db^3/12)(G)(0.26db^3)}}{2(bd^3/12)(\ell)}$$

Critical $f_b = Cb^2/\ell d$

where C is constant for a given species and grade of wood.

This result shows that critical flexural stress for lateral buckling is proportional to b^2 (square of beam thickness) and inversely proportional to both beam length and depth. It is interesting to contrast this critical stress for beams with critical buckling stress for a column. For a column, the critical stress is inversely proportional to the *square* of the length, rather than the length.

Equation 6-6 for "long" beams is based on the lateral buckling theory just presented. The allowable stress in Eq. 6-6 is a function of $b^2/\ell d$.

Example 6-6

A simply supported 6 × 18 beam with an unmodified $F_b = 1550$ psi and $E = 1,500,000$ psi is loaded with a uniform load of seven days duration. The beam will be in a humid location where moisture content is expected to remain about 20%. The span is 40 ft. The beam is laterally braced at its ends and at midspan only. Based on flexure, what uniform load can be superimposed?

Moisture modification (for both F_b and E) = 1.0 (see footnote 7 in Table B-3).
Load duration modification = 1.25 (from Fig. 2-14)
$C_F = (12/17.5)^{1/9} = 0.959$
For an assumed unit weight of 35 pcf, the beam weight is about 23 lb/ft.
$\ell_u = (20)(12) = 240$ in.
$\ell_e = 1.63(240) + 3(17.5) = 444$ in.
(Table 6-1)
$C_s = \sqrt{(444)(17.5)/(5.5)^2} = 16.0$

$C_k = 0.811\sqrt{\dfrac{1,500,000(1.0)}{1550(1.25)(1.0)}} = 22.6$

Because C_s exceeds 10 but is less than C_k, the beam is intermediate. By Eq. 6-3,

$$F_b' = (1.0)(1.25)(0.959)(1550) = 1858 \text{ psi}$$

By Eq. 6-5,

$$F_b' = (1.0)(1.25)(1550)\left[1 - \tfrac{1}{3}(16.0/22.6)^4\right]$$
$$= 1775 \text{ psi} < 1858 \quad \textit{Controls}$$

Allowable $M = F_b'S$
$\qquad\qquad = 1775(280.7)/12 = 41,500$ ft-lb
$w(40)^2/8 = 41,500$
Allowable $w = 208$ lb/ft

Subtracting the weight of the beam, the allowable superimposed load is 208 − 23 = 185 lb/ft.

Example 6-7

Repeat Ex. 6-6, but with $\ell_u = 44$ ft = 528 in.

$\ell_e = 1.63(528) + 3(17.5) = 913$ in.
$C_s = \sqrt{(913)(17.5)/(5.5)^2} = 23.0$
$C_k = 22.6$ (same as before)

Because C_s exceeds C_k, the beam is classified as long. By Eq. 6-3, $F_b' = 1858$ psi, as before. By Eq. 6-6,

$$F_b' = 0.438(1,500,000)/23.0^2$$
$$= 1242 \text{ psi} \quad \textit{Controls}$$

Allowable $M = F_b'S$
$\qquad\qquad = 1242(280.7)/12 = 29,050$ ft-lb
$w(44)^2/8 = 29,050$
Allowable $w = 120$ lb/ft

The allowable superimposed load is 120 − 23 = 97 lb/ft.

Example 6-8

Design a Douglas fir beam, spanning 22 ft, to support two concentrated loads of 0.77 kips each (from a stairway). Use either No. 1 or Dense No. 1 grade. Additionally the beam supports 100 lb/ft, not including self-weight. The first point load is 7 ft from the left support and the second is 10 ft from the left. Lateral support exists only at the location of the concentrated loads. Consider only flexure. Fig. 6-6 shows the solution in computation sheet form.

6-4. DESIGN OF BEAMS FOR SHEAR

Timber beams are relatively weak in shear parallel to the grain, so their shear strength needs to be calculated. Shear failure causes a long horizontal split at beam mid-depth (Fig. 2-11). To design to prevent shear failure, one compares the computed actual shear stress with the modified allowable shear stress. That is, the beam is satisfactory in shear if

$$f_v \leq F_v \qquad (6\text{-}7)$$

where $f_v = VQ/Ib$ (for rectangular members $f_v = 1.5V/A$). The basic allowable shear stress F_v is based on the assumption that a full-length split is present, so if end splits are known to be minor and will not grow in length, the allowable shear stress may be modified as shown by footnote 8 of Table B-3. Figure 6-7 shows how to measure the size of splits and checks when using footnote 8.

Some of the NDS provisions for shear are based on the 1933 work of Newlin, Heck, and March (5). Their experiments on checked beams showed horizontal shear stresses to be lower than usually assumed by virtue of the upper and lower halves of the checked beam each acting as independent beams. This "two-beam action" is the basis of NDS section 4.2.6. Because the experiments were for beams with severe (artificially created) splits, two-beam action cannot be expected in glued laminated beams that have relatively small checks. Keenan verified the lack of two-beam action in glulams by a finite element study (6). He also found from laboratory and finite element work

$$w = 0.10 + bm \ wt \ (est. \ 0.02) = 0.12 \ k/ft$$

Lateral support at end & conc. loads only. (Points x)

Try Dense No.1 D.fir (WWPA)

B & S

Basic $F_b = 1550 \ psi$

$E = 1.7 \times 10^6 \ psi$

$M \ max = A_1 = \frac{12}{2}[0.48 + 1.92] = 14.4 \ ft\text{-}k$; Assume $C_F = 0.96$

" $F_b' = 90\%$ of F_b

Approx. S req'd = $12 \times 14,400 / (0.90 \times 1550) = 124 \ in.^3$

Try 6×14 $(S = 167 \ in.^3)$ $\ell_u = 12(12) = 144"$; $\ell_u/d = 144/13.5 = 10.67 < 14.3$

$\ell_e = 1.63(144) + 3(13.5) = 275"$ (Table 6-1)

$C_k = 0.811\sqrt{1.7 \times 10^6/1550} = 26.9$

$C_s = \sqrt{275(13.5)}/5.5 = 11.1$; $10 < C_s < C_k$, $\therefore$ Intermediate

$F_b' = 1550[1 - \frac{1}{3}(11.1/26.9)^4] = 1535 \ psi$

$C_F = 0.987$; $F_b \ (mod) = 1550(0.987) = 1530 \ psi$ Controls

S req'd $= 12 \times 14,400/1530 = 113 \ in.^3$

Try No.1 grade 4×16 $(S = 135.7 \ in.^3)$ $E = 1.8 \times 10^6 psi$; $F_b = 1500 psi$

$\ell_e = 280"$; $C_k = 0.811\sqrt{1.8 \times 10^6/1500} = 28.1$

$C_s = \sqrt{280(15.25)}/3.5 = 18.7$ Intermediate

$F_b' = 1500[1 - \frac{1}{3}(\frac{18.7}{28.1})^4] = 1402 \ psi$ Controls

$C_F = 0.974$; $F_b(mod) = 1500(0.974) = 1460 \ psi$

S reg'd $= 12 \times 14400/1402 = 123 \ in.^3 < 135.7$ OK

Use 4×16, No.1 Grade

Fig. 6-6. Calculation sheet for Ex. 6-8.

that, contrary to popular assumption, shear strength was not increased by the application of compressive stress perpendicular to the grain as caused by the beam reaction.

Shear strength is affected by beam size and shear span (distance from reaction to first concentrated load). Researchers are attempting to develop ways of considering this size effect for shear when designing, just as the size effect for bending adjusts the allowable bending stress. Accumulating data from many sources, Keenan (reference 6) found that shear strength was a function of sheared area, A_s, defined as the product of beam width and the shear span. He proposed, for Douglas fir, an equation for the maximum actual shear stress at failure as

ℓ = LENGTH OF SPLIT ON WIDE FACE

ℓ = SIZE OF SHAKE

Fig. 6-7. Splits and shakes.

CRITICAL SECTION

FACE OF SUPPORT

(a) BEARING PRESSURE ON FACE OPPOSITE LOADED FACE

CRITICAL SECTION

(b) SUPPORT WITHOUT BEARING PRESSURE ON FACE OPPOSITE LOADS

Fig. 6-8. Critical sections for shear.

$$\tau_{\max}(\text{psi}) = 1382 - 486 \log \left(A_s / 4 \right) \quad (6\text{-}8)$$

Another researcher, Longworth, has discovered that the relationship between maximum shear stress at failure and sheared area is also affected by beam width (7). He found that failure stress was more properly a function of sheared volume.

Basic Method for Shear Calculations

The NDS suggests two methods of checking wood beams for shear. The first way is by methods of statics. There are two cases for solving for shear, V, by this method:

A. If the beam has loads applied to one surface and is supported by bearing on the opposite surface: Compute shear, V, at a critical section distance d (depth of beam) from the face of the support, locating the movable loads (if any) to cause largest possible V at the critical section. This case is illustrated in Fig. 6-8a.

Having computed the maximum shear force, V, at the critical section, calculate the maximum shear stress, f_v, and compare it with the allowable shear stress, F_v as modified (including footnote 8 if it applies). If f_v exceeds F_v (as

modified) but does *not* exceed $2F_v$ (modified by all except footnote 8) then the alternate method *may* be used, if desired, for sawn lumber.

B. If the beam is not supported by bearing pressure on the face opposite the loads, but rather is supported by bolts or other fasteners: Use a critical section *at the connection*, and locate the moving loads to cause the largest possible shear, V, at that point. Figure 6-8b shows this case. The design procedure for this case is described later in the section "Design of Joints in Shear."

Alternate Method for Shear

If case A above shows the beam to have insufficient shear strength, then for sawn lumber (but not glulams) the designer may choose to follow the two-beam provisions of NDS sections 4.2.6.2 and 4.2.6.3 instead. These sections are applicable to case A, above, only; they do not apply to case B. Nor do they apply to notched beams. (There is considerable disagreement among design professionals about the validity of the entire concept of two-beam action.)

For two-beam action on a simple span, shear is calculated differently for uniform loads than for concentrated loads. For uniform loads, end shear is taken to be the same as shear at distance d from the support, that is,

$$V = w\ell_c/2 - wd \qquad (6\text{-}9)$$

in which ℓ_c is the clear span between faces of supports. For ℓ_c/d ratios of about 3 to 35 this value is unnecessarily conservative, so Eq. 6-9 may be multiplied by C_v where

$$C_v = 0.95 + \frac{\sqrt{\ell_c/d}}{250} - 1.32 \, (d/\ell_c)$$
$$+ 11.5 \, (d/\ell_c)^3 \qquad (6\text{-}10)$$

For a concentrated load, P, located at distance x from the face of the support, the maximum shear force calculated by statics would be $P(\ell_c - x)/\ell_c$. This is multiplied by an algebraic function to reflect the experiments of Newlin et al. The maximum shear force, under two-beam action, due to a single concentrated load then becomes

$$V = \frac{P(\ell_c - x)\,(x/d)^2}{\ell_c[2 + (x/d)^2]} \qquad (6\text{-}11)$$

For two-beam action when there are moving concentrated loads rather than stationary ones, the loads should be positioned so as to cause the maximum V (from Eq. 6-11). This is usually done by trial and error using Eq. 6-11. But if there is a single moving load (or an outstandingly large load in a group of moving loads), placing that load at $3d$ or $\ell/4$ (whichever is closer) from the support will result in maximum or near maximum V, eliminating the need for trial and error.

To conclude that the beam is satisfactory in shear when V due to all loads is computed by this second method, the maximum shear stress computed for two-beam action is compared to the modified F_v. (Two-beam action is based on having severe checking, so the modification of footnote 8 of Table B-3 *cannot* be used here.)

End Notches

Sometimes at a support, the lower face of a beam is notched to bring the top of the beam to the proper elevation (Fig. 6-9). Notching a beam on the tension side is not recommended as the notch induces tensile stress normal to the grain; this, in combination with horizontal shear, tends to cause splitting. If an end notch on the tension face is used, the depth of this end notch must not exceed one-fourth the beam depth. In the shallow portion next to the notch, the shear stress is larger than $1.5V/A$, and is taken as:

$$f_v = \left(\frac{3V}{2bd'}\right)\left(\frac{d}{d'}\right) \qquad (6\text{-}12)$$

For notched beams the shear force, V, is usually considered to be the maximum shear at the face of the support, not the lesser shear at a distance d from the support. The NDS does not allow two-beam action to be considered for beams with end notches.

Design of Joints in Shear

Case B above describes shear in beams at end connections, where the beam is not supported by bearing on the lower face of the beam, but rather by bolts, lag screws, or other fasteners. The critical section is at the connector group, as shown in Fig. 6-8b. The shear stress at the critical section is calculated by Eq. 6-12, except that d' now equals the depth of the member less the distance from the unloaded edge of

Fig. 6-9. Notched beam.

the member to (1) the center of the nearest bolt or lag screw (for bolts or lag screws) or to (2) the nearest edge of the nearest connector (for split rings and shear plates).

For the special case when the joint is at least five times member depth from the end, the allowable shear stress may be increased 50%; actual stress is calculated by Eq. 6-12, except that the ratio d/d' in the equation is taken as 1.0.

Example 6-9

A hem-fir Select Structural grade 10×18 beam with clear span of 16 ft and a 4-in. bearing length at each end carries a dead load of 80 lb/ft plus its own weight (specific weight is about 30 pcf). The live loads are two 8-kip concentrated wheel loads separated by a distance of 5 ft; these loads may be in any location on the beam. The cumulative duration of the live loads is less than two months. Is the beam OK for shear?

$$\text{Beam weight} = (9.5)(17.5)(30)/144$$
$$= 35 \text{ lb/ft}$$
$$\text{Total uniform dead load} = 80 + 35 = 115 \text{ lb/ft}$$

Normally one would calculate the maximum shear,

(a)

(b)

Fig. 6-10. Beam for Example 6-9.

V, due to the two moving loads by placing one load at the face of the support as shown in Fig. 6-10a, in which case the maximum shear would be 14.42 k. However, the beam is supported by bearing on its bottom face, so the critical section is at a distance d from the face of the support. For maximum shear V at this location, the loads would be positioned as in Fig. 6-10b, with one load a distance $d = 17.5$ in. $= 1.46$ ft from the support face. Now the shear at the critical section calculates as 12.79 k.

$$f_v = 1.5V/A = 1.5(12,790)/(9.5)(17.5)$$
$$= 115 \text{ psi}$$
$$F_v = (1.15)(70) = 80 \text{ psi} < 115 \qquad \textit{No good}$$

Example 6-10

Reconsider the previous example, this time using the more liberal two-beam action provisions of the NDS. (It is permissible to use two-beam action because actual shear stress (115) is less than $2F_v$.)

Loads will be tried in several positions, using Eq. 6-11. First, place the left load 1 ft from the left support ($x = 1$ ft) and then the other load will be 6 ft from that support. Substituting into Eq. 6-11 with $P = 8$ k, $\ell_c = 16$ ft, and $d = 1.46$ ft: for $x_1 = 1$ ft, then $V_1 = 1.42$ k; using $x_2 = 6$ ft, $V_2 = 4.47$ k. That is,

For $x_1 = 1$ ft and $x_2 = 6$ ft, $V = 1.42 + 4.47$
$= 5.89$ k

Placing loads at other trial positions,

For $x_1 = 2$ ft and $x_2 = 7$ ft, $V = 3.39 + 4.14$
$= 7.53$ k
For $x_1 = 3$ ft and $x_2 = 8$ ft, $V = 4.41 + 3.75$
$= 8.16$ k
For $x_1 = 4$ ft and $x_2 = 9$ ft, $V = 4.74 + 3.32$
$= 8.06$ k
For $x_1 = 3.5$ ft and $x_2 = 8.5$ ft, $V = 4.64 + 3.54$
$= 8.18$ k   *Controls*

The additional V due to uniform load is

$$0.115 (16/2 - 1.46) = 0.75 \text{ k.}$$

For our $\ell_c/d = 11$, C_v (from Eq. 6-10) $= 0.85$, giving a total V of $8.18 + 0.85(0.75) = 8.82$ k.

$f_v = (1.5)(8820)/(9.5 \times 17.5) = 80$ psi

$F_v = (1.15)(70) = 80$ psi

OK under two-beam action

Example 6-11

Select a No. 1 hem-fir beam to carry a uniform dead load of 210 lb/ft and uniform snow load of 250 lb/ft on a clear span of 18 ft. Beam ends are supported on a concrete wall. The solution is shown by the designer's computation sheet in Fig. 6-11.

Example 6-12

Select a D. fir–larch beam to support a normal-duration live load of 230 lb/ft and a dead load of 40 lb/ft that includes an estimate of beam self-weight. The simple span for computing bending is 19 ft. (If the bearing length is not known, it will be conservative to use for shear the same span length as used in bending.) Notch the beam ends, if necessary, so that the distance from top of support to top of beam is 10 in. Choose the section based on flexure and check it for shear.

Assuming that the section will be in the Beam-and-Stringer size/use class, choose No. 1 grade.

LOADS: SNOW 250 LB/FT

DEAD 210

25 LB/FT - ASSUMED FOR BM. WT.

235 LB/FT

18'-0" CLEAR

L

$235/0.9 = 261 ; \dfrac{250+235}{1.15} = 422 > 261$

D + S CONTROLS

ALLOWABLE STRESSES: ASSUME B & S

ASSUME $C_F = 0.97$; $F_b = 1050 \times 1.15 \times 0.97 = 1171$ psi

$F_v = 70 \times 1.15 = 80$ psi

SPAN: ASSUME 3" REQ'D BEARING LENGTH AT EACH END.

FOR M AND Δ, L = 18.0 + 3/12 = 18.25'

$M_{D+S} = \frac{1}{8}(485)(18.25^2) = 20{,}190$ FT-LB

REQ'D $S = 12(20{,}190)/1171 = 207$ IN.³

TRY 6×16 (S = 220)($C_F = 0.972$)

CHECK SHEAR

$V = 485\left[\dfrac{18}{2} - \dfrac{15.5}{12}\right] = 3740$ LB

$f_v = 1.5(3740)/(5.5 \times 15.5) = 66$ psi $< F_v$ OK

VERIFY WEIGHT:

$38(5.5 \times 15.5/144) = 22.5$ LB/FT < 25 ASSUMED OK

USE 6×16

Fig. 6-11. Computation sheet for Example 6-11.

Basic F_b = 1300 psi (WCLIB)
Basic F_v = 85 psi

Since $270/1.0 > 40/0.9$, $D + L$ controls.
Assuming C_F = 1.0,

$$F_b = (1.0)(1300) = 1300 \text{ psi}$$
$$M = w\ell^2/8 = 270(19)^2/8 = 12,180 \text{ ft-lb}$$
$$S \text{ req} = M/F_b = 12,180(12)/1300 = 112 \text{ in.}^3$$

After scanning Table B-1, try a 6 × 12 (S = 121 in.^{3}; C_F = 1.0). The greatest end notch depth allowed in a 6 × 12 is $(1/4)(11.5) = 2.88$ in. Therefore, the 1.5-in. notch we need to make will be acceptable provided the shear stresses remain sufficiently low.

The maximum shear (at the notch, not at d from the support face) is

$$V = w\ell/2 = (270)(19/2) = 2570 \text{ lb}$$
$$f_v = \left(\frac{3(2570)}{2(5.5)(10)}\right)\left(\frac{11.5}{10}\right)$$
$$= 80.6 \text{ psi} < F_v = 85 \text{ psi} \quad OK$$

Use a 6 × 12, No. 1 grade.

Example 6-13

A 12-ft clear simple span beam (6 × 12 hem-fir) is loaded with a uniform normal-duration load of 785 lb/ft including self-weight. The beam has been well seasoned and no further drying is expected in service. For this reason, no checks (splits) are anticipated (none are present now). The basic F_v from Table B-3 is 70 psi. Check the beam for shear.

At a distance $d = 11.5$ in. from the support face,

$$V = (785)(12/2 - 11.5/12) = 3960 \text{ lb}$$
$$f_v = (1.5)(3960)/(5.5 \times 11.5) = 94 \text{ psi}$$

From footnote 8 of Table B-3, shear stress modification for no anticipated splits is 2.00.

$$F_v = (70)(2.00) = 140 \text{ psi} > 94 \text{ psi} \quad OK$$

Note that the NDS permits doubling the allowable shear stress in this case. However, using this increased allowable seems unwise in the authors' opinion. Conditions that may cause a beam to develop checks or splits in service just cannot be predicted with any certainty.

Example 6-14

The Douglas fir beam of Fig. 6-12 is to be supported at each end by two $\frac{1}{2}$-in. diameter bolts through two

Fig. 6-12. Beam for Example 6-14.

steel angles. What is the maximum allowable reaction based on the bolt value and based on shear in the beam?

Based on bolt value (double shear):

Allowable load per bolt = 430 lb
Allowable reaction = 2(430) = 860 lb

Based on shear in the beam:

$$F_v = 95 \text{ psi}$$
$$f_v = 1.5(R/bd')(d/d')$$
Allowable $R = (2/3)(95)(1.5 \times 8.25) \times$
$$(8.25/11.25) = 575 \text{ lb} \quad Controls$$

Clip-angle connections of this type are not advisable for heavily loaded members or with large numbers of bolts, because vertical shrinkage may cause splitting.

6-5. DEFLECTION

Besides having strength, a beam should also have adequate stiffness so that it will not deflect or sag too much. Excessive deflections can lead to broken windows, cracked plaster, ponding of water, or just bad appearance. Architectural or structural damage is not the only reason for limiting deflection. Also, the human factor needs to be considered. People feel uncomfortable walking across a floor that is saucer-shaped, no matter how safe and strong the floor is.

Because wood will creep under long-term loading (see Chapter 2), one cannot simply calculate live- and dead-load deflection by mechanics formulas and expect the result to be the maximum over the life of the beam. To account for creep over the long term, the immediate deflection due to long-term loads is magnified.

Assuming that all live loads are short-term loads, the NDS recommends separating the effects of the short-term (live) loads and long-term (dead) loads, calculating the total deflection over the long term as follows:

$$\Delta_{\text{Total}} = \Delta_{\text{Live}} + 1.5\,\Delta_{\text{Dead}}$$
$$= \Delta_L + 1.5\,\Delta_D \qquad (6\text{-}13)$$

This equation assumes that the additional deflection due to creep will equal one-half the initial, immediate dead-load deflection. This equation is applicable to either glulams or seasoned lumber. For unseasoned lumber, the factor 1.5 in the equation is replaced with 2.0. This

same method of calculating long-term deflection was in use as early as 1930 (8).

The designer must keep computed deflections within limits specified by authorities such as the Uniform Building Code (9) and the AITC (reference 3). The UBC deflection limits are reproduced in Table 6-2. According to this table, the UBC assumes additional deflection due to dead-load creep to be one-half of the immediate dead-load deflection for seasoned lumber—the same as the NDS. Notice also that the UBC does not consider dead-load deflection that takes place immediately to be of any consequence. This is because the immediate deflection due to dead load occurs before the fragile materials (glass, plaster, or gypsum board) are applied. However, designers in many offices include the immediate dead-load deflection, producing a more conservative design.

Table 6-3 shows deflection limits recommended by the AITC. Notice that the limits are less restrictive for industrial buildings than for commercial buildings. (In Table 6-3, applied load means all types of live load.)

The designer generally calculates beam deflection using simple formulas, a few of which are given in column a of Table 6-4. These for-

Table 6-2.   UBC Deflection Requirements.

Maximum Allowable Deflection for Structural Members[a]

Type of Member	Member Loaded with Live Load Only (L.L.)	Member Loaded with Live Load Plus Dead Load (L.L. + K.D.L.)
Roof member supporting plaster or floor member	$L/360$	$L/240$

[a]Sufficient slope or camber shall be provided for flat roofs in accordance with UBC Section 2305 (f).

L.L. = Live load
D.L. = Dead load
K   = Factor as determined below
L   = Length of member in same units as deflection

Value of K

Wood	
Unseasoned	Seasoned[b]
1.0	0.5

[b]Seasoned lumber is lumber having a moisture content of less than 16% at time of installation and used under dry conditions of use such as in covered structures.
Source: Reproduced from the 1985 edition of the Uniform Building Code, copyright © 1985, with permission of the publishers, the International Conference of Building Officials.

Table 6-3.   Recommended Limits for Deflection of Beams.

Type of Beam	Type of Load	
	Applied Loads Including Live, Snow, and Wind	Total of Dead Load Plus Applied Loads
	Fraction of Span Length	
Roof beams		
Industrial buildings	1/180	1/120
Commercial or institutional buildings		
With plaster below	1/360	1/240
No plaster below	1/240	1/180
Floor beams[a]	1/360	1/240
Bridge stringers		
Highway bridges	1/300 to 1/200	
Railway bridges	1/400 to 1/300	

[a]These limits for floor beams are intended to minimize plaster cracking and to ensure comfort to persons walking. They may not be adequate to eliminate levels of vibration that disturb building occupants. More rigid limits should be observed in such cases.

Table 6-4.   Maximum Deflections.

	(a) Bending Deflection	(b) Shear Deflection
Uniform load, simple span	$\dfrac{5w\ell^4}{384EI}$	$\dfrac{Kw\ell^2}{8AG}$
Center conc. load, simple span	$\dfrac{P\ell^3}{48EI}$	$\dfrac{KP\ell}{4AG}$
End conc. load, cantilever	$\dfrac{P\ell^3}{3EI}$	$\dfrac{KP\ell}{AG}$
Uniform load, cantilever	$\dfrac{w\ell^4}{8EI}$	$\dfrac{Kw\ell^2}{2AG}$

Note: $K = 1.20$ for a rectangular section and 1.11 for a circular section.

mulas are for flexural deflection only, that is, deflection due to the stretching and compressing of fibers parallel to the beam's axis. However, shear stresses cause additional deflection called *shear deflection*. Shear deflection occurs as plane cross sections become warped due to nonuniform distribution of the shear stresses.

Table 6-5.   Ratios of Elastic Properties and Shear Deflection Expressed as a Percentage of Total Deflection at Selected Span-Depth Ratios, for Rectangular Beams Centrally Loaded.

Species	$\dfrac{E^a}{E_{14:1}{}^b}$	$\dfrac{E}{G^c}$	Shear Deflection as a Percentage of Total at Midspan Span-to-Depth Ratio	
			(8:1)	(24:1)
Basswood	1.37	49.03	53.5	11.4
Andiroba	1.26	41.99	49.6	9.9
Hickory	1.24	39.79	48.3	9.4
Marishiballi	1.21	37.13	46.3	8.8
Balsa	1.20	36.08	45.8	8.6
Douglas-fir	1.18	29.29	40.7	7.1
Cedro	1.18	29.02	40.5	7.0
Angelique	1.15	24.87	36.8	6.1
Cumaru	1.15	24.35	36.3	6.0
Mahogany	1.10	22.83	32.1	5.6
Ceiba	1.08	19.18	31.0	4.8
Teak	1.02	10.12	19.2	2.6

[a] E = pure tangent modulus of elasticity in bending.
[b] $E_{14:1}$ = modulus of elasticity in bending at span-to-depth ratio 14:1.
[c] G = modulus of rigidity.
Source: Courtesy of Forest Products Research Society, *Forest Products Journal* (10).

Formulas for shear deflection are given by column b of Table 6-4.

It is common practice to assume that shear deflection is a minor part of the total deflection and to ignore it when designing. However, this may not be a good practice for all wood beams. Shear deflection contributes significantly to total deflection for beams with small span-to-depth ratios, box beams, and I-beams with thin webs. Also, shear deflection increases as the ratio of modulus of elasticity to shear modulus (E/G) increases. Table 6-5 shows how shear deflection is affected by both span-to-depth ratio and E/G ratio (10). For the first entry in the table, basswood, shear deflection for a deep beam is more than half the total deflection! It is improbable that a shear deflection this large would occur in a beam of species and span commonly used in construction. These species generally have an E/G ratio between 6 and 20 (11). As a rule of thumb, the designer should probably include shear deflection for a rectangular beam if the span-to-depth ratio is 15 or less.

Ponding

The accumulation of rainwater or snowmelt on roofs is a serious problem that should not be overlooked when designing roof decking, beams, and girders. Progressive deflection and rapid collapse as water accumulates is called *ponding*. The chance of ponding can be minimized by constructing building roofs so that they slope at least $\frac{1}{4}$ in. vertical per foot horizontal after long-term deflection. This does not completely eliminate the chance of ponding, however. Ponding can still occur due to inadequate capacity of drainage pipes in downpours or due to water confinement by parapet walls or ice dams. Beams and purlins designed for light loads generally have low stiffness and are most susceptible to ponding.

All members of a flat or nearly flat roof system must be designed with a great enough stiffness to prevent ponding. One simple rule of thumb is that any roof member should be stiff enough that a 5-psf load causes no more than $\frac{1}{2}$ in. deflection (12). Kuenzi and Bohannan (13) developed a method, verified experimentally, for estimating the effects of ponding. The method does not account for dead-load creep and resulting permanent set (which one may wish to consider, especially for unseasoned members). In this method, the before-ponding deflections and bending and shear stresses are multiplied by a magnification factor to estimate the increased deflections and stresses under ponding loads. This magnification factor, C_P (see reference 3) is given by

$$C_P = \frac{1}{1 - W'\ell^3/\pi^4 EI} \qquad (6\text{-}14)$$

where C_P = factor for magnifying stresses and deflections under all but ponding loads to determine stresses and deflections under all loads including ponded water

W' = ponding load due to a one-inch depth of water equaling 5 psf times the tributary area, lb

ℓ = span of member, in.

E = modulus of elasticity of member, psi

I = moment of inertia of member, in.^{4}

Example 6-15

Based on bending stress, a maximum permissible span for 2 × 6 Douglas fir T & G decking was determined in Example 6-5. What is the maximum permissible span for this same decking based on a deflection limit of $\ell/180$ due to dead load plus live load? Which controls?

Recall that total load, w, was 55 lb/ft per foot width. For controlled random lay-up,

$$\Delta = w\ell^4/[145E(2I/3)]$$
$$\approx w\ell^4/100EI$$
$$= (55/12)\ell^2/[100(1,700,000)(3.375)]$$
$$= 7.99 \times 10^{-9}\ell^4$$

Set equal to deflection limit of $\ell/180$

$$\ell/180 = 7.99 \times 10^{-9}\ell^4$$
$$\ell = 88.6 \text{ in. (based on deflection limit)}$$

The maximum span based on bending was 109 in., so the maximum permissible span of decking is 88.6

in. Actual spacing will be some convenient round number less than 88.6 in.

Example 6-16

A seasoned 2 × 12 Douglas fir floor joist (Select Structural grade) has a uniform dead load of 100 lb/ft including joist weight, and a live load of 60 lb/ft on a simple span of 16 ft. Find short- and long-term flexural deflection, and check against UBC deflection limitations. (Span/depth ratio = $(16)(12)/11.25 = 17$, so shear deflection can be assumed to be insignificant.)

$$\Delta_L = \frac{5(60)(16^4)(12^3)}{384(1,800,000)(178)} = 0.28 \text{ in.}$$
$$\Delta_D = \frac{5(100)(16^4)(12^3)}{384(1,800,000)(178)} = 0.46 \text{ in.}$$
$$(\text{or } \Delta_D = \Delta_L(100/60))$$

Total initial $\Delta = 0.28 + 0.46 = 0.74$ in.
Total long-term $\Delta = 0.28 + (1.5)(0.46)$
$= 0.97$ in.

For live load only,

UBC limit = $\ell/360 = (16)(12)/360 = 0.53$ in.
0.28 in. < 0.53 in.   *OK*

For live plus one-half dead ($\Delta_L + 0.5\,\Delta_D$),

UBC limit = $\ell/240 = (16)(12)/240 = 0.80$ in.
$0.28 + (0.5)(0.46) = 0.51$ in.
0.51 in. < 0.80 in.   *OK*

Example 6-17

A 6 × 16 unseasoned beam is used on a 14-ft span. Loads are a 2000-lb central concentrated live load and $w_D = 440$ lb/ft including beam weight. The average value of E from Table B-3 for the species = 1,400,000 psi. Assuming $G = E/13$, find the total (shear and flexural) long-term deflection.

Initial flexural deflection:

$$\Delta_L = \frac{2000(14^3)(12^3)}{48(1,400,000)(1.03)(1707)}$$
$$= 0.08 \text{ in.}$$
$$\Delta_D = \frac{5(440)(14^4)(12^3)}{384(1,400,000)(1.03)(1707)}$$
$$= 0.15 \text{ in.}$$

Notice that the average value of modulus of elasticity has been multiplied by 1.03 to obtain a pure bending modulus. When bending and shear deflections are calculated separately, it would be more nearly correct to use a pure bending modulus even though the difference is minor. Published values of E in Table B-3 are 97% of the true value so that, for average span/depth ratios, the designer will not have to calculate shear deflection separately. See NDS (reference 1) Appendix G.7 for further details.

Initial shear deflection (Table 6-4):

$$\Delta_L = \frac{1.2(2000)(14 \times 12)}{4(5.5 \times 15.5)(1,400,000 \times 1.03/13)}$$

$$= 0.01 \text{ in.}$$

$$\Delta_D = \frac{1.2(440/12)(14 \times 12)^2}{8(5.5 \times 15.5)(1,400,000 \times 1.03/13)}$$

$$= 0.02 \text{ in.}$$

Total long-term deflection which includes effect of creep (Eq. 6-13) is

$$(0.08 + 0.01) + (2)(0.15 + 0.02) = 0.43 \text{ in.}$$

Note that shear deflection causes 0.05 in., or almost 12% of the total long-term deflection for a beam of these proportions. If shear deflection were ignored, the deflection computed using $E = 1,400,000$ psi would be 0.40 in., only 7% less than the 0.43 in. value.

Example 6-18

No. 1 hem-fir 6 × 18 beams (basic $F_b = 1050$ psi, basic $F_v = 70$ psi, $E = 1,300,000$) are used at 8-ft centers to support an essentially flat roof system. Check the beams for the effect of ponding. The beams have been designed for dead load plus snow load. Span = 22.5 ft; dead load = 19 psf (which includes an allowance for weight of the beam); snow load = 30 psf; allowable total-load short-term deflection = $\ell/180$.

$D + S$ controls, since $19/0.9 < 49/1.15$

$$C_F = (12/17.5)^{1/9} = 0.96$$
$$F_b = (1.15)(0.96)(1050) = 1160 \text{ psi}$$
$$F_v = (1.15)(70) = 80 \text{ psi}$$

Under dead and snow loads (but not ponded water),

$$w = (19 + 30)(8) = 392 \text{ lb/ft}$$
$$M = (392)(22.5)^2/8 = 24,800 \text{ ft-lb}$$
$$f_b = M/S = (24,800)(12)/280.7 = 1060 \text{ psi}$$

For shear at a distance d from support,

$$V = w\ell/2 - wd = (392)(11.25)$$
$$- (392)(17.5/12)$$
$$= 3840 \text{ lb}$$
$$f_v = 1.5V/A = (1.5)(3840)/96.25$$
$$= 59.8 \text{ psi}$$
$$\Delta \text{ (total)} = \frac{5(392)(22.5)^4(12)^3}{384(1,300,000)(2456)} = 0.71 \text{ in.}$$

Calculating ponding magnification factor (Eq. 6-14),

$$W' = (5 \text{ psf})(22.5 \text{ ft})(8 \text{ ft}) = 900 \text{ lb}$$
$$C_P = \frac{1}{1 - \dfrac{(900)(22.5)^3(12)^3}{\pi^4(1,300,000)(2456)}}$$
$$= 1.06$$

Magnify stresses and deflections to find effect of dead, snow, and ponding (due to melted snow) loads combined.

$$f_b = (1060)(1.06) = 1120 < 1160 \quad OK$$
$$f_v = (59.8)(1.06) = 63.4 < 80 \quad OK$$
$$\Delta \text{ (total)} = (0.71)(1.06) = 0.75 \text{ in.}$$
$$\ell/180 = (22.5)(12)/180 = 1.5 \text{ in.} \quad OK$$

6-6. DESIGN FOR BEARING

The ends of a beam may bear directly on a supporting member, which could be either another wood member, a masonry wall, or a steel hanger, girder, or bearing plate. A beam may also be stressed in bearing when a member above transfers its load to the beam. A bearing failure is not a collapse; it is merely a crushing of fibers and generally not a serious failure. To prevent this crushing, the reaction, R, divided by the bearing area must not exceed the modified $F_{c\perp}$. In equation form,

$$R/A_b \leq F_{c\perp} \tag{6-15}$$

The bearing area equals the width of beam, b, times the bearing length, as seen in Fig. 6-13. Therefore, solving for bearing length in the above equation yields

$$\text{req } \ell_b = R/F_{c\perp}b \tag{6-16}$$

Fig. 6-13. Beam bearing.

The allowable stress in compression perpendicular to the grain is *not* to be modified for load duration (a recent revision to the design code), but it may need to be modified for moisture conditions or treated wood.

Having unloaded area adjacent to an area loaded perpendicular to the grain causes bearing strength to be higher than it would be if the entire block of wood were loaded. The code provides for this increase by raising the allowable stress for bearing areas that are both less than 6 in. long and not nearer than 3 in. to the end of a member. The allowable bearing stress in these cases is

$$F'_{c\perp} = F_{c\perp} \, (\ell_b + 0.375)/\ell_b \quad (6\text{-}17)$$

where ℓ_b is the length of bearing in inches measured along the grain of the wood. This is equivalent to saying that the effective bearing length is $\frac{3}{8}$ in. more than the actual bearing length. No allowance is made for the fact that pressure over the supporting area will probably not be uniform.

The design code does not specify a minimum bearing length, but common sense would indicate a minimum of 3 or 4 in. The bearing length should be large enough so that even when construction tolerances (location of support; beam length) are considered, the beam will not slip off the support.

Allowable stresses in compression perpendicular to the grain in Table B-3 are based on a deformation of 0.04 in. If deformation is critical, one may design sawn-lumber beams according to NDS Section 4.2.7, which limits the adjusted allowable stress, $F'_{c\perp}$, (based on a deformation of 0.02 in.) to

$$F'_{c\perp} = 5.60 + 0.73F_{c\perp} \quad (6\text{-}18)$$

In this equation, $F_{c\perp}$ is the value from Table B-3, as modified, but not for load duration.

Figure 6-14 shows two ways of framing timber beams into masonry walls. The method in Fig. 6-14a is known as a fire-cut, because floor joists framed in this manner can collapse in a fire without pulling the wall down along with them. The bearing plate in Fig. 6-14b may be used for larger beams to distribute the reaction over a larger area of masonry. The small space shown above the beam prevents wall load on the beam, prevents masonry from prying loose under beam deflection, and provides aeration to help prevent decay.

Bearing stresses may need to be considered for members other than beams. For example, the bearing capacity (perpendicular to grain) of a bottom plate (usually a 2 × 4) having 2 × 4 studs nailed to it can be checked using the equations of this section. The increase in allowable stress obtained from Eq. 6-17 would apply. The bearing capacity of the stud itself, parallel to grain, would also have to be considered (see Chapter 7).

Bearing at an Angle to Grain

Sometimes bearing stresses are applied at an angle (neither parallel nor perpendicular) to the grain. In this case, one needs to use the Hankinson formula, which accounts for the strength of wood being greater for end-grain bearing than for compression perpendicular to the grain. If θ is the angle between the load direction and the direction of the grain, then the allowable stress normal to the inclined surface is

Fig. 6-14. Two types of beam bearing.

$$F_n = \frac{(F_g)(F_{c\perp})}{F_g \sin^2 \theta + F_{c\perp} \cos^2 \theta} \quad (6\text{-}19)$$

In this expression, F_g is the modified allowable compressive end-grain stress (Table B-7), and $F_{c\perp}$ is the modified (but not for load duration) allowable compressive stress perpendicular to grain.

Example 6-19

Dense No. 1 D. fir–larch 4 × 16 beams, with a 12-ft simple clear span, have been designed to support a storage platform for sacks of cement. These heavy sacks and the beam's self-weight cause a uniform load of 1100 lb/ft. The wood is initially dry and remains dry in service. The beam ends will rest on steel connection brackets that have been embedded in masonry pilasters. To prevent a bearing failure of the Douglas fir, what length of connection bracket (bearing length) will be required?

Table B-3 shows $F_{c\perp} = 730$ psi. Because of dry conditions, there is no moisture modification of this value. Load-duration modification is not used for $F_{c\perp}$. Nor is the increase in allowable stress given by Eq. 6-17 applicable here.

The required bearing length is estimated to be 4 in. Assuming that the entire 12-ft, 8-in. length of beam can be loaded, the maximum reaction is

$$R = 1100(12.67)/2 = 6970 \text{ lb}$$

From Eq. 6-16,

$$\text{req } \ell_b = (6970)/[(730)(3.5)] = 2.73 \text{ in.}$$

In this case, it is an unnecessary refinement to recycle. Use a bearing length of $2\frac{3}{4}$ in. or more.

Example 6-20

A roof rafter (4 × 12 D. fir–larch No. 2) bears on a steel connection bracket atop a concrete wall as shown in Fig. 6-15a. As seen in Fig. 6-15b, the acute angle between the direction of the grain and direction of the load is 50°. Snow loading has controlled the rafter design. Find the allowable compressive stress on the horizontal bearing surface of the rafter.

From Table B-7, $F_g = 2020$ psi, and from Table B-3, $F_{c\perp} = 625$ psi. Using the Hankinson formula,

$$F_n = \frac{(1.15 \times 2020)(625)}{(1.15)(2020) \sin^2 50° + (625) \cos^2 50°}$$
$$= 895 \text{ psi}$$

Note that the end-grain allowable is modified for load duration, but the perpendicular-to-grain allowable is not.

Example 6-21

A temporary wood bridge is made necessary by a crane breakdown and is to carry wheeled bins of partially completed parts. Design the timber beams. No. 3 southern pine is available at a local lumber yard. The design is shown in computation sheet form by Fig. 6-16. A 3 × 8 is selected for bending. This section is then checked for shear and for end bearing and is found satisfactory.

6-7. FLOOR SYSTEM DESIGN

In the preceding sections design of bending members has been broken into small parts in order to concentrate on one aspect of the design at a time. Design is actually a more integrated process, usually involving the design of sys-

Fig. 6-15. Bearing at an angle to the grain for Example 6-20.

tems rather than member selection. True design involves choosing from many different layouts of the structural members before actually deciding upon sizes of members.

In the design of a complete floor system the planks, decking, or plywood to which the load is directly applied are designed first. Next the the joists supporting the first-designed members are chosen. Then the designer selects the beams that support the joists. Each member selected is, in some way, dependent on other members of the system. For example, the spacing chosen for joists will depend on how far the decking can span. These principles are best developed in an example.

Example 6-22

A floor system is to be designed for the storage building in Fig. 6-17. Floor live load = 100 psf and dead load = 10 psf plus weight of wood members. Assume both dead and live loads are of permanent duration since this is a storage building. Masonry walls are 12 in. thick. Dry conditions of use.

Alternates for the floor deck are: (1) planks, (2) T&G decking, and (3) plywood. Some alternates for the beam/joist system are:

1. Several large beams running N-S and spanning 18 ft clear, with joists E-W
2. Beams running N-S but with supports at walls and a center column (reducing beam span to one-half); joists E-W
3. Beams down the E-W centerline, supported on a

row of posts, with joists N-S having one end on the beam and one end on the masonry wall

Using steel beams in combination with wood joists is another option the designer may wish to consider.

Assuming we wish no obstructing columns, let us design components for the first of the beam/joist systems above.

Floor design: No. 2 D. fir–larch planks, 2-in. nominal, have been chosen for the deck material. If the designer calculates for a 12-in. width of floor, the actual width of plank will be immaterial (except as it may affect the allowable stresses). Based on bending strength, determine the allowable span of the planks. Using 40 pcf as the approximate weight of D. fir,

$$w = 100 + 10 + 40(1.5/12) = 115 \text{ psf}$$

Using a factor of 1.22 from footnote 4 of Table B-3,

$$F_b = 1250(0.9)(1.22) = 1370 \text{ psi}$$

$$M \text{ allow. per ft} = 1370(12 \times 1.5^2/6)/12$$
$$= 514 \text{ ft-lb/ft}$$

For four or more spans (Table A-9),

$$\text{Maximum } M = 0.107 \ w\ell^2$$

Based on bending,

$$\text{Allow } \ell = \sqrt{514/(115 \times 0.107)} = 6.46 \text{ ft}$$

Based on a deflection limit of $\ell/240$, determine the allowable span of the planks. Per foot,

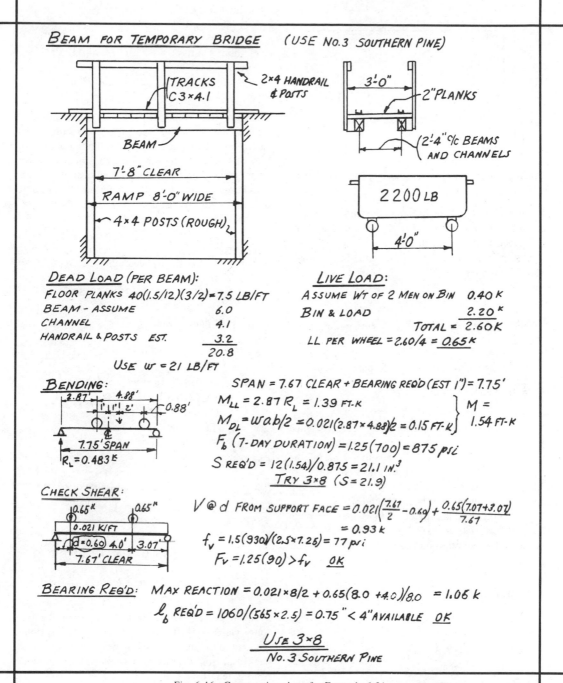

Fig. 6-16. Computation sheet for Example 6-21.

$$I = 12(1.5)^3/12 = 3.375 \text{ in.}^4/\text{ft}$$

For four or more spans (Table A-9),

$$\Delta = w\ell^4/154EI$$

But, since all load is long-term,

$$\Delta = 1.5 \; w\ell^4/154EI$$

$$\frac{12\ell}{240} = \frac{1.5(115)(\ell^4)(12^3)}{154(1.7)(10^6)(3.375)}$$

Allow $\ell = 5.29$ ft

Fig. 6-17. Plan view of storage building for Example 6-22.

Deflection rather than bending, controls. Planking cannot span more than 5.29 ft; that is, joists may not be spaced farther apart than 5.29 ft.

Check shear in planks. If joists are 2-in. nominal, plank clear span = $5.29 - 1.5/12 = 5.17$ ft maximum.

Critical $V = 115(5.17/2 - 1.5/12)$

$= 283$ lb/ft width

$f_v = 1.5(283)/(12 \times 1.5) = 24$ psi

$F_v = 0.9(95) = 86$ psi   *OK*

Joist design: To choose joists, the designer must know the spacing of the beams because this defines the length of the joists. The length (clear) of the building in the longitudinal direction is 78 ft. Try 11-ft beam spacing × 7 spaces = 77 ft. (Joists will cantilever slightly over beams at end walls.)

Assuming beams will be $9\frac{1}{2}$ in. wide, the clear span of the joists = 11.0 ft $- 9.5/12 = 10.21$ ft. Assume that the required bearing length of the joists is 2 in., then

Joist span = $10.21 + 2/12 = 10.38$ ft

Try No. 2 D. fir–larch.
Try joist spacing = 24 in c/c.

Estimated joist wt = 5 lb/ft
$w = 2(115) + 5 = 235$ lb/ft
$M = w\ell^2/8 = 3165$ ft/lb
F_b (repetitive) = $1450(0.9) = 1305$ psi
Req $S = 29.1$ in.³
Possible section: 2 × 12 ($S = 31.6$ in.³)

Try joist spacing = 36 in. c/c.

Estimated joist wt = 11 lb/ft
$w = 356$ lb/ft
$M = 4795$ ft-lb
F_b (nonrepetitive) = $1250(0.9) = 1125$ psi
Req $S = 51.1$ in.³
Possible section: 3 × 12 ($S = 52.7$ in.³)

There are many other joist spacings (up to a limit of 5.29 ft) that could be tried. The two sections considered above each use 1.0 fbm of lumber per square foot of floor, so choosing between them probably must be based on factors other than cost of the lumber. For example, the 2 × 12s are usually easier to obtain. *Tentatively use 2 × 12 joists at 24 in. c/c.*

Check joist bearing length:

$F_{c\perp} = 625$ psi
For $w = 235$ lb/ft, reaction = $235(11/2) = 1292$ lb
Req $\ell_b = 1292/(1.5 \times 625) = 1.38$ in.

Check joist deflection:

Span = $11 - 9.5/12$ (est) $+ 1.38/12$
$= 10.32$ ft
Δ(long term) = $1.5(5)\, w\ell^4/384EI$
$= \dfrac{1.5(5)(235)(10.32)^4(12)^3}{384(1.7 \times 10^6)(178)} = 0.30$ in.
$= \ell/413$ *OK*

Check joist shear:

Clear span = $11 - 9.5/12 = 10.21$ ft
Critical $V = 235(10.21/2 - 11.25/12)$
$= 979$ lb
$f_v = 1.5(979)/[1.5(11.25)] = 87.0$ psi
$F_v = 0.9(95) = 86$ psi   *OK* (1% overstress)

Use 2 × 12 joists at 24 in. c/c.

Beam design: Clear distance between the masonry walls is 18 ft. Use a steel plate to distribute the load to the masonry walls, where the plate is set back 1 in. from the interior face of the wall. Assume bearing length = 3 in.

Beam span = $18 + 2/12 + 3/12 = 18.42$ ft
w (from joists) = $11(235)/2.0 = 1292$ lb/ft
Beam wt (est) = $\underline{\quad 45 \quad}$
$w = 1337$ lb/ft
$M = 1337\,(18.42)^2/8 = 56{,}700$ ft-lb

Try No. 1 Douglas fir B & S (WWPA).

$$F_b = 0.9(1350) = 1215 \text{ psi}$$
$$\text{Req } S = M/F_b = 560 \text{ in.}^3$$

Try 12×18 ($S = 587$ in.^{3}).

$$C_F = (12/17.5)^{1/9} = 0.96$$
$$F_b = 0.9(.96)(1350) = 1166 \text{ psi}$$
Beam wt $= 38 \text{ pcf}(11.5)(17.5)/144 = 53 \text{ lb/ft}$
$> \text{est } 45$
Corrected $w = 1292 + 53 = 1345 \text{ lb/ft}$
Corrected $M = 57,040 \text{ ft-lb}$
Req $S = 587$ in.^{3} (12×18 OK)

Check beam bearing length:

$$\text{Req } \ell_b = 1345(9)/\big[625(11.5)\big]$$
$$= 1.68 \text{ in.} < 3 \text{ in. assumed} \quad OK$$

Check beam shear:

$$\text{Critical } V = 1345 \left(18.17/2 - 17.5/12\right)$$
$$= 10,260 \text{ lb}$$

$$f_v = 1.5(10,260)/(11.5 \times 17.5) = 76 \text{ psi}$$

$$F_v = 0.9(85) = 76 \text{ psi} \quad OK$$

Check beam long-term deflection

$$\Delta = \frac{1.5(5)(1345)(18.42)^4(12)^3}{384(1,600,000)(5136)}$$
$$= 0.64 \text{ in.}$$
$$= \ell/345 \quad OK$$

Use 12 × 18 beams.

REFERENCES

1. *National Design Specification for Wood Construction*, National Forest Products Association, Washington, DC, 1986.
2. *Standard for Tongue-and-Groove Heavy Timber Roof Decking*, AITC 112-81, American Institute of Timber Construction, Vancouver, WA, 1981.
3. American Institute of Timber Construction, *Timber Construction Manual*, 3rd ed., Wiley, New York, 1985.
4. Gerstle, K. H., *Basic Structural Design*, McGraw-Hill, New York, 1967.
5. Newlin, J. A., G. E. Heck, and H. W. March, "Wood-Beam Design Method Promises Economies," *Engineering News-Record*, **110**, May 1933, pp. 594–596.
6. Keenan, F. J., "Shear Strength of Wood Beams," *Forest Products Journal*, **24**(9), Sept. 1974, pp. 63–70.
7. Longworth, J., "Longitudinal Shear Strength of Timber Beams," *Forest Products Journal*, **27**(8), Aug. 1977, pp. 19–23.
8. Jacoby, Henry S., and Roland P. Davis, *Timber Design and Construction*, 2nd ed., Wiley, New York, 1930.
9. *Uniform Building Code*, International Conference of Building Officials, Whittier CA, 1985.
10. Biblis, E. J., "Shear Deflection of Wood Beams," *Forest Products Journal*, **15**(11), Nov. 1965, pp. 492–498.
11. Hoyle, R. J., *Wood Technology in the Design of Structures*, 4th ed., Mountain Press Publishing Co., Missoula, MT, 1978.
12. Haussler, R. W., "Roof Deflection Caused by Rainwater Pools," *Civil Engineering, ASCE*, **32**(10), Oct. 1962, pp. 58–59.
13. Kuenzi, E. W., and B. Bohannan, "Increases in Deflection and Stresses Caused by Ponding of Water on Roofs," *Forest Products Journal*, **14**(9), Sept. 1964, pp. 421–424.
14. American Institute of Steel Construction, *Manual of Steel Construction*, 8th ed., AISC, Chicago, 1980.

PROBLEMS

Assume normal duration of load and consider beam weight in problem solutions unless indicated otherwise.

6-1. A laterally supported beam of No. 1 hem-fir has a controlling bending moment of 8200 ft-lb (which includes an estimate of beam weight) from combined dead and wind loads. For bending strength, what is the shallowest satisfactory S4S section of 6-in. nominal thickness?

6-2. A No. 1 hem-fir beam, laterally supported, spans a clear distance of 20 ft and has superimposed dead load of 100 lb/ft and snow load of 150 lb/ft. Assuming the required bearing length at each end is 2 in., select the most economical section (the one of least cross-sectional area) considering bending alone. Do not neglect self-weight.

6-3. A laterally supported beam of 15-ft span carries at each one-third point concentrated loads that include 6 kips D, 4 kips normal L, and 2 kips S. Choose the most ecomomical No. 1 hem-fir section, considering bending only. (*Hint:* Select ignoring beam weight, then verify your choice considering it.)

6-4. What permanent uniform load can be put on a 12-in.-diameter log used as a beam if the span is 12 ft? Consider only bending stresses. Neglect log weight. Unmodified $F_b = 1500$ psi.

6-5. A uniform load of 400 lb/ft is applied to a log with unmodified $F_b = 1600$ psi. If the span is 12 ft, what is the required log diameter (based on flexure)?

6-6. A single 4 × 14 beam of No. 2 D. fir–larch spans 24 ft and has lateral support at its ends only. Considering flexure only, what is the allowable superimposed permanent uniform load per foot?

6-7. Consider a Select Structural D. fir–larch 6 × 22 beam that has a point load of normal duration applied at the center of the simple span. What is the smallest distance between lateral supports for which the beam is considered as "intermediate"? As "long"? Why is a "long" beam unlikely to be actually used?

6-8. A No. 1 hem-fir beam spanning 24 ft supports a permanent center point load of 1600 lb. There are lateral supports at the ends and center. Choose the most economical safe section whose smallest cross-sectional dimension is 6 in. nominal. Neglect beam weight.

6-9. Considering bending only, select the most economical No. 2 D. fir–larch section for the rafter of Fig. 4-13. The only loads are dead load and snow load of 30 psf of horizontal projection.

6-10. Roof rafters (repetitive members) are to be placed 24 in. c/c with a slope of 3/12. Snow load is 30 psf on a horizontal plane and dead load is 15 psf of sloping roof surface. Rafters have a horizontal span of 13 ft from ridge beam to exterior wall and are laterally supported. Based on flexure only, what is the smallest acceptable No. 2 hem-fir rafter?

6-11. A box beam (see Fig. 6-18) has No. 1 D. fir–larch 2 × 6 top and bottom flanges glued to $\frac{1}{2}$-in. plywood webs. Calculate the depth, d, required to resist an applied bending moment of 42 ft-kips assuming the contribution of the plywood to flexural strength is negligible. (Because one flange has nearly uniform tension and the other nearly uniform compression, the allowable stress will be the smaller of F_c or F_t.)

6-12. A 2 × 8 D. fir–larch beam of clear span 18 ft has an applied uniform load of 70 lb/ft. Assuming a specific weight of 38 pcf, compare the maximum shear stress at the critical section to the allowable shear stress.

Fig. 6-18. Beam for Problem 6-11.

6-13. Are the beams of Problem 4-28 satisfactory for bending and shear? Clear span is 15 ft 6 in.

6-14. California redwood beams (Select Structural) are used 6 ft apart on 18-ft spans to support the nearly flat roof over a laundry room, where temperature will be about 85°F and relative humidity will be 90%. The roof consists of 3-ply felt and gravel on 2-in. nominal wood decking with 3-in. rigid insulation. Snow load is 50 psf. Select the most economical section for flexure and shear. Consider beam weight.

6-15. Refer to Problem 4-7. Assume No. 2 D. fir–larch for 2- to 4-in. thicknesses and No. 1 (WWPA) if B&S size. Choose the best section for the interior floor beams, considering both flexure and shear.

6-16. For a 2 × 8 No. 2 D. fir–larch joist, simply supported at its ends, prepare a plot of the allowable uniform load (based on flexure and shear only) versus the span length of the joist.

6-17. Prove that the No. 2 D. fir–larch 4 × 16 beam of Fig. 6-19 is or is not satisfactory in shear. Use only the basic method: do not use two-beam action. All loads are permanent.

6-18. A 6 × 10 No. 1 hem-fir beam has a 10-ft clear span. A single concentrated load of 1800 lb may move to any position on the span. The beam is part of a temporary bridge that will only be used one month before being dismantled. Disregarding beam weight, is the beam satisfactory for shear?

Fig. 6-19. Beam for Problem 6-17.

Fig. 6-21. Beam for Problem 6-23.

6-19. A D. fir–larch 4 × 4 beam has a $3\frac{1}{2}$ × 3 cross section at the ends because it has been notched on the lower face. What is the maximum allowable shear, V?

6-20. Prove that the beam of Fig. 6-20 is or is not satisfactory in shear.

6-21. A nominal 6 × 12 beam of hem-fir is notched such that the cross section of the end measures 9-in. vertically. Center-to-center of required bearing lengths is 14 ft 2 in. The total (normal-duration) load on the beam is 230 lb/ft. Is the beam satisfactory for shear?

6-22. A 2 × 8 No. 1 D. fir–larch joist is simply supported and uniformly loaded. Calculate the length and load for which bending and shear capacity of the member are reached simultaneously.

6-23. The composite timber beam shown in Fig. 6-21 is joined by $\frac{1}{2}$-in. by 5-in. lag screws (allowable shear force = 440 lb/screw). For V = 1500 lb:
  (a) Calculate the shear stress at the joint.
  (b) Find the required lag screw spacing for this stress.
  (c) Calculate the shear stress at the neutral axis.
  (d) Is the section satisfactory?

6-24. A 2 × 8 No. 2 D. fir–larch beam carries 60 lb/ft snow load plus self-weight on an 18-ft span. It is supported at its ends by bolted connections to 4 × 4 D. fir posts. Design the

bolted connection, making sure that the beam will not be overstressed in shear. Give bolt size and locations.

6-25. Four-inch nominal decking having basic F_b = 2000 psi and E = 1,800,000 psi is arranged in a two-span continuous lay-up on a gently sloped roof. Deflection is limited to $\ell/240$ with roof beams spaced at 12 ft c/c. What is the allowable uniformly distributed two-month duration (snow) load?

6-26. Douglas fir Commercial grade WWPA decking, 3-in. nominal, is surfaced and used at 19% maximum MC. Lay-up is controlled random. Total load is 103 psf on the nearly flat roof, with snow load controlling. Deflection is limited to $\ell/180$. What is the maximum allowable span for the decking?

6-27. Simple-span 2-in. nominal decking (D. fir–larch WWPA Commercial grade) is to be used on a roof with slope of 5/12. Loads are S = 30 psf of horizontal projection and D = 10 psf of sloping surface. Deflection is limited to $\ell/240$. What is the maximum allowable decking span? Decking will be laid with its longitudinal axis parallel to slope, so the component of load parallel to the roof will cause compression. However, ignore the compression force for this problem.

6-28. Compute the total, long-term deflection (when snow load is present) for the section you chose for Problem 6-14. Does the section satisfy the AITC limits?

6-29. A 6 × 14 beam of No. 1 WWPA D. fir–larch supports a total uniform load of w kips per foot. (a) Find (as a function of w) the short-term deflection due to bending and due to shear for a simple span of 12 ft. Assume E/G = 13. What percent of the total deflection is shear deflection? (b) Repeat for a span of 24 ft.

Fig. 6-20. Beam for Problem 6-20.

6-30. An unseasoned 6 × 10 No. 2 D. fir–larch beam spans 18 ft. Assume $E/G = 13$. The live load is 100 lb/ft. The superimposed dead load is 150 lb/ft. (a) Determine the total long-term deflection, disregarding shear deflection. (b) Repeat including shear deflection. Remember to raise E by 3%.

6-31. A 4 × 16 No. 3 California redwood floor beam spans 15 ft. Assume $E/G = 13$; $L = 250$ lb/ft and D = 250 lb/ft plus beam weight.

(a) Determine total deflection over the long term, disregarding shear deflection.

(b) Repeat including shear deflection.

(c) What percent of total deflection is shear deflection?

(d) Under the UBC, is the beam satisfactory for deflection?

6-32. No. 2 D. fir–larch 4 × 12 beams for a mildly sloped roof system are to be checked for the effect of ponding. Dead load (including self-weight) is 18 psf and snow load is 25 psf. Beams are used at 4 ft c/c and span 16 ft. Allowable total-load deflection = $\ell/180$. Are the beams satisfactory?

6-33. Check the beams you chose in Problem 6-14 to determine whether they are satisfactory for resisting ponding if the roof is perfectly flat.

6-34. A southern pine 6 × 10 beam supports a uniform snow load of 170 lb/ft plus total dead load of 100 lb/ft. The beam is No. 2 SR grade. The clear distance between supports is 18 ft. Determine the required bearing length at each end of the beam. Check the beam for shear and flexure.

6-35. A No. 1 redwood 6 × 12 beam supports a 350-lb/ft uniform load of cumulative duration 15 hours. The clear distance between supports is 20 ft. Determine the required bearing length at each end of the beam if the

Fig. 6-23. Plan for Problem 6-39.

structure is to be located in a very humid climate.

6-36. For the 4 × 12 No. 2 D. fir–larch beam of Fig. 6-22, what bearing length, ℓ_b, is required?

6-37. A building floor consists of $\frac{1}{2}$-in. plywood resting on (but not rigidly connected to) No. 2 southern pine 2 × 12 planks (surfaced dry, used at 19% maximum MC), used flat at 12 in. c/c. Planks as long as 24 ft can be found easily. The planks, in turn, are supported by beams. Live load is 50 psf and the plywood weighs 1.5 psf. The planks are treated with fire-retardant chemicals. Limit live-load deflection of planks and beams to $\ell/240$. (a) How far apart may the beams be placed (nearest safe whole foot)? (b) Design an interior beam of No. 1 (WWPA) D. fir–larch, assuming a clear span of 17 ft 6 in., with ends supported on masonry walls.

6-38. Design a complete floor system for the storage building shown in Fig. 6-17. Loads are: $D = 10$ psf plus weight of wood members, and $L = 100$ psf. Dry conditions. The system should have a row of columns (you determine how many) on the east-west centerline with beams in an east-west direction and joists north-south. Use T & G decking for the structural floor.

6-39. A 12 × 22 ft platform is to be constructed of No. 2 Douglas fir–larch. Floor is to be planking (2 × 10s laid side by side). Live load is 100 psf. Design the platform as shown in Fig. 6-23. (a) Calculate maximum allowable joist spacing based on capacity of planking. (b) Choose a layout of joists so that all spacings (not necessarily five) will be equal and design the joists, J-1. (c) Design the beams, B-1.

Fig. 6-22. Beam for Problem 6-36.

7

Selecting Sawn-Timber Compression and Tension Members

7-1. WOOD COLUMNS

Columns are generally thought of as the vertical supporting members of buildings. However, there are other structural members that act as columns—the piers of a bridge or the compression chord of a truss, for example. Generally columns are compression members, but they can also have combined compression and bending or can even have tensile axial force under loading that causes uplift. For purposes of design, we define a column as a structural member whose primary loads are axial compression.

The short, thick wood column will fail by crushing of its fibers. The longer, more slender column will fail by buckling due to lateral instability. Columns that fail by crushing are known as *short* columns. Such columns are not necessarily short in length; rather, the larger of their two ratios of unbraced length to cross-sectional dimension is small. Columns that fail by elastic buckling are called *long* columns and have a larger ratio of length to lateral dimension. Today we realize that columns cannot be so easily compartmentalized into only two classes. Therefore, current design specifications define a third class—columns that have neither very small nor very large ℓ/d ratios. These are called *intermediate*. The intermediate column tends to fail by a combination of crushing and buckling.

Wood columns are available in a variety of shapes and types. Most often they are either rectangular, square, or round in cross section. They may be made of a single solid piece of wood or formed from several pieces of wood. In the latter category are glulam columns (discussed in Chapter 8), spaced columns, and built-up columns. Spaced columns are assembled from two or more long members that are separated at the ends and mid-length by spacer blocks. Built-up columns are formed by nailing or bolting two or more members to form either a solid cross section or other shape, such as a hollow, box-shaped section.

It is interesting to compare the specific strength, that is, the ratio of strength to unit weight, of both wood and steel columns. For geometrically similar (but not equal) sections of equal weight, the radius of gyration varies inversely as the square root of specific gravity (1). The strength of longer columns is directly dependent on the square of radius of gyration, so longer wood columns (having lesser specific gravity than steel) will have the superior specific strength over steel.

Figure 7-1 shows the results of comparing light column sections of wood and steel that are of equal weight and length. For the two sections compared, the advantage lies with steel only for very short lengths where strength is not a function of slenderness ratio, ℓ/r.

History of Wood Column Design

The historical development of column theory for columns of any material is a subject that

Fig. 7-1. Comparison of design strength of steel and timber columns of equal weight and length (wood $F_c = 1000$ psi).

alone could fill an entire book. Starting in the 1700s various linear, parabolic, and other formulas were proposed for design. A few examples will show how design practice for wood columns developed.

Henry Dewell, author of a 1917 timber design book (2), felt that the American Railway Engineering Association (AREA) recommendations for working stresses were the best available at that time. This association used a single linear formula for both long and intermediate column design, as follows:

$$F_c' = F_c(1 - \ell/60d) \qquad (7\text{-}1)$$

where F_c' is the allowable stress on the column (considering that it may buckle), F_c is the allowable stress on a small block of wood, and d is the least cross-sectional dimension. A maximum of 60 was allowed for ℓ/d, the ratio of column length to least lateral dimension d. For "short" columns (then considered to have ℓ/d less than 15) the working stress for column design was 75% of F_c. The AREA equations and allowable stresses were for use in designing railway bridges and trestles.

Dewell also proposed using a version of the Gordon-Rankine formula, which he developed by curve-fitting test data accumulated from other sources. The proposed column formula,

$$F_c' = \frac{5000/3.5}{1 + \dfrac{\ell^2}{1750d^2}} \qquad (7\text{-}2)$$

contained a factor of safety of 3.5 and was to be used for all columns having ℓ/d ratios between 10 and 60.

Today's *National Design Specification for Wood Construction* (3) gives three equations for column design—one each for short, intermediate, and long columns. This status was reached by what may appear to be an erratic process, the main steps of which are shown below.

In 1930 a technical bulletin (4) was published that outlined an empirical intermediate-column formula developed at the Forest Products Laboratory. By the mid-1940s the Southern Pine Association (5) was recommending this intermediate formula—the same formula found in the NDS today. The Southern Pine Association also recommended using the Euler formula for long columns, but with a combined reduction factor (which includes a factor of safety) of 2.50 (as compared to 2.74 in today's NDS Euler formula for long columns).

The 1947 NDS recommended the same intermediate-column formula as the NDS today. But following 1947 that formula was omitted from several editions of the National Design Specification; that is, NDS of those years did not recognize the need for an intermediate-column formula at all. Rather, in those editions the allowable compressive stress for a column was the smaller of: (1) the short-column value or (2) the stress obtained from Euler's formula (with reduction factor). Doing away with the intermediate formula does simplify design procedures, but may also markedly decrease the factor of safety. Table 7-1 shows, for columns in the intermediate range, the allowable stresses obtained by the simpler long-column formula (with upper limit of F_c) and the more complicated intermediate-column formula. The table is for Dense No. 1 P&T Douglas fir–larch. It shows that allowable stresses given by the Euler-type formula can be as much as 15% greater than those given by the intermediate-column formula. If using the Euler formula for columns of intermediate length is in error, it is an error on the unsafe side. (The formulas for intermediate and long columns will appear later in this chapter as Eqs. 7-5 and 7-6.)

Finally, in the 1977 edition (6), NDS revived the use of the intermediate formula. (This vacillation may have been due to the compo-

Table 7-1. Allowable Compressive Stress by Different Formulas (psi).[a]

ℓ/d	F_c' (long formula)	F_c' (interm. formula)	Percent by Which Long Value Exceeds Interm. Value
11	1200	1186	1
12	1200	1180	2
14	1200	1162	3
16	1200	1136	6
18	1200	1097	9
20	1200	1043	15
22	1054	970	9
24	885	874	1

[a]For Dense No. 1, Douglas fir–larch with $F_c = 1200$ psi and $E = 1,700,000$ psi.

sition of the governing committee and to differences of opinion on simplification versus refinement.) In view of the series of changes that have occurred in design methods for columns, it probably is foolish to assume that the present formulas and classifications represent the final word. Further improvements (or just changes) will certainly occur.

7-2. COLUMN DESIGN

Sections 3.6 through 3.10 of the *National Design Specification for Wood Construction* (reference 3) deal with column design. NDS Appendices A, G, H, and N can be especially helpful also.

We know that the deflected shapes of buckled columns will differ depending on end support conditions. In most cases the effective length (distance between inflection points) will differ from the actual unbraced length. Figure 7-2 shows both theoretical and recommended design values for coefficient of effective length, K_e. These coefficients can be multiplied by actual length or length between lateral supports to obtain effective column length ($K_e \ell = \ell_e$). The recommended values recognize that perfectly fixed ends are impossible; therefore, the recognized coefficients (for other than pinned ends) are slightly larger than the theoretical values.

Actual support conditions seldom are equivalent to one of these idealized end conditions (fixed, pinned, or free). Given a column whose ends are laterally supported but not rigidly fixed, designers often assume that actual (unbraced) length equals effective length. This assumption (that K_e equals 1.0) is conservative for columns in which sidesway is prevented.

Research by Neubauer (7) shows that assuming a K_e equal to 1.0 is overly conservative for flat-ended 2 × 4 studs. He first tested columns with the squared ends pressed directly between the upper and lower platens of the testing machine. Then he tested columns that had pointed ends formed from steel channels and angles. For intermediate and long columns with flat ends, the maximum load carried was two to

Buckled patterns					
Theoretical multiplier, K_e	1.0	0.5	0.7	2.0	1.0
Recommended value of K_e, for use when actual conditions approximate those shown.	1.0	0.65	0.8	2.1	1.2

Fig. 7-2. Effective-length coefficients for compression members.

Fig. 7-3. Weak- and strong-axis buckling.

four times as great as for pointed ends. Neubauer concluded that the increased strength should be recognized by a reduction in effective length.

Slenderness Ratio

Columns of all materials are classified by their *slenderness ratio*. A slenderness ratio of a column is the ratio of effective unbraced length to radius of gyration (ℓ_e/r). This ratio may differ for the two principal axes of the cross section. If so, the larger of the two slenderness ratios is the critical one. Radius of gyration, r, and cross-sectional dimension, d, are proportional to each other in a rectangular column ($d = r\sqrt{12}$). Because most wood columns are rectangular, the NDS classifies them, for simplicity, by ℓ_e/d ratio rather than ℓ_e/r ratio. A large slenderness ratio indicates a greater instability and tendency to buckle under lower axial load.

If the unbraced length of a column is the same in both directions, the column will buckle about the w-axis (Fig. 7-3a). In other words, its middle will move in the s-direction. Because of the tendency to buckle about the w-axis, this axis will be called the weak axis, and the s-axis will be called the strong axis.

There may be a different effective column length for buckling about each axis. An example to illustrate this is a plywood-sheathed wood shear wall like the one in Fig. 7-3b. The plywood is nailed to the narrow faces of the studs along the entire length of the stud. There-

fore, the stud cannot buckle about its weak axis and the effective column length for weak-axis buckling is essentially zero. (Studs are the vertical members in a light-frame wall, spaced at 16–24 in. on center and of small cross section. See Fig. 1-1.) When the studs are so braced, their controlling (larger) slenderness ratio will be the one about their strong axis.

Design Formulas

As mentioned earlier, solid wood columns are divided into three classes. The classification is based on the larger of the two ℓ_e/d ratios. Fig. 7-4 shows, for the three classes, the variation

Fig. 7-4. Allowable column stress as a function of ℓ_e/d ratio.

of the allowable compressive column stress, F_c', with the ratio ℓ_e/d. An ℓ_e/d ratio of 11 separates the short and intermediate classes. The intermediate- and long-column classes are divided by an ℓ_e/d ratio of

$$K = 0.671\sqrt{E/F_c} \qquad (7\text{-}3)$$

For short columns ($\ell_e/d \le 11$), the full allowable compressive stress parallel to the grain is used; that is,

$$F_c' = F_c \qquad (7\text{-}4)$$

For intermediate columns (ℓ_e/d ratio between 11 and K), the formula developed at the Forest Products Laboratory is used:

$$F_c' = F_c\left[1 - \frac{1}{3}\left(\frac{\ell_e/d}{K}\right)^4\right] \qquad (7\text{-}5)$$

This formula reduces the allowable stress to something below the full allowable compression parallel to the grain. At ℓ_e/d of 11, it gives an allowable stress that is approximately equal to F_c.

For long columns ($\ell_e/d \ge K$), Euler's formula is used in the following form:

$$F_c' = 0.30E/(\ell_e/d)^2 \qquad (7\text{-}6)$$

This long-column formula was derived from Euler's formula as follows. Euler's formula for the critical (ultimate) stress in pin-ended columns of any elastic material is:

$$P_{cr}/A = \pi^2 E/(\ell/r)^2$$

If this is divided by 2.74 to provide a factor of safety and to correct for variability in E, the formula for wood becomes:

$$F_c' = 3.60E/(\ell/r)^2$$

Making the replacement $r = d/\sqrt{12}$, the formula for rectangular columns is as shown by Eq. 7-6 above.

According to the NDS, the effective slenderness ratio, K, that separates the intermediate- and long-column classes is associated with an allowable column stress of two-thirds of the short-column allowable. One may derive the expression for K (Eq. 7-3) by equating two-thirds of the basic allowable stress to the Euler stress as given by Eq. 7-6 and then solving for ℓ_e/d (i.e., K).

In Eqs. 7-3 through 7-6, above, term F_c is subject to all applicable modifications. However, E is *never* modified for load duration; therefore the long-column formula has no duration modification. Also, allowable slenderness is limited by NDS by the requirement that in no case should the ℓ_e/d ratio exceed 50.

The values for modulus of elasticity found in the Appendix tables are average values, not 5% lower exclusion values. The combined factor of safety included in Eq. 7-6 anticipates the use of the average value of E with a coefficient of variation of 25%.

Design Procedure

Column design is a trial-and-error procedure. Worrying about what to use for a guess or estimate will only waste time. Instead, time is saved by making a quick estimate and then analyzing that trial section to get a much better second estimate. Following are the necessary steps for designing a solid column with axial load only (no applied bending moment):

1. Estimate whether the column is short, intermediate, or long.
2. Estimate the allowable stress based on your guess from step 1.
3. Use the estimate from step 2 to solve for the required area and select a trial section. Do not choose a section with $\ell_e/d > 50$.
4. Decide whether or not the trial section is satisfactory by either
   a. comparing the actual stress f_c for that section to the modified allowable stress F_c', or
   b. comparing the actual area of the trial section to the required area using the modified allowable stress F_c' computed for the trial section.
   A safe section has $f_c \le F_c'$, or actual area is greater than or equal to required area.
5. Try other sections and recycle.

After the section has been chosen, three other checks may be necessary:

1. If the member has holes for split rings or bolts, the compressive stress at the net section must be computed to see that it does not exceed the allowable short-column stress.

2. The column may have an end bearing area that is less than the column cross-sectional area. In this case the designer must check the end-grain stress in bearing parallel to the grain. This check insures that the fibers at the end of the column do not crush or broom. The bearing stress cannot exceed values listed in Appendix Table B-7. If the end-grain bearing stress exceeds 75% of the table values, then a bearing plate of metal or other rigid material must be used. These design values for end-grain bearing are quite large, so it is unlikely that end-grain bearing will control unless the column end has been reduced in cross section for some reason, or the column bears on a surface of lesser area than the column.

3. If the column or stud bears on a wood bottom plate or sill, the bearing capacity of the plate must be checked. The plate or sill may not be strong enough in compression perpendicular to the grain to support the load. Bearing capacity of certain grades of concrete floors (8) should be checked also. The load on a column may have to be limited to the capacity of the material the column must bear on. A few suggested working stresses are listed in Table 7-2.

All of the above assumes that the column is straight. If there is significant initial crookedness, the axial load will cause bending moment and the member must be designed as a beam-column. (Refer to Sections 7-6 and 7-7.)

Table 7-2. Allowable Bearing Stresses, F_p, on Concrete.[a]

28-day Compressive Strength f_c' (psi)	F_p (psi)
2500	750
3000	900
4000	1200

[a]When the supporting surface (area A_2) is wider on all sides than the area (A_1) loaded by the wood column or column base, the allowable bearing stress on the loaded area of concrete may be increased by a factor $\sqrt{A_2/A_1}$, but by not more than 2.0.

Example 7-1

A column with pinned ends is to be designed from No. 1 Douglas fir–larch. The unsupported length of the column is 13 ft 0 in. The total load, which includes snow load, is 75 kips ($D = 40$ k and $S = 35$ k). Find the best square wood column.

Determine which loading condition controls.

$40/0.9 = 44.4$ for dead (permanent) load
$75/1.15 = 65.2$ for dead plus snow load
$D + S$ controls

From Appendix Table B-3 (assuming P&T size/use group),

$$F_c = 1000 \text{ psi} \quad \text{and} \quad E = 1,600,000 \text{ psi}$$

The load duration factor for snow is 1.15. If the column is short, $F_c' = 1.15 \times 1000 = 1150$ psi. Guess that the column might be intermediate, so that the allowable is somewhat less than 1150—say 1050 psi. Then

$$A \text{ req} = 75,000/1050 = 71.43 \text{ in.}^2$$
$$d \text{ req} = 8.45 \text{ in.}$$

Try 10×10 nominal column with $d = 9.5$ in. Check to see if 10×10 is short, intermediate, or long.

$$\ell_e/d = 12 \times 13/9.5 = 16.4$$

This is greater than 11, so the column is not short.

$$K = 0.671\sqrt{E/F_c}$$
$$= 0.671\sqrt{1,600,000/(1.15 \times 1000)} = 25.0$$

Since $11 < 16.4 < 25.0$, the 10×10 column is intermediate.

$$F_c' = F_c\left[1 - \frac{1}{3}\left(\frac{\ell_e/d}{K}\right)^4\right]$$

$$= 1.15 \times 1000\left[1 - \tfrac{1}{3}(16.4/25.0)^4\right]$$

$$= 1079 \text{ psi}$$

So

$$A \text{ req} = 75,000/1079 = 69.5 \text{ in.}^2$$
$$d \text{ req} = 8.34 \text{ in. vs 9.5 in. provided}$$

The 10×10 is the smallest satisfactory section. It is not necessary to recycle. The next smaller section is an 8×8 with $d = 7.5$ in. It will have a slenderness ratio greater than 16.4 and therefore a smaller F'_c. We know we need a d of 8.34 in. or more, so it is obvious that the 8×8 would be too small. No information was given in the problem statement concerning the material the column bears on, so it cannot be checked in bearing.

Example 7-2

A 10×14 No. 1 D. fir–larch column is 22 ft long. About the strong axis it is braced only at its ends; however, about the weak axis it is attached to a bracing member at the 13-ft point. Two $\frac{3}{4}$-in. bolts installed in $\frac{25}{32}$-in.-diameter holes connect the brace to the column, as shown by Fig. 7-5a. The total load is 111 kips with snow load controlling. Is the column satisfactory for the design loads?

By Table B-3 (B&S), $F_c = 925$ psi and $E = 1,600,000$ psi. Load duration factor is 1.15. Assuming pinned ends,

About the weak axis, $\ell_e/d = 12 \times 13/9.5$
$= 16.4$

About the strong axis, $\ell_e/d = 12 \times 22/13.5$
$= 19.6$

The larger ratio controls.

(a)

(b)

Fig. 7-5. Column for Example 7-2.

$K = 0.671 \sqrt{E/F_c}$
$= 0.671 \sqrt{1,600,000/(1.15 \times 925)} = 26.0$

Since $11 < 19.6 < 26.0$, the column is intermediate.

$F'_c = 1.15 \times 925 \left[1 - \frac{1}{3} \left(19.6/26.0 \right)^4 \right] = 949$ psi
$f_c = 111,000/(9.5 \times 13.5) = 865$ psi

The actual stress, f_c, is less than the allowable stress F'_c, so the gross section is more than sufficient. Check stress on the net section at the drilled holes (Fig. 7-5b).

$f_c = P/A \text{(net)}$
$= 111,000/(9.5 \times 13.5 - 2(25/32)(9.5))$
$= 979$ psi $< (1.15 \times 925) = 1064$ psi

In this example, the holes are small relative to the size of the cross section, so the short-column allowable is not exceeded at the section containing the hole. However, if the holes were larger, the strength at the net section might control the design.

7-3. ROUND AND TAPERED COLUMNS

The strength of a round column, or of a round column with bending, has historically been assumed equal to that of a square member having the same area of cross section. This simplification is still allowed today because of ease of calculation (NDS Section 3.7.4). Therefore, to design a round column, one may find the minimum size of square column that would satisfy all requirements and then provide a round shape of equal area.

Alternatively, the designer may replace d in Eqs. 7-5 and 7-6 with $r\sqrt{12}$, where r stands for radius of gyration. For short columns the results of the two methods will not differ. For long columns, the first method results in an allowable stress 5% greater than the second, providing a factor of safety less than one would ordinarily have. For intermediate columns the first method gives an allowable compressive stress 1–5% greater. Therefore, the equivalent-square method is only slightly in error, but on the unsafe side.

Tapered columns can be found in varieties such as shown in Fig. 7-6. For tapered columns

(a)   (b)   (c)

Fig. 7-6. Tapered column types.

like (a) or (b) in the figure, a section at one-third of the way from the small end is assumed to be critical. The column is designed as if the entire column had this cross section. However, one must also check to insure that the compressive stress at the smallest cross section of the tapered column does not exceed the allowable stress for a short column. For columns that are not continuously tapered (such as Fig. 7-6c) or for columns that increase in dimension in one plane while decreasing in another plane, design is based on a critical dimension in each direction. In these cases the dimension d in each plane of the column is taken as the sum of the minimum d in that plane plus one-third the difference between the minimum d and maximum d in that plane.

Round sections are often also tapered. Examples are telephone poles or timber piles. Round, tapered columns are often designed as equivalent square columns, but again the design is based on the cross section that is one-third of the length from the small end.

Example 7-3

A round column is to be designed for a rustic lodge lobby. The wood to be used has $E = 1,600,000$ psi and $F_c = 1000$ psi. The column will be 17 ft long. What diameter (to the nearest half inch) is necessary to support a normal-duration load of 65 kips? Assume pinned ends. Use the equivalent square column method.

Whenever a trial-and-error solution is required (as is the case here), it is convenient to tabulate the solution. (Advantages of the tabulated solution are that it is compact and is easy for the designer or other person to check.) Figure 7-7 shows a designer's

computation sheet for this example. A 10.5-in.-diameter round column (untapered) would be the smallest satisfactory section.

Example 7-4

Find the allowable axial load of a round, tapered column of length 20 ft. The column has a diameter of 12 in. at the base, tapering to 8 in. at the top. The column has $E = 1,600,000$ psi and $F_c = 925$ psi. The load will be of more than ten years cumulative duration. Use the more accurate method.

The critical section will have a diameter = $8 + 0.33(4) = 9.33$ in. Find the radius of gyration for this cross section.

$$r = \sqrt{I/A} = D/4$$
$$r = D/4 = 9.33/4 = 2.33 \text{ in.}$$

Substitute $r\sqrt{12}$ for d

$$r\sqrt{12} = 8.07 \text{ in.}$$

Find slenderness ratio and K:

$$\ell_e/d = 12 \times 20/8.07 = 29.7$$
$$K = 0.671\sqrt{1,600,000/(0.9 \times 925)} = 29.4$$

Since $29.7 > 29.4$, this is a long column. Find allowable column stress:

$$F_c' = 0.30E/(\ell_e/d)^2$$
$$= 0.30(1,600,000)/29.7^2 = 544 \text{ psi}$$

(Note that the load-duration factor is not used in the long-column formula.) Find allowable load.

$$P(\text{allowable}) = (A)(F_c')$$
$$= (\pi \times 9.33^2/4)(544) = 37,200 \text{ lb}$$

7-4. SPACED COLUMNS

Spaced columns are composed of two or more individual members separated at the ends and middle by blocking as shown in Fig. 7-8. The end spacer blocks that separate the two (or more) members of a spaced column provide increased fixity of the individual pieces at the ends. This increased fixity causes the distance between inflection points in the buckled column to be much less than it would be for the elements acting as independent columns. Therefore, the strength of a spaced column of

$P = 65\,k$ ASSUME NORMAL DURATION

$E = 1,600,000\,psi \; ; F_c = 1000\,psi \; ; L = 17'$

HINGED ENDS, $\therefore \ell_e = 17(12) = 204''$

$K = 0.671\sqrt{\dfrac{1,600,000}{1000}} = 26.84$

IF LONG, $F_c' = \dfrac{0.3(1,600,000)}{(\ell/d)^2}$

IF INTERMEDIATE, $F_c' = 1000\left[1 - \dfrac{1}{3}\left(\dfrac{\ell_e/d}{26.84}\right)^4\right]$

FOR CIRCULAR SECTION —

S (SIDE OF EQUIV. SQUARE) $= \sqrt{\pi D^2/4}$

TRY D =	EQUIV S	A	ℓ_e/d	CLASS	F_c'	A req.
7″	6.20″	38.44 in.²	32.90	LONG	443	146.7 in.² NG
9″	7.98″	63.62	25.56	INT	726	89.5 NG
10″	8.86	78.54	23.02	"	820	79.3 NG
10.5″	9.31	86.59	21.91	"	852	76.3 OK

USE $10\tfrac{1}{2}''$ DIAMETER

Fig. 7-7. Computation sheet for Example 7-3.

two members is not merely twice that of a single member but, rather, may be many times that of a single member.

Frequently, compression members of trusses are spaced columns. In a truss the end spacer blocks are often other web or chord members of the truss. (See Chapter 10.)

A longer spacer block at the ends provides for better fixity, thereby affecting the buckled shape. (With the longer blocks, distance between inflection points in the buckled shape will decrease.) Better fixity leads to a higher strength; therefore, the code provides for two end-fixity conditions. If the centroid of connectors in the end block is within $\ell/20$ from the column end, this is known as *condition a*. If the centroid of connectors is between $\ell/20$ and $\ell/10$ from the ends (better fixity), it is *condition b*. A factor C_x for use in design formulas

CENTROID OF CONNECTORS

SPACER BLOCK

Fig. 7-8. Spaced column.

has a value of 2.5 for condition a and 3.0 for condition b.

Incorporating C_x into the equation for K for spaced columns causes the ℓ/d ratio that divides intermediate columns from long ones to increase, so that a section classified as long for a solid column might be classified as intermediate when used as an element of a spaced column. This fact, along with the fact that C_x appears directly or indirectly in design formulas, causes all spaced columns except short ones to have greater allowable stresses than their solid-column counterparts.

The NDS assumes that the end spacer blocks will be joined to the long individual column members with split-ring timber connectors (see Chapter 9), rather than nails or bolts. The British Standard Code of Practice (9) allows for the spacer blocks to be joined with nails, bolts, screws, timber connectors, or glue. The British Code considers gluing to produce the stiffest spaced columns and therefore allows spaced columns joined by gluing to have a lesser effective length than those joined by other methods. The NDS requires that when there is more than one spacer block at mid-length the connection be made with a timber connector; a single middle block may be joined with bolts, nails, or timber connectors. Spacer blocks must be as thick as or thicker than the individual pieces composing the column.

To find the load on an end-block split-ring connector, the designer chooses a value of spacer-block constant from Table 7-3. The required connector capacity is the spacer-block constant times the area of an individual column element. To choose size and/or number of split rings, this required capacity is then compared to the split-ring design values in Table C-17. The procedure is illustrated in Example 7-5.

The increased strength of a spaced column is considered to be effective only for buckling in the plane parallel to the small dimension, as in Fig. 7-9a. The column is designed in this plane as a spaced column. For buckling in the other plane (see Fig. 7-9b) the column is designed as a solid column. Therefore, every spaced-column design requires considering two cases—one for each plane of possible buckling—to see which case controls the design. For each of these cases the allowable load of an individual member is determined first. The allowable load for the spaced column is the product of the allowable load for one member times the number of members. Limits allowed for slenderness ratios in spaced columns are (refer to Fig. 7-8):

80 for ℓ_1/d_1
50 for ℓ_2/d_2
40 for ℓ_3/d_1

Design Equations

For short spaced columns the allowable compressive stress is the allowable F_c parallel to the grain, with all applicable modifications. Therefore, short spaced columns are designed in the same manner as short solid columns. For intermediate spaced columns the allowable stress is:

$$F'_c = F_c\left[1 - \frac{1}{3}\left(\frac{\ell_1/d_1}{K}\right)^4\right] \quad (7\text{-}7)$$

where ℓ_1 is the length of the spaced column (a spaced column may *not* have an effective length

Table 7-3.   Shear Constants for Designing Spacer-Block Connections in Spaced Columns.

Ratio ℓ_1/d_1 of Individual Element of Spaced Column	Shear Constant per End per Plane			
	Species Grouping			
	A	B	C	D
11 or less	0	0	0	0
	(INTERPOLATE for ratios between 11 and 60)			
60 or above (maximum ratio allowed = 80)	468	399	330	261

Fig. 7-9. Spaced-column buckling about two different axes.

that is less than the actual length) and d_1 is the thickness of one element, and where

$$K = 0.671\sqrt{C_x E/F_c} \qquad (7\text{-}8)$$

For long spaced columns the allowable is given by:

$$F_c' = \frac{0.30 C_x E}{(\ell_1/d_1)^2} \qquad (7\text{-}9)$$

In both of Eqs. 7-7 and 7-8, F_c is with modifications, but E in Eqs. 7-8 and 7-9 is *not* modified for load duration. It is a good idea to check the shorter spaced columns to see that the modified allowable compressive stress, F_c, is not exceeded on the net area. Removal of wood for the split-ring connectors could cause the actual stress on the net area to exceed this allowable value.

Example 7-5

Solve for the capacity of the spaced column shown by Fig. 7-10 under a load of seven days cumulative duration. The two members are Select Structural Douglas fir–larch 4 × 6s. Choose connectors.

Fig. 7-10. Spaced column for Example 7-5.

Assume actual length equals effective length.
Load duration factor = 1.25
$E = 1.8 \times 10^6$ psi; basic $F_c = 1400$ psi

Check end fixity class:

$\ell/20 = 144/20 = 7.2$ in.
6 in. < 7.2 in., so this is condition "*a*," and C_x
    = 2.5

Check about *a-a* axis:

Treat as a spaced column for buckling about this axis.
$\ell_1/d_1 = 144/3.5 = 41.14$

$$K = 0.671\sqrt{\frac{2.5(1,800,000)}{1400(1.25)}} = 34.03$$

So the column is "long".
$F_c' = [0.30(2.5)(1,800,000)]/(41.14)^2$
    = 798 psi

Check about other axis:

Treat as a solid column for buckling in this direction.

$\ell_2/d_2 = 144/5.5 = 26.18$

$K = 0.671\sqrt{1,800,000/[(1400)(1.25)]}$
$= 21.52$

The column is "long" in this direction also.

$F'_c = [0.30(1,800,000)]/(26.18)^2 = 788$ psi

The second axis checked controls.

Allowable $P = 2(788)(3.5 \times 5.5) = 30,300$ lb

Choose split rings:

$$\ell_1/d_1 = 41.4$$

For timber connectors, D. fir–larch is in group B woods, so by interpolation on Table 7-3,

Spacer block constant = 245 psi
Connector capacity req = $(3.5 \times 5.5)(245) =$
4716 lb on each plane.

Using Table C-17, one 4-in.-diameter ring has capacity = $5260(1.25) = 6575$ lb. Use one 4″ diameter split ring at each of the four end-block joints.

7-5. BUILT-UP COLUMNS

Columns made by nailing or bolting two or more pieces of wood together to form a solid column are known as *built-up* or *layered* columns. Sketches of some of the many possible cross sections appear in Fig. 7-11. Because of slip between adjacent surfaces as the fasteners deform, no built-up column will be as strong as a solid (i.e., one-piece) column of the same dimensions and quality of wood. Care must be taken so that the pieces are adequately spiked or bolted together. (Glued laminated columns are made of several pieces of wood, but they are not classed as built-up. Their glued surfaces allow for excellent transfer of shear, so that the glulam column has the same capacity as a solid column of the same quality of material.)

The percent of solid-column strength of a built-up column has been found to be a function of ℓ/d ratio (10, 11) as shown in Table 7-4. This table is based on experiments on columns with cover plates as in Fig. 7-11a or b and should be used only for columns with cover plates. Tests by Dewell (reference 2) indicated that built-up columns without cover plates were far inferior to those with cover plates. For columns without cover plates he recommended using a built-up column capacity of 80% of the mean (average) of the strengths computed (1) as a solid column and (2) as a summation of the capacities of the individual pieces composing the column.

Both of the guidelines concerning the strength of built-up columns mentioned in the previous paragraph were developed more than a half century ago. The most recent and comprehensive study of built-up columns was undertaken by Malhotra and Van Dyer (12). Their theoretical and experimental program applied to columns in elastic and inelastic ranges and to columns of all cross-sectional configurations, taking into account the effect of interlayer slip. The design procedures resulting from the study have a more rational basis than the method suggested by Table 7-4, but are fairly complex.

Because of its light weight and economy, two

(a)   (b)

(c)   (d)

Fig. 7-11. Built-up column cross sections.

Table 7-4. Capacities of Built-up Columns.

ℓ/d	Percent of Solid-Column Capacity
6	82
10	77
14	71
18	65
22	74
26	82

researchers singled out the box column (similar to Fig. 7-11b, except hollow) for study (13). They found that strength varied directly with the number of nails. The columns tested were formed from 1 × 4 or 1 × 6 boards, using sixpenny ringed nails. With optimum nailing (about 68 nails on an 8-ft-tall column composed of four 1 × 6s), each column had strength equal to or greater than that of a solid pinned column of the same net cross-sectional area.

Example 7-6

A one-story building has No. 2 hem-fir 4 × 8 interior columns that are 10 ft 0 in. long. The building owner wants to put an open-air restaurant on the roof, increasing the load on the columns. (a) What is the allowable normal-duration load on each existing column using the NDS? (b) If each column were built up by adding 2-in. nominal material on each face, as shown by Fig. 7-12, what normal-duration load could then be allowed? Assume that the elements would be well nailed together.

(a) For No. 2 hem-fir, $E = 1,400,000$ psi and $F_c = 875$ psi.

$$\ell_e/d = 12 \times 10/3.5 = 34.3$$
$$K = 0.671\sqrt{1,400,000/875} = 26.8$$
$$34.3 > 26.8, \text{ so the column is long.}$$

Find allowable column stress:

$$F_c' = 0.30(1,400,000)/34.3^2 = 357 \text{ psi}$$

Find allowable load for the 4 × 8 column:

$$P(\text{allowable}) = (F_c')(A)$$
$$= (357)(3.5 \times 7.25) = 9060 \text{ lb}$$

Fig. 7-12. Column cross section for Example 7-6.

Notice that although this situation may be controlled by snow load, assuming a normal duration of load did not affect the answer to the problem, because the equation for long columns is not modified for load duration (K would equal 25.0 if a snow duration factor of 1.15 were included).

(b) The added material will be from the same size/use group as the original column, so we do not look up new allowables.

For the built-up member, $\ell_e/d = 12(10)/6.5$
$= 18.5$.
$K = 26.8$ as before.
$11 < 18.5 < 26.8$, so the column is intermediate.

Find allowable stress for a *one-piece* column of this size:

$$F_c' = 875\left[1 - (1/3)(18.5/26.8)^4\right]$$
$$= 809 \text{ psi}$$

(If snow controls, $K = 25$ so $F_c' = 906$ psi.)
Find load allowed for a *one-piece* column of this size:

$$P \text{ allowable} = 809 \times (6.5 \times 10.25) = 53,900 \text{ lb}$$

For the built-up column, using Table 7-4, this value should be reduced to

$$P \text{ allowable} = 53,900(0.66) = 35,600 \text{ lb}$$

7-6. BEAM-COLUMNS

Many columns are subjected to bending in combination with axial compression loads. In fact, because of nonuniform bearing, misalignment, or member crookedness, the load on many columns probably does not pass through the centroid of the cross section. This causes the column to be stressed both in axial compression and flexure. Some designers (14) even feel that, because perfectly axial load is only a theoretical possibility, all wood columns should be designed for a minimum eccentricity.

Columns loaded in such a way that they must resist both compression and bending are called beam-columns. The top chord of a truss supporting roof loads between panel points (from joists or from roof sheathing nailed directly to the chord) is one example of a beam-column.

Another example is an exterior column carrying vertical gravity loads and also subjected to horizontal wind or seismic loads. A column whose vertical load is applied onto a side bracket is also a beam-column.

P-Δ Effect or Moment Magnification

Bending moments in beam-columns can result from either lateral forces, applied moment loads, or eccentrically applied end loads. Whatever the cause, these initial bending moments cause the member to deflect, and additional bending moment occurs equal to the product of the end load, P, and deflection Δ. Figure 7-13 illustrates the cause of this P-Δ effect, also referred to as *moment magnification*. The result of this action is an increase in bending moment and the flexural stress, f_b, that bending moment causes.

Beam-Column Design

Columns with bending moment as well as axial load are designed using interaction equations. The designer uses trial and error to find a satisfactory section. The NDS shows many forms of interaction equation, but they all reduce to simply

$$\frac{f_c}{F_c'} + \frac{f_b}{F_b'}(C_M) \leq 1 \qquad (7\text{-}10)$$

in which C_M is the magnification factor to show how the initial bending moment and the flexural stress, f_b, it causes are magnified by the P-Δ effect. The authors' preference is to base F_c' in the above equation on the maximum ℓ_e/d ratio. The NDS, however, permits using the ℓ_e/d ratio for the plane of bending.

Figure 7-14 shows Eq. 7-10 graphically. Expressed in words rather than symbols it says that "the percentage of allowable axial load stress used plus the percentage of allowable flexural stress used must not exceed 100%." To make Eq. 7-10 usable, all that is needed is a value for C_M. For short columns, C_M is unity; that is, there is negligible P-Δ effect for short columns ($\ell_e/d \leq 11$).

For longer columns, C_M (as implied by the NDS equations) varies with column slenderness and with cause of the initial unmagnified bending moment. A factor J is used to account for the effect of slenderness: J equals zero for "short" columns and 1.0 for "long" columns. For intermediate columns,

$$J = \frac{\ell_e/d - 11}{K - 11} \qquad (7\text{-}11)$$

Dimension d in Eq. 7-11 is the width or thickness parallel to the plane of bending, that is, normal to the bending neutral axis. (For bend-

Fig. 7-13. P-Δ effect.

Fig. 7-14. Interaction equation in graphical form.

ing about the strong axis, d is the larger cross-section dimension.)

For beam-columns with concentric end loads but either side loads or applied moment or both, use

$$C_M = \frac{1}{1 - Jf_c/F'_b}$$

so that the interaction equation (7-10) becomes

$$\frac{f_c}{F'_c} + \frac{f_b}{F'_b}\left(\frac{1}{1 - Jf_c/F'_b}\right) \leq 1 \quad (7\text{-}12)$$

For beam columns with only eccentric end load, use

$$C_M = (1 + J/4)/(1 - Jf_c/F'_b)$$

so that the interaction equation becomes

$$\frac{f_c}{F'_c} + \frac{f_b}{F'_b}\left(\frac{1 + J/4}{1 - Jf_c/F'_b}\right) \leq 1 \quad (7\text{-}13)$$

For columns with both eccentric end load and either (or both) side load or applied moments, use separate terms for the flexural stress due to eccentric end load and flexural stress due to other initial bending moments:

$$\frac{f_c}{F'_c} + \frac{f_{bs}}{F'_b}\left(\frac{1}{1 - Jf_c/F'_b}\right) + \frac{f_{be}}{F'_b}\left(\frac{1 + J/4}{1 - Jf_c/F'_b}\right)$$
$$\leq 1 \quad (7\text{-}14)$$

in which f_{bs} and f_{be} are flexural stresses (M/S) due to the initial unmagnified bending moments and where

f_{bs} is due to side load and/or applied moment
f_{be} is due to eccentric end load

The NDS equations for which the above are equivalent are: For Eq. 7-12:

$$\frac{f_c}{F'_c} + \frac{f_b}{F'_b - Jf_c} \leq 1 \quad (7\text{-}15)$$

For Eq. 7-13:

$$\frac{f_c}{F'_c} + \frac{f_c(6 + 1.5J)(e/d)}{F'_b - Jf_c} \leq 1 \quad (7\text{-}16)$$

For Eq. 7-14:

$$\frac{f_c}{F'_c} + \frac{f_b + f_c(6 + 1.5J)(e/d)}{F'_b - Jf_c} \leq 1 \quad (7\text{-}17)$$

The design equation which combines effects for many types of loads (Eq. 7-14), is to be used when all loads cause bending about the same axis. It can easily be adjusted for bending about both axes by putting x-subscripts on the two bending terms shown and adding two similar bending terms with y-subscripts (as illustrated in Example 7-10).

The authors feel that rearrangement of the NDS equations as shown by Eqs. 7-10 through 7-14 permit a better understanding of their nature. Further, in their rearranged form they are usable for columns of any shape, whereas the NDS arrangement is suited to rectangular columns only. Also, the rearranged equations are as easy (or easier) to use than the NDS form.

Special provisions exist for very light truss members (2 × 4 or smaller) that are subjected to combined flexure and compression. If such a member is stiffened by plywood nailed to its narrow face, it is allowed an increase in stress. These special provisions (NDS section 3.10.5) will be covered in Chapter 10.

Example 7-7

An 8 × 8 No. 2 hem-fir (WWPA) member, 12 ft long, has simply supported ends and carries vertical load of 3.2 k dead and 4.2 k snow, applied at the face of the column. Is it satisfactory?

Terms required for the interaction equation are computed first. Basic design values are $F_b = 525$ psi, $F_c = 375$ psi, and $E = 1,100,000$ psi.

$3.2/0.9 < 7.4/1.15$, so snow loading controls.
$K = 0.671\sqrt{1,100,000/(1.15 \times 375)} = 33.9$
$\ell_e/d = 12(12)/7.5 = 19.2 < K$ *Column is intermediate.*
$F'_c = (375 \times 1.15)[1 - \frac{1}{3}(19.2/33.9)^4]$
$\quad = 416$ psi
$f_c = 7400/7.5^2 = 132$ psi
$M = Pe = 7400(7.5/2) = 27,750$ in.-lb
$f_b = M/S = 27,750/70.3 = 395$ psi
$F'_b = 1.15(525) = 604$ psi
$J = (19.2 - 11)/(33.9 - 11) = 0.358$

Fig. 7-15. Beam-column for Example 7-8.

Substituting in Eq. 7-13,

$$\frac{132}{416} + \frac{395}{604}\left(\frac{1 + 0.358/4}{1 - (0.358)(132)/604}\right)$$

$$= 0.317 + 0.654(1.182) = 1.09 > 1.0$$

Not satisfactory

(Note that were it not for the magnification of bending stress f_b by a factor $C_M = 1.182$, the member would *seem* to be satisfactory.)

Example 7-8

An 8 × 12 column of No. 1 (B&S) Douglas fir-larch, graded under WCLIB rules, is used to carry a vertical floor load of 18 kips (normal duration) placed 4 in. off center as shown by Fig. 7-15. In addition, the column is subject to a uniform wind load of 400 lb/ft from the attached exterior wall system. Assuming the wall system does not provide lateral support, is the member satisfactory as a beam-column?

Reactions to the eccentric vertical load and wind load are computed first, and are shown by Fig. 7-15b. Then the shear diagram is used to solve for the maximum bending moment of 13.02 ft-k, shown by Fig. 7-15d. If the vertical load acts alone, maximum moment (at top end) is $Pe = (18\,\text{k})(4/12\,\text{ft}) = 6$ ft-k. There are two load conditions for which this member must be analyzed: (a) action as a beam-column under combined dead, live, and wind loads;

and (b) action as a beam-column under eccentric vertical load alone.

Case a:

Basic $F_b = 1300$ psi and basic $F_c = 925$ psi
Larger $\ell_e/d = 14(12)/7.5 = 22.4$

$$K = 0.671\sqrt{1,600,000/(1.33 \times 925)} = 24.2$$
$11 < 22.4 < 24.2$; member is intermediate, as a column.
$F'_c = 1.33(925)[1 - \frac{1}{3}(22.4/24.2)^4] = 929$ psi
Actual $f_c = 18.0/(7.5 \times 11.5) = 0.209$ ksi

For bending, laterally unsupported length = 14 ft = 168 in.

By Table 6-1, use $\ell_e = 1.84\ell_u = 309$ in.
$C_s = \sqrt{309(11.5)/7.5^2} = 7.95 < 10$
   Short, as a beam.
$F'_b = 1300 \times 1.33 = 1729$ psi

For strong-axis bending, $\ell_e/d = 14(12)/11.5 = 14.6$

$$J = (14.6 - 11)/(24.2 - 11) = 0.273$$

Unmagnified f_b:

$S_x = 165.3$ in.[3]
$f_{bs} = 12(9.57)/165.3 = 0.695$ ksi due to side load
$f_{be} = 12(3.45)/165.3 = 0.250$ ksi due to end load

Substituting into interaction Eq. 7-14,

$$\frac{209}{929} + \frac{695}{1729}\left(\frac{1}{1 - 0.273(209)/1729}\right)$$

$$+ \frac{250}{1729}\left(\frac{1 + 0.273/4}{1 - 0.273(209)/1729}\right)$$

$$= 0.80 < 1.0 \qquad OK \ for \ Case \ a$$

Case b:
Load duration factor = 1.0 (case does not include wind).
Maximum $M = Pe = 6.0$ ft-k
Larger $\ell_e/d = 22.4$

$$K = 0.671\sqrt{1,600,000/925} = 27.9$$
$11 < 22.4 < 27.9$, so intermediate column.
$F'_c = 925[1 - \frac{1}{3}(22.4/27.9)^4] = 797$ psi
$f_c = 209$ psi

$C_s = 7.95 < 10$, so short beam.

$F'_b = (1.0)(1300) = 1300$ psi
For strong-axis bending, $\ell_e/d = 14.6$
$J = 0.273$
$f_{be} = 12(6,000)/165.3 = 436$ psi

$$\frac{209}{797} + \frac{436}{1300}\left(\frac{1 + 0.273/4}{1 - 0.273(209)/1300}\right)$$
$$= 0.64 < 1.0 \qquad OK \ for \ Case \ b$$

Example 7-9

A roof truss with slope 4 : 12 supports snow load and dead loads due to decking, roofing, ceiling, and insulation. The top chord of the truss receives a uniform vertical load to the narrow face. Therefore, this chord has to carry both axial and bending loads. The top chord is continuous over three sloping 5-ft 7-in. spans, and the most heavily loaded of the three carries a compression load of 4300 lb and a uniform load of 125 lb per horizontal foot. Will a 2 × 6 of Douglas fir–larch, No. 2 (placed with its thickness horizontal) be sufficient for this member?

Assume, conservatively, that coefficient of effective length $k_e = 1$. Basic $F_b = 1250$ psi, $F_c = 1050$ psi, and $E = 1,700,000$ psi. Horizontal distance between panel points is 5.58 × (12/12.65) = 5.29 ft.
Determine moments and stresses.

$M = 0.1w\ell^2$ (Table A-9)
$M = 0.1(125)(5.29)^2 = 349.8$ ft-lb

$f_b = 12(349.8)/7.56 = 555$ psi
$f_c = 4300/(1.5 × 5.5) = 521$ psi
$F'_b = 1250(1.15) = 1438$ psi

(Note that decking nailed to the top (narrow) face prevents lateral buckling, so that the allowable bending stress, F'_b, need not be modified for instability.)

To find F'_c we will need to know if the member is short, intermediate, or long. The ℓ_e/d ratio we will use will be for column buckling in the strong direction. (Because of the nailed decking, the member is not able to buckle in the weak direction.)

$\ell_e/d = 5.58(12)/5.5 = 12.2$
$K = 0.671\sqrt{1,700,000/(1.15 × 1050)} = 25.2$
$11 < 12.2 < 25.2$, so member is intermediate as a column.
$F'_c = 1.15(1050)[1 - \frac{1}{3}(12.2/25.2)^4]$
$\quad = 1185$ psi

Determine J.

$$J = (12.2 - 11)/(25.2 - 11) = 0.085$$

Use interaction Eq. 7-12.

$$\frac{521}{1185} + \frac{555}{1438}\left(\frac{1}{1 - 0.085(521)/1438}\right)$$
$$= 0.84 < 1.0 \qquad OK$$

Based on combined stresses, a 2 × 6 is sufficient for the entire top chord, because the span analyzed was the most heavily loaded portion of the chord. If this truss used bolts or split rings as connectors, it would be necessary to check the stress at the net section. The axial load divided by net area would need to be compared to the allowable short column stress (modified for load duration).

Example 7-10

A No. 1 D. fir–larch 8 × 12 (graded under P&T rules), 16 ft long, is subject to an eccentric load (e = 7 in.) of 10 k causing uniform strong-axis bending and a side load of 0.9 k at mid-height which causes weak-axis bending. The member has lateral support at the ends only. Is the section satisfactory?

Basic values: $F_b = 1200$ psi; $F_c = 1000$ psi; $E = 1,600,000$ psi.
Compression:

$$K = 0.671\sqrt{1,600,000/1000} = 26.8$$

Using largest ℓ_e/d,

$(\ell_e/d)_y = 16(12)/7.5 = 25.6 < K$, so intermediate column
$F_c' = 1000[1 - \frac{1}{3}(25.6/26.8)^4] = 722$ psi
$f_c = 10,000/(7.5 \times 11.5) = 116$ psi

Strong-axis bending:

$\ell_u = 16(12) = 192$ in.
$\ell_e = 1.84\ell_u = 353$ in. (Table 6-1)
$C_{sx} = \sqrt{353(11.5)/7.5^2} = 8.5 < 10$, so short beam
$C_F = 1.0$
$F_{bx}' = F_b C_F = 1200$ psi
$(\ell_e/d)_x = 16(12)/11.5 = 16.7$
$J_x = (16.7 - 11)/(26.8 - 11) = 0.361$
$f_{bx} = Pe/S_x = 10,000(7)/165.3 = 423$ psi

Weak-axis bending:

$F_{by}' = 1200$ psi

(If this had been a 2× or 4× member, the basic F_b would be multiplied by a number greater than 1.0 to account for flatwise use.)

$(\ell_e/d)_y = 25.6$
$J_y = (25.6 - 11)/(26.8 - 11) = 0.924$
$M = P\ell/4 = (900)(16)/4 = 3600$ ft-lb
$f_{by} = M/S_y = 3600(12)/107.8 = 401$ psi

Interaction equation:

$$\frac{f_c}{F_c'} + \frac{f_{bx}}{F_{bx}'}\left(\frac{1 + J_x/4}{1 - J_x f_c/F_{bx}'}\right)$$

$$+ \frac{f_{by}}{F_{by}'}\left(\frac{1}{1 - J_y f_c/F_{by}'}\right)$$

$$= \frac{116}{722} + \frac{423}{1200}\left(\frac{1 + 0.361/4}{1 - 0.361(116)/1200}\right)$$

$$+ \frac{401}{1200}\left(\frac{1}{1 - 0.924(116)/1200}\right)$$

$$= 0.93 < 1.0 \quad OK$$

7-7. COLUMNS OR BEAM-COLUMNS WITH INITIAL CURVATURE

The preceding discussion of column design assumes the member is initially perfectly straight. Occasionally the designer may be called upon to investigate the capacity of a column whose

initial crookedness is not minor enough to disregard. In this case, some results from structural stability theory (15, 16) are of use.

A differential equation based on equilibrium of the initially bent column can be solved for the total deformation of the loaded column. Assuming that the initial shape of a pinned-end column is a half sine wave with mid-height deflection of a, the total mid-height deflection after load application is

$$\Delta_T = \frac{a}{1 - P/P_E} \qquad (7\text{-}18)$$

In this equation, P is the applied load and $P_E = P_{\text{Euler}} = \pi^2 EI/\ell^2$. Inspection of the denominator of Eq. 7-18 shows that as P approaches P_E (the Euler load), deformation becomes infinitely large. A perfect column will remain straight up to the Euler load; an initially deformed column begins to deform further as soon as any load is applied. The deformation is relatively insignificant at low loads, but increases rapidly as the applied load increases.

If the only load on the crooked column is an axial load, then the magnified bending moment due to initial crookedness is $P\Delta_T$. The designer can then use the interaction equation, Eq. 7-10, with no magnification factor, that is, C_M of 1.0.

For analysis of the initially curved column with eccentrically applied end loads, the designer can use superposition. The eccentrically applied end loads can be replaced by the equivalent set of loads consisting of axial load P and end moments Pe. The maximum deflection at mid-height of the beam-column is then

$$\Delta_{\max} = \frac{a}{1 - P/P_E} + e \sec \frac{\pi}{2}\sqrt{\frac{P}{P_E}} - e$$

$$(7\text{-}19)$$

where the first term is due to initial crookedness and the second and third to eccentricity of load. The maximum moment at mid-height of the beam-column is

$$M_{max} = P(e + \Delta_{max})$$

$$= P\left(\frac{a}{1 - P/P_E} + e \sec \frac{\pi}{2} \sqrt{\frac{P}{P_E}}\right)$$

$$(7\text{-}20)$$

For this case also, the designer can use Eq. 7-10. Similar expressions for bending moment can be developed for other load types on initially crooked columns.

Creep in Columns

For columns where dead load is a major portion of the total load, creep may be of concern to the designer. However, a recent study (17) shows that crookedness, rather than creep, is the primary cause of secondary bending stress in columns. Column curvature and material (modulus of elasticity) variability were found to be much more significant than creep. The study concluded that creep could safely be ignored if the interaction equation (rather than simple column equation) were used in design with an estimate (say $\frac{1}{2}$ in.) of expected initial crookedness.

7-8. TENSION MEMBERS

Wood is required to resist tensile stresses when it is used in trusses or in the tension flange of a plywood-lumber beam. Other tension members are the chords of horizontal and vertical diaphragms (Chapter 12). In the design of a tension member, the modified allowable tensile stress, F_t, must not be exceeded. The tensile stress in the member is calculated using the net area. This is the cross-sectional area remaining after bolt holes or projected connector holes are subtracted from the gross cross-sectional area of the member. Example 5-10 shows a joint for a member stressed in tension. This example points out that both the capacity of the wood in tension and the capacity of the bolts in shear must be checked.

Footnote 15 to Appendix Table B-3 applies often to tension members. This footnote contains a table of modifying factors that decrease the allowable tensile stress for 2- to 4-in.-thick

lumber wider than 6 in. A factor from the table is multiplied by the basic allowable tensile stress (along with any applicable load duration, moisture, temperature, or treatment factors) to obtain the modified allowable tensile stress.

Example 7-11

Because of poor soil conditions, a different support system has been chosen for the sign of Problem 5-12 and Fig. 5-24. Lateral bracing (dotted line in Fig. 7-16a) prevents movement parallel to the sign. The sign is continuously bolted to the upper part of the posts. Normal to the sign, diagonal braces will be used as shown in Fig. 7-16b. Design both the vertical and diagonal supports using No. 1 D. fir–larch. Design the connection at point A.

Recalling that the sign width is 16 ft and the wind speed is 70 mph,

$$p = C_e C_q q_s I = (1.2)(1.4)(13)(1.0)$$
$$= 21.8 \text{ psf}$$
$$w = (21.8)(16/2) = 174.4 \text{ lb/ft}$$
$$F_1 = 174.4(10 \text{ ft}) = 1744 \text{ lb}$$
Sign wt. $= 1400 \text{ lb}$
Est. post wt. $= (24 \text{ lb/ft})(18 \text{ ft}) = 430 \text{ lb}$
$$F_2 = 1400/2 + 430 = 1130 \text{ lb}$$

Using a free body of the vertical post of Fig. 7-16b,

$$\Sigma M_B = 0 = -F_1(13) + (F_{AC}/\sqrt{2})(7)$$
$$F_{AC} = 4580 \text{ lb (tension)}$$
$$F_{AB} = F_2 + F_{AC}/\sqrt{2} = 1130 + 3239 = 4369 \text{ lb}$$

The shear and bending moment diagrams for the post are shown in Figs. 7-16c and 7-16d.

Design of diagonals AC:

$F_{AC} = 4580 \text{ lb tension}$
For a 2 × 6, $F_t = 1000(1.33) = 1330 \text{ psi}$
Net area req. $= 4580/1330 = 3.44 \text{ in.}^2$
Assuming 2 rows of $\frac{3}{4}$-in. diam. bolts (hole diam. $= 0.81 \text{ in.}$),
Net area provided $= (1.5)(5.5 - 2(0.81))$
$\quad = 5.82 \text{ in.}^2 > 3.44 \quad OK$
Tentatively use 2 × 6 diagonal

Check diagonal for wind reversal:

$F_{AC} = 4580 \text{ lb comp.}$; basic $F_c = 1250 \text{ psi}$
$\ell_e/d = 7\sqrt{2}(12)/1.5 = 79.2 > 50 \quad NG$

Fig. 7-16. Sign support for Example 7-11.

Try 3 × 8

$\ell_e/d = 7\sqrt{2}(12)/2.5 = 47.5$
$K = 0.671\sqrt{1,800,000/(1250 \times 1.33)} = 22.1,$
  so "long"
$F'_c = 0.30(1,800,000)/47.5^2 = 239$ psi
$f_c = 4580/(2.5 \times 7.25) = 253$ psi   *NG*

Try 3 × 10

$\ell_e/d = 47.5; K = 22.1,$ so "long"
$F'_c = 239$ psi
$f_c = 198$ psi   *OK*
Use 3 × 10 for member AC

Design posts (beam-columns):
Try 6 × 10 (B&S, WCLIB). First check the bottom
8 ft of post and then the top 11 ft. (Buckling of the
lower portion is about the weak axis. For the top

part, the post can buckle only about its strong axis.)
For the bottom 8 ft, assume that the ends are free to
rotate, but not translate, so the coefficient of effective length = 1.0.

$K_e\ell_u/d = (1.0)(8)(12)/5.5 = 17.5$
$K = 0.671\sqrt{1,600,000/(925 \times 1.33)} = 24.2,$
  so intermediate column
$F'_c = 925(1.33)[1 - \frac{1}{3}(17.5/24.2)^4]$
  $= 1118$ psi
$f_c = 4369/(5.5 \times 9.5) = 83.6$ psi
$f_b = 10,460(12)/82.7 = 1518$ psi
$\ell_e = 1.63(8 \times 12) + 3(9.5) = 185$ in.
$C_s = \sqrt{(185)(9.5)/5.5^2} = 7.6,$ so short beam
  and $J = 0$
$F'_b = 1300(1.33) = 1729$

$$\frac{83.6}{1118} + \frac{1518}{1729} = 0.95 < 1.0 \qquad \textit{Bottom part is OK}$$

Check top

  Coeff. of effective length = 2.1 (Fig. 7-2)
  $K_e \ell_u/d = 2.1(11 \times 12)/9.5 = 29.2 > 24.2$, so
    long column
  $F'_c = 0.30(1,600,000)/29.2^2 = 563$ psi
  $P = 1400/2 + (24 \text{ lb/ft})(11 \text{ ft}) = 964$ lb

(To assume P as an end load is conservative. Actually it is distributed along the column length.)

  $f_c = 964/(5.5 \times 9.5) = 18.4$ psi
  $f_b = 1518$ psi
  $F'_b = 1729$ psi

(Weak-direction buckling is not possible; the sign is continuously bolted to the upper part of the columns.)

$\dfrac{18.4}{563} + \dfrac{1518}{1729} = 0.91 < 1.0$ *Upper part is OK*

Use 6 × 10 for post (larger dimension normal to plane of sign).

Design connection:

  In the brace, load is parallel to grain.
  For a main member of 2.5-in. thickness ($\frac{3}{4}$-in.
    bolts),

  $P = 2310(1.33) = 3072$ lb (double shear)

  $\frac{1}{2}P = 1536$ lb (single shear)

In the 6 × 10 post, load is at 45°.
For a main member of 5.5 in. thickness,

  $P = 2860 \qquad \frac{1}{2}P = 1430$

  $Q = 1880 \qquad \frac{1}{2}Q = 940$

By Hankinson's equation,

$$N = \frac{(1430)(940)(1.33)}{1430(0.707^2) + 940(0.707^2)}$$
$$= 1509 \text{ lb}$$

The lesser of 1509 or 1536 controls.
Number req. = $4580/1509 = 3.03$
Use four $\frac{3}{4}$-in.-diam. bolts (2 per row).

(A quick check shows spacing and edge distance re-
quirements can be met. It would also be possible to use three $\frac{3}{4}$-in. bolts in a single row.)

7-9. COMBINED TENSION AND BENDING

Occasionally wood members are stressed in combined axial tension and bending. Combined tension and bending might occur, for example, when the bottom chord of a truss supports ceiling joists directly. For combined tension and bending, the NDS has the following two requirements, the second of which is a lateral-stability criterion:

$$\frac{f_t}{F_t} + \frac{f_b}{F_b} \leq 1.0 \qquad (7\text{-}21)$$

in which F_t and F_b are modified allowables, except that F_b is not modified for lateral instability; and

$$\frac{f_b - f_t}{F'_b} \leq 1.0 \qquad (7\text{-}22)$$

For short beams it is unnecessary to check the second requirement.

In these interaction equations for combined tension and bending, net section properties should be used where the combined tension and bending moment occur at a section reduced by bolt holes or connector grooves. If the combination occurs where these holes or grooves are absent, use gross section properties. In any case, also check for tension alone and bending alone, using the appropriate section properties (net or gross) where the maximum values of tension or bending moment occur.

REFERENCES

1. Desch, Harold E., *Timber: Its Structure and Properties*, 5th ed., St. Martin's Press, New York, 1973.
2. Dewell, Henry, *Timber Framing*, Dewey Publishing Co., San Francisco, 1917.
3. *National Design Specification for Wood Construction*, National Forest Products Association, Washington, DC, 1986.
4. Newlin, J. A., and J. M. Gahagan, "Tests of Large Timber Columns and Presentation of the Forest Prod-

ucts Laboratory Column Formula,'' U.S. Dept. Agr. Tech. Bulletin 167, 1930.

5. Southern Pine Association, *Southern Yellow Pine*, 15th ed. New Orleans, LA, 1948.

6. *National Design Specification for Wood Construction*, National Forest Products Association, Washington, DC, 1977.

7. Neubauer, L. W., ''Full-Size Stud Tests Confirm Superior Strength of Square-End Wood Columns,'' *Transactions of the American Society of Agricultural Engineers*, **15**(2), 1972, pp. 346–349.

8. *Building Code Requirements for Reinforced Concrete*, ACI-318, American Concrete Institute, Detroit, MI, 1983.

9. Council for Codes of Practice, British Standards Institution, *British Standard Code of Practice: The Structural Use of Timber*, 1971.

10. U.S. Forest Products Laboratory, *Wood Handbook*, U.S. Dept. of Agriculture, Washington, D.C., 1940.

11. U.S. Forest Products Laboratory, *Wood Handbook*, U.S. Dept. of Agriculture, Washington, D.C., 1974.

12. Malhotra, S. K., and D. B. Van Dyer, ''Rational Approach to the Design of Built-up Timber Columns,'' *Wood Science*, **9**(4), 1977, pp. 174–186.

13. Neubauer, L. W., and V. K. Arora, ''Proposed Strength Formulas for Wood Box Columns,'' *Forest Products Journal*, **24,**(1) 1974, pp. 31–35.

14. Gurfinkel, German, *Wood Engineering*, 2nd ed., Kendall/Hunt Publishing Co., Dubuque, IA, 1981.

15. Chajes, Alexander, *Principles of Structural Stability Theory*, Prentice-Hall, Englewood Cliffs, NJ, 1974.

16. Timoshenko, S. P., and J. M. Gere, *Theory of Elastic Stability*, McGraw-Hill, NY, 1961.

17. Itani, R. Y., M. C. Griffith, and R. J. Hoyle, ''The Effect of Creep on Long Wood Column Design and Performance,'' *Journal of Structural Engineering, ASCE*, **112**(5), May 1986, pp. 1097–1114.

PROBLEMS

7-1. A nominal 8-in.-wide column, 8 ft long, is to be No. 1 grade D. fir–larch (WCLIB). The end fixity factor is 1.0. If the permanent load is 57 kips, determine the column size.

7-2. A No. 1 hem-fir nominal 10-in.-wide column is 8 ft 4 in. long. The end fixity factor is 1.0 and load is 90 kips. Determine the column size assuming dry use conditions.

7-3. Design a square column of D. fir–larch 9 ft long. Normal-duration load is 80 kips. Be sure to indicate the grade of lumber chosen.

7-4. Design a square column 23 ft long of No. 1 D. fir–larch for a 40-k load.

7-5. If the column chosen for Problem 7-1 rests on a 3- by $7\frac{1}{2}$-in. steel plate at its bottom end, is it satisfactory for end-grain bearing? If not, what size plate is required?

7-6. The column of Problem 7-1 will have a hole for a 7/8-in. bolt (7.5 in. long) near the upper end to connect it to a column cap. At the section where the hole is located, is the stress level acceptable? (The actual compressive stress on the net area must not exceed the modified F_c.)

7-7. The end of a 6 × 6 southern pine column carrying axial load of 40 k rests on a U-shaped steel bracket (post base). What bearing area must the post base provide to the bottom end of the column?

7-8. A 22-ft-long column has normal-duration axial load of 43 kips. It is laterally braced in all directions at both ends and the middle. Choose the column of least area using No. 1 D. fir–larch (WWPA).

7-9. Same as Problem 7-8 except the middle of the column is braced in one direction only.

7-10. Same as Problem 7-9, but with lateral supports, *one plane only*, at one-third points of column length. Choose the most economical section. (Caution: Allowable stresses for P&T and for B&S are different.)

7-11. The square column you selected for Problem 7-4 has one-plane bracing attached at its midheight. The cross section at that point will have holes for two $\frac{3}{4}$-in. bolts. Is the section still satisfactory? (Consider net area.)

7-12. A 10-ft-high stud wall carries an axial load due to $D + S$. Studs are 2 × 4 No. 3 grade D. fir–larch at 16 in. c/c. They are supported about the weak axis by sheathing nailed to the face. Find the allowable load per foot of wall.

7-13. An 8-in.-diameter round column is to be 12 ft long. The column has $E = 1,600,000$ psi and $F_c = 950$ psi. Assume pinned ends. Find the allowable concentric load, assuming the load duration to be seven days.

7-14. A round column (pinned ends) is to be designed for a ski lodge. The wood to be used has $E = 1,300,000$ psi and $F_c = 950$ psi. The column will be 20 ft long. What diameter is necessary to support a $S + D$ load of 65 kips?

7-15. Find the allowable axial load (of two months duration) for a round, tapered column, 15 ft long. The diameter at the bottom is 10 in., tapering to 7 in. at the top. ($E = 1,400,000$ psi and $F_c = 950$ psi.) Conditions of use will be wet.

7-16. Find the allowable permanent axial load for an 8-ft-long, hinged-end, tapered column. ($E = 1.6 \times 10^6$ psi and basic $F_c = 1000$ psi.)

The thickness is a constant 5.5 in., but the width is 11.5 in. at the bottom, tapering to 5.5 in. at the top.

7-17. A column has a rectangular cross section everywhere, but it is tapered so that it has a 4- by 6-in. cross section at the small end and 6- by 8-in. cross section at the large end (actual dimensions). For $\ell_e = 10$ ft, $E = 1,500,000$ psi, and basic $F_c = 1100$ psi, determine the maximum permissible axial load of normal duration.

7-18. A spaced column consisting of two 3 × 8s is to be 8 ft 2 in. long. Using No. 2 hem-fir and connector condition b, compute the allowable load P, assuming that snow is the controlling load.

7-19. Solve for the capacity of a spaced column consisting of two D. fir-larch 4 × 8s (Dense No. 1) under load of two months cumulative duration. The column length is 12 ft, and the centers of the connectors are 12 in. from each end. (Assume that the connectors are strong enough, whatever the load you solve.)

7-20. A 9-ft-long built-up column like Fig. 7-11a is composed of six 2 × 6s of No. 1 D. fir-larch. What axial load can it safely support? What load could a 9-ft-long 6 × 6 solid column of No. 1 D. fir-larch support?

7-21. A No. 1 P&T hem-fir (WWPA) column is 8 ft long. The 40-kip, permanent-duration end load is eccentric by 6 in. Find the most economical square section.

7-22. A nominal 10 × 12 No. 1 (WWPA) hem-fir column is 8 ft 6 in. long. A 3-in. eccentricity of the 35-kip end load causes bending about the strong axis. There is also a central side load of 1.5 kips applied to the smaller face. Load duration is normal. Is this section with pinned ends satisfactory?

7-23. Repeat Problem 6-27, this time considering compression. Use F_c for No. 2 Joists and Planks (2–4 in. thick, 5 in. and wider).

7-24. A 17-ft-long column supports a vertical $D + S$ load of 18 k and a uniform horizontal wind load of 175 lb/ft. Choose a square column of No. 1 D. fir-larch.

7-25. A No. 1, 8 × 12 D. fir-larch column has an effective length of 12 ft. A 4-in. eccentricity of the 20-kip end load causes bending about the strong axis. What uniform wind load that also causes strong-axis bending can be superimposed?

7-26. Repeat Problem 7-25, except the wind load causes weak-axis bending.

7-27. A door header of 14-ft span is to be designed using 2-in. nominal members of No. 2. D. fir-larch. It is expected that two or even three members side by side will be necessary. Bending loads are $D = 64$ lb/ft and $L = 120$ lb/ft. In addition, there is a wind load of 7 k axial tension. Choose the required members.

7-28. Find the allowable axial tension load (due to wind) in a 2 × 8 member if there are two rows of $\frac{3}{4}$-in.-diameter bolts at the net section. Lumber is No. 2 southern pine. Assume holes 1/32 in. larger than the bolt.

7-29. A 10 × 10 No. 1 D. fir-larch column has an effective length of 28 ft. Assuming that we wish to design for a minimum eccentricity of $\frac{1}{2}$ in., what is the allowable load for the column?

    (a) Use Eq. 7-13, assuming an end load that is eccentric by $\frac{1}{2}$ in.

    (b) Use Eq. 7-10, calculating f_b using the total deflection from Eq. 7-18 and $C_M = 1.0$.

    (c) What is the allowable load if $e = 0.0$?

7-30. Same as Problem 7-29, except $\ell_e = 17$ ft.

7-31. A 10 × 10 No. 1 D. fir-larch column has no initial crookedness, but has end loads that are eccentric by 8 in. Effective length is 25 ft.

    (a) Using a computer, if possible, graph C_M (from Eq. 7-13) vs. P.

    (b) To the same scale, graph

$$\sec\left(\frac{\pi}{2}\sqrt{\frac{P}{P_E}}\right) \text{ vs. } P.$$

8

Glued Laminated Members

Glued laminated timbers were used as early as the late 1800s. During and since World War II, they have been used extensively. Casein adhesives were the most frequently used until the early 1970s, when they were generally replaced by synthetic resins suitable either for exterior or interior use. Glued laminated members (glulams) have seen increasing use in both buildings and bridges. Structures that would not have been feasible using only sawn-timber members have proved practical and successful using glulams.

8-1. GLULAMS

Structural glued laminated timber is defined by the National Design Specification (1) as "an engineered, stress-rated product of a timber laminating plant, comprising assemblies of suitably selected and prepared wood laminations bonded together with adhesives." Obviously, glulams for reliable structural use are neither do-it-yourself items nor items that can be manufactured at the job site.

The American Institute of Timber Construction (AITC), whose specifications cover the manufacture, grading, and design of glulams, also helps to control quality of the finished product by inspecting and approving laminating plants and operations. Production standards are also defined by ANSI/AITC A190.1-1983 (2).

Most glulams are made by gluing together the wide faces of 2-in. nominal lumber, as shown by Fig. 8-1. One-inch nominal is also used but mostly for arches. In bending, glulams are most effective when manufactured so that the wide face of the laminations is parallel to the neutral axis, that is, with the applied loads normal to the wide face of the laminations, as in Fig. 8-1a. They may also be used with the loads parallel to the wide face of the laminations (Fig. 8-1b), although the strength, stiffness, and economy advantages of glued laminated timber may be lost when that is done. Glulams may also be used as tension members or as compression members. Laminations for members to be used primarily for axial loading are selected by criteria different from those for members to be used primarily for bending.

The inner laminations of wide members may consist of side-by-side pieces, ordinarily not glued along their edges. When the edges are not glued, the longitudinal joints of one lamination must be staggered relative to those of adjacent laminations, as shown by Fig. 8-1c, and the transverse distance between such joints in adjacent layers must be at least equal to the lamination thickness. If the member will be subject to torsion, however, the edge (longitudinal) joints must be glued, since torsion will cause longitudinal shear stresses along those joints.

The structural designer must know enough about glulams to be able to select grade, size, shape, and perhaps the species of the assembled glulam member, but ordinarily is not required to participate in selecting the lamination grades or the makeup of the glulam cross section. That part of the design is done by the laminator, who will produce members meeting the geometric and strength requirements specified by the structural designer. The laminator's service may also include design and manufacture of the hardware (connection material) needed to install the members.

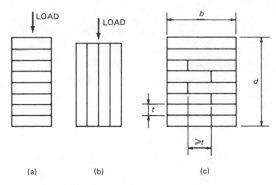

Fig. 8-1. Glulam cross sections.

Members may also be made using very thin laminations, much thinner than the thickness of a nominal 1-in. board. These are called *laminated veneer lumber* (LVL). Members of this type are available in certain proprietary products, although the NDS does not yet give specific requirements for their design. They are not considered to be glued laminated timber.

8-2. ADVANTAGES OF GLULAMS OVER SAWN TIMBERS

Glulams were originally developed to produce curved members such as large arches. Later they were used to solve the problems faced in obtaining sawn timbers of both large size and good quality. Yet as those problems were solved, other advantages were realized. Today we recognize not only that glulams may be superior to sawn timbers, but also that they provide for better utilization of our timber resources. Glulams have the following advantages over sawn timbers:

1. *Greater strength:* Lamination material is carefully selected, so that a glulam member can be of large size yet practically free from defects. More importantly, those defects that are present can be located (on the cross section) where they will do the least damage and can be dispersed along the length of the member. Thus, their effect at any given cross section is minimized and a higher modulus of rupture is obtained than would be possible in an equivalent size of sawn lumber. The highest grade of wood is used in the highest-stressed layers. In

a beam, the tension laminations are of highest quality, while those near the neutral axis are of lower quality.

2. *Larger cross sections:* Glulams can be assembled with larger section moduli and moments of inertia than are feasible with sawn timbers. With the disappearance of virgin forests, large sawn timbers are becoming more and more difficult to find.

3. *Longer spans:* These are possible because of the two above factors. For example, arched roof structures with spans of up to 500 ft have been built using glulams.

4. *Freedom from warping and checking:* Large sawn timbers almost always develop severe checks. These are caused by internal stresses that are largely the result of nonuniform drying. Large sawn timbers also are prone to warp, especially in a twisting pattern. By comparison, individual laminations of a glulam are generally only 2 in. in nominal thickness so that they can be fairly uniformly dried. They tend to remain straight and to have only minor checking. In the assembled glulam member no single lamination can warp independently, so the surface of each lamination is restrained to conform in shape with those to which it is glued.

5. *Diversity of shapes:* Glulams can be manufactured to many outlines, making possible architecturally attractive structures. Tapered beams, curved beams, arches, and rigid frames are all possible. Until glulams were developed, these shapes were normally possible only in structural steel or reinforced concrete (see Fig. 8-2).

6. *Varying section properties:* By tapering its depth or by changing the grade of lamination material used, the final member can be given a strength capacity that varies along the length according to the strength required.

7. *Camber:* Under dead load alone, a beam may have objectionable deflection. With sawn timbers, there is nothing that can be done to correct this. With glulams, however, it is a simple matter during assembly to bend the laminations before gluing, so that the resulting beam will have a slight upward bend (camber) to offset the downward deflection expected when the full dead load is applied.

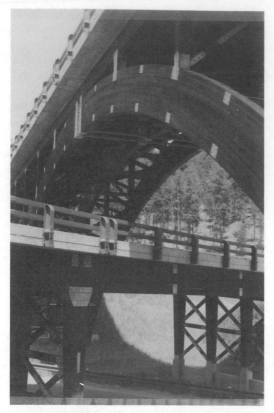

Fig. 8-2. Intersecting glulam highway bridges. (Photograph by Mike Douglas.)

8. *Fire resistance:* The fire rating of large glulam members is equal to that of heavy sawn timbers, equaling or often surpassing the ratings for structures of other materials. The fire resistance of glulam members is far greater than that of conventional light-frame construction, in which most members are of 2-in. nominal thickness.

9. *Economy:* Glulam members are often more economical than similar members of structural steel or reinforced concrete. Glulams may be more costly than sawn timbers (if timbers of the required size are available at all), but the glulams' other advantages often outweigh a slight increase in first cost. For a particular structure, of course, the answer must be obtained by comparative designs and is influenced by factors such as cost of transportation, erection cost, and the need for fireproofing.

As compared to sawn timbers, glulams do present some problems, but these are largely problems of manufacture only and do not often affect the structural designer. One such problem is that of residual stresses. If the swelling and shrinkage characteristics of adjacent laminations differ, then with changes of moisture content that may occur during use, the glulam can develop internal residual stress. Similarly, the presence of abnormalities such as cross grain and compression wood will cause unequal behavior by adjacent laminations and give rise to residual stress. Solving this problem is a task of the laminator's engineers, but not of the structural designer who selects the size and grade of glulam for a particular application.

Example 8-1

Compare the sizes of a glulam and a sawn-timber beam to carry a total normal-duration distributed load of 800 lb/ft on a simple, laterally supported span of 25 ft. Consider only flexural strength.

Solution:

$$\text{Maximum } M = 800(25^2/8) = 62{,}500 \text{ ft-lb}$$

For the glulam, assume western species, combination 22F-V3. (The meaning of this terminology will be explained later.) For this combination, Appendix Table B-5 shows an allowable flexural stress of 2200 psi. Ignoring the size factor,

$$S \text{ req} = 12(62{,}500)/2200 = 340.9 \text{ in.}^3$$

The approximate size of glulam required is 5.125 in. × 21 in. (Appendix Table B-2). The size factor for this depth is $C_F = 0.940$. The corrected allowable stress is $0.940(2200) = 2070$ psi, and the required S is now 362.3 in.^{3} The 21-in.-deep glulam has $S = 376.7$ in.^{3} *OK.*

For the sawn timber, assume D. fir–larch, B&S, Select Structural grade, which has an allowable flexural stress of 1600 psi. Ignoring the size factor, the required S is $12(62{,}500)/1600 = 468.8$ in.^{3} To provide this would take a 6 × 24 or an 8 × 20 timber. Try the 6 × 24, since it has the smaller area. For this member, the depth is 23.5 in. and the size factor is 0.928. The corrected allowable stress is $0.928(1600) = 1485$ psi. The required S is now 505 in.^{3} (vs. 506 provided). The 6 × 24 is still OK.

In this example the next-to-top grades of each were assumed. The comparison could be modified, of course, by choosing other grades. That the glulam should be more costly is suggested by the amount of wood consumed. The glulam is made by assembling 14 pieces of 2 × 6, which are later reduced to $5\frac{1}{8}$ in. wide. The 2 × 6s measure less than their name implies, but wood measuring a full 2 in. by 6 in. was used to manufacture them. Assuming the beam's overall length to be 26 ft, the total amount of wood required initially to produce the glulam is

$$14(2 \times 6/12)(26) = 364 \text{ fbm}$$

The sawn 6 × 24, on the other hand, used only 312 fbm.

The glulam used more wood, and its cost is further affected by the labor involved in its production. So, selection of the glulam over the sawn timber must be based on factors other than cost. For example, appearance, availability, or freedom from warping could justify a higher initial cost.

8-3. GLULAM PRODUCTION

Glulams are made using the following species of wood:

Softwoods

Douglas fir	Eastern spruce
Western larch	Hem fir
Southern pine	California redwood
Lodgepole pine	Softwood species (a species group)

Hardwoods

Hickory	Elm
Pecan	Sugar maple
Beech	Ash
Birch	Oak
Sweetgum	Poplar
Tupelo	Cottonwood

Of the species listed, Douglas fir and southern pine are probably the most commonly used. There is no reason why species other than those listed above cannot be used. All that is needed is to find suitable species and to develop appropriate standards for production. Other species are being experimented with and we may expect to see the above list expanded in the future. One reason that the list will expand is that good material for tension laminations is becoming increasingly difficult to find.

Researchers are seeking ways to improve the strength of tension-laminating material. One method under study is that of building up a tension lamination by gluing thin, clear (defect-free) strips of wood along each edge of a wider piece of wood. Thus far it has been discovered that the edge strips cause the resulting lamination to have significantly higher strength than a one-piece lamination of the same total width. Apparently the reason is that defects near the edge weaken a piece of lumber more than defects in the interior of the width. Edge strips prevent defects of the interior portion from acting as edge defects (3).

Adhesives

Adhesives for glulams must be able to withstand horizontal (longitudinal) shearing stresses due to load, in combination with stresses from causes such as shrinkage, swelling, or bending to produce a curved member. To be effective the glue must be stronger than the wood is in shear and in tension perpendicular to the grain. Performance requirements for glulam adhesives are spelled out by reference 2.

The AITC specifications (4) permit either dry-use adhesives or wet-use adhesives. Casein adhesives are for dry use only. They should not be used where the moisture content of the member will exceed 16%, either repeatedly or for long periods. Although permitted by the codes, casein glues are not often used.

Phenol, resorcinal, or melamine base adhesives are specified for wet use, but may be used for dry situations also. They should *always* be used where the glulam member will have over 16% moisture in service. This would include all cases of exposed (outdoor) members. It would also include members exposed to a moist environment: swimming pools, laundries, and manufacturing buildings where the processes cause high humidity.

Grades and Combinations

Wood for glulam laminations is graded under special rules. Grading may be either visual or by machine. Visual grading is based on the occurrence and spacing of defects such as knots and cross grain. Most glulams today are made from visually graded lumber. The trend will probably be toward greater use of machine-graded laminations, in which the strength is assumed to be a function of measured modulus of elasticity, E. Machine-graded lumber, of course, is also graded visually for edge defects.

The first and perhaps most critical step in glulam manufacture is to select the proper grades of laminations. For a bending member, for example, this involves finding a lamination of suitable tensile strength to serve as the outermost lamination on the tension side of the member. Normally this would be the bottom lamination. Wood just above the bottom layer is not so highly stressed. Thus the second layer need not be of such high quality as the bottom one. Layers near the neutral axis need only provide sufficient shear strength, the flexural stresses there being minimal. Laminations near the top surface, of course, must be strong enough to resist the high flexural compressive stresses.

The laminations are stacked according to the above reasoning, following criteria spelled out by the AITC specifications. Where moment reversals occur, as in continuous beams, it may be desirable to use those combinations that have the same allowable tensile flexural stress on both the top and bottom surfaces of the beam. One example of such a combination is western species 24F-V8.

Figure 8-3 shows the cross section of a typical glulam beam and indicates the grades of laminations used. Such groupings of laminations by strength are specified by the AITC specifications. These groupings are called combinations and are designated by terms such as *Combination 22F-V4*. (Appendix Table B-5 shows the various glulam combinations available for members subject primarily to bending.) In this system of nomenclature, the numerals preceding the first letter indicate the allowable flexural stress—the *22*, for example, meaning 2200 psi. The letter V indicates visual grading of the individual lamination; a letter E is used to show machine grading. The *4* of the combination number indicates the fourth com-

COMBINATION 22F

Fig. 8-3. Typical glulam makeup.

bination in that series that has the 2200 psi bending strength.

Combinations for members subject to primarily axial load are shown by Appendix Table B-6. Glulam columns are designed exactly as solid columns, except that F_c and E are from Table B-6.

When specifying a particular combination listed by the tables, the designer should verify that such material is actually available. Older publications listed allowable bending stresses as high as 3000 psi, but the present NDS gives a limit of 2400 psi for 24F—the strongest combination. If the designer specifies the allowable stresses required, rather than specifying a particular combination, the member desired is often easier to obtain.

Use of Mixed Species

In the interest of better utilization of the stronger woods, glulams may be made using stronger species for the outer laminations and weaker species for the interior ones. The weaker woods will have lower E-values than the stronger ones. Thus, when the laminator's engineers select wood species and grades to use in a glulam section, the effect of elastic properties, as well as strength, must be considered. This can be done by the transformed-section method—a method familiar to designers of reinforced concrete.

Figure 8-4a shows a glulam section made up of woods of different values of modulus of elasticity, E. In Fig. 8-4b, laminations having

Fig. 8-4. Transformed section.

(a) BUTT JOINT

(b) SCARF JOINT

(c) FINGER JOINT

Fig. 8-5. Types of end joint.

lower E-values are replaced by lesser areas of wood having the same E as the outer laminations. The section shown by Fig. 8-4b is the transformed section. Stresses in the outer laminations due to bending may be computed using the moment of inertia value of the transformed section in the flexure formula, $f_b = Mc/I_t$. To determine stresses in weaker laminations, the answer given by My/I_t for the transformed section must be multiplied by the ratio of $E(\text{inner})/E(\text{outer})$.

This type of computation is made only by those who design the makeup of the glulam section to conform to a particular combination designation. It is not done by the designer who selects a glulam section for a particular application.

End Joints

Laminations may be joined as needed to obtain sufficient length or to change the grade of lamination along the length of the member. Three types of end joint are used, as shown by Fig. 8-5.

Butt Joint. The butt joint is cheap and easy to make, but it is not a good structural joint. It is of no value whatever for transmitting tensile stress. It can transmit up to 80% of the allowable compression for the lamination. If the butt joint is used for compression laminations of a glulam beam, there will be appreciable deformation before the joint becomes effective. A butt joint can be made more effective by careful fitting of a steel bearing plate between the abutting ends. This is a costly procedure, however, and is rarely justified. As a general rule, the butt joint should not be used where the joined lamination must contribute significantly to strength.

Scarf Joint. The scarf joint of Fig. 8-5b can come close to being a full-strength splice. However, it wastes material and its manufacture is rather difficult to control. For a tension scarf joint, the slope of the joint must be nearly flat, not over 1:10. A compression scarf joint may have a steeper slope, up to 1:5. Both of these slope limits are for dry use. For wet use, a slope of not over 1:10 is recommended (5). Current standards from which glued laminated timbers are produced require qualification of the end joints to meet a strength criterion.

Finger Joint. The most commonly used joint today is the finger joint, shown by Fig. 8-5c. The finger joint can carry an appreciable percentage of the tensile capacity of the spliced lamination. The value of the joint factor depends on the exact geometry of the joint and on the manufacturing process. The long sloping edges of the ''fingers'' act like miniature scarf joints, but the ends of the fingers cannot be perfectly sharp. Finger joints may approach (but do not equal) the joint efficiency of well-made scarf joints.

Despite the fact that they are not quite as efficient as scarf joints, finger joints waste less lumber and are easier to produce. The finger joint is made by passing the pieces to be spliced though a machine whose rapidly rotating cutters form the ends to the overlapping finger pattern (see Fig. 8-6). The members are coated with glue and pressed together and the glue curing is accelerated. All of this—cutting, gluing, fitting together, and initial curing—can be done as the pieces move through the machine without stopping. Proof-testing of the finger-jointed splice may also be done as the lamination travels through the machine, and this has become common practice (6).

If scarf or finger joints are used, the strength-

Fig. 8-7. Glulam irregularities (before planing).

reducing effects of the knots are most often the controlling factors in the strength of a lamination.

Assembly and Gluing

After the laminations have been prepared, they are glued, assembled in forms to provide the proper curvature or camber, and clamped together until the glue is sufficiently cured to allow handling of the member. After gluing and clamping, the edges of the member are irregular, as shown in Fig. 8-7. The irregularities are caused by glue squeeze-out, width tolerances, and lateral crookedness of the laminations. To remove the irregularities, the members are smoothed by planing each side surface. The finished glulam will have smooth, planed sides, and its top and bottom surfaces will be the original smooth, wide faces of the boards used.

8-4. STANDARD WIDTHS

Glulam laminations are standard widths of sawn lumber. Wider laminations are harder to force into alignment with each other; thus greater irregularity is expected after gluing wide glulams than narrower ones. Consequently, a greater amount must be planed off from the wider members to produce the desired smooth side surfaces.

Table 8-1 shows the standard finished widths for glulam members made in the United States. Because of the wider and thicker standard lumber sizes then in use, glulams manufactured prior to about 1970 may be slightly wider, and their depths will be multiples of $1\frac{5}{8}$ in. rather than $1\frac{1}{2}$ in.

A glulam member is specified by its actual

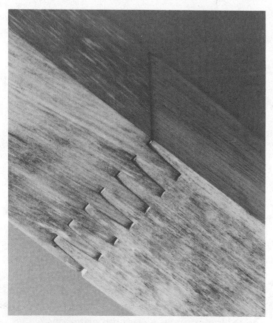

Fig. 8-6. Finger joint. Used to splice laminations for glulams. (Courtesy National Casein Co.)

Table 8-1. Standard Widths of Glulams.

Standard Actual Width (in.)	Nominal Width of Laminations (in.)	Width removed to Smooth Sides (in.)
$2\frac{1}{8}$	3	$\frac{3}{8}$
$3\frac{1}{8}$ [a]	4	$\frac{3}{8}$
$5\frac{1}{8}$ [a]	6	$\frac{3}{8}$
$6\frac{3}{4}$	8	$\frac{1}{2}$
$8\frac{3}{4}$	10	$\frac{1}{2}$
$10\frac{3}{4}$	12	$\frac{1}{2}$
$12\frac{1}{4}$	14	1
$14\frac{1}{4}$	16	Varies depending on how 16″ lamination is assembled

[a]Widths of glued laminated timbers made of southern pine are 3″ and 5″, respectively.

dimensions. Although a 2 × 6 piece of S4S lumber does not measure 2 in. by 6 in., a $6\frac{3}{4}$ × 15 glulam *does* measure what its name implies, $6\frac{3}{4}$ in. by 15 in. The width of southern pine laminated timbers in the nominal 6-in. and 4-in. width categories is $\frac{1}{8}$ in. less than other species because of the greater difficulty of holding the southern pine laminations in line while gluing. More lumber has to be planed off to produce members of the proper appearance.

8-5. LIMITS OF CURVATURE

Glulams are produced without softening the wood to permit easy bending. To avoid damaging the laminations as they are bent, the curvature must be limited. Thicker laminations cannot be bent as sharply as thinner ones. Curvature must also be limited so that high residual bending stresses are not present in the finished member. Residual stresses may also result from drying. Whatever their cause, residual stresses reduce the bending capacity of the member since the flexural and residual stresses are additive.

AITC's specification (see reference 4) recommends curvature limits. For laminations of 2-in. nominal thickness, the minimum radius of curvature (measured to the inner surface of the lamination) should not be less than 27 ft 6 in. (for southern pine, 18 ft 0 in.). For softwood laminations of 1-in. nominal thickness

($\frac{3}{4}$ in. actual) the minimum radius should not be less than 9 ft 4 in. (7 ft 0 in. for southern pine). These are the *minimum* radii. With thinner laminations, smaller radii may be used to meet architectural or other requirements. In no case, however, should the ratio of lamination thickness to inside-face radius of curvature, t/R_i, exceed $1/100$ for hardwoods or southern pine, or $1/125$ for other softwoods.

8-6. ALLOWABLE STRESSES AND MODIFICATIONS

Appendix Table B-5 gives allowable stresses for glulam members stressed primarily in bending. The table shows values for both visually graded and E-rated material, and for both western species and southern pine. The lumping of various species under a single name is justified by the fact that the laminations themselves are stress graded; thus the strength of the assembled glulam timber depends solely on the *combination* of lamination grades used and on the compatibility of mixed species for gluing.

The tabulated allowable bending stresses are based on a 5% exclusion value of tests of small clear specimens of each species, reduced to account for normal load duration, size effect, and the size and frequency of knots and other defects that affect the strength of a lamination. The table values are for use with members of 12-in. depth, carrying normal-duration loads, and under dry conditions. The table values also assume uniform load and a span-to-depth ratio of 21:1 (7); if these conditions are not met, the size factor must be adjusted, as will be explained later.

Appendix Table B-5 gives allowable stresses for bending about either axis, that is, bending due to loads normal to the wide face or due to loads parallel to the wide face of the laminations. Even though the table is for members stressed principally in bending, values are given also for allowable stresses due to axial loading that may occur simultaneously with the bending.

To use Table B-5, one must first find the portion of the table that covers the species and type of grading to be used (visually graded western species, for example). The combinations that may be available are listed under this classification. As explained earlier, the numeral at the beginning of the combination, when multiplied by 100, shows the allowable tensile flexural stress. However, this allowable is to be used only at locations where the tensile flexural stress is on the same side of the member as the tension lamination. An example of such designation would be *20F-V4*, in which the letter *V* indicates visually graded lamination material. For this combination the table shows an allowable flexural stress of 2000 psi, *provided the tensile bending stress is on the same face as the tension lamination*. If the beam is continuous, so that somewhere along its length flexural tensile stresses are on the opposite side (top rather than bottom, for example), the allowable tensile flexural stress at that location is only 1000 psi. Reading to the right in the table, various other allowable stresses are shown.

The table also gives allowable stresses for bending about the *YY*-axis. Notice that the allowables for bending about this axis are lower than those for the *XX*-axis. The reason is that, for bending about the *YY*-axis, even the lower-strength inner laminations are at the extreme fiber, where they are subject to the maximum flexural strains. For bending in this direction, most of the strength advantage of the glulam is lost.

The right-hand portion of Table B-5 gives allowable stresses due to axial loads. These allowables should be used for members that were graded primarily for bending but also have some applied axial load. For members whose loads are *primarily* axial, rather than bending, glulams graded for that type of use (Table B-6) should be specified.

At the bottom of each table is a line showing wet-use factors. These are factors by which the allowable stresses on the lines above must be multiplied whenever the moisture content in service will be 16% or more.

The AITC Manual (reference 7) shows values of allowable stress for glulams of various hardwoods. Before specifying hardwood glulams, the designer should make certain that there are hardwood-laminating plants in his or her locality.

Modifications

Most of the modifications to be considered are the same as for sawn timber, but there are a few differences.

Load Duration. Load-duration factors are the same as for sawn timbers.

Size Factor. The size factor must be used to modify allowable flexural stresses for *all* glulams whose depth in the plane of bending exceeds 12 in. (For sawn members, this factor need not be considered for members 4 in. or less in nominal thickness.) The basic size factor is given by the same equation (Eq. 4-12) as for sawn timbers. For glulam members, however, that basic factor may itself need to be adjusted. Because glulams are often deep, as compared to sawn beams, the size factor is more often important in the design of glulams.

The first adjustment to the size factor is for load type. If the load is not a uniformly distributed load, the basic size factor should be multiplied by

1.078 for a single concentrated load
0.968 for third-point loading

The second adjustment is for span-to-depth ratio. Equation 4-12 is based on a span/depth ratio of 21 : 1. For other ratios, the size factor should be adjusted according to the following:

ℓ/d	Adjustment Multiplier
7	1.063
14	1.023
21	1.000
28	0.984
35	0.972

If both adjustments are needed, they should be superimposed. Interpolation will usually be necessary to modify for span-to-depth ratio. For loadings that do not conform to the narrow cases defined above, some interpolation might be possible.

Current research suggests that, for glulams, a volume factor (considering depth, thickness, and span) may soon replace the size factor and its adjustments to better ensure safety in the design of large glulams.

The basic size factor (but not its adjustments) may be considered by use of an "effective section modulus," as shown by Appendix Table B-2. The effective section modulus is merely the product of the section modulus and the unadjusted size factor:

$$S_e = SC_F \qquad (8\text{-}1)$$

Fire Retardant. Treatment with fire retardant reduces the strength by between 10 and 25%. The effect varies with species and treatment method and according to whether the treatment is given before or after gluing. The NDS does not spell out the percent by which allowable stresses are to be reduced. Rather, it refers the designer to the manufacturer doing the fire-retardant treatment and drying.

Temperature. As explained in Chapter 2, the strength of wood is affected by abnormal temperatures. For combined high temperature and high MC, the effect is the greatest. The recommendations of Appendix C of NDS (reference 1) should be observed. Chapter 4 of the *Wood Handbook* (reference 5) gives further helpful information.

Slenderness. Lateral buckling may be considered by using Eqs. 6-3, 6-5, and 6-6 (or alternate forms recommended by AITC) to reduce the allowable flexural stress. The modulus of elasticity used in these equations should be the smaller of E_{xx} or E_{yy}. Size factor and slenderness are not cumulative.

Equations 6-3, 6-5, and 6-6 are from the NDS. Since the variability of glulam properties is less than for sawn timbers, AITC recommends using instead Eq. 8-2, below, to compute C_k (the value of slenderness factor that marks the division between intermediate and long beams) and Eq. 8-3 to determine the allowable bending stress for long beams. For intermediate length beams, Eq. 6-5 is used without change. These equations are suggested also by Appendix O of the NDS.

$$C_k = 0.956 \sqrt{E_{\min}/F_b} \qquad (8\text{-}2)$$

For long beams, having slenderness factor $C_s > C_k$

$$F_b' = 0.609\, E/C_s^2 \qquad (8\text{-}3)$$

Curvature Factor. In curved members, flexural stresses are not linear. Also, in curved glulams residual stresses are caused when the laminations are bent prior to their assembly. When bending moment is applied to the assembled glulam, the flexural stresses due to that bending moment are added algebraically to the already-present residual stresses. To account for both of these the quoted allowable is multiplied by

$$C_c = 1 - 2000\,(t/R_i)^2 \qquad (8\text{-}4)$$

in which t is the actual thickness of a lamination and R_i is the inside-face radius to which the beam is bent during manufacture.

Example 8-2

Use Appendix Table B-5 to find the allowable flexural and horizontal shear stresses for the member shown by Fig. 8-8. The member is of combination 22F-V4, western species, and the loading is mainly snow load. The member is to be used in a paper mill, where the atmospheric humidity is very high.

Appendix Table B-5 gives 2200 psi and 165 psi for the allowable stresses in flexure and horizontal shear, respectively. Modification factors for wet use (from the table) are 0.8 and 0.875, respectively.

The basic size factor is 0.956. The span-to-depth ratio is $\ell/d = 12(36.75)/18 = 24.5$. For this ratio, the basic size factor must be multiplied by about 0.992 (a value obtained by interpolation from the table in Section 8-6, above) to obtain an adjusted C_F

Fig. 8-8. Beam for Example 8-2.

Fig. 8-9. Cross section, floor system for Example 8-4.

$= 0.956(0.992) = 0.948$. The load-duration factor, applicable to both flexure and shear allowables, is 1.15 for snow load.

Thus, the modified allowables are:

$$F_b = 2200(0.8)(0.948)(1.15) = 1920 \text{ psi}$$
$$F_v = 165(0.875)(1.15) = 166 \text{ psi}$$

(*Note*: The size factor does not apply for shear.)

Example 8-3

What would be the allowable flexural stress F_b' if the beam of Example 8-2 were laterally supported at its ends only?

Modifications for wet use and for load duration still apply. Size factor, C_F, is ignored (unless considering it gives a lower allowable stress than considering lateral buckling).

Effective unbraced length (see Chapter 6) is

$$\ell_e = 1.63(12 \times 36.75) + 3(18) = 773 \text{ in.}$$
$$C_s = \sqrt{(773)(18)/6.75^2} = 17.48$$
$$F_b = 2200(0.8)(1.15) = 2024 \text{ psi}$$
$$C_k = 0.956\sqrt{1,600,000/2024} = 26.9$$

using E_{yy}, which is smaller than E_{xx}.

$$10 < 17.48 < 26.9 \qquad \text{beam is intermediate}$$
$$F_b' = 2024\left[1 - \frac{1}{3}\left(\frac{17.48}{26.9}\right)^4\right]$$
$$= 1900 \text{ psi} \qquad Controls$$

Using C_F, $F_b = 2024(0.948) = 1920 \text{ psi}$.

8-7. SUGGESTED DESIGN PROCEDURE

In most respects the design procedure for an untapered glulam is the same as for sawn-timber members. Perhaps the main points of difference are that the size factor is nearly always considered, curvature factors may have to be considered, and deflection will control more

often than for sawn-timber beams. A suggested design procedure for glulam beams is as follows:

1. Assume a combination (first make sure that glulams of that quality will actually be available).
2. Assume a weight per foot for the member, to be confirmed or corrected later in the design procedure.
3. Solve for the required section modulus, assuming a tentative value for the size factor. (Or solve directly for the required effective section modulus, if a table of effective S values is available.)
4. Solve for the required area to avoid overstress in horizontal shear.
5. Solve for the required moment of inertia to keep deflection within the allowable limit. (At this stage, consider only flexural deflection.) Long-term deflection for glulams is calculated by multiplying the deflection due to long-term load alone by 1.5 and adding the deflection due to short-term loads.
6. Select a section having the required S, A, and I.
7. Verify that all assumptions made are close enough that correcting them will not change the selection. (Or correct the assumptions in a further trial.)
8. If deflection controlled the selection, consider now whether including shear deflection will change the decision.

If the member is tapered or curved, additional steps will be required.

Example 8-4

Select a glulam section for the conditions shown by Fig. 8-9. This beam to be designed is one of many supporting a building floor system. The floor live load is 200 psf. Dead load is from 3-in. nominal decking, plus about 2 psf for floor finish. (Decking will give top lateral support.) The building will be used for long-term storage.

Assume that western species combination 20F-V4 will be available, for which the basic allowables are 2000 psi for bending and 165 psi for shear, and $E_{xx} = E_{yy} = 1,600,000$ psi. The load carried by one beam will equal the weight of dead load and live load in the area from the center of one spacing between beams to the center of the next space. Since the beams are equally spaced, that tributary area is equal to the beam span times the beam spacing. The load per foot for one beam is, therefore,

Beam weight (assumed): 40 lb/ft
Floor weight:
  Decking   7.29 psf (assume 35 × 2.5/12)
  Flooring   2   psf
             9.29 psf
  × 6 ft (spacing) =   56 lb/ft
Live load:
  6 (200) =   1200 lb/ft
  Total =   1296 lb/ft

The bending moment on one beam is

$$M = 1.296(25)^2/8 = 101.2 \text{ ft-k}$$

The allowable bending stress for permanent loading is

$$F_b = 2000(0.9)(\text{size factor})$$

Assuming a size factor of 0.95,

$$F_b = 1710 \text{ psi}$$
$$\text{Req } S = 12(101.2)/1.710 = 710.2 \text{ in.}^3$$

The maximum end reaction is $12.5(1.296) = 16.2$ kips. Ignoring (for now at least) the fact that shear at the critical section is less than the reaction,

$$\text{Req } A = 1.5 \times 16.2/(0.9 \times 0.165) = 163.6 \text{ in.}^2$$

Assume that the long-term deflection is to be limited to $\ell/240$, or 1.25 in., and that both dead and live loads are long term.

$$\text{Req } I = 5(1.296)(25^4)(12^3)/(384E \times 1.25/1.5)$$
$$= 8543 \text{ in.}^4$$

Table B-2 (Appendix) shows that a glulam measuring $6\frac{3}{4}$ by $25\frac{1}{2}$ in. ($A = 172.1$) will provide $S = 731.5$ and $I = 9327$. This section satisfies all three

requirements. However, we must confirm that the assumptions made were satisfactory.

For a span-to-depth ratio of 21, the size factor is 0.920. The actual ratio for the beam selected will be 11.8, so the corrected size factor is 1.036 times 0.920, or 0.953. Noticing that this combination is D. fir, which weighs about 35 pcf, the actual weight will be about $35(6.75)(25.5/144) = 42$ lb/ft. Both assumptions were close enough that no changes are needed in the computed bending moment or reaction. The $6\frac{3}{4}$- by $25\frac{1}{2}$-in. section is still satisfactory. Other, wider sections might also be found satisfactory, but—being shallower—they are more likely to be controlled by deflection.

Example 8-5

Repeat Example 8-4, but with a span of 60 ft.

In the previous example, design was controlled by the required section modulus. With $l = 60$ ft, we should expect deflection to be more important.

The bending moment is

$$M = 1.296 \times 60^2/8 = 583.2 \text{ ft-k}$$
$$R = 1.296 \times 60/2 = 38.9 \text{ k}$$
$$\text{Req } S = 12(583.2)/1.710 \text{ ksi} = 4093 \text{ in.}^3$$
$$\text{Req } A = 1.5(38.9)/(0.9 \times 0.165) = 392.9 \text{ in.}$$

Moment of inertia required to limit long-term total deflection to $\ell/240 = 3.00$ in. is

$$\text{Req } I = 5(1.296)(60^4)(12^3)/$$
$$(384 \times 1600 \times 3.00/1.5)$$
$$= 118,100 \text{ in.}^4$$

One possible section is $8\frac{3}{4} \times 55\frac{1}{2}$ in., but this large depth may be unacceptable. A wider, but shallower, section may be preferred. Try a $10\frac{3}{4} \times 51$-in. glulam, for which $A = 548.3$, $S = 4660$, and $I = 118,800$. Check the assumptions made.

Actual (approximate) beam weight = $35(10.75)(51)/144 = 133$ lb/ft, and the total load is now 1.389 k/ft. The basic size factor would be 0.851. The ratio of span to depth would be $12(60)/51 = 14.1$, and the adjusted size factor would be $0.851(1.023) = 0.871$. This great change in size factor suggests that a deeper section will be needed.

Try $10\frac{3}{4} \times 52\frac{1}{2}$ in. ($S = 4938$ in.^{3}, $I = 129,600$ in.^{4}).

Corrected values are:

$$w = 1.393 \text{ k/ft}$$
$$M = 1.393(60^2)/8 = 626.8 \text{ ft-k}$$

Fig. 8-10. Glulam with overhang for Example 8-6.

$C_F = 0.849(1.025) = 0.870$
Req $S = 12(626.8)/(2.0 \times 0.9 \times 0.870)$
    $= 4803$ in.^{3}
Req $I = 126,900$ in.^{4}

OK; use $10\frac{3}{4} \times 52\frac{1}{2}$ in.

Example 8-6

Figure 8-10 shows a simple glulam beam with an overhang at one end. Assume full lateral support. Assume that the dead load shown includes the beam weight; thus values shown for maximum positive and for maximum negative bending moment require no adjustment. The load duration is normal. Select a glulam size, using western species combination 20F-E3.

Allowable unit stresses shown by Table B-5 are:

Horizontal shear, 165 psi
For tension due to bending
    where tension is on bottom, 2000 psi
    where tension is on top, 1000 psi

No modifications are needed, except for the size factor for bending. Effective section modulus, $S(C_F)$, required:

For positive bending, $12(37.35)/2.00$
    $= 224.1$ in.^{3}
For negative bending, $12(14.4)/1.00$
    $= 172.8$ in.^{3}

From Appendix Table B-2, select a $6\frac{3}{4} \times 15$-in. section. The maximum shear (immediately left of the right support) is 8.27 k, and the shear at beam depth from the support face (assuming the support width to be $3\frac{1}{2}$ in.) is

$$8.27 - 0.8(1.25 + 1.75/12) = 7.15 \text{ k}$$

The computed shear stress is

$$1.5 \times 7150/(6.75 \times 15) = 106 \text{ psi}$$

which is well below the 165-psi allowable.
    Deflection must be checked at two locations. As

an approximation, the maximum total deflection in the 20-ft span is less than it would be for the same span without an overhang. That is, deflection is less than

$$(5/384)(800)(20^4)(1728)/(1,700,000 \times 1898)$$
$$= 0.893 \text{ in.}$$

or for live load alone, less than 5/8 of 0.893, or 0.558 in., which is only 1/430 of the 20-ft span. Satisfactory.

At the end of the overhang, the maximum deflection will occur when the overhang is fully loaded but the 20-ft span has only dead load. Computation by the conjugate beam method (not shown here) indicated a negligible downward deflection (0.06 in.) at that point. Perhaps more significant is the possible upward deflection at the right end when the overhang carries dead load only but the 20-ft span is fully loaded. This deflection is found to be about 0.4 in. upward. Whether or not this much deflection is objectionable will depend on the detail manner in which the beam is to be used. If the overhang is an interior balcony, not connected to anything else, the deflection may not be harmful. If it is an exterior balcony, water might not drain properly from the overhang. If the overhang is connected to other items, those items might be damaged by the deflection of the overhang.

8-8. BIAXIAL BENDING

Beams can be loaded to have simultaneous vertical and horizontal bending moment. If the allowable bending stresses for bending about the X-axis and for bending about the Y-axis were the same, one could merely compute the maximum combined flexural stress (i.e., the sum of M_x/S_x and M_y/S_y), and then compare it to the common allowable. For glulams, however, the allowable stress for bending about the Y-axis (perpendicular to the wide face of the laminations) is much less than for bending about the X-axis. No allowable value for the combined bending stress is given, so it is necessary to use an interaction equation to decide whether or not the member is satisfactory. One such equation, based on the idea that points representing bending stresses of members that are not overstressed plot below the diagonal line of Fig. 8-11, is

$$f_{bx}/F_{bx} + f_{by}/F_{by} \leq 1.00 \qquad (8\text{-}5)$$

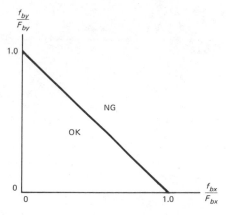

Fig. 8-11. Basis of interaction equation for biaxial bending.

In computing the allowable stresses for Eq. 8-5, note that the size factors, C_F, about the X- and Y-axes will usually be different.

Example 8-7

A $6\frac{3}{4} \times 24$-in. glulam, combination 20F-V9, is used on a 32-ft span with lateral support at the ends only. It has a diagonal concentrated load of 4.9 k at midspan. The load is inclined at 23° from the vertical. Self-weight of the beam is about 40 lb/ft. Is the beam satisfactory in biaxial bending?

Components of the load are

Vertical $P = 4.9 \cos 23° = 4.51$ k
Horizontal $P = 4.9 \sin 23° = 1.91$ k

Bending moments are

$$M_x = 4.51(32/4) + 0.040(32^2/8) = 41.2 \text{ ft-k}$$
$$M_y = 1.91(32/4) = 15.3 \text{ ft-k}$$

Flexural stresses are

$$f_{bx} = M_x/S_x = 12(41.2)/648 = 0.763 \text{ ksi}$$
$$f_{by} = M_y/S_y = 12(15.3)/182.2 = 1.008 \text{ ksi}$$

Allowable stresses:

$$\ell_e = 1.37(12 \times 32) + 3(24) = 598 \text{ in.}$$
$$C_{sx} = \sqrt{598(24)}/6.75 = 17.75$$
$$C_{kx} = 0.956\sqrt{1,400,000/2000} = 25.3$$
Beam is intermediate.
$$F'_{bx} = 2.0[1 - (1/3)(17.75/25.3)^4] =$$
$$1.838 \text{ ksi; } controls$$
$$C_F = 0.926; F_b(C_F) = 1.852 \text{ ksi} > F'_{bx}$$
$$F_{by} = 1,400 \text{ ksi}$$

Fig. 8-12. Types of cantilever construction.

Interaction equation:

$$0.763/1.838 + 1.008/1.400 = 1.135 > 1.0$$
$$NG$$

Had the beam proved satisfactory for biaxial bending, what other conditions should be investigated? For sure, shear should be checked. The authors feel that this should be checked using a similar interaction equation, but with the actual and allowable flexural stresses being replaced with computed and allowable shear stresses. Note from Table B-5 that the allowable shear stresses for the two directions of loads are unequal.

8-9. CANTILEVER BEAM SYSTEMS

Two examples of cantilever beam systems are shown schematically by Fig. 8-12. Figure 8-12a shows a two-span system, in which the left beam is extended a distance C beyond its interior support. From the right end of that cantilevered portion (overhang) another beam is supported, this beam having a span B. At the end of the overhang, the connection allows one member to rotate freely relative to the other. It is referred to as a *pinned connection* even though usually an actual pin is seldom used. (A detail for this type of connection may be found in Chapter 9.)

Why arrange beams in this manner? Why not merely use two simple beams, one of span A

and the other of span D? Reasons for preferring the cantilever system are:

1. Economy, caused by bending moments being less than for simple beams of spans A and D.
2. With the same size of member, longer feasible spans for the cantilever system than for the simple, one-span beams. Longer spans mean more usable floor area below.
3. Possible ease of construction as compared to a system that uses only single-span beams.

The advantages listed above are even more evident in the three-span system shown by Fig. 8-12b. The central beam, extending from one pin connection to the other, is called a *suspended span*. Another possibility, not shown here, is a three-span system in which a central span extends beyond its supports at each end, the cantilever overhangs each serving as supports for suspended end spans.

If the only load condition to be considered were full uniform load, and if the glulam to be used had the same strength for negative bending as for positive, then it would be easy to optimize the design by choosing the best length, C, for the overhang. However, the Uniform Building Code (8) and others require that partial live loads be considered. Thus, for each of the systems shown by Fig. 8-12, two load cases would have to be used by the designer:

Case 1—Full load on all spans

Case 2—Dead load on all spans, with live load either in the center span only or in both end spans

Dimension C that is most economical for case 1 will be different from the dimension that is best for case 2. Some compromise value will be the optimum.

Choice of the best length of overhang is affected by the glulam makeup. If the glulam has tension laminations on its bottom surface only, the allowable negative bending moment is less than the allowable positive. On the other hand, if the glulam is designed and fabricated with tension laminations on the top in portions of the length where negative bending moment is expected, and tension laminations on the bottom where positive bending moment is expected, the allowable positive and negative bending moments will be alike. The best overhang length, C, may be larger in the latter case than in the former. The larger overhang makes the system more effective, so for cantilever construction it may be advisable to specify combinations in which top and bottom allowable stresses are alike. These combinations can be identified from the third and fourth columns of Appendix Table B-5. For all other combinations, the allowable negative bending moment will be one-half the positive.

What are the disadvantages of cantilever beam systems? Possible larger deflection is one disadvantage, and with it come the potential problems of vibration and ponding. These disadvantages do not necessarily cancel the advantages, but they are factors that the designer must consider carefully. Larger deflection can be partially offset by initial camber, *which must be specified by the designer*. For the beam of Fig. 8-12b, for example, deflection would have to be computed at the interior of the end span, the end of the overhang, and the center of the suspended span. At those points, camber should be at least equal to $1\frac{1}{2}$ times the long-term dead load deflection. If the beam system must support a roof (for which ponding is a concern) the

UBC requires either that the roof be sloped to drain away all water after long-term deflection of all roof framing members or that the supporting beams be cambered additionally (beyond the camber required for 1.5 times dead load) to ensure that the roof will drain even under full design loads.

Where a cantilever beam system is used to support a floor or a roof, the structural floor or roof material (if attached to the glulams) usually has enough diaphragm stiffness to ensure lateral stability of the glulam over those portions of the length where the top side is in compression. For the portion that has negative bending moment, however, the compression (bottom) side is laterally unsupported. For that portion lateral stability must be checked, using the distance between the inflection point and the pin connection as the laterally unsupported length. If this distance proves to be too long (so that the required reduction of allowable stress is too much) it may be necessary to provide one or more intermediate lateral braces so as to reduce the laterally unsupported length.

Suggested Design Procedure

The first step, of course, is to try to find a satisfactory length of overhang, one that will give economy yet moderate deflections. This may have to be done by trial and error, completing alternate designs using the following procedure, or it may be dictated by other factors—architectural, for example.

1. Estimate the weight of the member. Usually, it will be best to keep the same member size throughout, although the suspended span can be made of a smaller section if appearance is not important.
2. Solve the maximum bending moments of each sign; do this for each span and for each load condition. It will often help to sketch the bending-moment diagrams.
3. Choose a tentative section.
4. Check any assumptions made and recycle as required.
5. Compute deflections and specify the required camber.
6. Check for bearing at the supports. At the

PLAN (PARTIAL)

CROSS SECTION

Fig. 8-13. Trial floor system for Example 8-8.

support next to the overhang, the reaction may be large enough that the compression stress perpendicular to the grain exceeds the allowable.

7. Design the connection at the "hinge."
8. Check horizontal shear at the cantilever end and at each side of all points of support.

Example 8-8

Choose a D. fir–larch glulam section for the cantilever beam floor system shown by the tentative layout of Fig. 8-13. Use a combination that allows equal positive and negative bending moments. Limit short-term total deflection at midspan to $\ell/180$. As instructed by the UBC, consider both full loading and partial loading. Dead load from the floor sheathing and joists is 20 psf; specified live load is 45 psf (after considering any allowable reductions).

The solution to this example is shown in designer's computation sheet form by Fig. 8-14. Three load cases are considered—all spans fully loaded, live load on end spans only, and live load on center span only. Deflection of the exterior span is estimated by computing the deflection for a simple span using the same section. (The actual deflection will be slightly less.) For the cantilever and the suspended span, it is solved by AISC handbook equations (9).

It is interesting to compare the section selected for the cantilever system with those that would be required if three simple beams were used.

For the left span, say 29.5 ft c/c req. bearing lengths,

$$M = 0.87(29.5^2/8) = 94.6 \text{ ft-k}$$
Req $SC_F = 12(94.6/1.6) = 709.5$
Use $6\frac{3}{4} \times 27$ in.

For the center span, say 33.5 ft,

$$M = 0.87(33.5^2/8) = 122 \text{ ft-k}$$
Req $SC_F = 12(122/1.6) = 915$
Use $6\frac{3}{4} \times 30$ in. (vs. $6\frac{3}{4} \times 25\frac{1}{2}$)

Using the cantilever beam system saves 216 fbm per bay (gross) over using simple beams. It also saves $4\frac{1}{2}$ in. of building height.

Example 8-9

Check lateral stability of the beam selected in Example 8-8. Are lateral braces required for the bottom of the glulam?

From Fig. 8-14, the designer's solution to the previous example, we can compute the length over which the bottom flange is in compression. Load case III produces the largest negative bending moment. Further, for case III the distance from the inflection point to the pin connection is larger than for either case II or case III. So case III controls. However, the NDS requires that lateral support be pro-

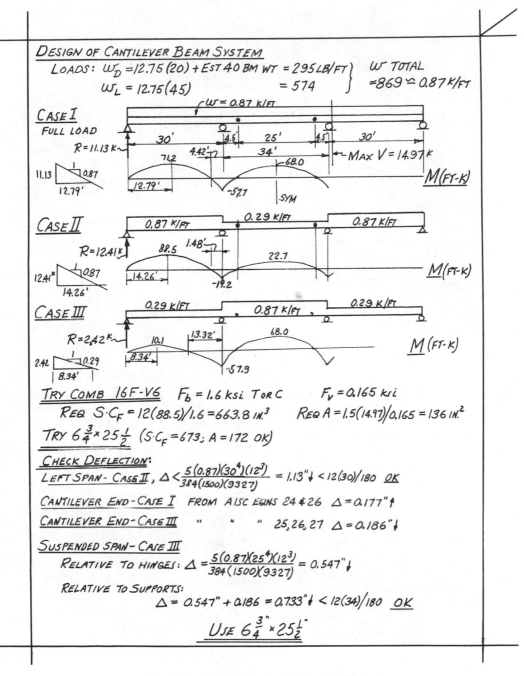

Fig. 8-14. Computation sheet for Example 8-8.

vided at the point of bearing on the interior column to prevent the beam from rotating laterally at that point. Thus the maximum laterally unsupported length over which the bottom of the beam is in compression is 13.32 ft. Ratio $\ell_u/d = 12(13.32)/25.5 = 6.27$. This is less than 14.3, so a conservative value of effective length (see Chapter 6) is

$$\ell_e = 1.63(12)(13.32) + 3(25.5) = 337 \text{ in.}$$

By Eq. 6-2, the slenderness factor is

$$C_s = \sqrt{337(25.5)}/6.75 = 13.73$$
$$C_k = 0.956 \sqrt{1,500,000/1600} = 29.27$$
$$F_b' = 1600[1 - (1/3)(13.73/29.27)^4]$$
$$= 1574 \text{ psi}$$

Fig. 8-15. Curved glulam pedestrian bridge. Deck is vertically laminated lumber. (Photograph by authors.)

Using the size factor, $F'_b = 0.92(1600) = 1470$ psi *Controls*.

$$f_b = 12(57,900)/731.5 = 950 \text{ psi} < F'_b \quad OK$$

Bracing is needed only where the beam crosses the column.

Example 8-10

How should the beams of Examples 8-8 and 8-9 be cambered?

Deflections were computed by the designer, as shown by the computation sheet in Fig. 8-14. For the left span, the computed downward deflection due to dead plus live loads was slightly less than 1.13 in. The portion due to dead load alone is slightly less than

$$(295/870)(1.13) = 0.38 \text{ in.}$$

(short-term deflection)

Camber at approximately the center of the left span should compensate for long-term dead-load deflection, or $1.5(0.38) = 0.57$ in. A camber of $\frac{5}{8}$ in. or more should be specified.

Dead-load deflection (short-term) of the suspended span, relative to the hinges, is

$(295/870)(0.547) = 0.185$ in. Camber of the suspended beam should be at least 1.5 times this amount, or 0.278 in. Say $\frac{3}{8}$ in.

Deflection of the cantilever end is very small in this particular instance. For this beam, camber of the cantilever end will be omitted. However, in designs of cantilever beam systems, deflection of this point may sometimes be important.

8-10. CURVED GLULAMS

The design of curved glulams (see Fig. 8-15) differs from that of straight ones in three ways:

1. Curvature limits must be checked.
2. The allowable flexural stress must be modified for curvature.
3. Radial stresses must be computed and provided for.

Curvature limits given in Section 8-5 should be followed to prevent residual stresses from being too high and to prevent damaging the laminations as they are bent to the desired shape during fabrication. The allowable bending stress for the glulam must then be reduced, so that the sum of the flexural stress due to loads and

Fig. 8-16. Plastic model of curved glulam. Top: unloaded. Bottom: with load tending to straighten the member. (Photograph by authors.)

the residual stress due to bending during fabrication does not exceed a reasonable total for the quality of wood used.

When bending moment is applied to a beam that is initially curved in the plane of bending, radial stresses (as well as flexural stresses) occur. These radial stresses may be either tensile or compressive.

For graphic evidence of tensile radial stress in a curved member, see Fig. 8-16, which shows what happens when a model of a curved glulam is bent. When the applied bending moment tends to straighten the bundle, the laminations try to move apart. Under the opposite direction of applied bending moment, however, the laminations are pressed more firmly together. This effect can easily be demonstrated by the reader by merely holding together firmly several sheets of paper in an initially curved shape. Bending the bundle one way separates the pages; bending it the other way compresses them together.

How can we compute the magnitude of these radial stresses? The free body in Fig. 8-17a is a short segment of a curved beam having bend-

ing moments that tend to straighten the segment. To the right is a diagram of stresses indicated by the flexure formula. (Curved-beam theory need not be used to compute the flexural stresses, since the radius of curvature is always large compared to the glulam depth.) Resultant forces T and C are equal to each other.

$$T = C = M/(2d/3) \qquad (8\text{-}6)$$

An element (shaded) from below the neutral axis of the segment is now removed as a free body as shown in Fig. 8-17b. Since the two forces T are not parallel to each other, their resultant is a radial force, F:

$$F = 2T \sin \alpha \qquad (8\text{-}7)$$

Force F has its largest value when the removed element extends all the way to the neutral axis, as shown.

An equal but opposite force must be applied to hold the removed element in equilibrium. This equilibrant force is normal to the interior surface of the removed element, and is equal

(a)

(b)

Fig. 8-17. Radial force in curved glulam.

to the product of the radial stress and the area of that surface. Thus,

$$f_r = F/(2R_c\alpha b) = 3M/(2R_c bd) \quad (8\text{-}8)$$

where R_c is the radius to the centerline of the glulam depth.

The above equations are for constant-section curved beams only; they do not apply to tapered curved beams, which are discussed later.

When the bending moment due to load tends to *straighten* the beam (as in Fig. 8-17), the radial stress is tensile. When the bending moment tends to increase the beam's curvature (i.e., opposite to that shown by Fig. 8-17), it causes compressive radial stress.

Radial tensile stresses are perpendicular to the grain of the wood lamination. In this direction the wood is weaker in tension than the glue that joins the laminations; therefore, it is the tensile strength of the wood itself that needs to be considered. The NDS allowable values for this radial *tensile* stress are:

Southern pine or California redwood
  For any load condition—
  One-third the usual allowable for horizontal shear.
Douglas fir or larch
  For wind or earthquake loadings—
  One-third the usual allowable for horizontal shear.
  For other loadings—15 psi

When the latter allowable (the 15-psi value) is exceeded for either Douglas fir or larch, radial mechanical fastenings may be used to resist the *entire* amount of the radial tension, but still the computed radial tensile stress must be limited to one-third the allowable horizontal shear stress. As another alternative for uniform-section curved members of Douglas fir or larch, *vertical grain* lumber may be used throughout the member (vertical-grain lumber is lumber in which the annual rings—as seen on the end of the piece—are at an angle of at least 60° with the wide face of the lamination). In this case, the ceiling value for the computed radial tensile stress is again only one-third the usual allowable for horizontal shear. While this latter alternative is permitted by the NDS, it is seldom applied.

The AITC (see reference 7) gives allowable values of radial *compressive* stress for some species. They are as follows:

Douglas fir–larch	560 psi
Douglas fir south	560 psi
Southern pine	560 psi
Hem-fir	375 psi
Western woods	255 psi
Canadian softwoods	255 psi

The above allowables for radial compression in softwoods are not necessarily the same as the allowables for compression normal to the grain; in some cases they are larger and in some cases smaller. All of the above allowables, whether for radial tension or for radial compression, are subject to modification for load duration, fire-retardant treatment, and service condition (wet use versus dry).

Suggested Design Procedure for Curved Glulams

Design is necessarily by trial and error. The member depth is one of the dimensions to be determined, but to do this the designer needs to know both the minimum radius and the centerline radius of curvature. Frequently only the outer radius is known when the design is commenced, so the other dimensions have to be estimated. A suggested procedure is as follows:

1. Make the usual estimate of member weight. Compute tentative bending moments and shears.
2. Ignoring curvature and size factors for the present, solve for an approximate required section modulus and choose a trial depth.
3. Use this trial depth (or slightly more) to compute both the minimum radius and the radius at the centerline of the member depth.
4. Determine the allowable flexural stress, including modifications for curvature and size and other applicable modifications.
5. Select a section for flexure and verify that it is satisfactory for shear.
6. Compute the radial stress and compare it to the allowable.

7. Verify all assumptions made above, and correct if necessary.
8. Recycle as required to step 3 (or to step 1 if the estimated member weight was badly in error).

The designer should be alert to the fact that a curved member may also have axial load. If this happens, axial load must be considered along with flexure.

Example 8-11

Select a glulam member for dry use to carry a normal-duration bending moment of 160 ft-k (including that due to the member weight) in a direction that tends to straighten the member. At the section having that moment, the member centerline is curved to a radius of 32 ft. Try combination 22F-V4, western species. Consider flexure only. The member is laterally supported. (Assume that no modification of the basic size factor is needed for load condition or ℓ/d ratio.)

With a normal allowable bending stress of 2200 psi, the required section modulus would be

$$\text{Req } S = 12(160)/2.2 = 873 \text{ in.}^3$$

which could be provided by a $6\frac{3}{4} \times 28\frac{1}{2}$-in. glulam. For a $28\frac{1}{2}$-in. depth the size factor is computed to be 0.908. The minimum radius for the $28\frac{1}{2}$-in. section would be

$$R_i = (12 \times 32) - 14.25 = 369.8 \text{ in. or } 30.8 \text{ ft}$$

This is larger than the permitted minimum radius of 27 ft 6 in. for $1\frac{1}{2}$-in. laminations (see Section 8-5). Also, ratio t/R_i is $1.5/369.8 = 0.0041$, which meets the requirement that t/R_i not exceed $1/125 = 0.008$ for Douglas fir or larch. By Eq. 8-4, the curvature factor to be used is

$$C_c = 1 - 2000(1.5/369.8)^2 = 0.967$$

The modified allowable flexural stress is

$$F_b = 2200(0.908)(0.967) = 1930 \text{ psi}$$

A revised value of section modulus is

$$\text{Req } S = 12(160)/1.93 = 994.8 \text{ in.}^3$$

A possible section would be $6\frac{3}{4} \times 30$ in., for which the actual S is 1012.5 in.^{3} If the minimum radius,

Fig. 8-18. Two types of mechanical means of resisting radial tension.

curvature factor, and size factor are checked again, this section is found to be satisfactory in bending.

Since the bending moment in this case tends to *straighten* the member, the radial stress is tensile. The next step is to make sure that this radial tensile stress does not exceed the allowable, given by the NDS as 15 psi. The computed actual radial stress will be

$$f_r = 3(12,000)(160)/(2 \times 384 \times 6.75 \times 30)$$

$$= 37.0 \text{ psi}$$

Since this exceeds the allowable, what can be done? There are two things that might solve the problem. One of these is to increase the beam area, bd, so that Eq. 8-8 gives a radial stress value of 15 psi or lower. That would be a rather expensive solution in this case, however, as it would more than double the size of the member.

The second alternative would be to use mechanical connections to hold the laminations together in the curved region. Figure 8-18 shows a short segment of curved beam with its laminations held together by either lag screws or rebars embedded in epoxy. Assuming that the connectors resist the entire radial tensile force, the load applied to each is merely the total of the radial stresses (at the neutral plane of the beam) on an area equal to the connector spacing (pitch) times the beam width. For the beam of this example, the connector force would be

$$37.0 \times 6.75p = 250p \text{ lb}$$

Try $\frac{3}{8}$-in. (no. 3) rebars embedded in epoxy. From the ACI specifications (10) we find that the allowable tensile stress for grade-60 rebars is 24,000 psi. For a no. 3 grade-60 rebar, the allowable tensile load is thus $0.11(24,000)$ or 2640 lb. Setting $250p$ equal to this allowable force per rebar, the maximum allowable pitch is $2640/250 = 10.6$ in. To use no. 3

rebars at 10-in. centers would be a practical solution. However, if lag bolts were used, the holes for these bolts (if long enough lag bolts could be found) would reduce the section modulus. In this case, the bending strength of the beam would have to be checked again using the net section.

Example 8-12

For a more realistic example, consider the building whose outline shows in Fig. 8-19. There are several ways in which the roof of this building could be supported. Five such schemes are shown by Fig. 8-20. The number of supporting members to be used depends on the strength and stiffness of the deck, and also on architectural requirements. Whether or not to use intermediate columns is also a matter to be decided by the architect or by the owner. Of these five, let's consider scheme II. In this scheme no intermediate columns are used and the roof support members are glulams with a short curved segment to give the roof a two-way pitch. Also, the architect has suggested a radius of 36 ft, measured to the top of the roof. Try western species combination 20F-V3, for which the basic allowable stresses are 2000 psi for bending (1000 psi where tension due to bending is on the top face) and 165 psi for shear.

Deflection of roof support members should be limited. Assume for this case that the desired limit for live-load deflection is $\ell/240$. Ponding need not be considered, since the roof has appreciable (3 : 12) pitch.

Roof loads consist of the deck weight, 9 psf for roofing and insulation and 30 psf of horizontal projection for snow. The weight of the beam itself, of course, must be estimated. Assume that it has already been decided that the roof deck will be nominal 3-in. hem-fir selected decking and that the insulation will be 3-in. thick.

How many beams should we use? This will depend, of course, on the strength and stiffness of the

Fig. 8-19. Building plan and elevation for Example 8-12.

roof deck. Under snow load, the basic allowable bending stress for the decking (a repetitive member) is $1600(1.04)(1.15) = 1914$ psi (Table B-3 and footnotes). Modulus of elasticity, E, is 1,500,000 psi.

Load per sq ft of roof surface is:

Decking (assume 30 pcf)

$30(2.5/12) =$ 6.3 psf

Roofing and insulation   9.0 psf

Total $w_d =$ 15.3 psf of roof surface

Converting this to pounds per foot of horizontal projection,

$15.3(12.37/12) = 15.8$ psf

Snow   30.0 psf

Total load $= 45.8$ psf of horizontal projection

Roof *deck* bending will be caused only by the component of this load that is normal to the roof surface. Per square foot of roof deck, this will be

$45.8 \times 12/12.37 = 44.4$ psf

Assuming that the roof deck will be continuous over two spans

Actual $M = 44.4\ell^2/8$
S furnished by 3-in. deck $= 12 \times 2.5^2/6$
                          $= 12.5$ in.^{3}/ft
Allowable $M = 12.5(1914/12) = 1994$ ft-lb/ft

Setting this allowable M equal to $44.4\ell^2/8$, the allowable ℓ based on strength of the decking is 18.95 ft. (That is, the glulam beams might be placed as far apart as 18 ft 11 in.) For two-span continuous decking, the deflection is $w\ell^4/185EI$. Equating this to the assumed total deflection of $\ell/180$,

$$\frac{12\ell}{180} = \frac{44.4\ell^4(12)^3}{185 \times 1,500,000 \times 12(2.5)^3/12}$$

From this, the maximum ℓ for which the deflection limit is not exceeded is found to be 15.56 ft. Deflection controls the solution and the supporting beams must be spaced at not over about 15 ft 6 in. apart. Looking at the floor plan of Fig. 8-19, we see that the end beams will be about 60 ft on centers.

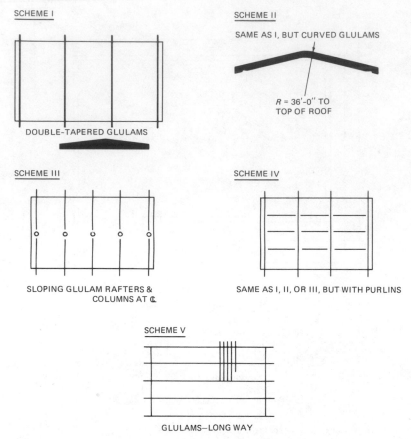

SCHEME I

DOUBLE-TAPERED GLULAMS

SCHEME II

SAME AS I, BUT CURVED GLULAMS

$R = 36'-0''$ TO TOP OF ROOF

SCHEME III

SLOPING GLULAM RAFTERS &
COLUMNS AT ℄

SCHEME IV

SAME AS I, II, OR III, BUT WITH PURLINS

SCHEME V

GLULAMS—LONG WAY

Fig. 8-20. Five framing schemes for building of Fig. 8-19 and Example 8-12. (Courtesy of Board of Regents, University of Colorado.)

This length could be divided into four 15-ft lengths. So try glulams on 15-ft centers.

Now that a tentative spacing for the glulam beams is known, we can turn our attention to the design of the glulams themselves. Each interior glulam will carry load from a 15-ft width of roof; the end ones (considering the effect of the roof overhang at the end of the building) will carry load from about 10.8 ft of roof. The load per foot of horizontal length for each interior glulam will be:

From roof: $45.8(15) = 687$

Glulam wt, estimated = $\underline{\quad 45\quad}$

Total $= 732$ lb/horizontal ft

For dead load alone, this total will be $732 - 30(15)$, or 282 lb/ft.

$$732/1.15 > 282/0.9$$

so the load combination that includes snow will control the design.

Assuming that the span will be 40.0 ft (from center to center of the walls) and considering the effect of the roof overhangs at the north and south walls, the midspan bending moment will be 139.0 ft-k. Using an allowable flexural stress of $1.15 \times 2000 = 2300$ psi, a first approximation for the required section modulus is:

$$\text{Req } S = 12(139.0)/2.3 = 725.2 \text{ in.}^3$$

Size factor and curvature factor were ignored in the first approximation above, so now they must be estimated to obtain a more accurate value of required section modulus. A section modulus of very slightly more than 725.2 would be provided by a $6\frac{3}{4}$- × $25\frac{1}{2}$-in. glulam, but when size and curvature factors are considered, this member will probably prove unsatisfactory. So for the next trial, let's consider the next deeper section, $6\frac{3}{4} \times 27$ in. For a 27-in. depth, the unadjusted size factor is 0.914; with adjustment for span-to-depth ratio it is 0.924. The assumed glulam weight is close enough. Allowing 1 in. for roof-

Fig. 8-21. Double-tapered glulam.

ing thickness and 3 in. for insulation, the inside (minimum) radius of the 27-in. glulam will be

$$12(36) - 1 - 3 - 2.5 - 27 = 398.5 \text{ in.}$$

For a 27-in. glulam and this minimum radius of curvature, the curvature factor will be 0.972. Thus the allowable stress for the second trial is

$$F_b = 2000(1.15)(0.924)(0.972) = 2066 \text{ psi}$$

The required section modulus is

$$\text{Req } S = 12(139.0)/2.066 = 807 \text{ in.}^3$$

The $6\frac{3}{4}$- × 27-in. section is satisfactory, its S being 820.

The maximum shear at distance d from the face of the support is about

$$V = (19.5 - 27/12)(732) = 12,630 \text{ lb}$$

$$f_v = 1.5(12,630)/(27 \times 6.75) = 104 \text{ psi}$$

$$< 1.15(165) \text{ psi} \quad OK$$

Negative bending at the support must be checked, since the flexural tension there is on the top surface, and the basic F_b is only 1000 psi:

$$M = 732 \times 4.50^2/2 = 7410 \text{ ft-lb}$$

$$f_b = 7410(12)/820.1 = 108 \text{ psi}$$

$$< F_b = 1.15(1000)(0.924) \quad OK$$

Check radial tension: The centerline radius using the 27-in. section will be

$$398.5 + 27/2 = 412 \text{ in.}$$

$$f_r = 3(12)(139,000)/(2 \times 412 \times 6.75 \times 27)$$

$$= 33.3 \text{ psi} > 15 \quad NG$$

Therefore, if this size of member is to be used, it must be mechanically reinforced for the entire value of radial tension. The procedure was illustrated in Example 8-11.

Check deflection: Since the roof is nearly flat, it will be fairly close to estimate the deflection by computing it for a simple beam of 40-ft span. (Because of the roof overhangs, the actual deflection will be slightly less than this amount.) Snow load deflection is

$$5(450)(40^4)(1728)/(384 \times 1,600,000 \times 11,070)$$

$$= 1.46 \text{ in.} < \ell/240 = 2.00 \text{ in.} \quad OK$$

Long-term dead load deflection will be about

$$1.5(1.46)(732 - 450)/450 \quad \text{or} \quad 1.37 \text{ in.}$$

8-11. TAPERED GLULAMS

Glulam beams are frequently tapered to provide a desired roof pitch. More often than not the bottom edge of the beam is horizontal and the top sloped. A design complication is that stress conditions are more complex than in a straight, parallel-edged beam. Consider a double-tapered beam such as shown by Fig. 8-21. Ignoring the local vertical compression that occurs at load points, the resultant stress at either surface, top or bottom, is parallel to that surface. It is a *principal* stress. Were the material isotropic, one might compute the principal stress and design the member merely by comparing the computed principal stress to an allowable stress value.

Wood is *not* isotropic, however, and the allowable stresses depend on direction relative to the wood grain. Consequently, an interaction equation must be used. The interaction equation used in designing tapered glulams is the spheroidal equation

$$\left(\frac{f_x}{F_x}\right)^2 + \left(\frac{f_y}{F_y}\right)^2 + \left(\frac{f_{xy}}{F_{xy}}\right)^2 \leq 1.0 \quad (8\text{-}9)$$

Terms in capital letters in the above equation are allowable stresses and those in lowercase are the computed or "actual" stresses. The definitions are:

f_x = computed bending stress at the extreme fiber of the side that slopes relative to the grain, obtained by the flexure formula and ignoring the taper; $f_x = M/S$.

f_y = the compressive or tensile stress perpendicular to the grain; it will be compressive if the taper cut is on the compression side of the member or tensile if the taper is on the tension side. The amount of f_y is given as

$$f_y = f_x \tan^2 \theta \qquad (8\text{-}10)$$

f_{xy} = a shear stress, computed as

$$f_{xy} = f_x \tan \theta \qquad (8\text{-}11)$$

F_x = allowable bending stress, with applicable modifications (size factor is not to be used)

F_y = allowable stress perpendicular to grain (tensile or compressive) with modifications

F_{xy} = allowable shear stress, with modifications

The definitions shown above for terms of Eq. 8-9 were obtained by elastic analysis of a small element of the tapered beam (11). They were derived assuming the material to be isotropic, but were later shown experimentally to apply to orthotropic materials also, including wood. Technically, the definitions shown above are limited to the specific case of a beam with a constant degree of taper. However, by replacing the tangent of the angle of slope by (dy/dx) in the expressions for f_{xy} and f_y, the definitions could be used for members in which the taper is not constant (12).

It is normal to think of shear stress as being highest at the neutral axis and zero at the extreme fiber, following the rule $f_v = VQ/Ib$. For a tapered beam, however, that is true only at the support of a simply supported beam or at the free end of a cantilever beam. Elsewhere, the location of maximum shear stress f_{xy}, is closer to the tapered face. Except for locations very close to the support, the maximum value of shear stress f_{xy} occurs at the extreme fiber of the tapered face. This effect is included in the definitions given above. For one particular condition—symmetrical double taper and a single midspan concentrated load—the shear stress distribution will be as shown by Fig. 8-22.

Critical Cross Section

To check a given design, all that is required is to compute values for all six factors in the interaction equation and then substitute them in the equation. If the sum of the squared terms does not exceed 1.0, the design is considered satisfactory. But at what cross section should these stresses be checked? Since the first term of the interaction equation usually is much larger than the others, the cross section that

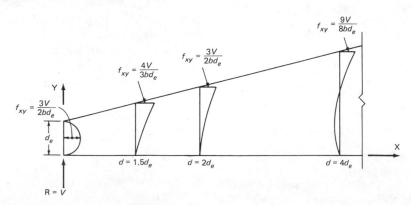

Fig. 8-22. Shear stress distribution in symmetrical double-tapered glulam with concentrated midspan load.

controls is usually the one for which the computed flexural stress, f_x, is the largest. Considering the double-tapered beam of Fig. 8-21, the flexural stress at any cross section is

$$f_x = M/S \qquad (8\text{-}12)$$

Under uniform load, the bending moment at any distance x from the left support is

$$M_x = w\ell x/2 - wx^2/2 \qquad (8\text{-}13)$$

Using d to designate the depth of the member at any point,

$$d = d_e + x(d_c - d_e)/(\ell/2) \qquad (8\text{-}14)$$

The flexural stress M/S at distance x from the support is f_x

$$= \frac{\dfrac{w\ell}{2}\left(\dfrac{\ell}{2}\right)\left(\dfrac{d - d_e}{d_c - d_e}\right) - \dfrac{w}{2}\left(\dfrac{\ell^2}{4}\right)\left(\dfrac{d - d_e}{d_c - d_e}\right)^2}{bd^2/6}$$

$$(8\text{-}15)$$

To find the maximum value of flexural stress at distance x from the support, the above expression is differentiated with respect to depth d and the derivative set equal to zero. After a little algebraic work it shows that, for a symmetrical double-tapered beam under uniform load, the maximum value of flexural stress will occur where the beam depth is

$$d = (d_e/d_c)(2d_c - d_e) \qquad (8\text{-}16)$$

For a symmetrical double-tapered beam with only a concentrated load at midspan, the maximum stresses will occur where the depth is twice the end depth.

The method shown above would be suitable for other shapes of beam or types of loading. In most cases, rather than deriving an equation as was done above, it is much easier merely to compute stresses and apply the interaction equation at several locations until the location giving the highest total is found. Such a successive-approximation solution is quite satisfactory.

Unusual Loadings or Shapes of Beam

In the years since the above method was derived, much progress has been made in application of the finite-element method. The authors suggest that problems of unusual loadings or shapes of glulam might be solved more reliably using finite-element analysis than by the above definitions. The nonisotropic nature of wood can easily be modeled in standard, available FEM programs. It would still be necessary, of course, to enter the resulting stress values into the interaction equation (Eq. 8-9) to determine whether or not the design is satisfactory.

Tapered-Beam Design Procedure

Normally, the slope of one face relative to the other is known. Usually the upper surface slopes to match the roof, while the lower face of the member is horizontal. Since shear will probably control at the end having the smallest depth, a good method involves finding the depth there first. A suggested sequence of steps is as follows:

1. Based on shear, choose a width and the minimum acceptable depth at the support. Since the depth there is not yet known, the designer must either calculate the shear, V, at the support or estimate the depth so as to be able to compute shear, V, at the critical section.
2. Based on the desired geometry (roof slope or architectural detail) determine a trial value for depth at the deepest point of the span.
3. Check the deflection for this beam. Both flexural and shear deflections should be considered and the total should not exceed the specified limit. (Shear deflection may, of course, prove to be a negligible part of the total.)
4. At the point of maximum flexural stress, if the location of that point is known, compute the stresses and use the interaction formula to determine whether the section is satisfactory. If the location of the maximum-stress section is not known, compute the stresses

Table 8-2.   Constant for Deflection of Tapered Beams with Uniform Loading.

Taper		C_{dt}
Double	When $0 < (d_c - d_e)/d_e \le 1.0$	$1 + (d_c - d_e)(0.66/d_e)$
Double	When $1.0 < (d_c - d_e)/d_e \le 3.0$	$1 + (d_c - d_e)(0.62/d_e)$
Single	When $0 < (d_c - d_e)/d_e \le 1.1$	$1 + (d_c - d_e)(0.46/d_e)$
Single	When $1.1 < (d_c - d_e)/d_e \le 2.0$	$1 + (d_c - d_e)(0.43/d_e)$

and use the interaction formula at sufficient, closely spaced locations along the span to be certain that the member is not over-stressed at any location.

5. Recycle as necessary to find a member that satisfies all strength and stiffness requirements.

6. Finally, determine the required camber.

Deflections of Tapered Beams

Of the above steps, the only one that may be laborious is computing the deflection. Tapered-beam deflections due to bending may be computed by any method that takes into account the variable moment of inertia—the moment area or conjugate beam methods, for example. Since this is a time-consuming procedure, an approximate method has been developed. In this method, deflection of the tapered beam is approximated as the deflection of an equivalent prismatic member. The depth of the equivalent member is determined by

$$d = C_{dt}d_e \qquad (8\text{-}17)$$

in which d_e is the end depth (small end, if single taper), and C_{dt} is a constant from Table 8-2. This approximation includes the effect of shear deflection (reference 7).

Example 8-13

Design a double-tapered glulam of western species combination 20F-V3 to support the roof of the building shown by Fig. 8-19. This type of roof support will be scheme I shown by Fig. 8-20. Assume that the required roof slope is $\frac{1}{2}$ in./ft. A curved glulam for this building (scheme II) was designed in Example 8-12. Because of the lesser roof slope, the total load is 725 (rather than 732) lb per horizontal foot.

Following the sequence of steps in the suggested procedure, above:

1. End shear, $V = 725(20) = 14,500$ lb
   Allowable $F_v = 1.15 \times 165 = 190$ psi
   Req $A = 1.5(14,500)/190 = 114.5$ in.^{2}
   At the end try $8\frac{3}{4} \times 13.5$ in. (area $= 118.1$ in.^{2})

2. For a roof slope of $\frac{1}{2}$: 12, the required center depth is

$$20(1/2) + 13.5 = 23.5 \text{ in.}$$

3. Check deflection:

   $(d_c - d_e)/d_e = (23.5 - 13.5)/13.5 = 0.74$
   So, $C_{dt} = 1 + (0.74)(0.66) = 1.49$
   Equiv $d = (13.5)(1.49) = 20.12$ in.
   Equiv $I = (8.75)(20.12^3)/12 = 5939$ in.^{4}

$$\Delta = \frac{5(725)(40^4)(1728)}{384(1,600,000)(5939)} = 4.39 \text{ in.}$$

This maximum deflection under total load is obviously much too high. To correct it we must increase either the depth or the width of the section, or both. The desired limit for total deflection is about $\ell/180$, or 2.67 in. After several trials, we find that a section measuring $8\frac{3}{4} \times 18.5$ in. at the support, and tapered to 28.5 in. deep at midspan will meet this limit. Deflection computations are as follows:
   Try $8\frac{3}{4} \times 18.5$ in. at support, tapered to 28.5 in. deep at midspan.

   $(d_c - d_e)/d_e = (28.5 - 18.5)/18.5 = 0.541$
   So, $C_{dt} = 1 + (0.541)(0.66) = 1.36$
   Equiv $d = (18.5)(1.36) = 25.16$ in.
   Equiv $I = (8.75)(25.16^3)/12 = 11,610$ in.^{4}
   $\Delta = 2.25$ in. < 2.67 in.; OK

4. Check interaction equation at point of maximum f_x. Maximum stress occurs where the beam depth (under uniform load) is

$$d = (18.5/28.5)(2 \times 28.5 - 18.5) = 25.0 \text{ in.}$$

By similar triangles, this depth occurs at 13 ft from the support, at which location the bending moment is

$$M = 725(20)(13) - 725(13)^2/2 = 127,200 \text{ ft-lb}$$

$$S = 8.75(25^2)/6 = 911 \text{ in.}^3$$

Terms for the interaction equation are:

$$f_x = M/S = 12(127,200)/911 = 1676 \text{ psi}$$

$$f_{xy} = 1676(10/240) = 70 \text{ psi}$$

$$f_y = 1676(10/240)^2 = 2.9 \text{ psi}$$

$$F_x = 2000(1.15) = 2300 \text{ psi}$$

$$F_{xy} = 165(1.15) = 190 \text{ psi}$$

$$F_y = 560 \text{ psi}$$

Note that F_y uses the allowable for *compression* perpendicular to the grain, since the tapered surface is on the compression side of the beam. Substituting in the interaction equation,

$$(1676/2300)^2 + (70/190)^2 + (2.9/560)^2$$

$$= 0.67 < 1.0 \quad OK$$

Check critical section using size factor:

$$C_F = (12/25)^{1/9} = 0.92$$

$$F_b' = (2000)(1.15)(0.92) = 2120 \text{ psi} > 1676 \text{ psi}$$

$$OK$$

5. Determine required camber. The total short-term deflection at midspan is computed above as 2.25 in. Of this total, $2.25(725 - 450)/725 = 0.85$ in. is dead-load deflection. The long-term dead-load deflection will be about 1.5 times this amount, or about 1.28 in. Camber of at least that much should be provided.

8-12. MEMBERS BOTH TAPERED AND CURVED

Glulam beams that are both double tapered and curved can be used for roof members. Usually, the top edge of these beams is formed by two straight lines meeting at the apex, and the bottom edge is curved. The result is a variable cross section, making design and analysis dif-

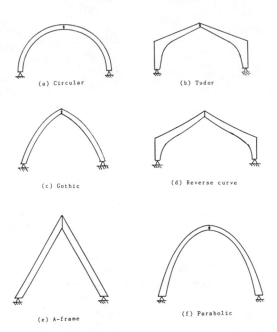

Fig. 8-23. Shapes used for three-hinged arches.

ficult. The complexity is somewhat reduced by a series of design aids (graphs and tables) available in the *Timber Construction Manual* (reference 7). From these aids, the designer can calculate trial values for end depth, centerline depth, and bottom slope, and can find approximate values of stress and deflection for the trial section.

A curved tapered member (such as the Tudor arch of Fig. 8-23) may be made up of two parts: a curved beam of constant section and a haunch that is attached mechanically.

8-13. THREE-HINGED ARCHES

The three-hinged arch is probably the most commonly used glulam member that is both tapered and curved. At first glance, a three-hinged arch looks like a rigid frame. If constructed in steel, a rigid frame would probably be used. The difficulty of making fixed-end connections in timber, however, makes the three-hinged arch a much better choice. Figure 8-23 shows several shapes that may be used. The advantage of one shape over another is mainly a question of aesthetics, although selecting a suitable shape can also affect economy.

Fig. 8-24. Three-hinged arch for Example 8-14.

The three-hinged arch is a statically determinate structure; therefore its analysis is quite simple. If the shape of the member centerlines is known and the location of all loads known, then the resulting bending moment, shear, and axial load can be computed at any cross section. With these known, the member can be checked at each cross section as a member subject to combined bending and axial load.

At the outset of the design process, however, *the designer does not yet know the shape of the member centerline.* Therefore, the design procedure is necessarily a trial-and-error procedure. An initial assumption must be made as to the shape. A procedure for doing this will be shown by Example 8-14. In many cases the architect will have a fairly good idea as to a shape that will satisfy aesthetic and functional requirements, and this shape may be used by the structural designer for the first trial analysis and design.

Design of a three-hinged arch will be slow and cumbersome, unless the designer follows an orderly procedure. The following is suggested as a sequence of design steps:

1. Determine the applied loads. This will include snow and wind loads, as well as roof dead loads. If these loads are brought to the arch member by roof decking, they will be uniformly distributed loads; if roof purlins are used, the loads will reach the arch member as concentrated loads. The weight of the arch member itself must be estimated (and later confirmed or revised).

2. Sketch *to scale* the estimated outline of the arch members, so that the location of the member centerlines can be estimated. At this step be certain that the limits on allowable curvature are not exceeded.

3. For each combination of loads to be considered, compute the arch reactions. These will include both horizontal and vertical reaction at each end, and also the force transferred from one segment of the arch to the other at the central "hinge." (The central connection may be actually pinned, or it may be merely a connection that is incapable of transferring bending moment from one segment to the other.)

4. Determine the area required at the base to resist shear. Setting the allowable longitudinal shear stress equal to the computed unit stress gives

$$\text{Req } A = bd = 1.5V/F_v \quad (8\text{-}18)$$

5. For the first trial, use a crown depth $d_c = 1.5b$ and tangent-point depth $d_t = 1.5d(\text{base})$.

Example 8-14

Select glulam sizes for the three-hinged arch shown by Fig. 8-24. The arches are spaced 12 ft c/c. The roof consists of 3-in. decking, with insulation and roofing, giving dead load of 17 psf of roof surface. Specified snow load for the locality is 30 psf of horizontal projection. Design wind speed is 80 mph and the importance factor is 1.0. Use UBC-specified load combinations.

Try western species combination 24F-V8, for which the basic allowable stresses are:

Bending, 2400 psi (either top or bottom)
Shear, 165 psi
Compression parallel to the grain, 1650 psi
E for bending, 1,800,000 psi
E for axial compression, 1,600,000 psi

UBC snow load reduction (Eq. 4-1) for each degree of slope above 20° is

$R_s = 30/40 - 1/2 = 0.25$ psf of horizontal projection

Reduced snow load $= 30 - 0.25(26.57 - 20)$
$= 28.4$ psf

Snow load on each arch is $12(28.4) = 341$ lb per horizontal foot.

For each square foot of horizontal projection, the roof area is $13.42/12$ ft.^{2}. Estimating the weight of the glulam to average 45 lb per horizontal foot, the total dead load is

$$12(17)(13.42/12) + 45$$
$$= 273 \text{ lb per horizontal ft}$$

UBC requires considering the following load cases:

Case I—$D + S$
 Case I-a, $D +$ full snow
 Case I-b, $D +$ unbalanced snow
Case II—$D + W$
Case III—$D + W + \frac{1}{2}S$
Case IV—$D + S + \frac{1}{2}W$

The designer feels that either case I-a or case I b will control the design, and considers those load cases first. The analysis for case I-a and a first trial design section are shown by Fig. 8-25a. Loads have already been computed above. From these, approximate arch reactions are calculated, based on the shape shown by Fig. 8-24. The approximate horizontal reaction at point A is then used to determine a tentative glulam size ($5\frac{1}{8} \times 9$ in.) based on shear at the column base. With this preliminary information, a shape for the first trial section is estimated, as shown in Fig. 8-25a for trial no. 1. The designer divided this trial section into seven segments (for the half arch) for checking stresses and for calculating deflections. The dimensions on the computation sheet are obtained by scaling. Exact dimensions will be needed, of course, for final construction drawings.

Now that a tentative shape for the arch centerline is available, corrected reactions are calculated, as shown in Fig. 8-25b. The vertical reactions are unchanged, but the horizontal reaction for case I-a reduces to 5477 lb. This corrected value of horizontal

reaction is then used in computing values of bending moment, shear, and axial load at each segment center. Figure 8-25b shows a table of these values (but not their computation) for case I-a.

To decide whether the trial section is satisfactory, interaction of combined bending and axial-load stresses is checked at each segment center. These calculations, tabulated in Fig. 8-25c, show trial section no. 1 to be satisfactory for combined stresses under load case I-a.

Finally, radial compressive stresses at the haunch are computed, and these too are satisfactory.

Having shown that the trial section is strong enough, the next step is to compute its deflection. Two deflections are of interest: vertical deflection at the hinge, and horizontal deflection at the top of the column. Unless a computer matrix analysis program is used, the easiest way to compute these deflections is by virtual work. A shortcut of this method permits a tabular solution in which actual writing of moment equations and integrating are avoided.

Figure 8-25d shows the deflection computation by this method for case I-a. Because for load case I-a both load and structure are symmetrical, only the left half of the arch need be considered. (For any unsymmetrical case, however, the entire arch would have to be considered: i.e., 14 segments would be used instead of only seven.)

Example 8-14 ends at this point after showing that the first trial section is satisfactory for load case I-a, but the design is by no means complete. All the other load cases (unless it can easily be seen that they will not control) must be checked by a similar procedure. If the trial section is found unsatisfactory for any of these cases, the trial section must be modified and analysis for that particular loading repeated. Because the procedure is lengthy, tables of standard solutions have been developed (see reference 7).

8-14. GLULAM COLUMNS

Column equations given in Chapter 7 are intended for use with sawn timbers. These equations may be used also for glued laminated columns. However, the variability of properties in glulams is less than found in sawn timbers, so the AITC suggests for glued laminated columns using Eqs. 8-19 and 8-20, below. Equation 8-19 defines K, the ℓ/d ratio beyond which the column is to be considered "long." Equation 8-20 gives the allowable average stress, P/A, for a long column.

$$K = 0.792\sqrt{E/F_c} \qquad (8\text{-}19)$$

Fig. 8-25a. Designer's calculations for Example 8-14.

$$F'_c = 0.418E/(\ell_e/d)^2 \qquad (8\text{-}20)$$

Appendix G of the NDS also suggests these equations for columns in which the coefficient of variation for modulus of elasticity is 0.11 or less. The allowable stress for intermediate-length columns is given by the same equation as for sawn-timber columns.

REFERENCES

1. *National Design Specification for Wood Construction*, National Forest Products Association, Washington, DC, 1986.
2. *Structural Glued Laminated Timber*, ANSI/AITC A190.1-1983, American Institute of Timber Construction, Vancouver, WA, 1983.
3. Johnson, J. W., ''Edge-Gluing Douglas Fir Lumber

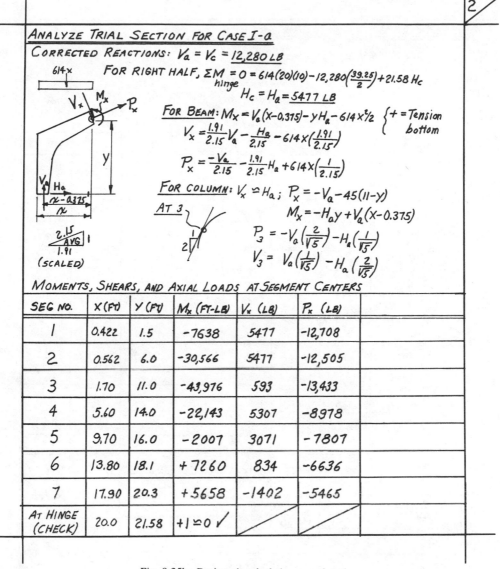

ANALYZE TRIAL SECTION FOR CASE I-a

CORRECTED REACTIONS: $V_a = V_c = 12,280\ LB$

FOR RIGHT HALF, $\Sigma M = 0 = 614(20)(10) - 12,280\left(\frac{39.25}{2}\right) + 21.58\ H_c$
hinge

$$H_c = H_a = 5477\ LB$$

FOR BEAM: $M_x = V_a(x - 0.375) - yH_a - 614\ x^2/2$ { + = Tension bottom

$$V_x = \frac{1.91}{2.15} V_a - \frac{H_a}{2.15} - 614 x\left(\frac{1.91}{2.15}\right)$$

$$P_x = \frac{-V_a}{2.15} - \frac{1.91}{2.15} H_a + 614 x\left(\frac{1}{2.15}\right)$$

FOR COLUMN: $V_x \simeq H_a$; $P_x = -V_a - 45(11 - y)$

$$M_x = -H_a y + V_a(x - 0.375)$$

AT 3

$$P_3 = -V_a\left(\frac{2}{\sqrt{5}}\right) - H_a\left(\frac{1}{\sqrt{5}}\right)$$

$$V_3 = V_a\left(\frac{1}{\sqrt{5}}\right) - H_a\left(\frac{2}{\sqrt{5}}\right)$$

MOMENTS, SHEARS, AND AXIAL LOADS AT SEGMENT CENTERS

SEG NO.	X (FT)	Y (FT)	M_x (FT-LB)	V_x (LB)	P_x (LB)	
1	0.422	1.5	-7638	5477	-12,708	
2	0.562	6.0	-30,566	5477	-12,505	
3	1.70	11.0	-43,976	593	-13,433	
4	5.60	14.0	-22,143	5307	-8978	
5	9.70	16.0	-2007	3071	-7807	
6	13.80	18.1	+7260	834	-6636	
7	17.90	20.3	+5658	-1402	-5465	
AT HINGE (CHECK)	20.0	21.58	+1 ≃ 0 ✓			

Fig. 8-25b. Designer's calculations, continued.

to Improve Strength Properties," *Forest Products Journal*, **28**(11), Nov. 1978, pp. 37–41.

4. *Design Standard Specifications for Structural Glued Laminated Timber of Softwood Species*, AITC 117-84, American Institute of Timber Construction, Vancouver, WA, 1984.

5. *Wood Handbook*, U.S. Department of Agriculture, Washington, DC, 1974.

6. Eby, R. E., "Proof Loading of Finger Joints for Glulam Timber," *Forest Products Journal*, **31**(1), Jan. 1981, pp. 37–41.

7. American Institute of Timber Construction, *Timber Construction Manual*, 3rd ed., Wiley, 1986.

8. *Uniform Building Code*, International Conference of Building Officials, Whittier, CA, 1985.

9. *Manual of Steel Construction*, 8th ed., American Institute of Steel Construction, Chicago, 1980.

10. *Building Code Requirements for Reinforced Concrete*, ACI 318-83, American Concrete Institute, Detroit, MI, 1983.

11. *Wood Structures, a Design Guide and Commentary*, ASCE Structural Division, American Society of Civil Engineers, New York, 1975.

12. Maki, A., and E. Kuenzi, "Deflection and Stresses of Tapered Wood Beams," Forest Products Laboratory Research Paper FPL-34, 1965.

3

CHECK STRESSES – LOAD CASE I-a

INTERACTION EQ: $\dfrac{f_c}{F_c'} + \dfrac{f_b}{F_{bx}'} \leqslant 1.0$

LATERAL SUPPORTS: COLUMN – BASE, ②, ③.

BEAM – HORIZONTALLY (BY DECKING) FULL LENGTH.

$K = 0.671\sqrt{1.6(10^6)/1.15(1650)} = 19.48$

LOCATION	1	2	3	4	5	6	7
DEPTH (in.) d	10.13	13.5	28	13.5	12.22	10.94	9.66
AREA (in.^{2})	51.9	69.2	143.5	69.2	62.6	56.1	49.5
I (in.^{4})	444	1051	9375	1051	779	559	385
S (in.^{3})	87.7	156	670	156	128	102	79.7
UNBRACED ℓ	72"	72"	—	—	—	—	—
ℓ/d	$14 < K$	$14 < K$	—	—	—	—	—
F_c'	1729	1729	1.15(1650) = 1898	1898	1898	1898	1898
C_F	—	0.987	0.910	0.987	0.998	—	—
$F_b' = 1.15(2400)C_F$	2760	2720	2512	2720	2750	2760	2760
$f_c = P/A$ psi	245	181	94	130	125	118	110
$f_b = 12M/S$ psi	1045	2351	788	1703	188	854	852
f_c/F_c'	0.142	0.105	0.050	0.068	0.066	0.062	0.058
f_b/F_b'	0.379	0.864	0.314	0.626	0.068	0.309	0.309
SUM	0.521	0.969	0.364	0.694	0.134	0.371	0.367
OK ?	✓	✓	✓	✓	✓	✓	✓

RADIAL COMPRESSION AT HAUNCH (PT. 3): $R_c \approx 9.33 + \frac{1}{2}(28/12) = 10.5'$

$f_r = \dfrac{3M}{2R_c\,bd} = \dfrac{3(12 \times 43,976)}{2(12 \times 10.5)(5,125 \times 28)} = 43.8\,psi < 650 \quad \underline{OK}$

Fig. 8-25c. Designer's calculations, continued.

PROBLEMS

Note: Unless otherwise noted, beams have full lateral support, dead load mentioned does not include weight of the beam, and live load is of normal duration.

8-1. A glulam beam is to carry live load of 700 lb/ft and superimposed dead load of 200 lb/ft on a 24-ft simple span. Select a section using combination 24F-V3 western species. Consider flexure and shear. Compare to a sawn timber of the best grade D. fir–larch.

8-2. Use the transformed-section method to calculate the allowable bending moment for a laterally supported glulam section made up of 2 × 8s, as follows, and trimmed to 6.75 in. wide after gluing: top two laminations, Select

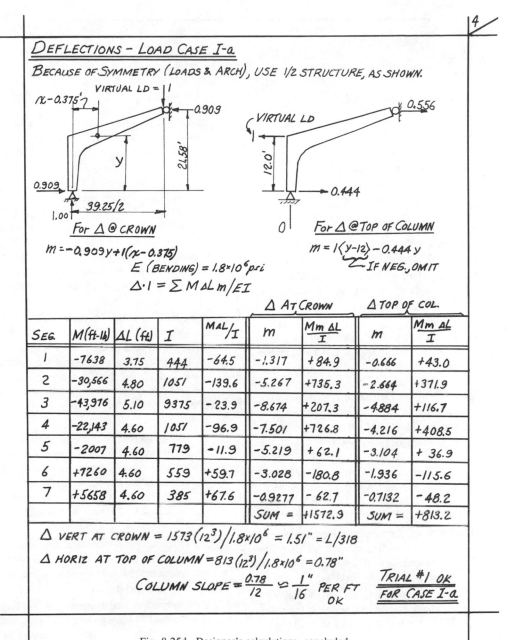

DEFLECTIONS – LOAD CASE I-a

BECAUSE OF SYMMETRY (LOADS & ARCH), USE 1/2 STRUCTURE, AS SHOWN.

For Δ @ CROWN

$$m = -0.909y + 1(x - 0.375)$$

$$E\ (BENDING) = 1.8 \times 10^6\ psi$$

$$\Delta \cdot 1 = \sum M \Delta L\, m / EI$$

For Δ @ TOP OF COLUMN

$$m = 1(y - 12) - 0.444y$$

IF NEG., OMIT

SEG.	M (ft-lb)	ΔL (ft)	I	MΔL/I	Δ AT CROWN m	Mm ΔL / I	Δ TOP OF COL. m	Mm ΔL / I
1	−7638	3.75	444	−64.5	−1.317	+84.9	−0.666	+43.0
2	−30,566	4.80	1051	−139.6	−5.267	+735.3	−2.664	+371.9
3	−43,976	5.10	9375	−23.9	−8.674	+207.3	−4.884	+116.7
4	−22,143	4.60	1051	−96.9	−7.501	+726.8	−4.216	+408.5
5	−2007	4.60	779	−11.9	−5.219	+62.1	−3.104	+36.9
6	+7260	4.60	559	+59.7	−3.028	−180.8	−1.936	−115.6
7	+5658	4.60	385	+67.6	−0.9277	−62.7	−0.7132	−48.2
					SUM =	+1572.9	SUM =	+813.2

$$\Delta\ VERT\ AT\ CROWN = 1573(12^3)/1.8 \times 10^6 = 1.51'' = L/318$$

$$\Delta\ HORIZ\ AT\ TOP\ OF\ COLUMN = 813(12^3)/1.8 \times 10^6 = 0.78''$$

$$COLUMN\ SLOPE = \frac{0.78}{12} \backsimeq \frac{1''}{16}\ PER\ FT\ OK$$

TRIAL #1 OK
FOR CASE I-a

Fig. 8-25d. Designer's calculations, concluded.

Structural southern pine; center six laminations, No. 2 D. fir–larch; bottom two laminations, Select Structural California redwood.

8-3. Glulam beams of 30-ft span are spaced 14 ft apart. Dead load is 8 psf of roof. Live load is 25 psf. Deflection limits are $\ell/360$ for live load alone and $\ell/240$ for total load. Consider long-term effect for dead load. Choose a section using combination 24F-V3 western species.

8-4. Choose a combination 20F-V4 western species glulam to carry vertical permanent load of 1.0 k/ft total (including beam weight) on a 20-ft simple span. Limit long-term total deflection to $\ell/360$: (a) with laminations horizontal; (b) with laminations vertical.

8-5. What is the allowable bending moment (strong-axis bending) for a $6\frac{3}{4}$- × 24-in. glulam of southern pine combination 22F-V4, subject to permanent uniform load and lat-

Fig. 8-26. Cantilever system for Problem 8-9.

erally supported only at the ends of a 30-ft simple span?

8-6. Select a 22F-V4 western species glulam to carry a 12-k normal-duration concentrated load at the center of a 28-ft simple span, with lateral support at the ends and at midspan only.

8-7. Try a wider section for the biaxially loaded beam of Example 8-7. Choose the most economical depth for that width section.

8-8. Choose a combination 22F-E1 southern pine glulam section, subject simultaneously to wind bending moments $M_x = 35.0$ ft-k and $M_y = 9.7$ ft-k.

8-9. Figure 8-26 shows a three-span cantilever beam system. Dead load is 0.4 k/ft, including the beam weight. Live load is 0.8 k/ft, placed in either a full-loading or a partial-loading pattern. Deflection limit under long-term loading is 1.5 in. Choose 20F-V4 western species glulam sections.

8-10. For a similar system to that of Fig. 8-26, but with spans of 20 ft, 25 ft, and 20 ft, choose the best location for the hinges and select glulams of the most appropriate western species 20F-V combination. Load is 1.6 k/ft total, normal duration, present all along the beams.

8-11. Rework the glulam design of Example 8-11, except that the bending moment is 24 ft-k. Use lag screws and check net section, if mechanical connectors are required.

8-12. A curved glulam beam with $\frac{3}{4}$-in. laminations, 24 in. deep by $8\frac{3}{4}$ in. wide, combination 20F-V3 western species, has an inside radius

of 15 ft 0 in. Normal-duration bending moment of 100 ft-k causes tension on the concave side. Is the beam OK in flexure? Is it OK for radial stress? Show proof.

8-13. Select a glulam section for the curved beam of Fig. 8-27. Design mechanical connectors if they are needed. Loads are: dead, 200 lb/ft of horizontal projection plus beam weight; snow, 600 lb/ft of horizontal projection. Use western species combination 22F-E1. What bearing length is required at the reaction? (*Note:* The bearing stress is not perpendicular to the grain.)

8-14. A western species combination 22F-V7 double-tapered glulam of 60-ft span is $8\frac{3}{4}$ in. wide, 30 in. deep at the ends, and 55 in. deep at the center. The controlling load includes snow and is 650 lb/ft, including the beam weight. Check the effect of combined stresses at distance $x = 15$ ft and $x = 20$ ft from the end.

8-15. A double-tapered glulam is to span 50 ft and have a top-surface slope of $\frac{1}{2}$ in. per ft. Beams are on 12-ft centers. Loads are: snow, 40 psf of horizontal projection; dead, 26 psf of roof plus beam weight. Use southern pine combination 22F-E1. Choose width and depths. What end-bearing length is needed?

8-16. Prove for a double-tapered glulam with a single concentrated load at midspan that the maximum flexural stresses occur where depth, d, is twice the end depth, d_e.

8-17. Design a wood roof system for a building 70 ft wide by 150 ft long. Use $\frac{1}{2}$-in. plywood on joists (at 16 in. c/c) that are parallel to the building length. Choose the number and spacing of beams. Either one row or two rows of interior columns will be acceptable. The roof is nearly flat. Sufficient slope to permit drainage and prevent ponding can be provided by varying the vertical position of joist hangers. Use glulams wherever possible. Limit long-term deflection to $\ell/200$. Snow load is 40 psf. Insulation and roofing weigh 12 psf. Suspended ceiling weighs 11 psf.

8-18. Design the glulam beam for scheme V of Fig. 8-20. Use a roof slope of 1 in. per ft.

Fig. 8-27. Curved beam for Problem 8-13.

GLULAMS @ 20′ C/C

10′

WALL

COLUMN

WALL

3′   30′   40′   40′   30′   3′

Fig. 8-28. Building for Problem 8-19.

8-19. For the building shown in cross section by Fig. 8-28, choose a size of 2-in. nominal roof joist. Then

(a) Choose glulams for the side-span beams.

(b) Choose glulams for the curved, center-span beams.

(c) Design the glulam columns.

Snow load is 30 psf. Roof is $\frac{1}{2}$-in. plywood on joists at 16-in. centers. Roofing, insulation, and ceiling weigh 17 psf. Exterior walls and diaphragm action by the plywood roof sheathing provide lateral bracing for the top end of the columns.

8-20. For the three-hinged arch of Example 8-14, analyze and check for load case I-b. If the trial section for Example 8-14 is found unsatisfactory, revise it and recalculate for case I-b. (Note that all 14 segments must be considered for this unsymmetrical load case.)

8-21. Same as Problem 8-20, but for load case II.

8-22. Choose a Douglas fir glulam, combination 3, for a column 19 ft long to support an axial load of 100 k.

8-23. Same as Problem 8-22, but with the load applied at the face of the column (i.e., eccentrically).

9

Bolts, Timber Connectors, and Special Weldments

Bolts for resisting direct shear or tension loads were covered in Chapter 5. This chapter introduces the use of bolts under more complex loading conditions.

A wide variety of metal connection devices makes it possible to use wood members more effectively and in situations where formerly only structural steel members were practical. Examples include beam and joist hangers, special welded steel assemblies, shear plates, split rings, and spiked grids. Many of these will be discussed in this chapter.

Some proprietary types of connector are selected by *catalog engineering*, that is, merely using allowable loads and dimension limits provided by the manufacturer's literature. These will not be covered in detail here. For other types, however, design either is entirely the responsibility of the structural designer or requires the designer's expertise to choose the connector number and arrangement.

The list of available connectors is long, but it will continue to grow, being limited only by the ingenuity of designers and the desire of manufacturers to meet new needs and develop new market opportunities.

9-1. BOLT GROUPS SUBJECT TO MOMENT

In wood construction, it is preferable to avoid applying bending moment to groups of fasteners, but occasionally it cannot be avoided. An example would be a bracket connection as shown by Fig. 9-2. To develop a method of analysis, refer to Fig. 9-3.

First, replace the eccentric load in Fig. 9-3 by an equal, centrally applied load and a moment equal to the original load times its eccentricity ($M = Fe$). The moment, acting alone, rotates the side member about the centroid of the fastener group. As it rotates, the side member pushes each bolt with a force, T. Force T applied to each bolt by the moment is in a direction normal to the radius, r, from that bolt to the center of rotation (Fig. 9-3b). Assuming that the bracket does not deform severely, the distance each bolt is moved by the moment is proportional to radius, r. Forces T are proportional to the distance moved; therefore, forces T are proportional to radii, r. Force T required to move each bolt is also proportional to the stiffness of the material that deforms to permit the bolt to move. This stiffness cannot be defined precisely, but is assumed to be proportional to the allowable shear load, S, for the bolt. This is a reasonable approximation since allowable bolt loads are based on slip rather than ultimate strength.

Putting the above in equation form, the ratio of forces, T, for two bolts is

$$\frac{T_1}{T_2} = \frac{r_1 \times S_1}{r_2 \times S_2} \tag{a}$$

or

$$T_2 = T_1(r_2 \times S_2)/(r_1 \times S_1) \tag{b}$$

The sum of the moments of all the T-forces about the center of rotation is equal to the applied moment Fe. Thus

Fig. 9-1. Hinged connection for glulam arch bridge. (Photograph by Mike Douglas.)

$$M = T_1 r_1 + T_2 r_2 + \cdots + T_n r_n \quad \text{(c)}$$

Next, substitute an expression like (b) for all except the first term of (c). Multiply the first term of (c) by $r_1 S_1 / r_1 S_1$. Then, collecting terms, expression (c) becomes

$$M = \frac{T_1}{r_1 S_1} \sum r_i^2 S_i \quad \text{(9-1)}$$

The summation term in this equation can be referred to as the polar moment of inertia of the strengths, S, about the center of rotation. Calling this summation J, Eq. 9-1 becomes

$$T_1 = M r_1 S_1 / J \quad \text{(9-2)}$$

Fig. 9-2. Bolted bracket.

Had the above been derived by writing expression (b) in terms of any other bolt force T, say T_i, then a more general form of the equation would result, as follows:

$$T_i = M r_i S_i / J \quad \text{(9-3)}$$

Each force, T, computed by Eq. 9-3 will be in the direction shown by Fig. 9-3b, normal to the distance, r, for each bolt. These are the components of bolt force due to $M = Fe$, acting alone. The central load, F, also must be accounted for, and its effect is a downward force, $V = F/n$, on each bolt. To find the total force on any bolt, add V and T for that bolt *vectorially*.

Since the strengths, S, are in both the numerator and denominator of Eq. 9-3, it is often simpler to use *relative* rather than actual strengths of the bolts in various directions. This is illustrated in Example 9-1.

The method shown above is not precise. For one thing, it uses a strength, S, based on the direction of force, T, due to moment only; whereas strength, S, in the direction of the resultant is not considered. Looking to the future, it may be that an ultimate strength method, similar to that now available for structural steel design, may be developed for wood.

Example 9-1

Figure 9-4 shows a bracket, consisting of two steel plates, connected by six $\frac{5}{8}$-in. bolts to a 6×12 D. fir–larch column. Is the connection satisfactory for the 3-k load shown?

This is a double-shear connection with steel side members. Length of bolt in the main member is $5\frac{1}{2}$ in. Basic allowable bolt loads from Table C-12 are $P = 1990$ lb and $Q = 1410$ lb. The eccentric load of 3 k is first replaced by its equivalent—a 3-k load acting at the centroid of the bolt group, and a clockwise moment, $M = 3(9) = 27$ in.-k.

Considering the effect of moment alone, the direction of forces, T, will be parallel to the grain for bolts nos. 2 and 5. For bolts 1, 3, 4, and 6, force T will be at an angle

$$\theta = \tan^{-1}(4/3) = 53.1°$$

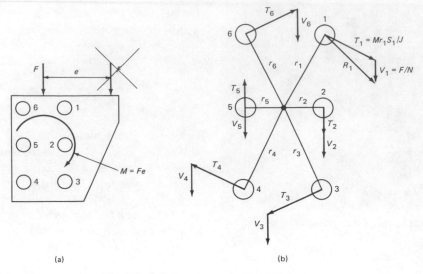

Fig. 9-3. Bolt forces on connection with moment.

By Hankinson's equation, and ignoring the increase of allowable P when steel side plates are used, the allowable force T_1 is

$$T_1 = \frac{1990 \times 1410}{1990 \sin^2 53.1° + 1410 \cos^2 53.1°}$$

$$= 1576 \text{ lb}$$

Rather than actual strengths, we will use their *relative* values. Letting $S_1 = 1.00$, the relative strengths for bolts 2 and 5 will each be $1990/1576$, or 1.263. Calculate the polar moment of inertia, J, of the relative strengths as tabulated.

Bolt	S	r (in.)	Sr^2
1	1.00	5	25.00
2	1.263	3	11.37
3	1.00	5	25.00
4	1.00	5	25.00
5	1.263	3	11.37
6	1.00	5	25.00

$$J = 122.74 \text{ bolts} \times \text{in.}^2$$

Bolt forces due to moment alone: Bolts 1, 3, 4, and 6 are alike:

$$T_1 = Mr_1S_1/J = 27(5)(1.0)/122.74$$
$$= 1.100 \text{ k/bolt}$$

(a)

(b)

Fig. 9-4. Connection for Example 9-1.

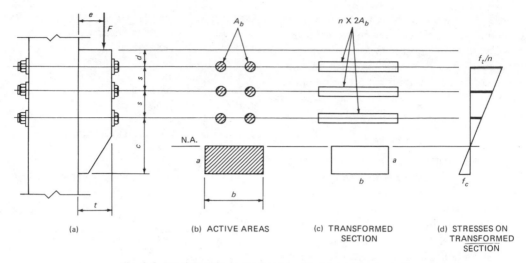

Fig. 9-5. Bracket; bolts in shear and tension (f_t = stress in steel).

T_2 (or T_5) = 27(3)(1.263)/122.74
      = 0.833 k/bolt

Due to direct vertical load alone:

Each bolt receives 3.0/6 = 0.500 k

Total forces are found by adding vectorially as shown by Fig. 9-3b for bolt nos. 1 and 2.
Resultant forces are:

R_1 or R_3 = 1.456 k (these are the most highly loaded bolts)
R_2 = 1.333 k

The allowable force on bolts 1 and 3 can be redetermined now that the direction of the resultant is known (37.2° to the grain).

Allowable $R_1 = \dfrac{1990 \times 1410}{1990 \sin^2 37.2° + 1410 \cos^2 37.2°}$

      = 1730 lb > 1456 lb   *OK*

For bolt 2, the allowable is

      1990 lb > 1333 lb   *OK*

As a refinement, one could recycle, using the above allowables to compute new relative strengths, S, and then solve for a revised J and new resultant bolt loads. However, the additional effort would bring about little change in the answers; further, the method itself is only approximate, so recycling would probably not improve the answer significantly.

9-2. CONNECTIONS WITH MOMENT— BOLTS IN TENSION AND SHEAR

Figure 9-5 shows a bracket consisting of a wood block bolted to the face of a wood column. Again, the actual load is replaced for analysis by an equal load at the shear plane and a moment $M = Fe$. The effects of the vertical load and of the moment are computed separately. Assume that each bolt receives an equal portion of the shear, V. The limit for V is the allowable bolt load shown by Table C-12, modified as necessary.

Figure 9-5b shows a cross section at the shear plane of the bolts. The shaded area at the bottom is the area over which the bracket is pressing against the column. The moment causes tension in the bolts above the compressive area. Two different materials are involved: The steel bolts are stretching, and the wood of the block and column are being shortened (horizontally) by the compressive stress. To analyze, the actual cross section is replaced by a transformed section, shown by Fig. 9-5c. In the transformed section, each square inch of steel is replaced by n times as much area of imaginary wood, n being the ratio of the modulus of elasticity of steel to that of wood perpendicular to the grain. Herein lies the greatest error of the method, since E perpendicular to the grain is normally not known. Fortunately, the procedure is not extremely sensitive to the value of n.

To find the centroid, which is the neutral axis for bending, merely equate the moment of the area above the neutral axis to the moment of the area below. Knowing the location of the centroid (dimension a on Fig. 9-5c), compute the moment of inertia of the transformed section, and then use the flexure formula to find stresses at any level, as shown in Fig. 9-5d. Notice that the stress value, My/I, computed at each bolt location must be multiplied by n to convert it from a transformed-section stress to a steel stress. The steel stress, multiplied by the area of the bolt shank, is the tensile force in one bolt at that level.

Knowing both the shear stress and the tensile stress in the bolts, how does one decide whether or not the combined stresses are acceptable? Wood design specifications (such as the NDS) do not answer this. The authors suggest, however, that the interaction equations given by the AISC Specification (1) for bolts in combined tension and shear could be used.

Example 9-2

For the bracket of Fig. 9-5a, assume that: load $F = 7.5$ k; eccentricity $e = 3$ in.; six $\frac{3}{4}$-in.-diameter bolts are used; bracket and column are both No. 1 D. fir-larch; dimensions are: $b = 9.5$ in., $c = 10$ in., $d = 2$ in., $s = 4$ in., and $t = 5\frac{1}{2}$ in.; and the column is a 10×10. What are the forces on the most highly stressed bolts? Is the connection satisfactory?

The 7.5-k vertical load is assumed to be shared equally by the six bolts, so that each receives a vertical shear of $7.5/6 = 1.25$ k. The allowable shear, P, per bolt is one-half the value shown by Table C-12, or

$$2860/2 = 1430 \text{ lb}$$

so the bolt group is satisfactory for the vertical load alone, located at the shear plane.

Modulus of elasticity perpendicular to the grain for structural woods varies with species, and values used range from about 4% to about 11% of the modulus parallel to the grain. Using E parallel to the grain of 1,600,000 psi (Table B-3), the transverse E is somewhere between 64,000 and 176,000 psi. Try an intermediate value, say 120,000 psi. Then ratio $n = 29,000,000/120,000 = 242$. For the transformed section, replace each pair of bolts by $2n$ times the area of the bolt shank, or $2(242)(0.442) = 214$

in.², represented by the broad areas of Fig. 9-5c. Dimension a is unknown. Distances of the bolt areas above the neutral axis are:

Top bolts $(18 \text{ in.} - a)$
Next pair down $(14 \text{ in.} - a)$
Bottom pair $(10 \text{ in.} - a)$

The total moment of the areas above the neutral axis, about the neutral axis (N.A.) is

$$214[(18 - a) + (14 - a) + (10 - a)]$$
$$= 8988 - 642a$$

The moment of the area below, about the neutral axis, is

$$9.5a(a/2)$$

Equating, $8988 - 642a = 4.75a^2$.

The solution to this quadratic equation is $a = 12.79$ in. This puts the neutral axis above the lowest pair of bolts, so the assumption used in the above equations was incorrect. Repeat the calculation, assuming the neutral axis to lie between the bottom and middle pairs of bolts. Then the equation of moments of areas about the neutral axis becomes

$$214[(18 - a + 14 - a)] = 4.75a^2$$

and the solution is $a = 13.87$ in., which is consistent with the assumption that a is between 10 and 14 in. The moment of inertia of the transformed section (about the neutral axis) is

$$9.5(13.87^3)/3 = \quad 8450$$
$$214(14 - 13.87)^2 = \quad 4$$
$$214(18 - 13.87)^2 = \quad \underline{3650}$$
$$I_t = 12,104 \text{ in.}^4$$

Stress on the top part of the transformed area is Mc/I_t, or

$$7500(3)(18 - 13.87)/12,104 = 7.68 \text{ psi}$$

This is a level of stress for the imaginary wood that was substituted for steel at the location of the top bolts. Stress in the steel is n times as large, or

$$f_s = 7.68 \times 242 = 1860 \text{ psi (tension)}$$

The AISC specification (1) gives an allowable ten-

Fig. 9-6. Connection reinforced by stitch bolts (shows two alternate positions for stitch bolts).

sile stress for A-307 bolts, when combined with shear, as

$$26 \text{ ksi} - 1.8f_v, \text{ but not over 20 ksi}$$

So the allowable tensile stress is

$$26,000 - 1.8 \times 1250/0.442 = 20,910 \text{ psi}$$

Use 20,000 max.

The computed tensile stress is much less than this allowable, so the bolts are satisfactory for combined shear and tension.

Using the same transformed section, we may estimate the compressive stress perpendicular to the grain at the bottom edge of the bracket as

$$f_{c\perp} = Ma/I_t = 22,500(13.87/12,104) = 26 \text{ psi}$$

which is well below the allowable shown by Table B-3, regardless of grade.

9-3. STITCH BOLTS

Bolts under load tend to split the wood they connect. Splitting may occur because of a wedging action as the bolts are pushed parallel to the grain, or it may result because the bolt-bearing pressures normal to the grain cause tensile stress in that direction. The allowable loads for bolts are established knowing that this tendency exists. Nevertheless, if the designer is concerned about the potential for splitting under the action of the bolts, or of splitting by other causes (drying shrinkage, for example), which might reduce the effectiveness of the bolts, then *stitch* bolts may be helpful.

Figure 9-6 shows a tension splice with metal side plates. Failure would likely involve splitting and formation of wood plugs sheared from the wood members. Vertical stitch bolts would help prevent that failure.

There is no doubt that the stitch bolts improve the ultimate strength of the connection. However, the authors are not aware of any quantitative information that would help determine appropriate increased allowable loads for the main bolts.

9-4. JOIST AND BEAM HANGERS AND FRAMING ANCHORS

These are, by far, the most commonly used of modern timber connectors. Figure 9-7 shows a sampling of the variety produced by one manufacturer.

Figure 9-8 shows a joist hanger and its typical application. In this type of hanger, the joist transfers its reaction to the hanger through bearing, but load transfer to the header is by shear in special nails through the hanger flanges. In other types of hanger, the outstanding legs project up and over to bear on the top surface of the header (the double member in Fig. 9-8), in which case the vertical element of the hanger acts as a tension member carrying the entire reaction. Standard manufactured hangers for 2-in. nominal members are made of sheet metal, often galvanized. Those for larger members are made from folded or welded steel plates. Catalog information normally gives

Fig. 9-7. A variety of commercial sheet metal joist hangers. (Photograph by authors.)

the structural designer all information needed to select the proper type and size of hanger.

Framing anchors are especially useful in light-frame construction. Used as in the top illustration of Fig. 9-9a, they perform the same function as joist hangers. In the top picture of Fig. 9-9b, they make a connection that otherwise would require toe nailing. Wind may cause severe uplift on roof rafters. Framing anchors often can transfer the required downward reaction from the rafter to the structure below, whereas toe nailing might be entirely inadequate. In the bottom illustration of Fig. 9-9c, framing anchors connect the wall studs to the bottom plate. Here, too, framing anchors replace toe nailing. The roof uplift must be transferred somehow all the way down into the foundation, so that uplift reaction (or a portion

of it) must also be transferred from stud to lower plate. Again, framing anchors do this job better than toe nailing.

A structural designer ordinarily would not be involved in designing these connectors unless working for a hanger manufacturer. In that case, the design process would be largely a matter of testing to determine allowable loads for new hangers or anchors. However, designers of wood structures must often determine *whether* to use framing anchors and, if so, the number and type required.

Unless shown otherwise, the usual load-duration modifications apply to the allowable load tables shown by the manufacturers' literature. Caution: The designer must make sure that using a duration factor larger than 1.0 does not overstress the metal parts of the connection. Load-duration factors do not apply to design in metals.

9-5. SPECIAL WELDMENTS

Figure 9-10 shows a few of many standard types of weldment. The two types of saddle shown are used in cantilever construction to transfer the reaction of a suspended span to the overhanging member. A tension tie is needed when expected longitudinal load as well as ver-

Fig. 9-8. Light-gauge metal joist hanger. (Courtesy of TECO Products & Testing Corp.)

Fig. 9-9. Framing anchors. (Courtesy of TECO Products & Testing Corp.)

(a)

(b)

(c)

light-gauge sheet metal as well as structural steel.) Figure 9-10g shows the connection of wood beams to the top of a wood column. The two arch-base details of Figs. 9-10e and f represent extremes for pinned-end arch construction. The first type would be used if stress conditions in the arch made it absolutely necessary to allow free end rotation. The second type would be used if some small amount of end moment could be tolerated. By no means, however, is it a fixed-end connection. Since the bolts through the arch are spaced as close together as possible ("minimum spacing") and the hole diameter is slightly larger than the bolt diameter, end rotation of the arch occurs with negligible end moment.

Finally, Fig. 9-10h shows a typical anchorage for a wood column to a concrete foundation. In another version of this connection, the U-strap has a hole in its center, and the entire assembly is placed over a single anchor bolt that is embedded in the foundation.

Most of the details in Fig. 9-10 are from the AITC *Timber Construction Manual* (2), which shows a large variety of other standard details. Hardware items of those details are not routinely stocked, but in many cases manufacturers can produce them from standard designs. If the situation is unusual in any way, however, the designer of the wood structure will probably have to design the special connector required. Figure 9-11 shows an example, which was designed for an architect by his structural engineering consultant. Figure 9-12 shows another special weldment.

The design of special weldments involves both timber design and structural steel design. The process can be broken into the following requirements:

1. Protect the wood members. This usually means providing either adequate bearing length or adequate spacing, end distance, and edge distance for bolts to carry the beam reaction. The member may be shorter than expected, and the members to which it connects may not be precisely located. So construction tolerance must be considered, providing added bearing length or (if the reaction is carried by bolts) added end distances.

tical must be carried from one member to the other. Thus, the detail in Fig. 9-10b would not be suitable in seismic zones. Ordinarily, and in the absence of the need to resist calculated longitudinal loads, we would prefer not to restrain to prevent slight longitudinal shrinkage. In that case, the simple saddle of Fig. 9-10b would have enough flexibility to permit some shrinkage. In either case, the vertical load causes horizontal forces in parts of the saddle. In the simple saddle of Fig. 9-10b, for example, forces F are each equal to $R(e/d)$.

Figures 9-10c and d show two of many types of beam hanger. (Beam hangers are made in

(a) SADDLE WITH TENSION TIE

(b) SIMPLE SADDLE AND FORCE DIAGRAM

SUPPORTED BEAM   LAG BOLT

(c) SEAT FOR SEASONED MEMBERS

0.01d SUGGESTED ALLOWANCE FOR SHRINKAGE

SEASONED MEMBER

0.15d MAX

HANGER

FASTENER LOCATED NEAR BEARING SEAT—PROVIDE MINIMUM EDGE DISTANCE FOR UPLIFT

(d) WELDED FACE HANGER

SHAPE CORNERS OF BEAM TO PERMIT SEATING

(e) TRUE HING ANCHORAGE FOR ARCHES

FASTENERS AT MINIMUM SPACING IN SLOTTED HOLES

ARCH

SHOE

PIN

BASE

WEEP HOLES OR SLOT IN LOWER PORTION OF BASE FOR DRAINAGE

ANCHOR BOLTS

(f) ARCH ANCHORAGE WHERE TRUE HINGE IS NOT REQUIRED

CONNECTIONS GROUPED NEAR CENTERLINE TO MINIMIZE SPLITTING

ARCH

SHOE

SLOT OR WEEP HOLES IN LOWER PORTION OF BOX SHOE FOR DRAINAGE

(g) SEAT FOR WOOD BEAM ON STEEL COLUMN

MACHINE BOLTS

STEEL U-STRAP WELDED TO STEEL COLUMN

(h) U-STRAP ANCHORAGE TO CONCRETE BASE

COLUMN

BOLTS

BEARING PLATE

STEEL U-STRAP

Fig. 9-10. Typical details. Figures a, c, d, e, f, and g (from reference 2) are shown courtesy of the American Institute of Timber Construction. Figures b and h are from an earlier edition of reference 2. Designers should check the latest edition of reference 2 for other details.

Fig. 9-12. Complex weldment for joining glulam arches in fabric-covered dome structure. (Photograph by authors.)

Fig. 9-11. Two views of a complex weldment. (Courtesy of V. Hairabedian, AIA.)

2. Protect the steel parts. This requires providing the proper bending strength for parts against which the wood members bear, providing the necessary tensile or compressive strength, and meeting the AISC requirements (see reference 1) for preventing local buckling of the steel.

3. Provide bolts or other devices to hold the wood parts in proper position in the weldment or to transfer member forces to the weldment.

4. Detail the weldment and the bolts or other

devices so as to allow the wood members to shrink (or swell) without restraint.

5. In all of the above, remember that load-duration modifications do *not* apply to the structural steel parts themselves.

Welded connections sometimes appear complicated. The complexity of analysis is reduced, however, by proper use of free bodies. This is shown by Example 9-4.

Example 9-3

Design a hanger to support an $8\frac{3}{4}$- × 33-in. sloping glulam purlin, combination 20F-V4, from a horizontal glulam girder, $8\frac{3}{4}$ × 42 in. Both members are Douglas fir. The 15-k vertical reaction is due to snow plus dead load. The following shows only major elements of the solution.

Bearing stress allowable (not to be modified for load duration) is that for compression perpendicular to the grain. On the compression face (top) this is 560 psi, while on the tension face (bottom) it is 650 psi. The bearing area required is

$$15,000/650 = 23.1 \text{ in.}^2$$

Bearing length must be $23.1/8.75 = 2.64$ in. or more. Tentatively, try a 5-in. bearing length (seat width).

A possible detail is as shown by Fig. 9-13. Allowing $\frac{1}{8}$-in. clearance for inserting the purlin into the hanger, dimension w should be $8\frac{7}{8}$ in. Assuming the vertical side members to be $\frac{3}{8}$-in. thick, the trans-

Fig. 9-13. Special hanger for Example 9-3.

verse steel seat the purlin bears on can be considered a simple beam of 9.25-in. span. If the purlin bears uniformly on the seat, the seat bending moment will be approximately

$$M = \left(\frac{15,000}{8.75 \times 5}\right) 9.25^2/8$$

$$= 3667 \text{ in.-lb/inch of plate width}$$

Allowing a 27-ksi bending stress (for A36 steel), the required section modulus is

$$S = 3667/27,000 = 0.136 \text{ in.}^3 \text{ per inch width}$$

$$bt^2/6 = 0.136; \qquad b = 1.0''$$

Required $t = 0.903$ in.

At first glance, this thickness seems excessive. What can be done about it? Four possibilities occur to the authors.

1. Consider the plate deflection when calculating the bending moment. (The plate deflection decreases the pressure near the center, concentrating it more near the edges of the supported glulam. This might deform the edges of the glulam. In any case, stress conditions would be difficult or impossible to compute accurately. Unless physical testing is done to verify the strength, it would be better to avoid this solution.)

2. Consider continuity of the vertical side plates with the bottom plate. (From Fig. 9-13c, however, we see that the vertical members have little flexural stiffness compared to the bottom plate. So continuity

will have little effect in reducing the bottom plate bending moment, and the moment computed above is not overconservative.)

3. Stiffen the bottom plate. (This could be done only if the projecting stiffener were not objectionable.)

4. Realize that this is a large hanger, and *use* a $\frac{15}{16}$-in. plate. (That is what will be done here.)

Each vertical side plate is subjected to the forces shown by Fig. 9-13b. Stress in the side plate on horizontal section 1-1 at the lower edge of the upper weld is $P/A + M/S$, where (from forces below the section)

$$M = 628(29.87 - 1.88) = 17,580 \text{ in.-lb}$$

$$f = 7500/5t + 17,580/(t \times 5^2/6)$$

$$= 5719/t \text{ psi}$$

Setting this equal to an allowable tensile stress of 22,000 psi, the required thickness t is 0.26 in.

From reference 1, the limit for ratio b/t (width to thickness) is $95/\sqrt{F_y} = 15.8$ for A36 steel. Using this limit, the minimum allowable t is $5.0/15.8 = 0.316$ in. To meet both requirements, use $t = \frac{3}{8}$ in.

The top angle is sized so that its bearing stress on top of the glulam girder will not exceed the allowable for the girder (560 psi) and thickness is determined so that the bending strength of its horizontal leg is adequate. The angle may be shimmed out from the face of the girder so that the angle fillet does not deform the edge of the girder. Lag screws shown carry no computed load, but merely keep the hanger in position on the girder.

Weld joining each vertical plate to the top angle carries the resultant of a 7500-lb vertical load and a 628-lb horizontal load.

Weld joining the bottom plate to the vertical plates must be adequate to transfer 7500 lb at each end of the bottom plate. Because the bottom plate acts as a beam, its ends rotate. Thus, a single weld at each end would be subject to a "wedge" action, severely overstressing the weld. To prevent this, use two welds as shown and assume the entire 7500-lb reaction to be resisted by the inner weld. The inner weld must be a groove weld (not a simple fillet weld) so that it does not interfere with the seating of the beam.

Something—bolts, lag screws, or even nails—should be used to prevent the rafter from sliding out of the hanger. Should the holes provided for this be in position x or position y?

Sad experience has shown that position x is bad. Glulam beams restrained there by nails suffered lon-

gitudinal splits, looking much like shear failures. The cause was vertical shrinkage, which raised the bottom of the beam off its intended seat. The reaction, then carried by the nails rather than by beam bearing, caused perpendicular-to-grain tension that split the glulam. Position y is the better position.

Example 9-4

Figure 9-14a shows two roof purlins, plus vertical and diagonal bracing members, all intersecting at and supported by a sloping rafter. All members are D. fir glulams. Each purlin transfers a 3-k vertical reaction to the rafter (dead + snow). Wind loads in the vertical and diagonal braces (reversible—one tension, the other compression) are 2.4 k and 4 k, respectively. For equilibrium, either purlin may be required to resist the horizontal component of the diagonal load. Purlins, vertical, and diagonal are all $5\frac{1}{8}$ in. wide. Design a welded and bolted connection detail.

Reference 2 does not suggest a detail for this situation, so the designer must plan the complete connection. One solution—not necessarily the best, and by no means the only one—is shown by Fig. 9-14b. The connection is in two separate parts, one on each side of the rafter. A cross plate could be welded between the two parts to join them together, but the resulting assembly might be more difficult to handle than two separate pieces.

The left portion is shown by Fig. 9-14c. Load applied to this part is the 3-k downward purlin force ($D + S$), plus a possible horizontal 3.2 k (horizontal component of the 4 k force in the diagonal) either way due to wind. From Table C-12, the normal-duration, double-shear value of one $\frac{5}{8}$-in. bolt is 1990 lb parallel and 1406 lb (interpolated) perpendicular to the grain. Modifying for steel side plates, the parallel-to-grain value becomes 1.69(1990) = 3363 lb. For the vertical reaction of the purlin, N req $= 3000/(1.15 \times 1406) = 1.86$ bolts.

For combined vertical reaction and horizontal wind load:

Resultant force = 4.39 k, at 43° to grain
By Hankinson's formula (Eq. 5-1) allowable normal-duration load per bolt = 2041 lb (double shear)
N req = $4390/(1.33 \times 2041) = 1.62$
Use two $\frac{5}{8}$-in. bolts, as shown

Axial load in the purlin can be tensile, so the required end distance in the purlin is $7D$, reduced by the ratio of actual bolt load to full allowable parallel to the grain, or

(a) MEMBERS

(b) POSSIBLE CONNECTION

(c) LEFT PORTION (d) WELD LOADS (e) RIGHT PORTION (f) WELD LOADS

Fig. 9-14. Connection for Example 9-4.

$(7 \times 5/8)(0.5 \times 3200)/(1.33 \times 3363)$
$= 1.57$ in.

but not less than one-half of $7D$, or 2.19 in. Use $2\frac{1}{2}$ in.

The two plates marked P1 are spaced $5\frac{1}{4}$ in. apart,

allowing $\frac{1}{8}$ in. more than the width of the purlin between them. Plates P1 are welded to plate P2. Allowing $\frac{1}{2}$-in. clearance between the end of the purlin and plate P2, each line of weld joining the plates must resist a 1.5-k vertical load at an eccentricity of 3 in. and a horizontal load of 1.6 k, as shown on the free-body diagram of plate P1 in Fig. 9-14d. The weld resists the applied loads with a vertical force, V, a horizontal force H, and a moment, M. A second load case ($D + S$ only) will not control.

How thick should the plates be? Thick enough that the bolt bearing value in the steel plates will be adequate, and preferably thick enough to meet the AISC width/thickness limit of $b \leq 15.8t$ to avoid local buckling of the edge of the A36 plate. For plates P1, $\frac{5}{16}$-in. thickness would be adequate for a 5-in. plate width. For simplicity, make plate P2 the same thickness.

The right portion, Fig. 9-14e, is slightly more complicated. The purlin connection to plates P3 is the same as for the left portion. The diagonal has either 4 k tension or 4 k compression, but no shear. The modified double-shear value of one $\frac{5}{8}$-in. bolt in the diagonal is $1.33(1.69)(1990) = 4470$ lb, so only one $\frac{5}{8}$-in. bolt is needed. The end distance required in the diagonal is $7(5/8)(4.0/4.47) = 3.91$ in. Rounding this up to 4 in. and plotting the detail to scale, we conclude that the plates P3 must be $5\frac{1}{2}$ in. wide. The b/t criterion suggests that plates P3 be $\frac{3}{8}$ in. thick. That criterion is conservative, however, and the authors would feel comfortable using $\frac{5}{16}$ in. for P3, as for P1.

Welds joining plates P3 to P4 are loaded eccentrically, by the diagonal as well as the purlin. Figure 9-14f shows one plate P3 as a free body. In designing the weld, two load cases must be recognized: the vertical 1.5 k ($D + S$) alone, and the combination of all three factors ($D + S + W$) shown.

The vertical member transfers 2.4 k wind load (either tension or compression) into plates P2 and P4. One $\frac{5}{8}$-in. bolt (value with wind 4470 lb) is adequate.

Loading for bolts joining plate P4 to rafter:

Under $D + S$
Purlin reaction = 3.0 k
Under $D + S + W$:

Purlin reaction	= 3.0 k down
Vertical component of 4 k tension in diagonal	= 2.4 k down
One-half of vertical member force	= <u>1.2 k</u> up
Sum	= 4.2 k on one shear plane

$3.0/1.15 = 2.61$ $4.2/1.33 = 3.16$

The loading including wind controls. Angle of load to grain = 78.2° (90 minus roof slope of 11.8°)

$$P = 1990 \times 1.33 \times 1.69/2 = 2236 \text{ lb}$$
$$Q = 1406 \times 1.33/2 = 935 \text{ lb}$$

By Hankinson's formula, allowable at 78.2° to the grain is 958 lb/bolt.

$$N \text{ req} = 4400/958 = 4.59$$

Try 6 bolts. These bolts must also pass through plate P2 on the other side of the rafter.

Bolts joining the left portion to the rafter must carry 3.0 k ($D + S$), or 4.2 k ($D + S + W$). The latter controls, and N req = $4200/958 = 4.38$. For symmetry, use six $\frac{5}{8}$-in. bolts passing through both plates, P2 and P4. These plates will have to be wide enough to satisfy bolt spacing and edge distance requirements. The lower edge of the rafter is the loaded edge, and the distance of the bolts from that edge must be at least $4D$, or $2\frac{1}{2}$ in.

As an alternate to connecting the purlins as shown, they could be supported on horizontal bearing plates welded between plates P1 and plates P3.

Example 9-5

In the glulam chapter, Example 8-8 covered the design of a cantilever beam system. The solution was in the form of a designer's computation sheet, shown by Fig. 8-14. Using loads and sections from that solution, design a saddle connection to support the suspended span.

Another designer's computation sheet, shown here by Fig. 9-15, shows a solution to this part of the problem. There being no longitudinal loads to contend with, the designer assumed first that a simple saddle (as in Fig. 9-10b) might suffice. This would require 7- or 8-in.-long lag screws to resist the horizontal forces, F. So the designer continued assuming a saddle in which the required forces F are resisted by end-grain bearing between the glulam and transverse steel bars. It was necessary to estimate the bar width in order to compute force F. For greater accuracy in designing the bearing plates, the designer might consider two-way plate action, using equations such as are developed in reference 3.

In the final step, the designer verified that $\frac{5}{8}$- $\times$ $4\frac{1}{2}$-in. bars would be satisfactory for the slanting side parts of the saddle. Notice that, since the allowable steel stresses for tension and for weak-axis bending are not alike, the designer used an interaction equation rather than merely adding $P/A + Mc/I$ to find the total stress.

Figure 9-15 does not show design of the necessary welds, although that certainly is a part of the problem. If appearance is of little concern, the detail of Fig. 9-10a might be used instead.

If the top seat of the saddle would interfere with

Fig. 9-15. Computation sheet for Example 9-5.

deck material, the designer would show that seat recessed into a dap in the top of the left glulam.

9-6. SHEAR PLATES AND SPLIT RINGS

Screws and bolts have a common fault—they slip until they come into bearing, following which they cause bearing deformation in the wood. Consequently, it is difficult to realize the full load capacity of wood members when bolts or screws are used. An early attempt to correct this inadequacy was the Kübler dowel, shown by Fig. 9-16. This double-tapered dowel of cast iron or oak was embedded, one-half in each of the two wood members being connected. A bolt passing through both wood members and the dowel held the pieces together, but did not transfer shear. The dowel (up to 4 in. wide) presented a large bearing area to the wood member and could transfer large loads from one member to the other without severe bearing de-

Fig. 9-16. Kübler dowel, a forerunner of shear plates and split rings.

Fig. 9-18. Split-ring connector. (Courtesy of TECO Products & Testing Corp.)

formation. The Kübler dowels probably inspired developers of modern split-ring and shear-plate connectors. These too, present fairly large bearing areas and thus avoid causing large bearing deformations in the wood.

Figure 9-17 shows two types of shear plate and Fig. 9-18 shows the split-ring connector. These connectors reached peak popularity during World War II, when material shortages made it necessary to use timber for long-span structures that until then had been practical only in structural steel. Today, the use of split rings has declined somewhat, but shear plates still find many applications.

9-7. SHEAR PLATES

Shear plates are of two varieties—pressed steel plates of $2\frac{5}{8}$-in. outside diameter, and malleable iron ones 4 in. in outside diameter. The smaller plates are used with $\frac{3}{4}$-in. bolts and the larger ones with either $\frac{3}{4}$- or $\frac{7}{8}$-in. bolts.

A shear plate is installed in a precut groove and dap so that its flat surface is flush with the face of the wood member and its flange is in

(a) Pressed steel   (b) Cast malleable iron

Fig. 9-17. Shear plates. (Courtesy of TECO Products & Testing Corp.)

the groove. Two nails through the small holes in the shear plate prevent it from dropping out during handling. Load is transferred from the member into the shear plate by wood bearing pressures against the sides of the flange. It is then transferred to the bolt by bearing of the shear plate against the bolt shank. The bolt then carries the load to the other member—again by way of a shear plate if the other member is wood.

Shear plates are useful for connecting wood members to steel members or for joining together wood members that have to be disconnected and reinstalled.

Table C-15 shows allowable loads for shear plates installed in seasoned wood that remains seasoned during use and is subject to normal-duration loads. These values must be modified for fire-retardant treatment, for moisture conditions (Table C-2), and for number in a row (Tables C-10 and C-11). Load-duration factors apply to the values in Table C-15, but the resulting modified allowable load must not exceed the allowable based on the strength of the steel bolt (see footnote 6 to Table C-15).

Modification for metal side plates is permitted only for 4-in. shear plates and for the parallel-to-grain allowables only, as follows: 18%, 11%, and 5% for species groups A, B, and C, respectively, but to not over the maximum shown by footnote 6 of Table C-15. No metal side plate modification is allowed for group D, nor for the allowable perpendicular to the grain, nor for $2\frac{5}{8}$-in. shear plates.

Interpolation on Table C-15 is permitted, but the lower thickness of wood listed in each block is the minimum thickness allowed, and the

Fig. 9-19. Net area of wood member with split rings (or shear plates) in two faces.

larger allowable load shown in each block (or by footnote 6) is the absolute maximum.

Table C-15 give allowables both perpendicular to and parallel to the grain. For allowable loads at other angles, first complete all modifications, *then* use Hankinson's formula (Eq. 5-1). Table C-16 gives end-distance and spacing requirements.

9-8. SPLIT RINGS

Split rings have tapered thickness, so special tools are needed to cut the grooves to receive them. The bolt merely holds the parts together firmly, but does not carry load from one member to the other. Each member must be grooved, so the split ring is not used to connect wood members to metal ones.

Figure 9-19 shows dimensions that are helpful in determining the net area remaining in a wood member after installation of a split ring (or shear plate) and its accompanying bolt. Table C-17 shows allowable loads for split rings installed in seasoned wood that remains seasoned in service and is subject to normal-duration loading. The table gives values for load both parallel to and perpendicular to the grain. For allowable loads at other angles, use Hankinson's equation. (All modifications apply equally to values parallel to and perpendicular to the grain, so Hankinson's equation may be solved first, if desired.) Modifications to con-

sider are: load duration, fire-retardant treatment, moisture (Table C-2), and number in a row (Table C-10).

Table C-16 shows end-distance and spacing requirements for split rings. Nomographs (reference 2) are helpful in computing spacings, end distances, and edge distances in joints where the axis (line through a row of connectors) is not parallel to the grain.

Example 9-6

What is the allowable load, F (including wind), for the connection of Fig. 9-20? The lumber is seasoned, but will be wet in service.

Fig. 9-20. Connection for Example 9-6.

Load is parallel to the grain in the 2 × 8 member and at 53.1° to the grain in the 4 × 10. From Table C-1, the members are of species group B. From Table C-17, the unmodified allowables for one $2\frac{1}{2}$-in. split ring installed on one side of the member are:

Diagonal member, 2730 lb, parallel to the grain
Vertical member, 2730 lb parallel to grain and 1940 perpendicular to grain
By Hankinson's equation, allowable at 53.1° to grain in the vertical member = 2166 lb

The unmodified value in the vertical member controls. Modifying for load duration and for wet service,

$$\text{Allowable } F \text{ (2 rings)}$$
$$= 2(2166)(1.33)(0.67) = 3860 \text{ lb}$$

based on the connection strength. Net area of the diagonal is

$$1.5(7.25) - 0.375(2.92) - 1.125(9/16)$$
$$= 9.15 \text{ in.}^2$$

The allowable tensile stress, by footnotes 6 and 15 of Table B-3, is

$$F_t = 0.8(0.84)(1000)(1.33)$$
$$= 894 \text{ psi (on net section)}$$

Allowable F based on tensile strength of the diagonal is

$$F = 9.15(894) = 8180 \text{ lb}$$

The controlling allowable F is 3860 lb.

Example 9-7

Design a double-shear splice using $2\frac{1}{2}$-in. split rings to develop the maximum allowable tension in Select

Structural southern pine 3 × 10s, surfaced dry and used at not over 19% MC.

If one row of rings is used, the net area of the member spliced will be its gross area minus the projected area of two half-rings and of the $\frac{9}{16}$-in. bolt hole between bottoms of ring grooves. That is, it will be the shaded area of Fig. 9-19, or

$$23.125 - 2(2.92 \times 0.375)$$
$$- (2.5 - 0.75)(9/16) = 19.95 \text{ in.}^2$$

For Select Structural southern pine, the 1150 psi allowable shown by Table B-3 is modified by a factor of 0.8 (footnote 15) to give 920 psi allowable on the net area. So the splice will have to be designed for 19.95(920) = 18,350 lb. The two side members, combined, must have at least as much net area as the main member, so they will have to be 1.25 in. thick or more. Use 2 × 10 side pieces.

Load is parallel to the grain in both main and side members. The allowable load per $2\frac{1}{2}$-in. ring in species group B is 2730 lb in the 2-in. side member and also in the 3-in. main member. The only modification needed is for the number of connectors in a row. Approximate required $N = 18,350/2730 = 6.72$ rings.

Try eight rings per end of the splice (one row in each face of the main member, four rings per row). Use Table C-10: $A_1/A_2 = 2.5/3.0 = 0.83$, and $A_1 = 23.125$ in.^{2}. Interpolating, find a factor of $3.83/4 = 0.96$.

$$N \text{ required } = 18,350/(0.96 \times 2730) = 7.0 \text{ rings}$$

Use eight rings per end of splice, as shown by Fig. 9-21.

Example 9-8

What spacing and end distance are needed for the splice of Example 9-7?

$2\frac{1}{2}''$ SPLIT RINGS   2 × 10 X 4'-8$\frac{1}{2}''$
(ONE ROW ON ℄)

Fig. 9-21. Solution to Examples 9-7 and 9-8.

Fig. 9-22. Connection for Example 9-9.

Appendix Table C-16 gives spacing requirements for connectors loaded to 100% and to 50% of their full allowable values. In this case only 7.0 rings are required but we are using 8, so the rings are loaded to only 87.5% of their allowable. Interpolating between the 3.5 in. shown for 50% load and 6.75 in. for 100%, the required spacing for this splice is

$$3.50 + (6.75 - 3.50)(87.5 - 50)/50$$
$$= 5.94 \text{ in.}$$

Use a 6-in. spacing.

Similarly, for end distance, the table shows requirements for 100% and 62.5% full load. The interpolated requirement is 4.58 in. Use 5-in. end distance. This applies to the inner ends of the members

being spliced, as well as the outer ends of the side pieces.

Example 9-9

Figure 9-22 shows the intersection of members at a bottom-chord joint of an existing roof truss. All members are species group C. The new owner of the building wants to add a materials-handling system that will apply an additional 1.5-k load just next to the panel point, as shown by Fig. 9-22. Will the connection be satisfactory under the increased loads?

From record prints of the original construction drawings, you find that the original member forces under dead plus snow loads are as shown by Fig. 9-22b. Assume that you have computed member forces due to the new conveyor load; they are shown in Fig. 9-22c. The total member forces are as in Fig. 9-22d.

Figure 9-22e shows a cross section of the truss joint. There are a total of eight split rings, two in each plane between members. It is easiest to see what is happening in a joint of this type if we use an "exploded" view, as in Fig. 9-22f. This exploded view isolates a portion of each member, showing it as a free body with *all* forces that act on that portion—a set of forces that must be in equilibrium.

Start with the vertical. The member is being pushed downward with a 5.83-k load. That force is resisted by four rings labeled *B*. The force carried by each ring must be 5.83/4, or 1.46 k.

Next, consider a short portion of one member of the bottom chord. It is pulled to the right by 3.93 kips (one-half of 7.86 k) and downward with 0.75 k from the proposed conveyor load. To keep this free body in equilibrium requires a 4.00-k force at 11° to the horizontal, or 2.00 k in each of the two rings *A*.

This completes the solution of forces in the rings. To find the ring forces it was not necessary to consider the diagonal, but we do now know the forces applied to it, as shown by the center picture of Fig. 9-22f. (As a general rule, consider the inside and outside members first, and the intermediate ones last.) Now that we know the force in each ring, we must compare it to the allowable for each ring and in each member in which the ring is installed.

For rings *A*:

In diagonal—load is at 32° to the grain.
   Allowable parallel to grain = 1760 lb
   Allowable perpendicular to grain
      Edge distance = 2.75 − 0.70 (scaled)
         = 2.05 in.
      For *ED* = 1.75 in. the allowable is 1040 lb;
         for 2.75 in. it is 1250 lb
      Interpolating, for *ED* of 2.05 in., allowable
         = 1103 lb
   By Hankinson's equation, at 32° to grain
   Allowable = 1508 lb

In bottom chord—does not control (by inspection, Table C-17).
Modify for snow. Controlling allowable in ring *A* = 1.15(1508) = 1734 lb.

The calculated load is 2.00 k, so rings *A* will be overloaded, and regardless of the outcome of checking rings *B*, the connection is unsatisfactory for the increased load. Just to illustrate the procedure, however, here is the check for rings *B*:

In diagonal—load-to-grain angle is 47°.
   By Hankinson's equation, allowable at 47° to grain = 1335 lb
In vertical, allowable = 1760 lb (does not control).
Modified for snow, allowable per ring = 1.15(1335) = 1535 lb.
   Actual load = 1460 lb. Rings *B* are OK (but rings *A* are not).

9-9. SPIKED GRIDS AND GRID PLATES

Figure 9-23 shows three types of spiked grid that are used mainly in pole construction, railroad or highway trestles, and marine construction. The flat and circular grids are for connecting one sawn timber to another, and the single-curved rectangular grid is for connecting a sawn timber to a round pole. The grids illustrated are cast in malleable iron. In one manufacturer's standard, the rectangular grids are about $4\frac{1}{8}$ in. square. Overall thickness for the flat grid is one inch, so that the spikes penetrate about $\frac{7}{16}$ in. into the wood members. To reach similar penetration, the curved grid has an overall thickness of $1\frac{3}{8}$ in. The circular one is $3\frac{1}{4}$ in. in diameter and is 1.20 in. deep.

A $\frac{3}{4}$- or 1-in. bolt passes through the members and the central hole in the grid. The bolt

(a) CIRCULAR

(b) FLAT

(c) SINGLE CURVE

Fig. 9-23. Spiked grids. (Courtesy of TECO Products & Testing Corp.)

is in a $\frac{1}{16}$-in. oversize hole, so under light loads it merely holds the parts together; it does not resist shear unless the grid connection itself slips through bearing deformation of the wood.

It requires considerable pressure to embed the spikes fully in each member. The force needed is more than can be applied merely by tightening the bolt. Manufacturers provide special installation tools to embed the spikes. After the installation tool is removed, the permanent bolt is installed and tightened.

Table C-18 shows allowable loads for one manufacturer's spiked grids. Load duration factors for wood do not apply to the allowables shown. Instead, the allowable may be increased by 20% for load combinations that include wind or earthquake.

Minimum grid spacings allowed are 7 in. parallel, and $5\frac{1}{2}$ in. perpendicular to the grain, for all three types of grid. Minimum end distances are 7 in. under full load or 5 in. (absolute minimum) under 85% of full load. Similarly, edge distances (for any load/grain direction) are $3\frac{5}{8}$ in. and $2\frac{3}{4}$ in. minimum.

9-10. NAILER PLATES AND TOOTHED PLATES

Nailer plates are sheet metal plates, frequently galvanized, with an array of holes through which nails may be driven. For standard sizes of nail (covered by tables such as Appendix Table C-4) use the tabulated values, multiplied by the proper modifiers, including the 25% increase for metal side plates. The object of the nailer plate is to make a connection more compact than it would be with only wood members. But, as load is picked up from a large number of nails, this may have the undesirable effect of overstressing the steel of the nailer plate itself. The chance of overstressing is greater when the nailer plate is used in a shear or compression connection than in a tension connection. The designer should be alert to this; it does no good to provide a connection that is excellent from the standpoint of the wood, only to have the steel parts fail.

Nailer plates are frequently used with special nails, shorter than those of Table C-4, and often with fluted shanks. This is usually the case when nailer plates are used with 2-in. nominal material.

Toothed plates have their own "nails" built in. The teeth are formed by bending the steel from long triangular punched holes in the sheet metal plates so that an array of closely spaced teeth project from one side of the plate. Since they are used mainly for wood truss construction, they are usually called *truss plates*. Installation is by pressing the plate firmly so that all teeth are driven into the wood. Chapter 10 covers design using truss plates.

9-11. DRIFT PINS AND DOWELS

Figure 9-24 shows fasteners that may be used to connect heavy-timber cap members to wood piles, stringers to cap members, and the like. Each type is driven into predrilled lead holes. For drift bolts and drift pins the hole diameter is $\frac{1}{16}$ in. less than that of the fastener. For spiral dowels (twisted steel rods), the hole diameter is about 75% of the outside diameter of the spiral. Drift pins, being merely round steel rods, can be made in a wide variety of sizes. Standard sizes of spiral dowels are from $\frac{1}{4}$ in. to $\frac{5}{8}$ in. and lengths available are by $\frac{1}{2}$-in. incre-

Fig. 9-24. Drift pins and dowels.

ments up to 8 in. and by 1-in. increments for dowels over 8 in. long.

Drift pins, drift bolts, and spiral dowels have significant strength in either lateral load or withdrawal and can be used in either side grain or edge grain. Withdrawal from end grain, while not allowed by the NDS for nails or lag screws, is a common direction of loading for these fasteners, and tests show that the end-grain withdrawal value is about one-half of that for withdrawal from side grain.

Allowable loads for these fasteners are given by equations involving the same parameters as the equations for nails. Reference 4 shows the equations and gives tables of allowable loads.

REFERENCES

1. *Specification for the Design, Fabrication and Erection of Structural Steel for Buildings*, American Institute of Steel Construction, Chicago, IL, 1978.
2. American Institute of Timber Construction, *Timber Construction Manual*, 3rd ed., Wiley, New York, 1986.
3. Timoshenko, S., and S. Woinowsky Krieger, *Theory of Plates and Shells*, 2nd ed., McGraw-Hill, New York, 1959.
4. *Western Woods Use Book*, Western Wood Products Association, Portland, OR, 1979.

PROBLEMS

All lumber sizes are nominal, S4S, unless otherwise noted. Loads are normal duration, unless otherwise noted.

9-1. Refer to the eccentrically loaded connection of Fig. 9-4. What would be the allowable load at $e = 9$ in. if four bolts were used in each vertical row and the spacing in the rows were still 4 in.?

9-2. Design a bracket, as shown by Fig. 9-5, to support a 9-k normal-duration load that is $2\frac{1}{2}$ in. from the face of an $8\frac{3}{4}$- $\times$ 9-in. glulam column, combination 11 Douglas fir. Use the same species in the bracket. (Solution is by trial and error. Assume a number and size of bolts and dimensions b, c, and s. Analyze the assumed joint. If not satisfactory, revise and analyze again.)

9-3. Complete the computations of Example 9-5 (Fig. 9-15) by designing the necessary welds.

9-4. Design a hanger to support rafters framing to opposite sides of a ridge beam. The rafters, 4 $\times$ 16 sawn D. fir, No. 1, slope at 4 in. in 12 in. (down away from the ridge beam), and have reactions of 3.5 k each (1.0 dead plus 2.5 snow). The ridge beam is a $6\frac{3}{4}$- $\times$ 30-in. glulam, western species combination 20F-E1.

9-5. Figure 9-25 shows roof framing and bracing members and their loads. Design a connection weldment to use with bolts (and shear plates, if necessary).

9-6. Design a split-ring connection to splice together two 2 $\times$ 12 tension members of Select Structural D. fir–larch. Use a double-shear connection. Design the splice to carry the maximum allowable tensile load for the main members. Use $2\frac{1}{2}$-in. rings.

9-7. Repeat the above problem using shear plates and metal side plates. Use $2\frac{5}{8}$-in. shear plates and $\frac{3}{4}$-in. bolts.

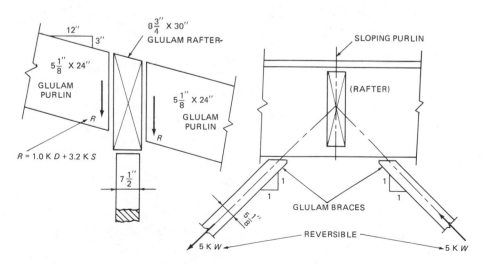

Fig. 9-25. Diagram for Problem 9-5.

Fig. 9-26. Truss joint for Problem 9-10.

9-8. A 2 × 6 diagonal brace is connected to an 8 × 8 column using 2½ in. split rings. The brace slopes 6 in./ft and has a load (from wind) of 4.5 k. The diagonal is Select Structural D. fir and the column is No. 1 D. fir. Design the connection and show a detail of the joint, including edge and end distances and spacing.

9-9. Design a split-ring connection to carry a 4.8-k load (1 k dead plus 3.8 k live) from a vertical hanger into a horizontal beam. The beam is a 6 × 12 of No. 1 D. fir, and the vertical hanger consists of two No. 1 redwood 2 × 6s, one on each side of the beam. The hanger is 4 ft from the end of the 12-ft span. Live-load duration (cumulative) is two months.

9-10. Figure 9-26 shows a top-chord joint of a truss connected with split rings. The members are all D. fir–larch, Select Structural grade. Revised use of the building causes member loads to increase above the original design loads. Is the split-ring connection satisfactory for the new increased loads?

9-11. A horizontal 8 × 14 dense D. fir timber is to be attached to the vertical piling of a trestle bent. The vertical load to be transferred to each pile is 2000 lb dead load and 5000 lb live load. The D. fir coast region wood piles are approximately 15 in. in diameter where the 8 × 14 is attached. Choose a size and number of single-curve spiked grids to connect the 8 × 14 to one pile.

10

Timber Trusses

For long or moderate spans, wood trusses are often lighter, more economical, and more practical than wood beams. In addition, a structure using trusses usually is stiffer than one using beams of the same span; that is, it has smaller deflections.

The wood truss is not a recent invention—wood trusses have been used for centuries for both buildings and bridges. For buildings, an early recorded use was during the first century B.C. (1). Recent improvements in methods of connecting members together to form a truss, however, have led to new uses for wood trusses and an increase in their popularity. This is particularly true in light-frame construction.

10-1. WOOD TRUSS TYPES

The variety of geometrical patterns that may be used is almost unlimited. Those shown by Fig. 10-1, however, have proved practical for wood trusses and are frequently used. Figure 10-1a shows two examples of the Pratt truss, a patented configuration when it was first developed. When the Pratt truss is fully loaded, its web diagonal members are in tension, while the verticals are in compression.

The sign of web member forces is the opposite for the Howe truss: Web verticals are in tension and diagonals in compression. Note that is it not the direction of member slope that distinguishes the Pratt truss from the Howe; rather, it is the sign of force in the web members.

The Howe truss of Fig. 10-1c has diagonals in both directions. If its members were all rigidly connected, this truss would be highly statically indeterminate. However, its interior verticals can carry only tension and the diagonals

are so connected that they can resist compression only. Under a particular moving-load position, one of the diagonals in a panel will act in compression and the other will be inactive. When the load moves so as to change the sign of shear in the panel, the first diagonal becomes inactive and the second receives compression.

Figure 10-1d shows a Warren truss. Its web diagonals alternate in both direction and sign of stress. Whether or not there are vertical members is immaterial. Verticals are included only if needed to reduce the laterally unsupported length of chord members or to shorten the length of floor system members that bring load to the truss panel points.

In its action, the bowstring truss of Fig. 10-1e is both a tied arch and a truss. Under uniform top-chord loading, the curved top chord acts (in compression) as an arch and the bottom chord as a tie connecting the ends together, the web members carrying little or no load. Under unsymmetrical load, however, the web members carry appreciable load and the system acts more as a truss and less as an arch. The bowstring truss is useful for both wood and structural-steel construction.

The fan, scissors, and Fink trusses of Fig. 10-1f, g, and h are also well suited to wood construction. The scissors truss is often used exposed for interiors of buildings with steeply sloped roofs (see Fig. 10-2).

Wood trusses are often categorized into one of three broad groups: light-frame trusses; heavy-timber trusses; and split-ring connected trusses.

Light-frame trusses are made using members of 2-in. nominal thickness, usually connected by metal nailer plates or toothed truss plates,

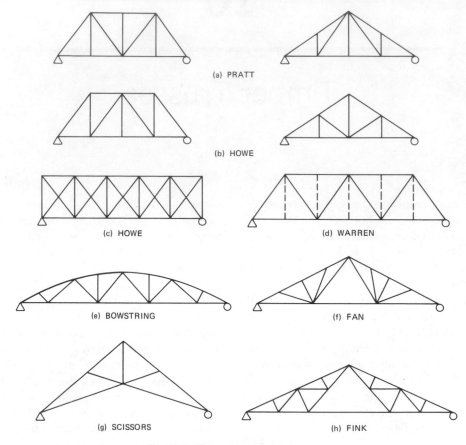

Fig. 10-1. Truss geometrical patterns.

or occasionally by nailed or glued plywood gusset plates. Whereas a few years ago sawn lumber was almost always used for sloping roof rafters, ceiling joists, and floor joists, today light-frame trusses have largely replaced the sawn-lumber members for these uses.

Light-frame trusses are especially economical for roof framing in buildings where spans are small or moderate. For floors, trusses with horizontal top and bottom chords also are often found economical. Floor trusses may provide better rigidity than sawn-lumber joists, and the openings between their members provide for passage of pipes, conduits, and ducts without cutting, as might be required if sawn-lumber joists were used.

Light-frame trusses, assembled in a fabrication shop, are complete and ready to install when shipped to the building site. Their advantages of stiffness, light weight, and low cost are partially offset by their greater potential for fire damage than heavier sawn timbers and glu-lams. (See Fig. 10-3.)

Heavy-timber trusses have member thicknesses up to 8 or 10 in. With truss members of such large size, connections pose a major problem. In large older structures, heavy truss connections were made using end bearing whenever possible to transfer load from one member to another and by using steel rods for selected tension members. This will be shown by a later example.

Timber connectors make possible a variety of large timber trusses that would otherwise be impractical. The larger members of these trusses are *spaced*; that is, they are built up using two or more separated elements of 2, 3, or 4-in. lumber. Lighter members are single pieces. At joints, the various elements overlap and are connected by split rings installed between adjacent elements. Such trusses are economical in their use of material and were es-

Fig. 10-2. Interior of church building with scissors trusses formed from peeled logs. (Photograph by authors.)

Fig. 10-3. Light-frame trusses for industrial structure. (Courtesy of Lumbermate Company, St. Louis, Missouri.)

Fig. 10-4. Weldment at conjunction of glulam truss and transverse bracing. (Photograph by authors.)

pecially popular in times of material shortage during World War II.

For best appearance (as in exposed trusses) solid members rather than spaced multiple-element members should be used. For these members glulams are often advantageous. It is attractive as well as practical to join the members of exposed trusses with bolted steel gusset plates on each face of the members (see Fig. 10-4). Shear plates may be used on the inner surfaces of the gusset plates if the bearing strength of the bolt on the wood is not sufficient.

"Hybrid" trusses using mixed materials are becoming increasingly popular. Examples include floor members whose top and bottom chords are wood, while the web members are metal. Top and bottom chords of these trusses are also being fabricated with finely laminated wood (called laminated veneer lumber, or LVL) rather than sawn lumber.

10-2. LIGHT-FRAME TRUSSES, ANALYSIS

Specifications by the Truss Plate Institute (2, 3) define current practice for design, fabrication, and erection of light-frame trusses. The first of these, TPI-85, *Design Specification for Metal Plate Connected Wood Trusses*, now includes design guides for three standard truss configurations, shown by Fig. 10-5. As other arrangements are adequately studied and tested, design criteria for those also may be added to future editions of TPI-85. The second specification, PCT-80, *Design Specification for Metal Plate Connected Parallel Chord Wood Trusses*, covers wood trusses whose parallel top and bottom chords are made from 2 × 4s, placed with the 4-in. nominal dimension horizontal. (PCT-80 defines these chord members as "4 × 2.")

Each of these specifications defines in detail how to determine the bending moments in the top and bottom chord members. After axial load and bending moment are determined for each member, actual member design is by the criteria of the NDS (4).

Member axial forces are computed assuming all members to be pin-connected at each truss joint. Actually, however, the top and bottom chords are usually continuous over two or more panels. When chord members receive transverse (usually vertical) load at locations other than panel points, they must be designed for either combined bending and compression or combined bending and tension. Having chord members continuous over two or more panels (rather than ending members at each panel point) makes fabrication easier, but makes analysis more difficult.

In common use are three methods of analysis to find values of member bending moments:

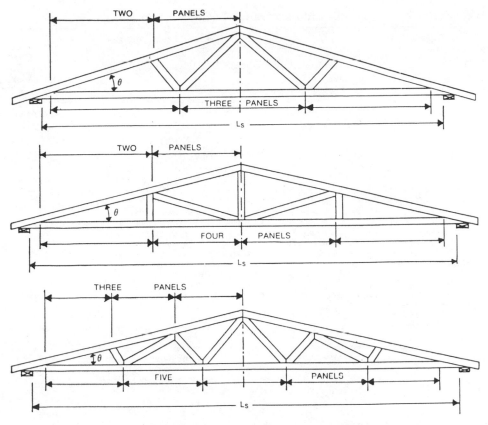

Fig. 10-5. Truss configurations covered by TPI-85. (Courtesy of Truss Plate Institute.)

1. Computer solution of the truss by matrix methods, which provides values for both axial load and bending moment. One such method (5) was produced by researchers at Purdue University. General-purpose systems such as SAP IV or SAP80 (6, 7) could also be used for analysis.

2. A simplified, empirical method given by TPI-85. In this method, a moment length factor, Q, is determined from a table such that QL represents the horizontal component of length between inflection points. Factors Q and L are shown by Tables 10-1, 10-2, and 10-3 for the *top* chords of different types of truss. Using the appropriate Q and L, the bending moment for top-chord segments with uniform vertical load is

$$M = w(QL)^2/8 \qquad (10\text{-}1)$$

Some of the tabulated expressions for Q seem to be involved and complex. Their object is

simply to account for the lack of complete fixity of the chord members at the heel joint. Since the effective length, QL, is squared in Eq. 10-1, it is important to consider all factors that affect the location of the inflection points. For segments of *bottom* chords having vertical uniform load, the same equation applies, but with factors Q and L from Table 10-4.

3. A simplified empirical method in which end moments are determined by simple moment coefficients (8).

In each of simplified methods 2 and 3 above, axial-load member forces are determined separately, assuming all joints to be pin-connected. Specification PCT-80, however, requires that no member be designed for less than 375 lb of axial load and that the vertical component of axial load for any diagonal web member be not less than one-sixth of the total load on the simple span.

Specification TPI-85 does not suggest how

Table 10-1. Top-Chord Factors for Triangular and Scissor Trusses.

Panel Point Moment		Mid-Panel Moment	
Q	L	Q	L
One panel Not Applicable	**One panel** Not Applicable	**One panel** 0.90	**One panel**[c] $L_i + S_a$
Two panels 0.90	**Two or three panels**[a] Largest of: $0.9L_i$, or $\dfrac{(L_i + L_a)}{2}$, or $0.9L_a$	**Two panels**[b] $0.58 \, (\cot \theta)^{0.23}$	**Two or three panels**[c, d] Largest of: $0.9(L_i + cS_a)$, or $\dfrac{(L_i + L_a)}{2} + cS_a$, or $0.9 \, (L_a + cS_a)$
Three panels 0.85		**Three panels**[b] $0.53 \, (\cot \theta)^{0.36}$	

[a] If S_t exceeds 24 in., add excess to end (heel) panel L_i or L_a (see Figure 10-7).
[b] $Q = \alpha \, (\cot \theta) \, \beta$ but shall not be less than 0.74; α and β are constants derived from PPSA analysis.
[c] $S_a = S_t - B$ but not less than zero; cS_a shall be added only to the length of the end (heel) panel.
[d] $c = 0.5$ for two panels; $c = 0.33$ for three panels; if neither L_i nor L_a are end (heel) panel lengths, then $cS_a = 0$.
Note: L_i = horizontal length of panel being investigated.
L_a = horizontal length of larger panel adjacent to panel of length L_i.
Source: Courtesy of Truss Plate Institute.

Table 10-2. Top-Chord Factors for Monopitch and Monoscissor Trusses.

Panel Point Moment		Mid-Panel Moment	
Q	L	Q	L
One panel N/A	**One panel** N/A	**One panel** 1.00	**One panel**[c] $L_i + S_a$
Two panels 0.90	**Two or More panels**[a] Largest of: $0.9 \, L_i$, or $\dfrac{(L_i + L_a)}{2}$,	**Two panels**[b] 0.90	**Two or more panels**[c, d] Largest of: $0.9 \, (L_i + cS_a)$, or $\dfrac{(L_i + L_a)}{2} + cS_a$,
Three or four panels 0.90		**Three or four panels**[b] $0.58 \, (\cot \theta)^{0.23}$	
Five or six panels 0.85	or $0.9 \, L_a$	**Five or six panels**[b] $0.53 \, (\cot \theta)^{0.36}$	$0.9 \, (L_a + cS_a)$

[a] If S_t exceeds 24 in., add excess to end (heel) panel L_i or L_a (see Figure 10-7).
[b] $Q = \alpha(\cot \theta)\beta$ but shall not be less than 0.74; α and β are constants derived from PPSA analysis.
[c] $S_a = S_t - B$ but not less than zero; cS_a shall be added only to the length of the end (heel) panel.
[d] $c = 0.5$ for two panels; $c = 0.33$ for three panels; if neither L_i nor L_a are end (heel) panel lengths, then $cS_a = 0$.
Source: Courtesy of Truss Plate Institute.

Table 10-3. Top-Chord Factors for Chord Segments Without Heel Joints, for Top Chords with Slopes of Less than 1.5/12.

Panel Point Moment		Mid-Panel Moment	
Q	L	Q	L
		One panel 0.95	**One panel** L_i
	Not Applicable[a]	**Two panels** 0.90	**Two or more panels** L is largest of: $(L_i + L_a)/2$, or $0.9 \, L_a$, or $0.9 \, L_i$
		Three or more panels 0.85	

[a] As these factors are derived from TPI-74, only mid-panel members are considered critical.
Source: Courtesy of Truss Plate Institute.

Table 10-4. Bottom-Chord Factors.

Q	L
One panel 1.0	One panel L_i
Two or more panels 1.0	Two or more panels L is largest of: $0.9 L_i$, or $\dfrac{(L_i + L_a)}{2}$, or $0.9 L_a$

Source: Courtesy of Truss Plate Institute.

truss deflections should be computed, but does require that live-load deflection of roof trusses not exceed $\ell/360$ when plaster is used on the ceiling below, nor $\ell/240$ when a flexibile ceiling such as drywall or a suspended ceiling is used. If there is no finished ceiling, the deflection limit is $\ell/180$ (ℓ is the truss span). The authors' suggestion for calculating deflection is presented later in this chapter.

Example 10-1

Compute design axial forces and bending moments for members of the light-frame wood roof truss of Fig. 10-6. The trusses will be installed at 24-in. centers. Tentatively, the designer is considering six equal top-chord and five equal bottom-chord panels as shown. (This is the third of the standard configurations covered by TPI-85.)

Loads are:
Applied to the top chord—

Snow	25 psf of horizontal projection
Sheathing	2.5 psf of roof surface
Roofing	6 psf of roof surface

Applied to the bottom chord—

| Ceiling | 8 psf |
| Insulation | 3 psf |

Member sizes are not yet known, so we must estimate the truss weight. Luckily, truss weight is a small part of the total, so a precise estimate is not needed. Assuming that all members will be 2 × 4s, and using Appendix Table B-1, a reasonably conservative weight for the truss is estimated to be about

5 lb per horizontal foot. Assume 3 lb/ft of this as top-chord load and the other 2 lb/ft as bottom-chord.

For a roof pitch of 4 in. per ft, the sloping length per foot is 12.65 in., or 1.05 times the horizontal distance. Loads per *horizontal* ft applied to the top chord of one truss are:

Snow 2 × 25		= 50
Truss weight		= 3
Plywood sheathing	1.05 × 2 × 2.5 =	5
Roofing 1.05 × 2 × 6		= 13
Total (per horizontal ft)		= 71 lb

The horizontal length of each top-chord panel is $26/6 = 4.333$ ft.

Distributed load along the top chord is transferred to the truss at panel points. Panel point loads are estimated treating the chord in each 4.333-ft length as a simple beam, that is, ignoring continuity. Thus, a full top panel point load is $4.333 \times 71 = 308$ lb. End panel points receive one-half as much from top-chord loading.

Bottom-chord loads are:

Ceiling 2 × 8	=	16 lb/ft
Insulation 2 × 3 =		6
Truss weight	=	2
Total	=	24 lb/ft

The panel length for the bottom chord is $26/5 = 5.2$ ft, so a full bottom panel point load is $5.2 \times 24 = 125$ lb. The end bottom-chord panel point in this case is also a top-chord panel point, so the end downward load is the sum of loads from a half top panel and a half bottom panel, or $308/2 + 125/2 = 216$ lb. Total end reactions (not merely the net force at the end panel point) are needed to compute the end-bearing requirement, and are 1236 lb each.

Figure 10-6b summarizes the panel point loads for full, symmetrical snow load and dead load and shows the axial forces in each member due to these loads. For some members, though, full snow load *may* not cause the worst condition, so the truss is also analyzed assuming snow load on only one-half of the roof. Figure 10-6c shows the panel point loads and axial forces for this case. (Truss analysis was by the method of sections, as shown in any book on structural analysis.)

Both top and bottom chords have transverse load, so they will have bending moments as well as axial forces. Web members are all assumed to be pin-connected to the chords. Having no transverse loads, the web members have no bending moments.

For the top chord, for negative bending moment at the panel point, Table 10-1 gives factor $Q = 0.85$

(a)

(b) LOAD CASE I: *D* + FULL SNOW (LB)

(c) LOAD CASE II: *D* + SNOW ON LEFT ONLY (LB)

Fig. 10-6. Truss for Examples 10-1 and 10-2.

for a three-panel member. For length, L, we must use the largest of the three choices shown:

1. Horizontal length of the panel being considered times 0.9
2. Average horizontal length of the panel being considered and the longer adjacent panel
3. Horizontal length of the longer adjacent panel times 0.9

This truss has equal-length top panels, so the second of the above choices is largest, and $L = 4.333$ ft.

Negative bending moments at the end of top-chord panels are

$$M = 71(0.85 \times 4.333)^2/8 = 120 \text{ ft-lb}$$

To compute positive bending moment at the mid-length of top-chord panels requires knowing dimensions S_t and S_a, dimensions of the *scarf*, that is, the diagonal cut, shown by Fig. 10-7. We do not yet know these dimensions, but can make a fairly good guess. Later, if our guess proves wrong, we must correct it and compute again. For now, knowing the pitch of the top chord and assuming 4-in. nominal

Typical Heel Detail

Typical Scissor Heel

Fig. 10-7. Heel joints. (Courtesy of Truss Plate Institute.)

depths of members, we can either sketch the heel joint to scale or calculate to see that S_t will be 10.5 in. and, if $B = 3.5$ in., S_a will be 7-in. Now, for the panel next to the heel joint, using $c = 0.33$, the largest of the three choices for L in Table 10-1 is

$$\tfrac{1}{2}(L_i + L_a) + cS_a$$
$$= 0.5(4.333 + 4.333) + (0.33 \times 7/12)$$
$$= 4.53 \text{ ft}$$

Factor Q is

$$0.53(\cot \theta)^{0.36} = 0.53(12/4)^{0.36} = 0.787$$

and

$$\text{Positive } M = 71(0.787 \times 4.53)^2/8 = 113 \text{ ft-lb}$$

For the bottom chord, from Table 10-4, factor Q is 1.0 and length L is the largest of three choices shown, which is 5.2 ft. The critical bottom-chord bending moment, at the panel points, is

$$M = 24(5.2^2/8) = 81 \text{ ft-lb}$$

Example 10-2 will show how top- and bottom-chord members are designed for these combined axial loads and bending moments.

10-3. LIGHT-FRAME TRUSSES, MEMBER DESIGN

Compression Members

The equations for allowable compressive stress given by Table 10-5 for light-frame truss mem-

bers are the same as those of Chapter 7, except for the additional buckling stiffness factor, C_T.

For the top-chord compression members, buckling may be possible either in the plane of the truss or perpendicular to that plane. For buckling in the plane of the truss, the effective buckling length, L', equals QL as found in Tables 10-1, 10-2, and 10-3, except that for sloping members the horizontal length, L, is multiplied by sec θ (Fig. 10-8).

Buckling perpendicular to the plane of the truss is prevented when roof or floor sheathing is attached all along the length of the top chord. If sheathing is not attached, the effective length for buckling of the top chord is the distance between purlins (dimension L_p in Fig. 10-8) or between points that are otherwise laterally braced.

For compression web members, the effective length for buckling in the plane of the truss is 0.8 times the member length, L_w. For buckling perpendicular to the plane of the truss, the effective length is $0.8L_w$ or (if a brace is attached) the length L_p shown for a web member in Fig. 10-8.

Bottom chords are seldom compression members, but if they are stressed in compression, their effective lengths are as shown in Fig. 10-8.

Factor C_T in the equations of Table 10-5 is normally 1.0, but may be determined by Eqs. 10-2 and 10-3 below, provided:

1. The member is 2×4 or smaller.
2. Continuous 3/8-in. or thicker plywood is attached to the narrow face of the chord.
3. The member has both bending and axial compression.
4. The truss is used under dry conditions only.

If *all four* of these conditions are met, then

$$C_T = 1 + 2300L'/(kE) \qquad (10\text{-}2)$$

for wood seasoned to 19% or less MC when the plywood is attached, or

$$C_T = 1 + 1200L'/(kE) \qquad (10\text{-}3)$$

Table 10-5.   Formulas for Determining F'_c and K.

L'/d Ratio	Visually Graded Lumber	Machine-Stress-Rated Lumber[a]
$L'/d \leq 11$	$F'_c = F_c$	$F'_c = F_c$
$11 < L'/d < K$	$F'_c = F_c\left[1 - \dfrac{1}{3}\left(\dfrac{L'/d}{K}\right)^4\right]$ where $K = 0.671\sqrt{\dfrac{EC_T}{F_c}}$	$F'_c = F_c\left[1 - \dfrac{1}{3}\left(\dfrac{L'/d}{K}\right)^4\right]$ where $K = 0.792\sqrt{\dfrac{EC_T}{F_c}}$
$L'/d \geq K$	$F'_c = \dfrac{0.30\,EC_T}{(L'/d)^2}$	$F'_c = \dfrac{0.418\,EC_T}{(L'/d)^2}$

[a]See Paragraph G.9, Appendix G, *National Design Specification for Wood Construction*, 1986 Edition.
Source: Courtesy of Truss Plate Institute.

for wood that is either partially seasoned or unseasoned when the plywood is attached. In the equations for C_T:

Effective buckling length L' must be not more than 96 in.

Factor k is 0.819 for machine-stress-rated lumber and 0.589 for visual graded lumber.

Just as for compression members designed under NDS, the maximum allowable ratio L'/d is 50.

Allowable bending stresses, F'_b, are found using the equations in Table 10-6. These are the same as the NDS equations in Chapter 6. TPI-85 does not say whether the basic allowable F_b should be that for single-member use, or that for repetitive members. The repetitive-member allowable would seem reasonable for top-chord design, provided all conditions spelled out by the NDS for such members are met.

For members with combined bending and axial compression, interaction equation 7-12 is used, in which factor J is

$$J = \frac{L'/d - 11}{K - 11} \qquad (10\text{-}4)$$

Member	Effective Dimension (d)	Effective Buckling Length (L')
Top Chord	h	See 3.2.3
	b	L_p
Bottom Chord	h	L_1, L_2
	b	L_p
Web	h	$0.8\,L_w$
	b	L_p or $0.8\,L_w$

Where sheathing or ceiling is nailed to a compression chord (either top or bottom), lateral support of the supported axis may be assumed to be continuous and $L'/d = L_p/b$ may be neglected.

Fig. 10-8. Effective buckling lengths. (Courtesy of Truss Plate Institute.)

10-6. Formulas for Determining F_b'.

C_s Value	Visually Graded Lumber	Machine-Stress-Rated Lumber
$C_s \leq 10$	$F_b' = F_b$	$F_b' = F_b$
$10 < C_s \leq C_k$	$F_b' = F_b\left[1 - \dfrac{1}{3}\left(\dfrac{C_s}{C_k}\right)^4\right]$ where $C_k = 0.811\sqrt{\dfrac{E}{F_b}}$	$F_b' = F_b\left[1 - \dfrac{1}{3}\left(\dfrac{C_s}{C_k}\right)^4\right]$ where $C_k = 0.956\sqrt{\dfrac{E}{F_b}}$
$C_s > C_k$	$F_b = \dfrac{0.438\,E}{(C_s)^2}$	$F_b' = \dfrac{0.609\,E}{(C_s)^2}$

Source: Courtesty of Truss Plate Institute.

Factor J is based on L'/d for the plane of bending; J must be not less than zero, nor larger than 1.0, and is taken to be zero for buckling perpendicular to the truss plane.

Tension Members

Tension members are much easier to design. The NDS interaction equations of Chapter 7 apply to truss tension members with bending. Note that *both* Eqs. 7-21 and 7-22 must be used, since a large bending moment, even when acting together with axial tension, can cause compressive stresses on one extreme fiber. The maximum allowable slenderness, L'/d, for tension members is 80.

10-4. LIGHT-FRAME TRUSSES, CONNECTION DESIGN

Connections for light-frame trusses can be made using nailed or glued plywood gusset plates, metal nailer plates, or toothed metal plates commonly known as *truss plates*. For economic reasons, the toothed truss plate is now used almost exclusively. Truss plates are used in pairs, one plate on each side of the members being joined. They are pressed into the wood, so that the many projecting teeth penetrate the wood, acting as small nails, but in a regular, closely spaced array.

Standard design values for toothed plates are usually not available as they are for other connectors. Rather, individual manufacturers determine (for their own type of plate) the necessary values for truss design. Specification

TPI-85 spells out detailed test procedures the manufacturer must follow in determining those design values. These include a procedure for finding load-carrying capacities per tooth, per nail, or per square inch of contact area for each plate. Other test procedures are for the tensile and shear strength of the plate itself (9, 10).

The test procedure in TPI-85 tells how a basic design value for lateral resistance of a particular plate is determined. This basic design value can be expressed in various ways: pounds of lateral load transferred per tooth, pounds per nail, pounds per square inch of net contact area, or pounds per square inch of gross contact area. Capacity is measured both parallel and perpendicular to the grain and design values are determined for each direction. Design values found from a test on one species of lumber may be used for any other species having the same or higher specific gravity. The basic design value for lateral resistance is whichever is lower: 1/3 of the ultimate load reached in the test; or (1/1.6) times the load at *critical slip*. Critical slip is 0.015-in. movement of the plate relative to one piece of wood; if two members are tested end to end, it is 0.030-in. wood to wood.

Metal plate truss joints can reach their limit of usefulness by slipping, by the teeth breaking or pulling out of the wood, or by failure of the steel plate. The plate itself may determine the limit of load by buckling, by tensile fracture, or by plate shear failure. The truss designer must proportion the plates to prevent these types of failure.

Diagonal end cuts in either tension or compression members often reduce the net area

Fig. 10-9. Net widths, members with diagonal end cuts. (Courtesy of Truss Plate Institute.)

of the wood member. Figure 10-9 shows three such cases. Where diagonal cuts occur, the member stress is increased, and that increased stress must be limited to the modified allowable axial stress for the wood.

Basic Design Value for a Truss Plate

The basic design value for lateral resistance is used to determine how many teeth or how many square inches of contact area (net or gross, according to the manner of test evaluation used) are needed to transfer the member force. The basic design value must be modified for load duration (using the usual factors). It is reduced by 20% if the lumber has over 19% MC when the plates are installed. It is also reduced if the lumber has been pressure treated with fire-retardant chemicals. (TPI-85 does not say by what percent, but NDS appears to suggest 10%.)

If the plate basic design value was determined using the *Net Plate Area Method*, then in truss joint design, portions of the contact area between the plate and the wood must be ignored. These portions are a $\frac{1}{2}$-in. end distance parallel to the grain and $\frac{1}{4}$-in. edge distance (for each edge) perpendicular to the grain of the member. But if the basic design value were determined by the *Gross Plate Area Method*, these areas need not be excluded. In any case, these areas need not be excluded in the design of the "heel" joint, since reduction factors for that joint will account for end and edge distances.

Connection of Tension Members

A plate used to connect tension members will itself have tensile stress. Truss plate tests are conducted to find the load a perforated plate can carry and that which a similar solid plate can carry. The ratio of these loads is called the tensile effectiveness ratio, R. The allowable tensile load for a plate is then

$$T = RA_g F_{st} \qquad (10\text{-}5)$$

in which F_{st} is the allowable tensile stress for the steel.

The tension connection must have sufficient net contact area (or gross contact area or number of embedded teeth) to transfer the entire force in the member ending at that plate. For lightly loaded members, a minimum member force of 375 lb must be used.

Connection of Compression Members

Up to 50% of a compression member's load may be transferred by direct bearing of one member against the other; the rest must be transferred by the truss plate. Also, the plate area must be sufficient to transfer at least 375 lb of force per member. Truss plates in a compression connection might be able to buckle. To prevent this requires keeping the distance between adjoining ends of the connected members as small as possible. Specification TPI-85 requires that the member ends bear firmly against the abutting member.

Heel-Plate Connections

Figure 10-7 shows typical heel details. The heel joint is the most difficult light-frame truss joint to design rationally. Since the two member forces do not intersect within the limits of the plate area, they cause moments that tend to rotate the truss plates.

Specification TPI-85 provides an empirical design method to compensate for the effects of this rotating moment. The members are assumed to have axial load only, but the basic plate design values (allowable lateral resistance) for plates that are *not* eccentrically loaded are multiplied by a reducing factor

$$R_F = 0.85 - 0.05(12 \tan \theta - 2.0) \quad (10\text{-}6)$$

but not less than 0.65 nor larger than 0.85.

Example 10-2

Select member sizes and design connections for the truss analyzed in Example 10-1. Use the gross contact area method for connection design. Design values for lateral resistance for truss plates vary with the manufacturer and generally are not published, so we will assume a value of 90 lb per square inch of gross plate contact area.

If top chord members are well connected to a fairly rigid roof material that can transfer load from a weak truss to adjacent stronger ones (see the NDS definition), it seems reasonable to assume the top chord to be repetitive.

Unmodified design values given by the local building code are:

Bending, 1650 psi (repetitive)
Bending, 1400 psi (single member use)
Compression parallel to grain, 975 psi
Compression perpendicular to grain, 565 psi
Tension, 825 psi
Shear, 90 psi
Modulus of elasticity, 1,600,000 psi

Dead load is only about 30% of the total load per square foot, so the load combination $D + S$ controls, and all allowable stresses (except compression perpendicular to grain) can be modified by a load-duration factor of 1.15.

The top chord is laterally braced by closely spaced nails connecting the roof plywood (a rigid dia-phragm) to the truss. The chord has combined axial load and bending, so its design is by trial and error.

Top-chord selection: The maximum bending moment in the end panel of the top chord is 120 ft-lb, and the simultaneous axial load in that panel is 3226 lb compression. For bending alone,

$$S \text{ req} = 12(120)/(1.15 \times 1650) = 0.759 \text{ in.}^3$$

Try a 2 × 4 (for which $S = 3.063$).

Horizontal buckling is prevented. For vertical buckling, the effective length of each top chord panel is

$$QL \sec \theta = 0.85(4.333)(1.05) = 3.87 \text{ ft}$$
$$L'/d = 12(3.87)/3.5 = 13.27$$

By Eq. 10-2, buckling stiffness factor C_t is

$$1.0 + (2300 \times 12 \times 3.87)/(0.589 \times 1,600,000)$$
$$= 1.113$$

and factor K (Table 10-5) is

$$0.671\sqrt{EC_t/F_c}$$
$$= 0.671\sqrt{1,600,000(1.113)/(975 \times 1.15)}$$
$$= 26.7$$

L'/d is greater than 11 but less than K, so the fourth-degree parabolic equation of Table 10-5 is used to determine the allowable compressive stress.

$$F_c' = 1.15(975)\left[1 - \frac{1}{3}\left(\frac{13.27}{26.7}\right)^4\right] = 1098 \text{ psi}$$
$$F_b' = 1.15(1650) = 1898 \text{ psi}$$

Terms for use in interaction equation 7-12 are:

$$f_c = P/A = 3226/5.25 = 614 \text{ psi}$$
$$f_b = M/S = 12(120)/3.063 = 470 \text{ psi}$$
$$J = (13.27 - 11)/(26.7 - 11) = 0.145$$

Substituting in the interaction equation,

$$\frac{614}{1098} + \frac{470}{1898}\left(\frac{1}{1 - 0.145(614/1898)}\right)$$
$$= 0.819 < 1.0 \quad OK$$

Use 2 × 4 (assuming that satisfactory connections can be designed).

Bottom-chord selection: Try a 2 × 4 for the bottom chord also. The maximum axial tension is 3060 lb, and the bending moment in this same member (from Example 10-1) is 81 ft-lb. The modified allowable tensile stress is 1.15(825) = 949 psi. The material attached to the bottom chords is not stiff enough to transfer load from a chord that is weak in bending to adjacent stronger ones, so the bottom chord is a single-use member (not repetitive). Thus the modified bending allowable stress would be 1.15(1400) = 1610 psi. Other allowables are the same as for the top chord. Two interaction equations must be used—Eqs. 7-21 and 7-22. Terms needed for these equations are:

$$f_t = P/A = 3060/5.25 = 583 \text{ psi}$$

$$f_b = M/S = 12(81)/3.063 = 317 \text{ psi}$$

By Eq. 7-21,

$$583/949 + 317/1610 = 0.81 < 1.0 \quad OK$$

By Eq. 7-22,

$$317 - 583 = \text{negative}, < F_b = 1610 \quad OK$$

Both top and bottom chords can be made of 2 × 4 material.

Web member selection: For simplicity, try to find the minimum size that can be used for all members. Consider first the compression web members. In this truss, the longest compression diagonal also has the largest axial force (522 lb), so it will control the design for compression. After choosing a size for that member, we can then check to see if it is also satisfactory for the 668-lb maximum tension. It would probably be practical to use 2 × 4 members throughout, although a manufacturer might elect to use a smaller size for the web members if, in total, it would reduce the cost and if the narrower width did not make the truss-plate contact areas too small.

For our example design, try 2 × 4. Figure 10-8 shows the assumed length of web members to be from the near edge of the top chord to the near edge of the bottom chord, that is, the net length of the web member. For the compression diagonal, this net length is 3.369 ft minus approximately 3.5 in., or 36.9 in. Buckling effective length, L', is 0.8(36.9) = 29.52 in. Ratio L'/d will be 29.52/1.5 = 19.68, which is less than the limit of 50. Then K, found using $C_t = 1.0$, is 25.3. By Table 10-5, the allowable compressive stress is 984 psi, while the actual stress, P/A, is only 522/5.25 = 99 psi. Obviously, even a 2 × 2 could be used for this compression

member, provided it did not cause difficulties in connection design.

The longest tension web member is 5.053′ minus about 3.5″, or about 57 inches long. $L'/d = (0.8)(57)/1.5 = 30$, which is well within the limit of 80 for tension members. The axial stress is 668/5.25 = 127 psi, much less than the modified allowable 825(1.15) = 949 psi.

Connections: Design of two of the connections for lateral resistance only will be illustrated here. Other connections will be the subject of problems at the end of the chapter. (The shear and axial load values of each plate would also need to be checked in a complete design.)

Heel joint: Figure 10-10a shows the heel joint drawn to scale. Because top and bottom chords intersect beyond the limits of the area in which the truss plates can be placed, the heel-plate connection to each member is eccentrically loaded. The horizontal component of the top-chord compression and the tensile force in the bottom chord are a couple, tending to rotate the truss plate counterclockwise. To compensate for this, TPI-85 allows the connection to be designed as though each member applies axial load only, but requires that the design value for the truss plate be reduced by the factor in Eq. 10-6. For this particular connection, the reduction factor is

$$0.85 - 0.05[12(4/12) - 2.0] = 0.75$$

The modified allowable per square inch of plate contact area is

$$0.75(90)(1.15) = 77.6 \text{ psi}$$

and the gross contact areas required, per side of the joint, are

Top chord: $3226/(2 \times 77.6) = 20.79 \text{ in.}^2$

Bottom chord: $3060/(2 \times 77.6) = 19.72 \text{ in.}^2$

The scale drawing of the joint is used to determine the outline of plate that will best provide the above areas. After trying several sizes of rectangular plate, it was decided that the 4- × 13-in. plate shown as a dashed line in Fig. 10-10a would be satisfactory. However, there are many different satisfactory solutions. The one shown requires a diagonal cut at the corner so that the plate does not interfere with the roof plywood. This may be objectionable, but to eliminate the diagonal cut would require using a type of plate with a higher rating per square inch of area. Or, it might be possible to put a rectangular plate in

(a) HEEL JOINT

(b) JOINT *U*2

(c) ECCENTRIC LOAD EFFECT

Fig. 10-10. Connections for truss of Examples 10-1 and 10-2.

a skew position to get the required areas without such a cut.

Net areas of each wood member at this joint must be checked. The net area is defined as the product of member thickness and dimension h_1, as shown on Fig. 10-10a. With the plate edge $\frac{1}{2}$ in. from the edge of the 2 × 4, each member has $h_1 = 3$ in., so the net area of either member is $1.5(3) = 4.5$ in.^{2}. For the top chord, stress on this net area is $3226/4.5 = 717$ psi. This is less than the allowable F_c of $1.15(975) = 1120$ psi, so the top chord is satisfactory for net area. The bottom chord, also, can be shown satisfactory for stress on the net area.

Interior top-chord joint: Figure 10-10b shows, to scale, the second interior top-chord joint. This joint is highly statically indeterminate, since the members do not intersect at a common point and part of the load of the compression diagonal is transferred by direct bearing rather than by way of the truss plates. If this bearing force is ignored (or if poor fabrication leaves a gap between the chord and the end of the compression diagonal) then the teeth connecting the plates to the chord are eccentrically loaded. In practice this eccentricity is commonly ignored, and TPI-85 gives no guide as to how (or whether) to consider it. The authors believe it can be considered, however, and in the following they show how:

For the tension diagonal, each plate must carry one-half of whichever is larger, the member load or

375 lb. For lateral resistance, the required gross contact area per plate is

$$596/(2 \times 90 \times 1.15) = 2.88 \text{ in.}^2$$

For the compression diagonal, up to 50% of the 522-lb axial force may be considered transferred by end bearing; the plates must transfer the balance or 375 lb, whichever is larger. For this case, end bearing can be used for $522/2 = 261$ lb. The plates must be designed for 375 lb, or 188 lb per plate. The contact area required per plate is

$$375/(2 \times 90 \times 1.15) = 1.81 \text{ in.}^2$$

For the top chord, the plates must transfer the maximum *difference* between chord forces on opposite sides of the joint. In this case this maximum difference at this joint, for either case I or case II, is $3063 - 2297 = 766$ lb. If the chord were *spliced* at this joint, the two plates (one on each face of the spliced member) would have to carry the entire chord force minus up to 50% of that force assigned to end bearing.

The common design method merely provides enough plate contact area to carry this 766 lb. However, the truss members at this joint are not concurrent. In the absence of guiding rules by TPI-85, the authors would prefer to consider the plate-to-chord joint eccentrically loaded, as shown by Fig. 10-10c.

Try two 5- $\times$ $4\frac{1}{2}$-in. plates, as shown by Fig. 10-10b. This provides more than enough contact area to connect the two diagonal web members. The contact area with the top chord measures 4.5 in. by 2.25 in. Assuming that this area is subject to $766/2 = 383$ lb per plate and a moment of $383e = 383(1.125) = 431$ in.-lb per plate, the largest shear stress *per square inch of contact area* (with the top chord) is the vector sum of P/A and Md/J, where J is the polar moment of inertia of the contact area and d is the distance from the centroid of the area to the point farthest from the centroid. (This is exactly the same as the elastic method for analyzing eccentrically loaded bolted joints in structural steel.)

$$J = I_x + I_y = (1/12)[(4.5(2.25)^3 + 2.25(4.5)^3] = 21.36 \text{ in.}^4$$

$$d = \sqrt{2.25^2 + 1.125^2} = 2.52 \text{ in.}$$

The maximum shear stress between one plate and the top chord, computed as suggested above, is the vector sum of

$$P/A = 383/(2.25 \times 4.5) = 38 \text{ psi}$$

and

$$Md/J = 431(2.52)/21.36 = 51 \text{ psi}$$

or 76 psi, shown graphically on Fig. 10-10c. This is less than the allowable of $1.15(90) = 104$ psi, so the plate size selected should be satisfactory.

10-5. BRACING AND ERECTION OF LIGHT-FRAME TRUSSES

Light-frame truss material is thin. Thus, light-frame trusses are easily damaged in handling and erection, and their member strength is often controlled by buckling. Obviously, bracing is very important. Reference 11 emphasizes this, giving valuable recommendations for bracing light-frame trusses. The bracing serves two purposes: (1) It holds the trusses in the desired position, keeping them parallel and preventing tipping; and (2) it may also reduce the effective length of compression members. Bracing is needed when the finished building is subject to

design loads, but is often even more important during construction. While the truss designer may specify the type of bracing required for strength of the truss members, it is often the *building* designer who decides and specifies the complete bracing system.

Figure 10-11 shows recommended bracing for a series of light-frame trusses. Bracing to keep the trusses upright, preventing them from tilting sidewise, is as important during construction as it is in the finished building. This bracing usually consists of 2-in. nominal diagonal members, sloping at about 45° and attached to the web members of three consecutive trusses, as shown by Fig. 10-11b. Being attached to all three trusses, the diagonal brace does two things: (1) it prevents each truss from tilting, and (2) it reduces to one-half the effective length of the web member in the intermediate truss. Continuous longitudinal (horizontal) 2-in. members attached to the other face of this braced member transfer the bracing effect to similar members in adjacent trusses along the building length. Additional diagonal bracing sets are used at about 20-ft intervals along the building.

Lateral support at mid-length of a web member cannot be provided by the horizontal brace alone, unless some detail prevents that brace from moving longitudinally. Without such a detail, the horizontal member merely forces all members to which it is attached to buckle in the same direction, as shown by Fig. 10-11c. To prevent this, the brace member must either be connected to a diagonal brace (Fig. 10-11b) or have its end restrained as in Fig. 10-11d.

After plywood sheathing is nailed to the top chords of roof or floor trusses, the chords are continuously laterally braced against buckling normal to the plane of the truss. However, significant loads may be applied during construction, before the sheathing is attached. To prevent buckling at this stage, top-chord bracing is needed. Figure 10-12 shows a recommended bracing system. The bracing, rather than being attached as shown, is preferably attached to the *bottom* surface of the top chord, so that it does not have to be removed as the sheathing is attached to the top surface.

Fig. 10-11. Bracing between light-frame trusses. Courtesy of Truss Plate Institute.

Fig. 10-12. Recommended top-chord bracing for light-frame trusses. (Courtesy of Truss Plate Institute.)

10-6. HEAVY-TIMBER TRUSSES

Before structural steel became practical, heavy-timber trusses were commonly used for bridges and buildings. Today, building designers often select exposed, heavy-timber trusses for their attractive appearance. Such trusses are not only attractive, but because of the large size of their members, they also have good fire resistance. Glued laminated members rather than sawn timbers are used in most modern heavy-timber trusses. Glulams as truss members offer the same advantages over sawn timbers as they do for beams (Chapter 8).

Types of Heavy-Timber Truss

For our purpose, there are two main types of heavy-timber truss: (1) those with steel or (on very old trusses) wrought-iron vertical tension members but without gusset plates, and (2) those with steel gusset plates at every joint. The first of these we will refer to as old-style timber trusses.

10-7. OLD-STYLE HEAVY-TIMBER TRUSSES

Old-style heavy-timber trusses usually were of Howe truss configuration, shown in Fig. 10-1b. When the Howe truss is fully loaded, its web verticals are tension members and its web diagonals compression members. For this reason, the web verticals could be merely threaded metal rods.

Connections for Old-Style Heavy-Timber Trusses

Connections for the Howe truss were relatively simple. The vertical rods were connected using just large washers and nuts. Tightening the nuts compressed the ends of the diagonals firmly against the chords, holding the whole assembly together. The compression members transferred their forces to the chords by end bearing of wood against wood, or of wood against cast-iron joint blocks.

Figure 10-13 shows several examples of connections recommended by Henry Dewell (12) for old-style heavy-timber trusses. They are shown and discussed here because, in building renovation or in historic preservation work, today's structural designer may have to investigate existing old-style trusses to determine their suitability for a new use.

Figure 10-13a shows a simple heel joint, in which the entire load of the top chord was transferred to the bottom-chord member by end bearing. Note that the axes of the two members met at a point directly above the reaction, so that the reaction and two member forces were

Fig. 10-13. Old-style heavy-timber truss connections. (From reference 12.)

apparently concurrent. Four steps were involved in designing this joint:

1. The allowable stress in compression at 30° to the grain of the bottom chord was determined so that the 5-in. dimension could be verified as adequate. (Of course, if the top chord were of a different species or grade than the

bottom chord, it is possible that bearing parallel to the grain of the top chord might control.)

2. The length of bottom chord extension to the left was determined, so that the bottom chord would not be "sliced off" by the 49,000-lb horizontal component of the top-chord force.

3. The net area below the bottom-chord notch

was checked to see whether the bottom chord was adequate to resist the 49,000-lb axial load. For this purpose, Dewell considered the force to be centrally located on the net section. (The net section at the bolt hole was also considered.)

4. The exact location of the support was determined to avoid applying a moment to the heel joint. Because of the tapered ends, the top-chord force of 56,500 lb is actually applied at a point not on the centroidal axis of the top chord. (Its horizontal component is 49,000 lb acting to the left.) Also, the bottom-chord tension acts through the center of the remaining net width after the slot is cut. Thus the two 49,000-lb forces form a couple with a 4-in. arm. To resist the moment of this counterclockwise couple, the support was moved 7 in. to the left of the intersection of top- and bottom-chord centerlines.

Force transfer from chord to chord on the long diagonal cut was apparently ignored. However, if there were any force transfer there, the two members would be forced apart, and the two bolts shown would resist this tendency.

Figure 10-13b shows a similar antique connection, but in this case neither cut at the end of the top chord was perpendicular to the grain. Thus, each surface was assumed to take a component of the chord force, each surface having a different allowable compressive stress, computed by Hankinson's equation. Other features of the design are similar to those for Fig. 10-13a.

Figure 10-13c shows a heel joint using a "bolster" to serve as a bearing block, so that the top-chord member is in bearing parallel to the grain. A short steel dowel keeps the top chord and the bolster in proper alignment. Pipe pins transfer the necessary shear force from the bolster to the bottom chord, and vertical bolts tie the bolster to the bottom chord. Note that with this connection, there is no need to shift the support point from the intersection of member centerlines. Bottom-chord net area is whichever is less, net area at a bolt hole or at a groove for a pipe pin.

Figures 10-13d and e show two types of intermediate joint for top and bottom chords of

Howe trusses. In the first detail the ends of the diagonal bear against the sides of notches cut into the chords. The top chord also is notched where a horizontal bearing surface is needed under the washer and nut of the threaded vertical member. The second detail uses bolster blocks to transfer the diagonal member's load to the chords. In this detail, the bolster blocks are keyed into $\frac{3}{4}$-in. daps of the chord members. A component of the diagonal's force is transferred to the chord where the bolster bears against the end of the dap. A single bolt through each bolster block and chord holds the parts together until assembly is complete.

Of these latter two details, that of Fig. 10-13e is preferable. The joint of Fig. 10-13d, having the diagonal inserted into skew-cut chord notches, required skillful carpentry to ensure good contact between members over the entire intended bearing surfaces. The joint using the bolster was much easier to fabricate so as to get contact over the entire cross section. All that should be added to the detail of Fig. 10-13e are short dowels to key the ends of the diagonal member to the bolsters.

The vertical rods of old-style trusses often had *upset* ends, that is, the rod ends were forged to a larger diameter. Threads cut into the upset portion would not reduce the net area at the threads to less than the gross area of the unthreaded shank. Dewell cautions investigators to make certain that the upset portion was not attached by welding. At the time that these trusses were popular, welding was not nearly as reliable as it is today. Nevertheless, some fabricators, lacking the expensive equipment needed for upsetting, would often substitute welded rods of lesser quality.

Engineers or architects investigating trusses of this type would benefit by reading the 1917 book by Dewell (12) to learn from his extensive experience with these trusses and his remarkable insight into how these connections function.

Example 10-3

Assume that you are investigating an old building to see if it can carry a 55-psf live load, plus the code-required 20 psf for partitions, applied to the existing

Fig. 10-14. Truss for Example 10-3.

floor system. Figure 10-14, a scale drawing prepared from your field notes, shows one of several old-style trusses in the building. These trusses are 18 ft 0 in. c/c. Member sizes shown are all "full" dimensions, that is, the member is the size its name implies. The truss is some 80–90 years old, but your thorough examination of the truss shows all materials to be in good condition, with no moisture damage and no splits. The vertical rods (not upset) are wrought iron. The bottom layer of floor planks is laid diagonally, so that the floor serves as an effective diaphragm.

For this example check only the top chord, the vertical rods, the bottom chord at joint L_2, and the connection at joint L_1.

First, you must determine suitable allowable stresses. If the vertical rods were steel, allowable stresses in use for structural steel when the truss was built would probably be satisfactory. But these rods are wrought iron and today's building codes do not show allowable stresses for wrought iron. So you

have shored up the truss carefully and have removed and tested one of the vertical members, intending to replace it with a modern steel rod. Your test of the removed rod shows a 30,500-psi yield strength and 50,000-psi tensile strength. Comparing to the AISC code for structural steel, you note that for rods of A36 steel, the allowable tensile stress is about 0.53 times the yield strength and 0.33 times the minimum ultimate strength. These same percentages, applied to the wrought iron, suggest 16 ksi based on yield or 16.5 ksi based on ultimate. With these comparisons as a guide, you decide, conservatively, to allow 15-ksi tensile stress on the net area of the threaded rod. (For today's threaded steel rods, gross area is now used.)

For the wood members, determining an allowable stress is not nearly so easy. (See references 1 and 13.) Species, grade and condition must all be considered. Grade is probably unknown. Even if a grade name is known, the grading system used when this truss was built probably differs from that of today. Even worse, there may have been no grading system at all. A suitable solution would be to employ a qualified timber grader to grade the lumber of this truss according to today's standards. Assume that you have done this and, using your grader's answers, have decided from the Supplement to NDS (reference 4) that the following unmodified allowable stresses should be used:

Bending, 1300 psi
Compression parallel to grain, 970 psi
Compression perpendicular to grain, 520 psi
Tension, 900 psi
Modulus of elasticity, 1,300,000 psi

The above are unmodified allowables for dry use under normal-load duration. The truss is some 80–90 years old and the building was in use for most of that time. It seems reasonable to assume that, by now, the entire 10-year period of full design loading has occurred and that continued use of the building will extend loading beyond normal duration. Therefore, you decide that it would be wise to treat any further loading as permanent, multiplying each of the above allowables for the wood by a duration factor of 0.9 (except for $F_{c\perp}$).

Your next step is to compute the loads. The floor load is

Live load	= 55	psf
Double board floor, about $35(1.5/12)$ =	4.4	
Specified partition load	= 20	
Total	= 79.4	psf

Weight of joist = 2 × 12 (full dimensions)$(35/144)$ = 5.8 lb/ft
Reaction of one joist = $9(79.4 + 5.8)$ = 767 lb

The load per foot of top chord is the sum of two joist reactions plus the self-weight of the chord itself. Estimated truss weight (pounds per foot of truss):

Top chord, $10(8)(34/144)$	= 18.9
Bottom chord, $8(8)(34/144)$	= 15.1
Wood web members, $1.38(8)(8) \times$ $(34/144)$	= 20.9
Wrought-iron rods, $1.23(3.4)(7/5.5)$	= 5.3
Total, per foot of truss	= 60.2 lb

(3.4 is the weight, in pounds, of a 12-in. length of wrought iron or steel with a 1.0-in.^{2} cross-sectional area.)

Dividing the last two items equally between top

and bottom chords, the top chord will receive a total load of $32 + (2 \times 767)$ = 1566 lb/ft and the bottom chord only 28 lb/ft. For practical purposes, all the load (say 1590 lb/ft) can be treated as top-chord load. Thus, a top interior panel point will receive $5.5(1590)$, or 8740 lb, and end ones each one-half as much. Figure 10-14 shows the truss loads and the member forces.

Actual top-chord uniform load of about $(2 \times 767) + 19$ = 1553 lb/ft causes vertical bending moments in the chord. (Web member and bottom-chord weights do not cause top-chord bending.) About $0.1w\ell^2$ is a reasonable value to use for bending moments in a continuous member of several spans, as this top chord. Thus, the estimated bending moment is

$$M = 0.1(1553)(5.5^2) = 4700 \text{ ft-lb}$$

Check the top chord in panel $U_2 U_3$: Computed actual stresses on the net area are:

$$f_b = 12 \times (4700)/(8 \times 8.5^2/6) = 585 \text{ psi}$$
$$f_c = 32,050/(8.5 \times 8) = 471 \text{ psi}$$

The $1\frac{1}{2}$-in. notch is on the edge of the member where the bending stress is compressive. The web member may not tightly fill the notch, however, so the net section rather than gross was used above.

Allowable stresses are computed next. The top chord is laterally braced horizontally by the floor system nailed to it. Vertically, it is free to buckle over a 5-ft 6-in. length. Ratio ℓ/d is $12(5.5)/10$ = $6.6 < 11$, so the member is considered to be a short column. The allowable compressive stress is F_c = $0.9(970)$ = 873 psi. Factor K is

$$K = 0.671\sqrt{1,300,000/873} = 25.9$$
$$J = (\ell/d - 11)/(K - 11) = \text{negative}$$
$$\text{Use } J = 0.0$$

By interaction Eq. 7-12,

$$471/873 + 585/(1300 \times 0.9) = 1.04 \simeq 1.0$$

which is marginal, but most designers would accept as satisfactory.

Check wrought-iron vertical: Figure 10-14 shows the largest tensile force to be 13.11 k. For a $1\frac{1}{4}$-in.-diameter tension rod with U.S. Standard threads the net (root) area is 0.969 in.^{2} (14). Stress on this net area is

$$13,110/0.969 = 13,529 \text{ psi}$$

This does not exceed the allowable of 15,000 psi, so the rods are satisfactory.

Check bottom chord at L_2: Net area of the bottom chord at the notch is

$$5.30(8) = 42.40 \text{ in.}^2$$

The bolt hole is at about the same location as the notch, so the net area is further reduced to

$$42.40 - (1.3125 \times 5.30) = 35.44 \text{ in.}^2$$

The tensile stress on the net area is

$$36,050/35.44 = 1017 \text{ psi}$$

which is considerably above the allowable of $0.9(900) = 810$ psi. Panel 1-2 of the bottom chord will also show overstress. If the truss is to carry the planned load, the central four panels of the bottom chord will have to be reinforced.

Check joint L_1: Assuming only the square-cut end of the diagonal to be effective, the member load of 17.78 k is transferred on an area of 4×8 in., or 32.0 in.^{2}. The compressive stress is $17,780/32.0 = 556$ psi. For the web diagonal, this is less than $F_c = 0.9(970)$, and obviously less than the end bearing value, F_g, so it is satisfactory. For the bottom chord, the stress is at an angle of $47.5.°$ to the grain. Hankinson's equation gives an allowable compressive stress of 638 psi > 556 psi. So the joint is satisfactory. (*Note:* F_g in end bearing could be used, if it were known, in place of F_c in Hankinson's equation also.)

10-8. MODERN GUSSETED HEAVY-TIMBER TRUSSES

Members of modern heavy-timber trusses are usually connected by bolts passing through the members and two steel gusset plates, one on each face of the members. For this to be practical, thickness normal to the truss plane should be alike for all members. Where the bearing strength of the bolt on the wood is not adequate, bolts may be supplemented by shear plates embedded in each face of the timber members. Advantages of the truss with double-plane gusset plates are:

1. Design flexibility. It is not necessary to conform to particular truss configurations, as was the case with the old-style trusses.

2. Less carpentry skill is needed to produce good-quality trusses than was the case with the old-style truss. Members do not have to be notched, nor do ends of web members need to be cut to precise angles to obtain end bearing against other members.

3. Double-gusseted trusses can be of very attractive appearance. For this reason, they are frequently exposed in building interiors. The truss configuration used can be selected for aesthetics, rather than just utility, although the simplest pattern is often also the most attractive.

In general, member design is for axial load only. The exception would be members having significant bending moment. Bending can result from three causes: (1) floor or roof loads applied to chords between panel points; (2) self-weight, for members with fairly large horizontal components of length; and (3) secondary stresses. Secondary bending stresses occur because the truss members are *not* pin connected, as assumed in analysis. Where members are long and slender, these secondary stresses are usually negligible. However, if the truss members have low length/depth ratios, secondary stresses may be significant. Matrix computer analysis giving end bending moments as well as axial forces is so simple and inexpensive today that a designer can easily consider all causes of bending. If significant bending moment is present, the members would be designed for combined axial load and bending.

When choosing a top-chord size, a designer must consider how the top chord is braced laterally (against movement normal to the plane of the truss). If the structural roof material has appreciable diaphram stiffness and is attached to the chord at very short intervals, the chord will be prevented from buckling normal to the truss plane. However, it still can buckle vertically over the length between panel points. In this case, its effective length can be taken as the panel length.

If the structural roof or floor material is not connected directly to the top chord, then it is usually connected to purlins, that is, to beams spanning from truss to truss. In this case, at best, the chord would be laterally supported only at points where the purlins are attached to the truss. Even then, if the purlins are not attached to a competent roof diaphragm, the top chord may not be adequately laterally supported. In this case the effective length of the top chord for buckling normal to the truss plane would be the distance between points at which an adequate bracing system frames into the truss. Usually, this distance is more than a single panel length. So the chord in this case would have two effective lengths—a single panel length for buckling vertically and a longer length (two or three panels, for example) for buckling normal to the plane of the truss. Figure 10-15 illustrates this.

Web compression members ordinarily are unbraced in either direction. Thus, their effective length for determining allowable stress is their actual length. The authors consider the actual length to be that measured along the web member's centerline from the center of one chord to the center of the other. Bottom-chord members and web tension members must be designed for their *net* area. Connections for

gusseted trusses are designed by the methods shown in Chapters 5 and 9.

10-9. TRUSSES CONNECTED BY SPLIT RINGS

Split-ring trusses, many of which are still in use, often had fairly large spans and were used for bridges, buildings, towers, and other structures. Figure 10-16 shows an example of such a truss, and Example 9-9 illustrated the analysis of one of its split-ring joints. The same procedure could be used for analyzing the rest of the joints.

In the designing of these trusses, an important step was to determine how the various layers of adjoining members were to be "stacked." This problem is shown by Fig. 10-16. In this truss, the largest number of pieces used per member was two (for both chords and web diagonals). The vertical members had just one element each and the supporting column had three elements. All pieces

Fig. 10-15. Effective lengths for top-chord buckling.

Fig. 10-16. Timber roof truss using $2\frac{1}{2}''$ split-ring connectors.

overlapped each other as shown, so that no element had to be bent to a different position in the stack at one end than at the other. In the designing of the members for strength, it might be that a single element would have been good enough; however, if two were needed so that the truss could be assembled without any members being bent or being off-center in the stack, then two elements were used.

Compression members having more than one element were designed as spaced columns (see Chapter 7), or the individual elements were treated as separate solid columns. In the truss of Fig. 10-16, the top chord was designed as a spaced column, while the compression verticals were single-element solid columns.

Having to design such a truss is unlikely today, but if the need does arise, a suggested design sequence is:

1. Determine member forces for load combinations to be considered.
2. Choose a species and stress grade.
3. Plan a tentative stacking arangement.
4. Design the top chord, considering the problems of connections, so that end- and edge-distance requirements will not be violated.
5. Design the other members.

6. If the tentative stacking arrangement is still OK, design the joints.

Analysis of existing split-ring trusses is a more likely task than new design. In analyzing existing trusses, it is important to determine the condition of all members. Split-ring connections having more than one bolt often caused members to split as moisture content changed. In the rush of wartime construction, rings were sometimes inadvertently omitted. Through shrinkage, connections may have loosened. And, of course, the wood may have deteriorated. Allowable tensile stresses used in wartime were often higher than those tolerated today. And, finally, the trusses may already have been loaded for periods exceeding the ten-year normal duration, so future loading would wisely be treated as permanent when allowable stresses are selected (references 1, 13).

10-10. TRUSS DEFLECTIONS AND CAMBER

Truss deflections should be limited for the same reasons that beam deflections are limited. However, computing deflections for trusses is slightly more trouble than for beams. As with beams, both immediate and long-term deflections must be considered. Trusses deflect because of (1) changes in member length, and (2) slip at the end connections of members. (See Fig. 10-17.)

If a computer matrix analysis method is used, the portion of deflection due to member elongation or shortening and member bending is easily determined for every joint. The computer program would probably have to be modified, however, so that it would include the effect of joint slip.

If the deflection at only a single point is needed (this is usually the case) then the virtual work method is the easiest to use. (See any book on structural analysis.) For trusses, this requires making a table showing, for each truss member, the following:

1. Force F (axial load) due to the loading considered (lb)
2. Cross sectional area, A (in.^{2})
3. Length L (in.)
4. Modulus of elasticity, E (psi)
5. Member elongation, FL/AE (in.), either positive or negative. Modify member elongation, FL/AE, by adding or subtracting to cover slip of the end connections. The resulting total, $FL/AE + S$, is the change of distance between joints at opposite ends of the member.
6. Force, u, in each member due to a unit (virtual) load in the direction and at the location of the deflection to be computed.

The desired deflection, Δ, is then given by

$$1 \times \Delta = \sum \left[(FL/AE + S)(u) \right] \quad (10\text{-}7)$$

F and S may each be either positive or negative, depending on whether the member force, F, is tension or compression. Slip, S, may involve slip at one end only or both ends, depending on where the actual member ends are. Virtual force, u, may be of either sign. The product for each member in Eq. 10-7 may also have either sign. For hybrid trusses, in which some members are metal and others are wood, appropriate values of E must be used for each member. Equation 10-7 considers only the effects of axial force and end slip in each member. If deflection at mid-length of a member is desired, bending deflection for that member must be added to the deflection of the truss joints.

Fig. 10-17. Light-frame trusses under construction load of over 100 psf. Truss deflection about one in. Note square panel at midspan for passage of duct work. (Photograph by authors.)

The deflection solved will be the immediate deflection, not considering creep. Long-term deflection could be estimated by multiplying the immediate deflections by a factor, as was done in Chapter 6 for beams.

For heavy timber trusses, reference 15 recommends using the difference between bolt diameter and the drilled hole diameter for joint slip. Dimensions S to use in calculating deflection for a truss with bolted joints would be twice this amount for members that could have such slip at each end. For split rings or shear plates, reference 15 suggests that slip be taken as

For unseasoned lumber, 0.07 in. × load ratio, or
For seasoned lumber, 0.05 in. × load ratio

The load ratio is the actual load for one connector divided by its allowable load.

For light-frame trusses, TPI-85 offers no help in determining deflections, except for full-scale testing. The authors believe, however, that a fair estimate might be obtained using the above method, assuming that dimension S is found using either the "critical slip" of 0.015 in. at each end, or perhaps using about two-thirds of that amount to represent probable slip under working loads. Greater reliability is obtained for light-frame trusses by performing full-scale tests, and this is often a practical method since the trusses are mass-production items.

Reference 3 gives a beam-type equation that may be used to estimate deflection of trusses with parallel top and bottom chords.

Camber

Bottom-chord camber offsets the bad appearance of sagging when the truss deflects. If the top chord is horizontal, it too should be cambered. If the amount of camber is not specified by code, it should be at least equal to the long-term dead-load deflection expected. Trusses supporting a flat roof should be cambered even more to prevent ponding. To provide camber, web member lengths must be adjusted to fit the geometry of the cambered position of the panel points. Continuous top- and bottom-chord

members are bent into the cambered position. Since the camber dimension is small, stresses due to this bending are usually ignored in design. In any case, these bending stresses are relieved when the truss deflects under load.

REFERENCES

1. *Evaluation, Maintenance and Upgrading of Wood Structures, A Guide and Commentary*, American Society of Civil Engineers, New York, 1982.
2. *Design Specification for Metal Plate Connected Wood Trusses*, TPI-85, Truss Plate Institute, Madison, WI, 1985.
3. *Design Specification for Metal Plate Connected Parallel Chord Wood Trusses*, PCT-80, Truss Plate Institute, Madison, WI, 1981.
4. *National Design Specification for Wood Construction* (and Supplement), National Forest Products Association, Washington, DC, 1986.
5. "A Computerized Wood Engineering System: Purdue Plane Structures Analyzer," USDA Forest Service Research Paper FPL 168, Forest Products Laboratory, Madison, WI, 1972.
6. Bathe, K. J., E. L. Wilson, and F. E. Peterson, *SAP IV, A Structural Analysis Program for Static and Dynamic Response of Linear Systems*, College of Engineering, University of California, Berkeley, CA, 1974.
7. Wilson, E. L., and A. Habibullah, *SAP80 Structural Analysis Programs*, Computers and Structures, Inc., Berkeley, CA, 1986.
8. Hoyle, R. J., Jr., *Wood Technology in the Design of Structures*, Mountain Press Publishing Co., Missoula, MT, 1978.
9. *Standard Method of Test for Tensile Strength Properties of Steel Truss Plates*, ASTM E 489-81, American Society for Testing and Materials, Philadelphia, PA, 1981.
10. *Standard Test Method for Shear Resistance of Steel Truss Plates*, ASTM E67-80, American Society for Testing and Materials, Philadelphia, PA, 1980.
11. *Commentary and Recommendations for Bracing Wood Trusses*, BWT-76, Truss Plate Institute, Madison, WI, 1976.
12. Dewell Henry D., *Timber Framing*, Dewey Publishing Company, San Francisco, 1917.
13. Lanius, R. M., Jr., "Evaluating Residual Strength and Repair of Structures," in Meyer, R. W., and R. M. Kellogg, *Structural Uses of Wood in Adverse Environments*, Van Nostrand Reinhold, New York, 1982.
14. *Specification for the Design, Fabrication and Erection of Structural Steel for Buildings*, American Institute of Steel Construction, Chicago, IL, 1978.
15. Milbradt, K. P., "Timber Structures," in Gaylord, E. H., Jr., and C. N. Gaylord, *Handbook of Structural Engineering*, McGraw-Hill, New York, 1979.

PROBLEMS

10-1. Try redesigning the truss of Examples 10-1 and 10-2 (Fig. 10-6) using four equal panels rather than six for the top chord, and three rather than five for the bottom chord. Can all NDS requirements be met?

10-2. Design truss-plate connections for the other joints of the truss of Example 10-2, using the gross area method and 90 psi for allowable lateral resistance.

10-3. Design truss-plate connections for the joints of the truss of Example 10-2 (except the heel joint), but use the net contact area method and an allowable lateral resistance of 95 psi.

10-4. Choose a truss configuration and design a light-frame roof truss for a 30-ft span, without overhangs, with pitch of 4 in. in 12 in. Use 24-in. truss spacing. Assume snow load of 25 psf and roof dead loads the same as for Example 10-1. Assume that the ceiling attached to the bottom chord is accessible through a "scuttle," for which case TPI-85 specifies a minimum total bottom-chord load of 10 psf. However, for our case, the ceiling and insulation already weigh 11 psf. Would it be prudent to use a total bottom-chord load larger than 11 psf to allow for the chance that the building owner would store things in the space above the ceiling? (This is a type of decision a designer may have to make. You make the decision, and design accordingly.)

10-5. In one area of the old building of Example 10-3 (Fig. 10-14) the truss span is only 30 ft 0 in. rather than 33 ft 0 in. Floor load (psf), truss member sizes, and truss spacing are the same as for Example 10-3. Check strength of members for this truss.

Fig. 10-18. Truss for Problem 10-6.

10-6. Design the double-gusseted scissors truss shown by Fig. 10-18. Trusses are 18 ft apart. The roof is 3-in. wood decking, with roofing weighing 6 psf and insulation 3 psf. Purlins (at each top-chord panel point) weigh 340 lb each. Snow load is 35 psf of horizontal projection for a level surface. Determine required member sizes, using glulams, combination 10, D. fir.

10-7. For the truss of Problem 10-6, design joints *A* and *B*, using bolts and steel gusset plates on each face.

10-8. For the truss of Problem 10-6, design joint *C* using bolts and steel gusset places on each face. *Hint:* One of the long sloping members can be continuous through joint *C*.

10-9. What would be the total long-term deflection for the truss you designed for Problem 10-4?

10-10. Find the total long-term deflection for the truss of Example 10-2.

11

Plywood and Similar Wood Products

One of the most important wood products for structural use is plywood. It is used for roof and floor sheathing, concrete formwork, webs of wood beams, and even the hulls of boats. It can be used to resist gravity loads or to resist lateral loads as in plywood diaphragms and shearwalls. Because of the way plywood is manufactured, it has more nearly isotropic properties and greater dimensional stability (resistance to change in dimension with moisture content) than sawn lumber. The dimensional change of plywood is only about one-tenth of that for solid timber over any given range in temperatures and relative humidities (1). Other manufactured wood products such as particleboard and insulation board have found general acceptance as building materials.

11-1. PLYWOOD PRODUCTION AND CLASSIFICATION

Plywood consists of thin wood sheets (called veneers or plies) assembled and glued together with the grain directions of adjacent layers arranged at right angles. The plies vary in thickness from 1/16 in. to 5/16 in. Structural plywood has an odd number of layers. In four-ply panels the two central plies have their grain in the same direction and therefore form a single layer. The standard nominal thickness of panels varies from 1/4 in. to $1\frac{1}{8}$ in.

The logs from which plywood is made are moistened before the manufacturing process begins by storing them in ponds, spraying them with water, or putting them in steam vats. They are then dropped into a vat where they are kept

in boiling water for six hours, so that the veneer will not break apart as it is peeled off. After the bark is removed, a saw cuts the logs typically into sections just over 8 ft long. Then a giant lathe rotates the logs, peeling off a continuous sheet of veneer, which is stored on a rack. When the veneer comes off the peel rack, it goes through a machine that clips out any bad veneer and strings it into continuous rolls. After the veneer is seasoned in a dryer to a moisture content of 5% or less, it is graded. The plies are assembled, glued, and pressed. Under heat and pressure, the glue hardens quickly and the sheets of plywood are complete. After the panel is inspected for quality of face and back, it is stamped with the grade trademark.

The standard plywood panel size is 4 × 8 ft. Other than 4 × 9 and 4 × 10 panels, odd sizes are so uncommon as to be uneconomical. Nearly all plywood in the United States conforms to the U.S. Department of Commerce Product Standard PS 1-83 (2). The American Plywood Association (APA), a manufacturers' association, publishes much information concerning the proper use and design properties of plywood. Most notable is the Plywood Design Specification (3), which contains allowable stresses and other design information.

Veneer Grades

The veneer is always seasoned to eliminate residual stresses that would occur if it were glued and hot-pressed before being allowed to dry. To classify plywood by its strength properties, the veneer itself must be graded. This grading process occurs just before the plies are assembled. The five veneer grades, in descending order of quality, are as follows:

N   No knots; may have small patches; sanded smooth; for natural finish

A   No knots; may have small patches; sanded smooth; for painting

B   Solid surface; small round knots; patches and round plugs are allowed

C   Plugged Improved C grade

C   Small knots, knotholes, patches; lowest grade allowed for exterior use

D   Larger knots, knotholes

The veneer grades that interest the structural designer the most are grades C and D. Many interior panels have grade C veneer on the face and grade D for inner and back plies. This is why these panels are sometimes called C-D interior. They are also referred to as APA Rated Sheathing, Exposure 1 or Exposure 2.

There are two main types of plywood—interior and exterior. Interior plywood is further categorized as either Exposure 1, Exposure 2, or simply Interior. Exposure 1 is made with an exterior glue that adds shear strength to the panel. Exposure 1 plywood is recommended if the conditions in use will be dry but prolonged exposure to moisture is expected during construction. Exposure 2 panels may be used where only moderate delays in providing protection may be expected. The third type of interior plywood is not used for strictly structural purposes.

Exterior plywood is bonded with waterproof adhesives. It has superior moisture resistance and should be used for panels exposed to wet conditions or high equilibrium moisture content. Grade D veneer (the lowest grade) is not allowed in Exterior plywood. The large open knotholes in grade D veneer could allow moisture to seep in between the plies, causing them to separate.

Plywood Grades

Table 11-1 shows plywood grades. The first three rows of interior plywood and the first three rows of exterior plywood each contain four grades of plywood that are for structural use. Most of these are unsanded. These four grades as classified by APA for structural use are:

1. APA Rated Sheathing (also called C-D Interior or C-C Exterior)—a general-use structural plywood that has all C-grade veneers if exterior and C face with D inner and back veneers if interior.
2. APA Structural I Rated Sheathing (also called Structural I C-D Interior or Structural I C-C Exterior)—very strong panel using all Group I species wood and only exterior glue; veneers same as above.
3. APA Structural II Rated Sheathing (also called Structural II C-D Interior or Structural II C-C Interior or Structural II C-C Exterior)—similar to Structural I, but Group 3 species wood is allowed.
4. APA Rated Sturd-I-Floor—for combination subfloor-underlayment (see Section 11-7).

Each of these comes in Exterior type, Exposure 1 type, and Exposure 2 type, except for Structural I and II for which Exposure 2 is not available. (The Structural II grade is not very common and is difficult to obtain.) Sturd-I-Floor is the only engineered grade that has been sanded at all; it is merely touch-sanded (somewhat, but not thoroughly smoothed).

The other plywood grades listed in Table 11-1 are appearance grades. These grades are sanded so that they will have a pleasing appearance for cabinets, shelving, and the like. The appearance grades are generally not used in structural work, with the exception of APA B-B Plyform. This special-use exterior panel has sanded surfaces and is mill-oiled to serve as formwork for concrete. Allowable stresses are given for the appearance grades for the instances where appearance and strength are both important.

Conformance to Standards

Plywood was first manufactured under a product standard in 1966. A product standard contains voluntary requirements for producing and

Table 11-1. Guide to Use of Allowable Stress and Section Properties Tables—Interior.

Plywood Grade	Description and Use	Typical Trademarks	Veneer Grade Face	Back	Inner	Common Thicknesses	Grade Stress Level (Table 3)	Species Group	Section Property Table
APA RATED SHEATHING EXP 1 or 2[3]	Unsanded sheathing grade for wall, roof, sub-flooring, and industrial applications such as pallets and for engineering design, with proper stresses. Manufactured with intermediate and exterior glue (1). For permanent exposure to weather or moisture only Exterior type plywood is suitable.	APA RATED SHEATHING 32/16 15/32 INCH SIZED FOR SPACING EXPOSURE 1 000 PS 1-83 C-D NER-108	C	D	D	5/16, 3/8, 15/32, 1/2, 19/32, 5/8, 23/32, 3/4	S-3 (1)	See "Key to Span Rating"	Table 4A (unsanded)
APA STRUCTURAL I RATED SHEATHING EXP 1 or APA STRUCTURAL II[2] RATED SHEATHING EXP 1	Plywood grades to use where strength properties are of maximum importance, such as plywood-lumber components. Made with exterior glue only. STRUCTURAL I is made from all Group 1 woods. STRUCTURAL II allows Group 3 woods.	APA RATED SHEATHING STRUCTURAL I 24/0 3/8 INCH SIZED FOR SPACING EXPOSURE 1 000 PS 1-83 C-D NER-108	C	D	D	5/16, 3/8, 15/32, 1/2, 19/32, 5/8, 23/32, 3/4	S-2	Structural I use Group 1 / Structural II See "Key to Span Rating"	Table 4B (unsanded) / Table 4A (unsanded)
APA RATED STURD-I-FLOOR EXP 1 or 2[3]	For combination subfloor-underlayment. Provides smooth surface for application of carpet. Possesses high concentrated and impact load resistance during construction and occupancy. Manufactured with intermediate and exterior glue. Touch-sanded (4). Available with tongue and groove.(5)	APA RATED STURD-I-FLOOR 20 oc 19/32 INCH SIZED FOR SPACING T&G NET WIDTH 47-1/2 EXPOSURE 1 000 UNDERLAYMENT PS 1-83 NER-108	C plugged	D	C & D	19/32, 5/8, 23/32, 3/4, 1-1/8 (2-4-1)	S-3 (1)	See "Key to Span Rating"	Table 4A (touch-sanded)
APA UNDERLAYMENT EXP 1, 2 or INT	For underlayment under carpet. Available with exterior glue. Touch-sanded. Available with tongue and groove.(5)	APA UNDERLAYMENT GROUP 1 EXPOSURE 1 000 PS 1-83	C plugged	D	C & D	1/2, 19/32, 5/8, 23/32, 3/4	S-3 (1)	As Specified	Table 4A (touch-sanded)
APA C-D PLUGGED EXP 1, 2 or INT	For built-ins, wall and ceiling tile backing, NOT for underlayment. Available with exterior glue. Touch-sanded.(5)	APA C-D PLUGGED GROUP 2 EXPOSURE 1 000 PS 1-83	C plugged	D	D	1/2, 19/32, 5/8, 23/32, 3/4	S-3 (1)	As Specified	Table 4A (touch-sanded)
APA APPEARANCE GRADES EXP 1, 2 or INT	Generally applied where a high quality surface is required. Includes APA N-N, N-A, N-B, N-D, A-A, A-B, A-D, B-B, and B-D INT grades.(b)	APA A-D GROUP 1 EXPOSURE 1 000 PS 1-83	B or better	D or better	C & D	1/4, 11/32, 3/8, 15/32, 1/2, 19/32, 5/8, 23/32, 3/4	S-3 (1)	As Specified	Table 4A (sanded)

(All rows above are grouped under the vertical heading: INTERIOR OR PROTECTED APPLICATIONS.)

(Table 11-1 continued on next page.)

marketing a product. The revised version of this standard is Product Standard PS 1-83 (see reference 2). Until recently, most grades of plywood were manufactured under PS 1-83 or its predecessors. Now some plywood is instead manufactured under APA performance standards as set forth in *Performance Standards and Policies for Structural-Use Panels* (4). A performance standard describes how a panel must perform in end use rather than how it must be manufactured. Performance-rated panels must undergo specific tests for strength and bond integrity as detailed in reference 4 to gain trademark privilege. Besides certain plywood, structural composites (veneer faces bonded to reconstituted wood cores) and nonveneered panels (waferboard, oriented strand board, and structural particleboard) are produced under performance standards.

All Structural I and Structural II plywood sheets conform to PS 1-83 because their strength properties are of maximum importance. Sturd-I-Floor may or may not conform to PS 1-83; however, because it is a flooring panel, its performance (maximum span) is generally its significant feature. Ordinary APA Rated Sheathing (other than Structural I or II) may or may not meet all the provisions of the product standard. For engineered applications where design stresses and section properties need to be known, only panels conforming to PS 1-83 should be specified.

Grade Stamps

Performance-rated panels are acknowledged by the National Evaluation Service, a body sponsored by major building codes, under report number NER-108 (previously NRB-108). Panels stamped with NER-108 may be used for spans shown on the trademark stamp. APA Rated Sheathing conforming to the product

Table 11-1. (*Continued*)   Guide to Use of Allowable Stress and Section Properties Tables—Exterior.

Plywood Grade	Description and Use	Typical Trademarks	Veneer Grade			Common Thicknesses	Grade Stress Level (Table 3)	Species Group	Section Property Table
			Face	Back	Inner				
APA RATED SHEATHING EXT(3)	Unsanded sheathing grade with waterproof glue bond for wall, roof, subfloor and industrial applications such as pallet bins.	APA RATED SHEATHING 48/24 23/32 INCH SIZED FOR SPACING EXTERIOR 000 PS 1-83 C-C NER-108	C	C	C	5/16, 3/8, 15/32, 1/2, 19/32, 5/8, 23/32, 3/4	S-1	See "Key to Span Rating"	Table 4A (unsanded)
APA STRUCTURAL I RATED SHEATHING EXT or APA STRUCTURAL II(2) RATED SHEATHING EXT	"Structural" is a modifier for this unsanded sheathing grade. For engineered applications in construction and industry where full exterior-type panels are required. STRUCTURAL I is made from Group 1 woods only.	APA RATED SHEATHING STRUCTURAL I 24/0 3/8 INCH SIZED FOR SPACING EXTERIOR 000 PS 1-83 C-C NER-108	C	C	C	5/16, 3/8, 15/32, 1/2, 19/32, 5/8, 23/32, 3/4	S-1	Structural I use Group 1 / Structural II See "Key to Span Rating"	Table 4B (unsanded) / Table 4A (unsanded)
APA RATED STURD-I-FLOOR EXT(3)	For combination subfloor-underlayment where severe moisture conditions may be present, as in balcony decks. Possesses high concentrated and impact load resistance during construction and occupancy. Touch-sanded (4). Available with tongue and groove.(5)	APA RATED STURD-I-FLOOR 20 oc 19/32 INCH SIZED FOR SPACING EXTERIOR 000 UNDERLAYMENT PS 1-83 NER-108	C plugged	C	C	19/32, 5/8, 23/32, 3/4,	S-2	See "Key to Span Rating"	Table 4A (touch-sanded)
APA UNDERLAYMENT EXT and APA C-C PLUGGED EXT	Underlayment for floor under where severe moisture conditions may exist. Also for controlled atmosphere rooms and many industrial applications. Touch-sanded. Available with tongue and groove.(5)	APA C-C PLUGGED GROUP 2 EXTERIOR 000 PS 1-83	C plugged	C	C	1/2, 19/32, 5/8, 23/32, 3/4	S-2	As Specified	Table 4A (touch-sanded)
APA B-B PLYFORM CLASS I or II(2)	Concrete-form grade with high reuse factor. Sanded both sides, mill-oiled unless otherwise specified. Available in HDO. For refined design information on this special-use panel see APA Design/Construction Guide: Concrete Forming, Form No. V345. Design using values from this specification will result in a conservative design.(5)	APA PLYFORM B-B CLASS I EXTERIOR 000 PS 1-83	B	B	C	19/32, 5/8, 23/32, 3/4	S-2	Class I use Group 1; Class II use Group 3	Table 4A (sanded)
APA MARINE EXT	Superior Exterior-type plywood made only with Douglas Fir or Western Larch. Special solid-core construction. Available with MDO or HDO face. Ideal for boat hull construction.	MARINE · A-A · EXT APA · 000 · PS1-83	A or B	A or B	B	1/4, 3/8, 1/2, 5/8, 3/4	A face & back use S-1 / B face or back use S-2	Group 1	Table 4B (sanded)
APA APPEARANCE GRADES EXT	Generally applied where a high quality surface is required. Includes APA A-A, A-B, A-C, B-B, B-C, HDO and MDO EXT.(5)	APA A-C GROUP 1 EXTERIOR 000 PS 1-83	B or better	C or better	C	1/4, 11/32, 3/8, 15/32, 1/2, 19/32, 5/8, 23/32, 3/4	A or C face and back use S-1 / B face or back use S-2	As Specified	Table 4A (sanded)

(Left margin spanning label: EXTERIOR APPLICATIONS)

(1) When exterior glue is specified, i.e. Exposure 1, stress level 2 (S-2) should be used.
(2) Check local suppliers for availability of STRUCTURAL II and PLYFORM Class II grades.
(3) Properties and stresses apply only to APA RATED STURD-I-FLOOR and APA RATED SHEATHING manufactured entirely with veneers.
(4) APA RATED STURD-I-FLOOR 2-4-1 may be produced unsanded.
(5) May be modified to STRUCTURAL I. For such designation use Group 1 stresses and Table 4B section properties.

Source: Courtesy of American Plywood Association.

standard will have PS 1-83 contained in the trademark stamp. Examples of typical grade/trademark stamps are shown in Fig. 11-1.

Species Group Number

The term *group number* in plywood refers to the wood species group from which the plywood is manufactured. The 70 species used in the United States for plywood are divided into five groups, four of which are used in the engineered grades of plywood. Species Group 1 contains the strongest varieties of wood. Mostly softwoods are used in plywood manufacture.

To find the allowable stresses for a particular plywood panel, the designer must know the species group for the outer plies of the panel. Except for Structural I plywood, the inner plies do not necessarily have to belong to this same group. (Plies of Structural I must all be from species Group 1.) Allowable stresses for all panels except Structural I are based on having inner plies of the weakest species (Group 4).

Span Rating

The span-rating system indicates the performance of the plywood when used as a roof or

```
━━━━━━━━━━━━━━━━━━
      APA
  RATED SHEATHING
  48/24  23/32 INCH
   SIZED FOR SPACING
    EXPOSURE 1
━━━━━━ 000 ━━━━━━
  PS 1-83  C-D  NER-108
━━━━━━━━━━━━━━━━━━
```

```
━━━━━━━━━━━━━━━━━━
      APA
  RATED SHEATHING
   STRUCTURAL I
  48/24  23/32 INCH
   SIZED FOR SPACING
    EXPOSURE 1
━━━━━━ 000 ━━━━━━
  PS 1-83  C-D  NER-108
━━━━━━━━━━━━━━━━━━
```

Fig. 11-1. Plywood trademark stamps. (Courtesy of American Plywood Association.)

floor sheathing, without the need for further engineered design. It gives the allowable span when the face grain is placed perpendicular to the supports (span and grain in the same direction). Usually two nominal thicknesses have the same span rating. Generally the thinner of these is easier to obtain.

The span rating consists of two numbers separated by a slash, such as 32/16. The number on the left is the maximum spacing (in inches) of supports for plywood as roof sheathing, and the one on the right is the maximum spacing for floor supports. Table 11-2 shows how to relate span rating to species group.

Building codes generally accept the span rating, so long as edge support requirements of the codes are met. For roof sheathing, edges parallel to the span usually need to be supported by either blocking, tongue-and-groove edges, or H-shaped clips; otherwise a reduced span must be used. These edge supports may be unnecessary if the actual span is smaller than the span rating. If the roof is to carry diaphragm loads (see Chapter 12) in addition to gravity loads, blocking (Fig. 6-1) may be the only permissible edge support.

Because such a large percentage of plywood is used as sheathing in light-frame construction, the span rating recommendations make it unnecessary to perform design calculations in common sheathing situations. However, the designer needs to understand the directional properties of plywood in order to be able to design for such applications as gusset plates and beam webs.

Allowable Stresses

Allowable stresses for plywood are shown by Table 11-3. These allowable stresses depend on (1) species group of the face ply, (2) condition of use (wet or dry), and (3) grade stress level

Table 11-2.   Key to Span Rating and Species Group.
For panels with "Span Rating" as across top, and thickness as at left, use stress for species group given in table.

Thickness (in.)	Span Rating (APA RATED SHEATHING grades)							
	12/0	16/0	20/0	24/0	32/16	40/20	48/24	
					Span Rating (STURD-I-FLOOR grades)			
					16 o.c.	20 o.c.	24 o.c.	48 o.c.
5/16	4	3	1					
3/8			4[3]	1				
15/32 & 1/2				4[3]	1[1]			
19/32 & 5/8					4[3]	1		
23/32 & 3/4						4[3]	1	
7/8							3[2]	
1-1/8								1

(1) Thicknesses not applicable to APA RATED STURD-I-FLOOR.
(2) For APA RATED STURD-I-FLOOR 24 oc, use Group 4 stresses.
(3) For STRUCTURAL II, use Group 3 stresses.

Source: Courtesy American Plywood Association.

Table 11-3.   Allowable Stresses for Plywood (psi) conforming to U.S. Product Standard PS 1-83 for Construction and Industrial Plywood. Stresses are based on normal duration of load, and on common structural applications where panels are 24″ or greater in width.

Type of Stress		Species Group of Face Ply	Grade Stress Level[1]				
			S-1		S-2		S-3
			Wet	Dry	Wet	Dry	Dry Only
EXTREME FIBER STRESS IN BENDING (F_b)	F_b & F_t	1	1430	2000	1190	1650	1650
TENSION IN PLANE OF PLIES (F_t)		2, 3	980	1400	820	1200	1200
Face Grain Parallel or Perpendicular to Span (At 45° to Face Grain Use 1/6 F_t)		4	940	1330	780	1110	1110
COMPRESSION IN PLANE OF PLIES	F_c	1	970	1640	900	1540	1540
		2	730	1200	680	1100	1100
Parallel or Perpendicular to Face Grain		3	610	1060	580	990	990
(At 45° to Face Grain Use 1/3 F_c)		4	610	1000	580	950	950
SHEAR THROUGH THE THICKNESS[3]	F_v	1	155	190	155	190	160
Parallel or Perpendicular to Face Grain		2, 3	120	140	120	140	120
(At 45° to Face Grain Use 2 F_v)		4	110	130	110	130	115
ROLLING SHEAR (IN THE PLANE OF PLIES)	F_s	MARINE & STRUCTURAL I	63	75	63	75	—
Parallel or Perpendicular to Face Grain (At 45° to Face Grain Use 1-1/3 F_s)		ALL OTHER[2]	44	53	44	53	48
MODULUS OF RIGIDITY (OR SHEAR MODULUS)	G	1	70,000	90,000	70,000	90,000	82,000
Shear in Plane Perpendicular		2	60,000	75,000	60,000	75,000	68,000
to Plies (through the thickness)		3	50,000	60,000	50,000	60,000	55,000
(At 45° to Face Grain Use 4G)		4	45,000	50,000	45,000	50,000	45,000
BEARING (ON FACE)	$F_c \perp$	1	210	340	210	340	340
Perpendicular to Plane of Plies		2, 3	135	210	135	210	210
		4	105	160	105	160	160
MODULUS OF ELASTICITY IN BENDING IN PLANE OF PLIES	E	1	1,500,000	1,800,000	1,500,000	1,800,000	1,800,000
		2	1,300,000	1,500,000	1,300,000	1,500,000	1,500,000
Face Grain Parallel or		3	1,100,000	1,200,000	1,100,000	1,200,000	1,200,000
Perpendicular to Span		4	900,000	1,000,000	900,000	1,000,000	1,000,000

(1) See Table 11-1 for guide.
  To qualify for stress level S-1, gluelines must be exterior and only veneer grades N, A, and C are allowed in either face or back.
  For stress level S-2, gluelines must be exterior and veneer grade B, C-Plugged and D are allowed on the face or back.
  Stress level S-3 includes all panels with interior or intermediate (IMG) gluelines.

(2) Reduce stresses 25% for 3-layer (4- or 5-ply) panels over 5/8″ thick. Such layups are possible under PS 1-83 for APA RATED SHEATHING, APA RATED STURD-I-FLOOR, UNDERLAYMENT, C-C Plugged and C-D Plugged grades over 5/8″ through 3/4″ thick.

(3) Shear-through-the-thickness stresses for MARINE and SPECIAL EXTERIOR grades may be increased 33%. See Section 3.8.1 for conditions under which stresses for other grades may be increased.

Source: Courtesy of American Plywood Association.

(shown by third column from the right in Table 11-1.)

Example 11-1

Find the grade stress level of $\frac{1}{2}$-in. APA Rated Sheathing, Exposure 1.

See Table 11-1 for interior plywood. Footnote 1, referred to in the Grade Stress Level column, indicates that S-2 is the proper grade stress level.

11-2. CROSS-SECTIONAL PROPERTIES

Because a plywood layer has its grain at right angles to the grain in adjacent layers, the section properties are not as readily found as in sawn lumber. It is important that the designer understand how the grain direction affects section properties to be used in design.

Wood is much stronger and stiffer parallel to the grain than it is perpendicular to the grain.

Fig. 11-2. Plywood panel with face grain perpendicular to supports.

In a three-layer plywood panel the grain of the outer plies runs parallel to the long dimension of the panel, while the grain of the middle layer (crossband) runs perpendicular. Since the outer plies contain roughly two-thirds of the wood and these outer plies are farthest from the neutral axis, it would be advantageous for bending stresses to act in the same direction as the grain of the outer plies. This is the reason that plywood in flatwise bending should almost always be placed with the face grain perpendicular to the supports, as shown in Fig. 11-2.

Positioning the panel as in Fig. 11-2 (rather than turned 90°) so that the bending stress is acting in the same direction as face grain is called placing it in the "strong" direction. For some panels the effective moment of inertia is more than ten times greater when they are correctly placed (face grain perpendicular to the supports) than when face grain is parallel to the supports. To understand the structural behavior of plywood, it is imperative to know the *direction* of stress and its relation to *direction* of face grain.

A few terms dealing with grain direction of the layers must be defined. An inner ply (in a five- or seven-layer panel) whose grain is in the same direction as the face grain is called a *center ply*. Inner plies whose grain runs perpendicular to the face grain are called *crossbands*. The outer plies can be called the face or back interchangeably, unless they are of different grades, in which case the face is the outer ply of higher grade while the back is the outer ply of lower grade.

The effective section properties for design are shown in Table 11-4. It is important to note that *the properties listed are for a one-foot-wide strip of plywood* (see Fig. 11-2). Because these effective properties are published, the designer does not have to be concerned with the multilayer quality of the wood. These properties have been adjusted for orientation of wood fiber and for grade of veneer in the different layers and are based on minimum manufacturing requirements. There are two sets of cross-sectional properties listed—the properties for computing parallel-to-face-grain stress (in columns 4–7) and those for computing perpendicular-to-face-grain stress (in columns 8–11). In most cases, the largest stresses will act parallel to the face grain, so the parallel properties will be used. The perpendicular properties are used if the stress is acting perpendicular to the face grain, that is, if the plywood is bent in the weaker direction. Sanding can remove considerable surface wood, so the table has different properties listed for sanded, touch-sanded, and unsanded panels. Usually the structural designer will be using the data for unsanded panels.

The moments of inertia listed in the table of section properties are for stiffness (deflection) calculations only, not for calculating bending stress. They were calculated using a transformed section to account for the fact that, for a thin peeled layer, E perpendicular to the grain is only $1/35$ of its E parallel to the grain. The listed moments of inertia are for use when the panel is loaded perpendicular to the plane of the panel (flatwise bending).

The effective section modulus, KS, is for use

Table 11-4. EFFECTIVE SECTION PROPERTIES FOR PLYWOOD.

Table 11-4A. Face Plies of Different Species Group from Inner Plies
(Includes all Product Standard Grades except those noted in Table 11-4B.)

Nominal Thickness (in.)	Approximate Weight (psf)	t_s Effective Thickness For Shear (in.)	Stress Applied Parallel to Face Grain				Stress Applied Perpendicular to Face Grain			
			A Area (in.²/ft)	I Moment of Inertia (in.⁴/ft)	KS Effective Section Modulus (in.³/ft)	Ib/Q Rolling Shear Constant (in.²/ft)	A Area (in.²/ft)	I Moment of Inertia (in.⁴/ft)	KS Effective Section Modulus (in.³/ft)	Ib/Q Rolling Shear Constant (in.²/ft)
UNSANDED PANELS										
5/16-U	1.0	0.268	1.491	0.022	0.112	2.569	0.660	0.001	0.023	4.497
3/8 -U	1.1	0.278	1.866	0.039	0.152	3.110	0.799	0.002	0.033	5.444
15/32&1/2 -U	1.5	0.298	2.292	0.067	0.213	3.921	1.007	0.004	0.056	2.450
19/32&5/8 -U	1.8	0.319	2.330	0.121	0.379	5.004	1.285	0.010	0.091	3.106
23/32&3/4 -U	2.2	0.445	3.247	0.234	0.496	6.455	1.563	0.036	0.232	3.613
7/8 -U	2.6	0.607	3.509	0.340	0.678	7.175	1.950	0.112	0.397	4.791
1 -U	3.0	0.842	3.916	0.493	0.859	9.244	3.145	0.210	0.660	6.533
1-1/8 -U	3.3	0.859	4.725	0.676	1.047	9.960	3.079	0.288	0.768	7.931
SANDED PANELS										
1/4 -S	0.8	0.267	0.996	0.008	0.059	2.010	0.348	0.001	0.009	2.019
11/32-S	1.0	0.284	0.996	0.019	0.093	2.765	0.417	0.001	0.016	2.589
3/8 -S	1.1	0.288	1.307	0.027	0.125	3.088	0.626	0.002	0.023	3.510
15/32-S	1.4	0.421	1.947	0.066	0.214	4.113	1.204	0.006	0.067	2.434
1/2 -S	1.5	0.425	1.947	0.077	0.236	4.466	1.240	0.009	0.087	2.752
19/32-S	1.7	0.546	2.423	0.115	0.315	5.471	1.389	0.021	0.137	2.861
5/8 -S	1.8	0.550	2.475	0.129	0.339	5.824	1.528	0.027	0.164	3.119
23/32-S	2.1	0.563	2.822	0.179	0.389	6.581	1.737	0.050	0.231	3.818
3/4 -S	2.2	0.568	2.884	0.197	0.412	6.762	2.081	0.063	0.285	4.079
7/8 -S	2.6	0.586	2.942	0.278	0.515	8.050	2.651	0.104	0.394	5.078
1 -S	3.0	0.817	3.721	0.423	0.664	8.882	3.163	0.185	0.591	7.031
1-1/8 -S	3.3	0.836	3.854	0.548	0.820	9.883	3.180	0.271	0.744	8.428
TOUCH-SANDED PANELS										
1/2 -T	1.5	0.342	2.698	0.083	0.271	4.252	1.159	0.006	0.061	2.746
19/32&5/8 -T	1.8	0.408	2.354	0.123	0.327	5.346	1.555	0.016	0.135	3.220
23/32&3/4 -T	2.2	0.439	2.715	0.193	0.398	6.589	1.622	0.032	0.219	3.635
1-1/8 -T	3.3	0.839	4.548	0.633	0.977	11.258	4.067	0.272	0.743	8.535

Table 11-4B. Structural I and Marine

Nominal Thickness (in.)	Approximate Weight (psf)	t_s Effective Thickness For Shear (in.)	Stress Applied Parallel to Face Grain				Stress Applied Perpendicular to Face Grain			
			A Area (in.²/ft)	I Moment of Inertia (in.⁴/ft)	KS Effective Section Modulus (in.³/ft)	Ib/Q Rolling Shear Constant (in.²/ft)	A Area (in.²/ft)	I Moment of Inertia (in.⁴/ft)	KS Effective Section Modulus (in.³/ft)	Ib/Q Rolling Shear Constant (in.²/ft)
UNSANDED PANELS										
5/16-U	1.0	0.356	1.619	0.022	0.126	2.567	1.188	0.002	0.029	6.037
3/8 -U	1.1	0.371	2.226	0.041	0.195	3.107	1.438	0.003	0.043	7.307
15/32&1/2 -U	1.5	0.535	2.719	0.074	0.279	4.157	2.175	0.012	0.116	2.408
19/32&5/8 -U	1.8	0.707	3.464	0.154	0.437	5.685	2.742	0.045	0.240	3.072
23/32&3/4 -U	2.2	0.739	4.219	0.236	0.549	6.148	2.813	0.064	0.299	3.540
7/8 -U	2.6	0.776	4.388	0.346	0.690	6.948	3.510	0.131	0.457	4.722
1 -U	3.0	1.088	5.200	0.529	0.922	8.512	5.661	0.270	0.781	6.435
1-1/8 -U	3.3	1.118	6.654	0.751	1.164	9.061	5.542	0.408	0.999	7.833
SANDED PANELS										
1/4 -S	0.8	0.342	1.280	0.012	0.083	2.009	0.626	0.001	0.013	2.723
11/32-S	1.0	0.365	1.280	0.026	0.133	2.764	0.751	0.001	0.023	3.397
3/8 -S	1.1	0.373	1.680	0.038	0.177	3.086	1.126	0.002	0.033	4.927
15/32-S	1.4	0.537	1.947	0.067	0.246	4.107	2.168	0.009	0.093	2.405
1/2 -S	1.5	0.545	1.947	0.078	0.271	4.457	2.232	0.014	0.123	2.725
19/32-S	1.7	0.709	3.018	0.116	0.338	5.566	2.501	0.034	0.199	2.811
5/8 -S	1.8	0.717	3.112	0.131	0.361	5.934	2.751	0.045	0.238	3.073
23/32-S	2.1	0.741	3.735	0.183	0.439	6.109	3.126	0.085	0.338	3.780
3/4 -S	2.2	0.748	3.848	0.202	0.464	6.189	3.745	0.108	0.418	4.047
7/8 -S	2.6	0.778	3.952	0.288	0.569	7.539	4.772	0.179	0.579	5.046
1 -S	3.0	1.091	5.215	0.479	0.827	7.978	5.693	0.321	0.870	6.981
1-1/8 -S	3.3	1.121	5.593	0.623	0.955	8.841	5.724	0.474	1.098	8.377
TOUCH-SANDED PANELS										
1/2 -T	1.5	0.543	2.698	0.084	0.282	4.511	2.486	0.020	0.162	2.720
19/32&5/8 -T	1.8	0.707	3.127	0.124	0.349	5.500	2.799	0.050	0.259	3.183
23/32&3/4 -T	2.2	0.739	4.059	0.201	0.469	6.592	3.625	0.078	0.350	3.596

Source: Courtesy of American Plywood Association.

in bending-stress calculations for flatwise bending. This effective section modulus is not the same as I/c because KS for the perpendicular condition is calculated ignoring the outermost tension ply. (This ply adds little strength for bending that causes flexural stress normal to the face grain.) Also the constant K has been partially determined by experiment.

Sometimes loads are applied in the same plane as the plane of the panel rather than normal to the face. This is known as edge loading. The plywood web of a wood I-beam and a plywood diaphragm are examples of edge loading. If the panel is edge loaded, then the moment of inertia is found by $I = bh^3/12$, where b equals the effective thickness of either the parallel plies or the perpendicular plies (crossbands), whichever have their grain in the direction of bending stress due to the edge loading. For example, if the edge loading is at 90° to the face grain, then b is the effective thickness of the *parallel* plies; the crossbands are considered worthless in this situation because stress is perpendicular to their grain direction.

To find the effective thickness of the parallel plies, divide the area of the parallel plies (given in Table 11-4 in units of square inches per foot) by 12 in. To find the effective thickness of the perpendicular plies, divide the area of the perpendicular plies by 12. Moment of inertia calculated by the above equation may be used for either deflection or bending-stress calculations of edge-loaded panels.

The modulus of elasticity listed in the table of allowable stresses (Table 11-3) is a conservative figure that includes an allowance for average shear deflection. Therefore, if shear deflection and bending deflection are being calculated separately in a design, the modulus can be increased by 10% in calculating the bending deflection. (This 10% allowance for shear deflection is identical in intent to the 3% allowance for shear deflection that is made in determining published E values for sawn lumber.)

Example 11-2

For a 5/16-in. Structural I unsanded panel, show how moment of inertia for stress parallel to the face grain could be determined for bending by loads acting normal to the panel face. Compare the answer to the value shown in Table 11-4B.

Figure 11-3 shows the 5/16-in. panel makeup for a one-foot strip and the transformed section. To ensure that the moment of inertia of this transformed section is a minimum expected value, three conservative assumptions are made:

1. The overall thickness, t, is taken to be the nominal thickness of the panel minus one-half of the thickness tolerance. (PS 1-83 gives $\pm 1/32$ in. for unsanded panels up to 13/16-in. nominal; for thicker panels the tolerance is $\pm 5\%$ of nominal thickness.)
2. The thickness of the ply (or plies) having stress parallel to the grain is computed using the tabulated minimum area for that direction (Table 11-4B).
3. The thickness of plies not parallel to the bending stress is whatever must be added to the minimum for parallel-to-stress plies (assumption 2) to add up to the total, t, of assumption 1.

For this 5/16-in. panel, minimum total thickness, t, is

$$5/16 - 0.5(1/32) = 0.297 \text{ in.}$$

Fig. 11-3. Transformed section for Example 11-2.

For stress parallel to face grain, the minimum thickness of each face and back ply is one-half of the tabulated minimum area of 1.619 in.^{2}/ft, or

$$t_1 = t_3 = 0.5(1.619/12) = 0.0675 \text{ in.}$$

The inner-ply thickness is

$$t_2 = 0.297 - 2(0.0675) = 0.162 \text{ in.}$$

The 12-in. actual width of inner ply is replaced in Fig. 11-3b by 1/35 of 12 in., the fraction 1/35 being the ratio of E perpendicular to the grain to E parallel. For this transformed section moment of inertia, I, is

$$I = 2[(1/12)(12)(0.0675^3)$$
$$+ 12(0.0675)(0.1147^2)]$$
$$+ (1/12)(12/35)(0.162^3)$$
$$= 0.0220 \text{ in.}^4$$

This is identical to the value published in Table 11-4B. The procedure illustrated is for the simplest case—a panel of three plies of the same species. For all other cases, the computation is more involved. Luckily, minimum moments of inertia for the designer's use have been computed by the APA.

Example 11-3

An unsanded strip of $\frac{1}{2}$-in. plywood whose inner and face plies are from different species groups is edge loaded with load perpendicular to the face grain. (This causes bending stress parallel to the face grain.) The strip is 24 in. wide as shown in Fig. 11-4. Find I and S.

The area of the parallel plies is 2.292 in.^{2}/ft, so the thickness of the parallel plies is 2.292/12 = 0.191 in. Assuming the crossbands to contribute nothing to stiffness,

$$I = 0.191(24)^3/12 = 220 \text{ in.}^4$$
$$S = I/c = 220/12 = 18.3 \text{ in.}^3$$

Fig. 11-4. Edge-loaded plywood for Example 11-3.

11-3. PLYWOOD DESIGN FOR BENDING AND AXIAL LOAD

Unless the panels are span rated and used as common floor or roof sheathing, plywood will be chosen by allowable-stress design. The same NDS (5) load-duration factors apply in plywood that are used for sawn lumber. Preservative-treated panels do not have a reduction in allowable stresses. However, fire-retardant treated panels have a reduction in allowable stresses that must be obtained from the company providing the treatment, because standards have not been developed.

Flexure

For flatwise bending, one uses the effective-section modulus KS to calculate bending stress. The maximum bending moment and maximum bending stress depend on the number of spans over which the panel is continuous. Appendix Table A-9 shows values of maximum bending moment and maximum shear for fully uniformly loaded beams with one, two, three, or four spans. The bending stress for plywood is calculated using bending moment and section properties each for a *one-foot wide* strip, as shown in Fig. 11-2.

Tension or Compression

To detemine the allowable axial force in tension or compression, the allowable stress is multiplied by the area, using the area of only those plies whose grain is in the same direction as the tensile or compressive stress. Plies with grain at right angles to the stress are considered ineffective.

If the tensile stress is at 45° to the face grain, the allowable stress is one-sixth F_t listed in Table 11-3. The area to use in conjunction with this allowable is the full area based on nominal thickness if the panel is Structural I where plies are all from the same species group. If the panel has plies from different species groups in its makeup, then an adjustment must be made to account for the varying strengths of the plies. An adjusted area can be approximated by using 70% of the gross area. The same areas are used for compression at 45°, except that the allowable is one-third of F_c listed in Table 11-3.

The allowable tensile or compressive load at some angle other than 0° (direction of the face grain), 45°, or 90° (direction of crossbands) can be found by interpolating between the area-stress product (load) for 45° and the area-stress product for either 0° or 90°.

Section 3.3.3 of the Plywood Design Specification (reference 3) recommends reducing the allowable load to 50% of the computed value for plates 8 in. or less in width, which have a greater possibility of a critical defect. (Full computed load is recommended for plates 24 in. or wider, interpolating to find the value for plates between 8 in and 24 in. wide.) This reduction is based on the idea that a defect in a very small piece of plywood could cause collapse.

Example 11-4

Find the allowable bending stress and modulus of elasticity for the following plywood panel: $\frac{1}{2}$-in. APA Rated Sheathing, Exterior, span rating of 32/16, wet use. ("Wet" allowable stresses are used if the equilibrium MC in service will be 16% or greater.)

Using the Key to Span Rating and Species Group (Table 11-2), this panel belongs to species Group 1. The first row of Table 11-1 for Exterior plywood shows that this panel is Grade Stress Level S-1. For species Group 1, Grade Stress Level S-1, wet conditions, the allowables shown by Table 11-3 are: F_b = 1430 psi; E = 1,500,000 psi.

Example 11-5

The lower chord of an interior wood truss has a tension splice with 5- × 14-in. plywood gusset plates on each side as shown in Fig. 11-5. The plywood plates are oriented in the strong direction, i.e. with face grain parallel to the tensile stress. What is the allowable load for these two 1/2" Structural I, Exp. 1 plates based on the tensile strength of the plywood?

For one gusset plate,

A of parallel plies per ft = 2.719 in.^{2}
A of parallel plies per 5 in. = 2.719(5)/12
 = 1.133 in.^{2}
F_t = 1650 psi (S-2, Group 1, dry)

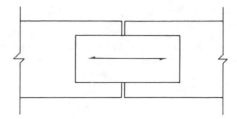

Fig. 11-5. Gusset plate for Example 11-5.

Using the 50% reduction for plates smaller than 8 in.,

Allowable T = 0.50(2)(1650)(1.133)

 = 1870 lb for a pair of plates

Example 11-6

Determine the allowable tensile load parallel to an edge on a 10- × 10-in. plywood plate whose face grain is at 45° to the edges of the plate. The plywood is $\frac{3}{4}$-in. Exterior APA Rated Sheathing, 48/24.

Adjusted gross area = (0.70)(3/4)(10) = 5.25 in.^{2} Table 11-2 shows species Group 1; Table 11-1 shows grade stress level S-1. Choose F_t for S-1 stresses, species Group 1, and wet conditions.

$F_t/6$ = 1430/6 = 238 psi

T = (238)(5.25)

 = 1250 lb, for a plate larger than 24 in.

With the reduction for small plates the allowable load is

$T = 1250 - 625(24 - 10)/(24 - 8) = 703$ lb

Example 11-7

Find the allowable compressive load on a 6-in.-wide plywood plate if the stress is at 25° to the direction of the face grain (wet conditions of use). The plate is $\frac{1}{2}$-in. Structural I, Exterior.

Allowable C at 45°
 = [(0.70)(1/2)(6)][(1/3)(970)] = 679 lb.

Table 11-4 shows that area of parallel plies = 2.719 in.^{2}/ft. Allowable C at 0° = [2.719(6/12)][970] = 1319 lb. By interpolation, Allowable C at 25° is

$1319 - (1319 - 679)(25/45) = 963$ lb

Example 11-8

A construction project is under a tight time schedule. The structural grade of plywood needed by the contractor is not available in his city. For a higher price, however, appearance-grade exterior panels are available. The stamp on the panel indicates "A-C Group 1 Exterior." Will this $\frac{1}{2}$-in. panel with supports perpendicular to face grain 24 in. c/c be sufficient for 70 psf in flatwise bending (dry use; permanent duration of load)?

Table 11-1 shows that Grade Stress Level S-1 should be used. Table 11-3 shows $F_b = 2000$ psi, so

$$\text{Modified } F_b = 0.9(2000) = 1800 \text{ psi}$$

Table 11-4A (sanded) shows $KS = 0.236$ in.^{3}/ft. For a one-foot width, $w = (70 \text{ psf})(1 \text{ ft}) = 70$ lb/ft. Assuming two spans,

$$f_b = M/KS = (w\ell^2/8)/0.236$$
$$= [(70)(2)^2(12)/8]/0.236$$
$$= 1780 \text{ psi} < F_b \quad OK$$

Example 11-9

Sheets of plywood (4 × 8 ft) are each cut into four equal pieces (2 ft dimension parallel to face grain), forming 24-in. deep web members for a lumber-flanged box beam. Plywood is $\frac{15}{32}$-in. APA Rated Sheathing Exposure 1. Calculate the amount that one web will contribute to the total moment of inertia of the beam.

The moment of inertia of the web will be relatively small, as only the center ply is effective in bending. It is the only ply that has grain in the direction of stress. The thickness of this ply = 1.007/12 = 0.084 in.

$$I \text{ (one web)} = bh^3/12$$
$$= 0.084(24)^3/12 = 97 \text{ in.}^4$$

11-4. PLYWOOD DESIGN FOR ROLLING SHEAR

Because plywood is built up in layers, two types of shear must be considered. Rolling shear is the first of these. Rolling shear occurs when the loads are normal to the surface of the plywood, as in floor sheathing. The layers of veneer tend to slide with respect to one another just like horizontal shear in a beam. A rolling-shear failure occurs only in a layer perpendicular to the face grain (a crossband). This is because shear strength is low across the grain.

Rolling shear can possibly best be visualized if one thinks of a panel whose crossband consists of a layer of toothpicks all oriented perpendicular to the face grain. The toothpicks represent the wood fibers. When the load is applied and the layers slide with respect to one another, the toothpicks are rolled and pushed past one another. Rolling shear takes its name from the rolling of these fibers, as shown in Fig. 11-6a. It should be noted that a rolling-shear failure could occur in the outer plies if the panel were placed across the supports in the weak direction (outer grain parallel to the supports).

It is conventional to use a subscript s when referring to rolling-shear stress, to prevent confusing it with shear through the thickness. The allowable rolling-shear stress, F_s, is less than half of the allowable for shear through the thickness. References 6 and 7 give further information on the modulus of rigidity in rolling shear for plywood.

Rolling-shear stress is calculated from $f_s =$

(a) Rolling shear

(b) Shear through the thickness

Fig. 11-6. Plywood in shear.

VQ/Ib, with the constant Ib/Q being given in Table 11-4. The maximum shear V is based on a one-foot width, as in most plywood design. For certain component design (such as stressed-skin panels and flange-web joints in box beams) the rolling-shear allowable is reduced 50% to account for stress concentrations.

Example 11-10

A $\frac{5}{8}$-in. Structural I, Exposure 1 standard plywood panel is loaded in flatwise bending with 70 psf. The face grain is perpendicular to the supports. For dry use with the supports at 32 in. c/c, find the actual and allowable rolling-shear stresses.

All Structural I plywood uses species Group 1 wood, so it is not necessary to use the Key to Span Rating. Table 11-1 shows that it is Grade Stress Level S-2 and is unsanded. Remember that if the face grain is perpendicular to the supports, the bending stress is parallel to the face grain. For Structural I, $\frac{5}{8}$ in. unsanded Table 11-4 shows (for stress parallel to grain) $Ib/Q = 5.685$ in.^{2}/ft. Also, 32 in. is one-third of the 8-ft panel length, so we have a three-span condition. Maximum shear for a three-span beam can be found in Table A-9. Assuming 2-in. nominal supports, $\ell_c = 30.5$ in.

$$V = 0.6w\ell_c = 0.6(70 \text{ lb/ft})(30.5/12 \text{ ft})$$

$$= 107 \text{ lb (per ft of width)}$$

Actual $f_s = V/(Ib/Q) = 107/5.685 = 18.8$ psi

Allowable $F_s = 75$ psi (Table 11-3)   *OK*

11-5. PLYWOOD DESIGN FOR SHEAR THROUGH THE THICKNESS

Plywood can also be stressed in shear on a plane that is perpendicular to the panel surface, that is, by loads in the plane of the panel. This is known as *shear through the thickness* (see Fig. 11-6b) because it affects the total thickness of a panel rather than just a crossband. Loading in the plane of the plywood (as in a diaphragm) can cause shear through the thickness. This same type of shear occurs in plywood-lumber built-up beams made using dimension lumber flanges and plywood webs. The plywood web of these beams is subjected to this type of shear, with the maximum shear stress at the neutral axis. Of course, this is similar to the horizontal shear that would occur in a steel wide-flange beam.

A load perpendicular to the panel surface and concentrated over a small area may punch out a core of wood; hence the name *punching shear*. Rolling of fibers past one another is not involved with this type of failure. Punching-shear calculations should be made with shear-through-the-thickness (rather than rolling-shear) allowable stresses.

When making calculations for shear through the thickness, the designer should *not* use the nominal thickness of the panel, but instead should use the effective thickness for shear as listed by Table 11-4. This effective thickness for shear is necessary to compensate for several factors. One such factor is that the shear strength of a panel increases with the number of glue lines, because the glue adds shear strength. Another factor has to do with the reduced shear strength of inner plies in mixed-species panels.

For shear at 45° to the face grain, the allowable stress, F_v, for shear through the thickness may be increased 100%. Where plywood is continuously glued (instead of mechanically fastened) to framing, other increases are allowed. If a panel is glued to continuous framing around all its edges, the increase is 33%. If the gluing is to framing at only the two edges parallel to the face grain, the increase is only 19%. Increases are cumulative with duration factor.

Example 11-11

A 5000-lb condenser is located on a plywood roof. The condenser has four legs, each with a 4- × 4-in. base. The plywood is $\frac{7}{8}$-in. Exterior APA Rated Sheathing, 48/24. Find the punching shear stress and compare with the allowable.

Table 11-2 indicates that species Group 3 stresses should be used. Table 11-1 shows that it is an unsanded panel and grade stress level S-1 is to be used. The effective thickness for shear is found to be 0.607 in. from Table 11-4.

$$A = (\text{eff. } t)(\text{perim}) = 0.607(16) = 9.7 \text{ in.}^2$$

$$V = 1250 \text{ lb/leg}$$

$$f_v = 1250/9.7 = 129 \text{ psi}$$

$$F_v = 120 \text{ psi (Table 11-3, assumed wet)} \qquad NG$$

11-6. ROOF SHEATHING

Although lumber and particleboard are also used, plywood is the most common roof sheathing. Table 11-5, reproduced from the UBC (8), shows allowable spans for plywood roof sheathing oriented in the strong direction (face grain perpendicular to supports). These allowable spans could also be determined directly from the span rating of the particular panel as discussed in Section 11-1. The table also gives allowable total and live roof loads when the specified span is used. This table deals with vertical loads only and does not list requirements for diaphragm action.

To choose plywood roof sheathing for gravity loads, Table 11-5 will generally be all the designer needs. Should an unusually large snow load exist so that the table cannot be used, the roof sheathing can be chosen based on bending moment, shear, and deflection. Use deflection limits of $\ell/180$ for total load and $\ell/240$ for live load alone.

The UBC has another table (not shown here) that gives allowable loads for roof sheathing oriented in the weak direction. Preframed roof systems often use plywood in the weak direction. In the preframed system, panels have stiffeners preattached before the assembly is lifted into position. The expense of having to use thicker panels is more than offset by the speed of construction.

Unless special means are taken, panel edges parallel to the span will be unsupported. Differential movement of these free edges in adjacent panels can occur. To prevent this movement, panel free edges may be nailed to lumber blocking fitted between the roof joists. Tongue-and-groove edges and panel clips (H-shaped

Table 11-5. Allowable Spans for Plywood Subfloor and Roof Sheathing Continuous over Two or More Spans and Face Grain Perpendicular to Supports[a].

Panel Span Rating[c]	Plywood Thickness (in.)	Roof[b]				Floor Maximum Span[d] (in.)
		Maximum Span (in.)		Load (psf)		
		Edges Blocked	Edges Unblocked	Total Load	Live Load	
12/0	$\frac{15}{16}$	12		135	130	0
16/0	$\frac{5}{16}, \frac{3}{8}$	16		80	65	0
20/0	$\frac{5}{16}, \frac{3}{8}$	20		70	55	0
24/0	$\frac{3}{8}$	24	16	60	45	0
24/0	$\frac{15}{32}, \frac{1}{2}$	24	24	60	45	0
32/16	$\frac{15}{32}, \frac{1}{2}, \frac{19}{32}, \frac{5}{8}$	32	28	55	35[e]	16[g]
40/20	$\frac{19}{32}, \frac{5}{8}, \frac{23}{32}, \frac{3}{4}, \frac{7}{8}$	40	32	40[e]	35[e]	20[g,h]
48/24	$\frac{23}{32}, \frac{3}{4}, \frac{7}{8}$	48	36	40[e]	35[e]	24

[a] These values apply for C-C, C-D, Structural I and II grades only. Spans shall be limited to values shown because of possible effect of concentrated loads.

[b] Uniform load deflection limitations $\frac{1}{180}$ of the span under live load plus dead load, $\frac{1}{240}$ under live load only. Edges may be blocked with lumber or other approved type of edge support.

[c] Span rating appears on all panels in the construction grades listed in footnote a.

[d] Plywood edges shall have approved tongue-and-groove joints or shall be supported with blocking unless $\frac{1}{4}$-in. minimum thickness underlayment, or $1\frac{1}{2}$ in. of approved cellular or lightweight concrete is placed over the subfloor, or finish floor is $\frac{25}{32}$-in. wood strip. Allowable uniform load based on deflection of $\frac{1}{360}$ of span is 165 psf.

[e] For roof live load of 40 psf or total load of 55 psf, decrease spans by 13% or use panel with next greater span rating.

[f] May be 24 in. if $\frac{25}{32}$-in. wood strip flooring is installed at right angles to joists.

[g] May be 24 in. where a minimum of $1\frac{1}{2}$ in. of approved cellular or lightweight concrete is placed over the subfloor and the plywood sheathing is manufactured with exterior glue.

[h] Floor or roof sheathing conforming with this table shall be deemed to meet the design criteria of Section 2516.

Source: Reproduced from the 1985 edition of the Uniform Building Code, copyright © 1985, with permission of the publishers, the International Conference of Building Officials.

metal devices) are other means of providing edge support. Notice from Table 11-5 that most panels require edge support to achieve the full nominal allowable span (as given by span rating).

Example 11-12

A roof framing system consists of trusses at 24 in. c/c. The roof has a slope of $1:1$. Roof $DL = 9$ psf of sloping surface and roof $SL = 30$ psf horizontal. Design plywood for sheathing loads only (not diaphragm loads).

Panel orientation: For greatest strength, plywood will be placed with face grain perpendicular to the top chords of the roof trusses.

Glue type: Although the UBC allows intermediate glue (unless the roof panel is exposed on the underside as at the edge of an overhanging roof), the authors recommend using only panels with exterior glue.

Loads:

UBC snow load reduction per degree of pitch over
  $20° = S/40 - 0.5 = 30$ psf$/40 - 0.5 = 0.25$
  psf/deg over 20°
$45° - 20° = 25°$
$(25)(0.25) = 6$ psf
Snow load $= 30 - 6 = 24$ psf of horizontal projection
Design snow load $= 24/1.414 = 17.0$ psf of roof surface
Total load $= 9 + 17 = 26$ psf of roof surface

The component of this which is perpendicular to the roof surface and which causes bending is 18.4 psf.

Panel options: Using Table 11-5, any Rated Sheathing that is $\frac{15}{32}$ in. or $\frac{1}{2}$ in. thick with a span rating of 24/0 will be sufficient to carry 18.4 psf total load. Table 11-5 shows these thicknesses will not require blocking (unless required for diaphragm loads). *Use $\frac{15}{32}$-in. or thicker C-C Exterior or C-D Interior Exposure 1, 24/0.*

Another panel, a $\frac{3}{8}$-in. panel with span rating of 24/0, will also work. If this panel is used, blocking or T&G edges will be required according to the table. This panel is Structural I Rated Sheathing (see Table 11-2). *Alternatively, use $\frac{3}{8}$-in. Structural I Exterior or Interior 24/0.*

Example 11-13

A roof framing system with beams supporting 2-in. nominal joists at 24 in. c/c is desired for a roof with a gentle slope. The structure is located in a ski area at Steamboat, Colorado, where measured snow loads have reached 175 psf. Roof dead load is 12 psf. Design plywood for vertical loads.

Loads are too great for Table 11-5 to be useful. Assume a Structural I Exterior panel and dry use, so $F_b = 1.15(2000) = 2300$ psi, $F_s = 1.15(75) = 86$ psi, and $E = 1,800,000$ psi. Check bending of a 12-in. width of plywood.

$$w = (175 + 12)/12 = 15.58 \text{ lb/in.}$$

$$\text{Req } KS = M/F_b$$

$$= 0.107(15.58)(24^2)/2300$$

$$= 0.417 \text{ in.}^3/\text{ft}$$

$\frac{19}{32}$ in. req for bending (see Table 11-4)

Check shear

$$\text{Req } (Ib/Q) = 0.607w\ell_c/F_s$$

$$= 0.607(15.58)(22.5)/86$$

$$= 2.47 \text{ in.}^2/\text{ft}$$

$\frac{5}{16}$ in. req for rolling shear

Check deflection

$$\ell/240 = 24/240 = 0.10 \text{ in.}$$

$$\text{Req } I = 0.0065w\ell^4/0.10E$$

$$\text{where APA recommends } \ell$$

$$= \ell_c + 0.25 \text{ in.}$$

$$\text{Req } I = \frac{(175/12)(22.75)^4}{154 \times 0.10(1,800,000)}$$

$$= 0.141 \text{ in.}^4/\text{ft}$$

$\frac{19}{32}$ in. req for deflection under LL alone (total load won't control)

Use $\frac{19}{32}$-in. Structural I C-C Exterior.

11-7. FLOOR SHEATHING

Many plywood floors consist of a bottom structural layer of APA Rated Sheathing called the subfloor. A layer of plywood called the underlayment may be applied over the structural subfloor to provide a smooth surface for linoleum, tile, carpet, or other finish floor. If the joints in the two plywood layers are offset, edge support (such as blocking) is not required by the Uniform Building Code (8). The rightmost column of Table 11-5 gives allowable spans (strong orientation) for Rated Sheathing used as subfloor. When used at their listed spans, the panels in the table are good for at least 165 psf based on an $\ell/360$ deflection limit. (See footnote d of Table 11-5.)

Staggered underlayment panels in the two-layer floor system are considered to have only two structural functions: preventing differential movement of free edges and resistance to dents. Typical thicknesses of underlayment for use over subfloor are $\frac{1}{4}$, $\frac{11}{32}$, and $\frac{3}{8}$ in. Underlayment properties can be found in Table 11-1. For areas to be covered with vinyl tile or linoleum rather than carpet, underlayment with "sanded face" should be specified. As an alternative to underlayment, lightweight concrete or wood strip finish floor may be applied over the subfloor with no edge support required for the plywood floor sheathing.

Sturd-I-Floor is a thicker panel that can be used in place of subfloor and underlayment to make a single-layer floor system. It comes in thicknesses up to $1\frac{1}{8}$ in. In its strong direction, the $1\frac{1}{8}$-in. thickness will span 48 in. and is known as Sturd-I-Floor 48 oc (or 2-4-1). Sturd-I-Floor 48 oc can carry a total service uniform load of 65 psf. The proper thickness of Sturd-I-Floor can be chosen simply from span rating. Tongue-and-groove edges are specified unless square-cut edges are to be blocked with lumber.

The thicker panels of Underlayment grade plywood are also allowed as the structural floor in single-layer floor construction. Table 11-6 gives allowable spans for either Sturd-I-Floor or Underlayment when used in a single-plywood-layer system.

Example 11-14

The floor system for a junior high school consists of steel girders that support wood I-joists at 24 in. c/c. The school has movable partions between classrooms and flooring is linoleum. Floor $DL = 6$ psf, partition $DL = 20$ psf, and $LL = 40$ psf. Choose plywood floor sheathing for vertical loads.

The allowable load for subflooring listed in Table 11-5 = 165 psf. The total floor load for this case is

Table 11-6. Allowable Span for Plywood Combination Subfloor-Underlayment.[a] Plywood Continuous over Two or More Spans and Face Grain Perpendicular to Supports.

Span Rating	Plywood Thickness (in.)	Maximum Span (in.)
16 oc	$\frac{1}{2}, \frac{19}{32}, \frac{5}{8}$	16
20 oc	$\frac{19}{32}, \frac{5}{8}, \frac{23}{32}, \frac{3}{4}$	20
24 oc	$\frac{23}{32}, \frac{3}{4}, \frac{7}{8}$	24
48 oc	$1\frac{1}{8}$	48

[a] Applicable to Underlayment grade and C-C (plugged). Spans limited to values shown because of possible effect of concentrated loads. Allowable uniform load based on deflection of $\frac{1}{360}$ of span is 125 psf, except allowable total uniform load for $1\frac{1}{8}$-in. plywood over joists spaced 48 in. on center is 65 psf. Plywood edges shall have approved tongue-and-groove joints or shall be supported with blocking, unless $\frac{1}{4}$-in. minimum thickness underlayment is installed, or finish floor is $\frac{25}{32}$-in. wood strip. If wood strips are perpendicular to supports, thicknesses shown for 16-in. and 20-in. spans may be used on 24-in. span.

Source: Reproduced from the 1985 edition of the Uniform Building Code, copyright © 1985, with permission of the publishers, the International Conference of Building Officials.

66 psf, so the table may be used. The right-hand column of Table 11-5 shows only one panel that may be used with a 24-in. span. A minimum $\frac{23}{32}$-in. thickness with span rating 48/24 is necessary for the subfloor (strong orientation, of course). Interior type will be specified.

A $\frac{1}{4}$-in. interior underlayment with joints offset from subfloor joints will be specified to be placed over the subfloor to ensure a smooth surface for the linoleum. Blocking will not be required (unless needed for lateral loads).

Use $\frac{23}{32}$-in. or thicker Interior Rated Sheathing 48/24 for subfloor and $\frac{1}{4}$-in. or thicker Interior Underlayment.

Alternatively, a single-layer floor system could be used. This could be either Sturd-I-Floor 24 oc or Exterior Underlayment 23/32 C-C plugged (Table 11-6).

11-8. WOOD-BASED FIBER AND PARTICLE PANELS

Plywood is by far the most common wood sheet or panel for structural use. Other wood panel products are manufactured using wood particles or wood pulp. The manufacturing process causes these panels to have properties different from the original wood particles. Most of these fiber and particle panels (particleboard, insulation board, and hardboard) are used in light-

Table 11-7. Allowable Loads for Particleboard Roof Sheathing.[a,b,d]

Grade	Thickness (in.)	Maximum on-Center Spacings of Supports (in.)	Live Load (psf)	Total Load (psf)
2-M-W	$\frac{3}{8}$[c]	15	45	65
and	$\frac{7}{16}$	16	105	105
2-M-F	$\frac{7}{16}$[c]	24	30	40
	$\frac{1}{2}$	16	110	150
	$\frac{1}{2}$	24	40	55

[a] Panels are continuous over two or more spans.
[b] Uniform load deflection limitation: $\frac{1}{180}$ of the span under live load plus dead load and $\frac{1}{240}$ of the span under live load only.
[c] Edges shall be tongue-and-groove or supported with blocking or edge clips.
[d] Roof sheathing conforming with this table shall be deemed to meet the design criteria of Section 2516.
Source: Reproduced from the 1985 edition of the Uniform Building Code, copyright © 1985, with permission of the publishers, the International Conference of Building Officials.

frame construction, their suitability and strength having been determined over time. With the advent of performance standards, some of these panels can now be rated based on their performance in standard tests.

Particleboard

Particleboard is made of discrete wood particles and synthetic resin adhesives bonded together under heat and pressure. Sometimes particleboard is known as flakeboard, chipboard, waferboard, or shavings board, indicating the size and shape of particles used. Wood chips and particles for making particleboard may come from waste materials of sawmills and plywood and woodworking plants. Most particleboard is compressed in flat-platen presses, although a small amount is extruded through a long, wide, and thin heated die.

Wood particles may range in size from an inch or more in length to very fine. Panels may consist of a single layer or may be multilayer with higher-density top and bottom surfaces. Particles may be either uniform, graduated, or mixed.

Particleboard production began in Europe around 1941. Current consumption of particleboard is much greater in Europe (particularly West Germany and Switzerland) than it is in either the United States or Canada (reference 1).

One of the main structural uses of particleboard is for horizontal and vertical diaphragms (see Chapter 12 for definition). The Uniform

Building Code now allows particleboard to be used as a lateral-force-resisting diaphragm. Other structural uses are subflooring and roof sheathing. Table 11-7 shows allowable loads for particleboard when used as a roof sheathing. Allowable spans for particleboard subfloor and combined subfloor-underlayment are given in Table 11-8.

Because particleboard can be made in very large sizes (up to 8 ft by 40 ft), special structural uses have developed for large particle-

Table 11-8. Allowable Spans for Particleboard Subfloor and Combined Subfloor-Underlayment.[a,e]

Grade	Thickness (in.)	Maximum Spacing of Supports[b]	
		Subfloor	Combined Subfloor-Underlayment[c,d]
2-M-W	$\frac{5}{8}$	16	16
	$\frac{21}{32}$	19.2	19.2
	$\frac{3}{4}$	24	24
2-M-F	$\frac{3}{4}$	19.2	19.2
2-M-3	$\frac{3}{4}$	19.2	19.2

[a] All panels are continuous over two or more spans.
[b] Uniform deflection limitation: $\frac{1}{360}$ of the span under 100 psf minimum load.
[c] Edges shall have tongue-and-groove joints or shall be supported with blocking. The tongue-and-groove panels are installed with the long dimension perpendicular to supports.
[d] A finish wearing surface is to be applied to the top of the panel.
[e] Floor sheathing conforming with this table shall be deemed to meet the design criteria of Section 2516.
Source: Reproduced from the 1985 edition of the Uniform Building Code, copyright © 1985, with permission of the publishers, the International Conference of Building Officials.

board panels. Huge combination subfloor-un-derlayment particleboard panels for mobile homes and for factory-built homes can be handled mechanically in the factory. Separate standards (9, 10) have been developed for this mobile home decking and factory-built housing decking.

Fiber-Based Panels

Fiber building boards have been in production longer than most other wood-base fiber and particle panel materials. Fiberboards are made of wood fibers interfelted together, the matting under heat and pressure producing a bond. They are manufactured from wood pulp (fibers suspended in water), and no adhesives are used in their manufacture (except in the case of tempered hardboard). Additives may be used during manufacture to make the panels more resistant to fire, moisture, or decay.

Although other manufacturing methods are possible, the pulp or slurry is usually laid out on a mesh screen during manufacture. After the water has been removed, the impression of the screen is still apparent on the back of the final board. The two broad types of fiberboards are rigid insulation board and hardboard. Insulation board is light in weight and not compressed during manufacture, whereas the denser hardboard is subject to intense pressure at high temperature. Insulation board is typically thicker than hardboard.

Sheathing-quality insulation board, the most common material for sheathing houses in the United States, is manufactured in three grades (11): regular-density (18 pcf), intermediate (22 pcf), and nail-base (25 pcf). It provides good resistance to heat flow and to racking (shearing deformation). Most building codes approve its use for sheathing wood stud walls and to serve as vertical diaphragms, provided it is properly placed and nailed and provided edge-blocking is used.

Hardboard varies in density from 33 to 90 pcf. Its strength (in all directions) is greater than the perpendicular-to-grain strength of natural wood. It has a myriad of uses—house siding, furniture backs, floor underlayment, core material for thin veneers, beam webs, facing for

concrete forms for architectural concrete, door panels of autos, and so on. However, few of these applications are strictly structural. Hardboard does qualify under building codes as an approved material for wall bracing in conventional, light-frame construction; it can be used in lieu of let-in bracing (Fig. 15-1) or plywood or other sheathing.

In Europe and Australia, hardboard has been used for beam webs since the 1930s (12). However the only composite beams made commercially in the United States with wood product webs have plywood webs rather than hardboard webs. Permissible design stresses for structural use of tempered hardboard are now available in the British Standard Code of Practice, CP: 112 (13).

Designing with fiberboard or particleboard is difficult because design properties vary with the manufacturer. For example, modulus of rupture for hardboard can vary from 1,900 to 12,500 psi. As knowledge of these products increases, there may be greater standardization. If so, the designer will be able to design without knowing in advance who the producer will be.

11-9. OTHER NONPANEL WOOD PRODUCTS

Impregnated and Densified Woods

Work has been done on the treatment of wood with chemicals, especially for securing dimensional stability, but also to improve strength. Several kinds of compressed wood were developed before World War II and were used for propellers and other aircraft parts.

Impreg is treated with resin and cured without compressing. Compreg is similarly treated with resin, but is highly compressed. Often formed from veneers, compreg is used for bowling alley approaches, tool handles, forming dies, pulleys, fan blades, and so on.

Staypak is wood compressed at high temperature and moisture content. This process causes the lignin to flow, relieves internal stresses, and increases the wood's density. Staypak is not impregnated with resin, so it can be made from solid wood as well as from veneer. It is less brittle than either impreg or compreg and has higher tensile and flexural strength properties. Staypak is not being manufactured at the present time.

Other products are made from wood treated with plastic-forming monomers which are polymerized under radiation, and from wood treated with polyethylene-glycol.

Currently, laminated veneer lumber (LVL) is a very popular wood product, commercial production having started in the early 1970s. Thin veneers are laid up with all grain parallel, forming large thick sheets that are eventually cut into sizes approximating dimension lumber. The predominant end uses of this lumber are flanges of plywood-webbed beams and chords of light trusses with steel tubular web members. References 14 and 15 give further details.

REFERENCES

1. Desch, H. E., *Timber—Its Structure and Properties*, 5th ed., St. Martin's Press, New York, 1973.
2. *U.S. Product Standard PS 1-83 for Construction and Industrial Plywood*, National Bureau of Standards, 1983.
3. *Plywood Design Specification*, American Plywood Association, Tacoma, WA, 1986.
4. *Performance Standards and Policies for Structural-Use Panels*, American Plywood Association, Tacoma, WA, 1984.
5. *National Design Specification for Wood Construction*, National Forest Products Association, Washington, D.C., 1986.
6. Lee, W. C., and E. J. Biblis, "Predicting Rolling Shear Modulus of Plywood," *Forest Products Journal*, **29**(4), Apr. 1979, pp. 31–34.
7. Polka, L. C., and J. Hejjes, "Effect of Moisture Content on the Mechanical Properties of Douglas Fir Plywood in Rolling Shear," *Forest Products Journal*, **27**(4), Apr. 1977, pp. 49–53.
8. *Uniform Building Code*, International Conference of Building Officials, Whittier, CA, 1985.
9. *Standard for Particleboard for Mobile Home Decking*, NPA 1-70, National Particleboard Association, Oct. 1970.
10. *Standard for Particleboard for Decking for Factory-Built Housing*, NPA 2-70, National Particleboard Association, 1971.
11. *Wood Handbook*, U.S. Forest Products Laboratory, U.S. Department of Agriculture, 1974.
12. McNatt, J. D., "Hardboard-Webbed Beams: Re-

search and Application," *Forest Products Journal*, **30**(10), Oct. 1980, pp. 57–64.
13. *British Standard Code of Practice*, CP:112.
14. Kunesh, R. H., "Micro Lam: Structural Laminated Veneer Lumber," *Forest Products Journal*, **28**(7), July 1978, pp. 41–44.
15. Laufenberg, Theodore, "Exposure Effects upon Performance of Laminated Veneer Lumber and Glulam Materials," *Forest Products Journal*, **32**(5), May 1982, pp. 42–48.

PROBLEMS

11-1. What would be the allowable bending stress (dry use) for the following plywood panels? (a) $\frac{3}{8}$-in. APA Rated Sheathing 24/0, Exp 1; (b) $\frac{5}{16}$-in. APA Rated Sheathing 12/0, Exp 1; (c) $\frac{7}{8}$-in. APA Rated Sheathing 48/24, Exp 2; (d) $\frac{3}{4}$-in. Structural I, Exp 1.

11-2. What are the modulus of elasticity and modulus of rigidity for dry $\frac{5}{8}$-in. APA Rated Sheathing 32/16, Exp 1?

11-3. If the face grain is perpendicular to the supports, what is the moment of inertia to use in figuring deflection for $\frac{1}{2}$-in. APA Rated Sheathing 24/0, Exp 2?

11-4. Determine the moment of inertia for bending stress perpendicular to face grain of the plywood considered in Example 11-2. Compare to the published value in Table 11-4B.

11-5. A truss tension-member splice has a nailed 6- × 17-in. plywood gusset plate on each side. The tensile force is parallel to the 17-in. dimension and also parallel to the face grain of the plywood. The plates are $\frac{15}{32}$-in. C-D Interior Exp 1, 32/16. What is the allowable load for the member based on the tensile strength of the plywood? Based on the rolling-shear strength of the plywood?

11-6. Figure 11-7 shows two plywood gusset plates joining the 2 × 4 members of a wood truss. Select a thickness of Structural I Exterior plywood to satisfy stress require-

Fig. 11-7. Plywood gusset plates for Problem 11-6.

Fig. 11-8. Plywood for Problems 11-7 and 11-9.

ments. Run face grain parallel to the bottom chord. Suggested sequence of steps: (a) Assuming a critical section on line a-a, make a free-body diagram of the portion of one gusset plate above a-a; (b) solve for the equilibrant shear and in-plane moment on line a-a; (c) solve for shear stress and flexural stress along section a-a; and (d) select the thickness.

11-7. A horizontal 4- × 8-ft plywood panel is loaded with 95 psf. The supports for the panel are at 32 in. c/c with the panel having its face grain perpendicular to the supports, as shown in Fig. 11-8. The panel is $\frac{1}{2}$-in. Structural II Exp 1, 32/16. Find the allowable rolling-shear stress and compare it to the actual stress.

11-8. A rough plywood-lumber I-shaped beam (dry use) consists of four 2 × 4s with a plywood central web of $\frac{1}{2}$-in. Structural I Exp 1 plywood. Total depth is 24 in. and the 2 × 4s are surfaced to dimensions of 1.5 in. by 3.5 in. The 2 × 4s are placed with their larger-dimension vertical. Considering shear through the thickness at the neutral axis, what is the allowable total shear, V, for the beam? The load is of permanent duration.

11-9. For a dry 4- × 8-ft plywood sheet ($\frac{3}{8}$-in. Structural I Exp 1), supported as shown in Fig. 11-8 and loaded with 35 psf, what is the actual bending stress? What is the allowable bending stress? Is the panel satisfactory for bending?

11-10. A roof framing system consists of joists at 24 in. and plywood sheathing. Loads on the gently sloping roof are $D = 12$ psf and $S = 50$ psf. Choose plywood for vertical loads.

11-11. Choose plywood for Problem 11-10 if the design snow load is increased to 80 psf.

11-12. Determine the correct thickness if the sheathing in Problem 11-10 is particleboard instead of plywood. Is blocking required?

11-13. A floor system consists of joists at 16 in. c/c, a plywood subfloor, and plywood underlayment; $D = 5$ psf and $L = 40$ psf. Choose subfloor and underlayment.

11-14. If the subfloor in Problem 11-13 is of particleboard, what grade and thickness are required?

11-15. A floor system has $\frac{3}{4}$-in. Structural I plywood supported by 2 × 8 No. 1 D. fir joists at 12 in. c/c. The joists are 12 ft long and have a span of 11 ft 8 in.

   (a) What is the allowable normal-duration load per square foot if the plywood face grain is perpendicular to the joists and the plywood is merely nailed to the joists?

   (b) What is the allowable normal-duration load if the plywood face grain is *parallel* to the joists and the plywood is glued to the joists to create a composite action? (In calculating properties of the transformed composite section, neglect plies whose grain is perpendicular to the length of the joists.)

   (c) Calculate the shear strength of the glue necessary to ensure composite action.

12

Diaphragm Action and Design

12-1. LATERAL LOADS ON STRUCTURES

In addition to designing for gravity loads, the designer must proportion a structure to resist lateral loads caused by wind or earthquake. After every major earthquake, such as the Mexico City one in September 1985, there is renewed interest in improving wind and seismic provisions of building codes and in understanding better the effect of dynamic loads on structures. Designing for lateral loads is an important consideration in most structures. For residential wood structures, however, lateral-load resistance is assured by conformance to code rules for structural details.

In a rectangular wood building, the most common elements resisting lateral force are horizontal and vertical diaphragms. A diaphragm is a structural member (wall, floor, or roof section) that resists lateral load in the plane of the member, thereby acting as a thin, deep beam. In a concrete building, the concrete floor slabs are horizontal diaphragms and the elevator core and/or shearwalls are the vertical diaphragms.

In wood structures, although the roof and floor sheathing are designed for gravity loads, they can also carry lateral load at minor additional cost due to nailing or blocking requirements. Plywood diaphragms can eliminate the need for diagonal bracing members.

Figure 12-1 shows the outline of a rectangular wood building under wind loading on its larger face. The plywood or other sheathing on the windward face transfers the wind load to the studs which, in turn, act as vertical beams to divide the load between the foundation and the chord at roof level. The chords (also known as top plates) are boundary members perpendicular to the applied lateral load. They consist of dimension lumber around the outer boundary of the roof.

The roof (horizontal diaphragm) acts like a beam, resisting both shear and bending moments. This "beam" differs from the usual beam because its loads are not vertical, but horizontal. If we were considering a multistory building, then the floors would also be horizontal diaphragms. The chords of Fig. 12-1 act as the beam's flanges. They alone are considered to resist the bending moment in the horizontal diaphragm caused by the wind. The roof sheathing and the framing members that support it act as the beam's web, resisting only the shear.

The end reactions of the horizontal diaphragm become lateral loads at the top of the two small walls in Fig. 12-1. These walls act as shearwalls. The shearwall load (called R in the figure) is applied along the length of the strut (a boundary member parallel to the applied lateral load) at the top of the shearwall. The shearwalls act as cantilever beams with their fixed ends at the foundation. Because these "cantilever beams" have small span-to-depth ratios, shear deformation is more significant than flexural deformation.

A rectangular building must be designed for lateral loads in both the transverse and the longitudinal directions. After members were chosen for wind or seismic loads normal to the large wall of Fig. 12-1, then the same design procedure would need to be repeated for load normal to the small wall. With wind in this second direction, the functions of the chords and struts would reverse. Also, the two longer walls would now be the shearwalls.

Fig. 12-1. Lateral loading of a rectangular building.

Figure 12-2 shows the shear and bending-moment diagrams for the roof diaphragm of the rectangular building. The variation of shear and moment at various locations along the span is no different than for a simple beam. However, for design of the chords, bending moment is resolved into a couple consisting of a tension force in one chord and an equal compression force in the other. In an ordinary beam, the variation of shear over the depth is parabolic. In a horizontal diaphragm, however, the shear has been found to be nearly uniform. Therefore, shear per unit width of diaphragm is

$$v = V/b \qquad (12\text{-}1)$$

(In this chapter shear force per unit width of plywood will be called *unit shear*, expressed as pounds per foot, for example.)

Wood diaphragms, whether at service or limit states, have been the subject of considerable analytical and experimental research.

Fig. 12-2. Shear and bending-moment diagrams for roof diaphragm.

References 1 and 2 contain bibliographies of literature published on wood diaphragms and are a good starting point for those interested in lumber or plywood diaphragms. Many journal articles report on diaphragm strength and stiffness as affected by such things as composite action, nail slip, adhesives, and openings.

Connections

Attention to connections is important in the design of lateral-force resisting systems. Without proper connections, the various parts of the system cannot accept or transfer load. For example, columns must be properly connected to the footing, beam-type members (joists, girders, and beams) must be properly connected at their ends (for both gravity and lateral loads), and studs must be connected to wall plates and sills for lateral loads parallel and perpendicular to the wall.

Section 8.8.5.5 of the NDS (3) permits a 30% increase in allowable values for nails used in lateral-load resisting diaphragms, in addition to all other modifications including the duration-of-load increase. However, this 30% increase applies only to nailing of sheathing to framing members and cannot be applied to other diaphragm nailing such as splices and joist-to-top-plate toe nails.

12-2. HORIZONTAL-DIAPHRAGM WEB DESIGN

The roof or floor diaphragm is generally sheathed in plywood, although it may be sheathed with lumber (4) or particleboard. Resistance of the sheathing to shear caused by lateral loads depends on four things: sheathing thickness and layout, nailing type and spacing, provision for blocking, and width of framing members. Resistance to shear actually is also a function of framing spacing, although this factor is ignored in design. Often plywood thickness is chosen by requirements for gravity loads, and then proper nailing and blocking are used so that lateral loads also can be carried. If lateral-load shears are large, panel thickness may not be controlled by gravity loads.

A plywood diaphragm may fail by nail heads pulling through the panel face, by nails pulling out through panel edges, by nails causing framing members to split, or by buckling of the plywood. Generally shear-through-the-thickness failure is not a factor. Sometimes staples rather than nails are used; staples are less likely than nails to cause splitting.

Table 12-1 (reproduced from 5) shows allowable shears in horizontal plywood diaphragms nailed to either D. fir–larch or southern pine framing. Notice from the table that plywood other than Structural I is assumed to have the same strength as a Structural I panel one size thinner. For example, $\frac{15}{32}$-in. C-D has the same strength as $\frac{3}{8}$-in. Structural I. The allowable shear given in Table 12-1 depends on the load case, that is, the layout of the plywood and framing members. The six load cases sketched in the table are based on whether the lateral load is parallel to or perpendicular to continuous panel joints and to unblocked edges.

A diaphragm with lateral load perpendicular to continuous panel joints (like case 1 of Table 12-1) is stronger than a diaphragm with load parallel (like cases 3–6). Case 1 is stronger than case 2 because unblocked edges are perpendicular to load in case 1. Experiments (6) indicate, too, that diaphragms with an orientation of continuous panel joints perpendicular to eccentric load have greater torsional resistance than if continuous joints are parallel to load. Orientation of face grain is not a consideration in finding the correct load case to use. Case 2 is often the panelized system, a partially preframed diaphragm that can be erected quickly.

When plywood joints in the direction parallel to applied force are offset (Case 1), the allowable nail spacing along interior edges is larger than at boundary members.

For *unblocked* diaphragms, nails should be spaced at 6 in. c/c along supported plywood edges (including diaphragm boundaries; i.e., chords and struts) and at 10 in. c/c (12 in. for roofs) for floors along intermediate framing members. Plywood panels with T&G edges or panel clips, although having some edge support, are considered to be unblocked because there is no member into which nails can be driven to resist shear.

Blocked diaphragms will be stronger than

Table 12-1. Allowable Shear in Pounds per Foot for Horizontal Plywood Diaphragms with Framing of Douglas Fir–Larch or Southern Pine[a].

Plywood Grade	Common Nail Size	Minimum Nominal Penetration in Framing (in.)	Minimum Nominal Plywood Thickness (in.)	Minimum Nominal Width of Framing Member (in.)	Blocked Diaphragms — Nail spacing at diaphragm boundaries (all cases), at continuous panel edges parallel to load (cases 3 and 4) and at all panel edges (cases 5 and 6) / 6 / Nail spacing at other plywood panel edges 6	4 / 6	2½[b] / 4	2[b] / 3	Unblocked Diaphragm — Nails spaced 6″ max. at supported end / Load perpendicular to unblocked edges and continuous panel joints (case 1)	Other configurations (cases 2, 3, & 4)
Structural I	6d	$1\frac{1}{4}$	$\frac{5}{16}$	2	185	250	375	420	165	125
				3	210	280	420	475	185	140
	8d	$1\frac{1}{2}$	$\frac{3}{8}$	2	270	360	530	600	240	180
				3	300	400	600	675	265	200
	10d	$1\frac{5}{8}$	$\frac{15}{32}$	2	320	425	640	730[b]	285	215
				3	360	480	720	820	320	240
C-D, C-C, Structural II and other grades covered in U.B.C. Standard No. 25-9	6d	$1\frac{1}{4}$	$\frac{5}{16}$	2	170	225	335	380	150	110
				3	190	250	380	430	170	125
			$\frac{3}{8}$	2	185	250	375	420	165	125
				3	210	280	420	475	185	140
	8d	$1\frac{1}{2}$	$\frac{3}{8}$	2	240	320	480	545	215	160
				3	270	360	540	610	240	180
			$\frac{15}{32}$	2	270	360	530	600	240	180
				3	300	400	600	675	265	200
	10d	$1\frac{5}{8}$	$\frac{15}{32}$	2	290	385	575	655[b]	255	190
				3	325	430	650	735	290	215
			$\frac{19}{32}$	2	320	425	640	730[b]	285	215
				3	360	480	720	820	320	240

[a] These values are for short-time loads due to wind or earthquake and must be reduced 25% for normal loading. Space nails 10 in. on center for floors and 12 in. on center for roofs along intermediate framing members.
Allowable shear values for nails in framing members of other species set forth in Table No. 25-17-J of UBC Standards shall be calculated for all grades by multiplying the values for nails in Structural I by the following factors: Group III, 0.82 and Group IV, 0.65.

[b] Framing shall be 3-in. nominal or wider and nails shall be staggered where nails are spaced 2 in. or $2\frac{1}{2}$ in. on center, and where 10d nails having penetration into framing of more than $1\frac{5}{8}$ in. are spaced 3 in. on center.

Source: Reproduced from the 1985 edition of the Uniform Building Code, copyright © 1985, with permission of the publishers, the International Conference of Building Officials.

Note: Framing may be located in either direction for blocked diaphragms.

those with no blocking, because all four plywood edges are nailed. The nail spacing used at each edge will determine the exact allowable shear. However, minimum requirements for use as a blocked diaphragm are that all four panel edges be nailed at 6 in. c/c. Nails are spaced along intermediate framing members as for unblocked diaphragms. Blocking of a diaphragm is really a means of splicing the sheets of plywood so that they will have better continuity.

The diaphragm must be designed for both load cases—that of transverse lateral load and that of longitudinal lateral load. Not only will each cause a different shear per foot in the diaphragm, but also allowable shears will be different because plywood layout (case number) will be different for different directions of load. (For example, if wind from the south causes load parallel to continuous panel joints, wind from the west may cause load perpendicular to continuous panel joints.)

Derivation of Table Values

There may be occasions for which Table 12-1 does not cover a specific design problem. In such instances it is helpful to know how to determine the allowable shear without using the table. Based on research results (7), the allowable shear in *blocked* diaphragms is calculated as the smaller of (1) the allowable shear, V_{cp}, based on plywood shear stress and (2) the allowable shear, V_{np}, based on lateral fastener load at the boundary. The product of the following items will give V_{cp}:

1. Allowable stress for shear through the thickness of the plywood (Table 11-3 or reference 8)
2. A factor of 1.33 for load duration (assuming wind or seismic)
3. A factor of 12 (in./ft)
4. The effective thickness for shear for the plywood (Table 11-4)

The product of the following items will give V_{np}:

1. Lateral load design value for the particular nail
2. A factor of 1.30 for diaphragm construction by NDS Section 8.8.5.5

3. A factor of 1.33 for load duration
4. The number of boundary nails per foot, except use 2.25 (experimentally derived) when there are actually 2 nails per foot
5. A factor of 0.89 if 2-in. nominal lumber is used, *or* if two rows of fasteners are used in 3-in. nominal lumber, *or* if three rows are used in 4-in. nominal lumber
6. A factor of 0.90 if the diaphragm consists of D. fir framing members and non-Structural-I plywood
7. A factor of 0.85 if nails are spaced 2 in. c/c at boundary
8. For 10d nails only, an additional factor of 0.85 if nails are spaced 2 in. or $2\frac{1}{2}$ in. c/c at boundary when boundary members are single 2-in. nominal members
9. For non-Douglas fir lumber, a factor of 0.82 for lumber Group III and a factor of 0.65 for lumber Group IV

For *unblocked* diaphragms, calculate the allowable shear for "basic" nailing (nailing 4 in. c/c at the boundary and 6 in. c/c at interior panel edges). Then, if the unblocked diaphragm pattern is case 1, use 67% of this value; if cases 2 through 6, use 50%. These reduction factors for unblocked diaphragms have been established by test, and no direct analytic procedure has been developed.

The ability to calculate allowable shear is very useful for highly loaded diaphragms that are not covered by Table 12-1. For instance, if a two-layer diaphragm is made by nailing the first layer to the boundary with nails at 4 in. c/c and the second layer (offset) is nailed over the first also with boundary nails at 4 in. c/c, and also penetrating into the framing member, then there are six fasteners per foot. Once this is known, the allowable shear, V_{np}, based on lateral fastener load at the boundary, can now be calculated. For these highly loaded diaphragms, an additional check for shear-through-the-thickness failure at discontinuous panel joints must be made for case 1 and 2 diaphragms. Reference 7 describes this check.

Deflection of Diaphragms

Deflection of a horizontal diaphragm (and resulting deflection of shearwalls) is generally not critical. According to reference 2, "Many offices do not determine the diaphragm deflection and building officials seldom ask for it." In order to limit deflection, the UBC limits the span-to-width ratio of plywood and particleboard horizontal diaphragms to 4 : 1. If the designer feels special conditions necessitate deflection calculations, references 9 and 10 give a procedure that includes the effects of nail deformation and slip at chord splices.

Continuity

When there are three or more shearwalls (rather than two), it is difficult to assess the degree of continuity of the horizontal diaphragm over the supporting shearwalls. In buildings where the horizontal and vertical diaphragms are both wood frame, the walls are probably more rigid than the floors or roof, because the walls are likely to have several layers of sheathing (siding, plywood, and a finish layer of interior gypsum board). Therefore, the shearwalls are generally considered to be rigid, unyielding supports. Assuming rigid supports, if bending deformations predominate, a horizontal diaphragm over three supports (two equal spans) will transmit five-eighths of the total lateral force to the center wall (analogous to a continuous beam reaction). If shear deformations predominate, then the distribution will be one-half to the center wall (as for two simple beams). The conservative approach would be to calculate both the simple-beam and continuous beam shears and use the larger value at all locations both for determining load transmitted to the vertical diaphragms and for determining the unit shear in the horizontal diaphragms at various locations. Experiments (reference 6) seem to indicate that having continuity over a center support does not change the shear resistance of a horizontal diaphragm.

Nail Capacity

The allowable lateral load given in the NDS for a particular nail size is based on nailing through lumber into lumber. When nailing through plywood into lumber, a too-large nail in a thin plywood will not be able to reach its full design capacity (reference 7). Maximum effective nail sizes for specific thicknesses of plywood are given in Section 5-3.

12-3. HORIZONTAL-DIAPHRAGM CHORD DESIGN

The chords are designed for the axial tension or compression forces that make up the bending-moment couple. The axial force at any point in the chord (Fig. 12-2) can be determined by dividing the moment at that point by the lever arm of the couple (distance between chords, b). That is,

$$T = C = M/b \qquad (12\text{-}2)$$

The chord force varies with M along the chord's length. Because it is easier (and conservative), the chord and its connections are usually designed for the maximum force in the chord, even though the force may be considerably less at the location under consideration. A load duration factor of 1.33 is used in chord and connection design, because the chord force is due to either earthquake or wind.

If the shearwalls in a building with plywood horizontal diaphragms should be masonry or concrete rather than wood, then the chord may consist of continuous horizontal reinforcing bars in the wall at diaphragm level. The rebar alone is assumed to take the entire tension load, and the masonry or concrete the compression load if the "chord" happens to be in compression. (Allowable stresses for steel, concrete, and masonry are all raised by a factor of 1.33 for wind or seismic load.) For force transfer from the diaphragm to the chord, connections with anchor bolts must be well designed.

In ordinary light-frame construction, the chord consists of a double top plate (two pieces of dimension lumber) as shown in Fig. 12-3. Using two pieces allows one plate member to be continuous at the location where the other is being spliced, assuring (with adequate connections by either nails or bolts, depending on magnitude of force) that the chord force can be transferred from one piece to the other. If chord forces are large, three top plates may be necessary. It is best, of course, to locate splices as

(a) NORTH WALL UNDER NORTH WIND

(b) NORTH WALL UNDER WEST WIND

Fig. 12-3. Wall-to-roof connections.

far as possible from the position of maximum moment.

Sometimes the designer may object to using double or triple top plates just to provide continuity through a splice. In this case, the designer may use a steel splice strap top and bottom to transfer the chord force across the splice.

Figure 12-3 shows that the plywood sheathing and the chord do not touch one another, emphasizing the need for good connection design. Since we do not know the direction of the lateral load, each chord must be designed for tension or compression, whichever is more severe. Usually tension is critical, because F_t generally is less than F_c. Also, a portion of compressive force can be transferred across a splice by end bearing, but a tension force cannot.

The examples that follow are for either wind forces alone or earthquake forces alone. In actual design practice, the designer must determine specified values of both wind loads and seismic loads, and design for the more severe of these in combination with dead and other live loads. (See Section 4-6.)

Example 12-1

A one-story wood frame warehouse whose plan view is shown in Fig. 12-4a is 13 ft high. Based on vertical loads, $\frac{15}{32}$-in. C-C Exterior plywood has been chosen tentatively for sheathing of the nearly flat roof. Blocking was not necessary for vertical loads. Determine roof diaphragm requirements for wind pressure of 35 psf on a vertical surface, assuming the plywood layout as shown by Fig. 12-4b. Design the chords.

(a) PLAN VIEW

(b) PLYWOOD LAYOUT

Fig. 12-4. Plan for Examples 12-1 through 12-4.

Wind pressure on north wall: Wind pressure on the upper one-half (6.5 ft) on the wall will be transferred to the roof diaphragm, with the rest going to the foundation. Horizontal load applied to the diaphragm is

$$w = (6.5 \text{ ft})(35 \text{ psf}) = 228 \text{ lb/ft}$$

Recalling the shear diagram of Fig. 12-2,

$$\text{Max } V = (228)(88)/2 = 10{,}000 \text{ lb}$$

$$\text{Max unit } v = 10{,}000/50 = 200 \text{ lb/ft}$$

The plywood layout shown in Fig. 12-4b is like case 3. Table 12-1 shows that for $\frac{15}{32}$-in. C-C plywood with 2-in. nominal framing members, an unblocked diaphragm has an allowable unit shear (if $8d$ nails are used) of

$$180 \text{ lb/ft} < 200 \text{ lb/ft} \quad NG$$

Blocking will be required. (Even with $10d$ nails, blocking would still be required.) Blocking can be installed over the entire roof or, alternatively, can be installed only in regions of large shear (shear > 180 lb/ft) near the east and west edges of the diaphragm. The designer must weigh the expense of blocking over the entire roof against the confusion and errors that might occur if blocking is specified only in certain regions.

By similar triangles on the shear diagram, determine distance x, measured from the building centerline to where no blocking is necessary.

$$x = 44(180/200) = 39 \text{ ft}$$

Blocking is necessary within $44 - 39 = 5$ ft of the west wall and within 5 ft of the east wall. By Table 12-1, a blocked diaphragm has allowable shear of 270 lb/ft. *Use $8d$ nails at 6 in. c/c at framed edges and at 12 in. c/c at intermediate framing members (field nailing). Install blocking within 5 ft of the west wall and within 5 ft of the east wall.*

Wind pressure on west wall: The plywood layout is now case 1 because the load is no longer parallel to the continuous panel joints.

Max $V = 228(50)/2 = 5700$ lb

Max unit $v = 5700/88 = 65$ lb/ft

Obviously this will not control over the wind load from the north. It should be noted here that it would have been better to use a plywood layout with the continuous panel joints running east-west rather than north-south. Because the critical (larger) unit shear will be due to wind against the longer wall, continuous joints should be parallel to the longer wall.

Chord design (for wind from north): Recalling the moment diagram of Fig. 12-2,

$$\text{Max } M = w\ell^2/8 = (228)(88)^2/8$$

$$= 221,000 \text{ ft-lb}$$

By Eq. 12-2 and assuming the chords to be $50 - 6/12 = 49.5$ ft apart,

$$T = C = 221,000/49.5 = 4460 \text{ lb}$$

(Chord force actually varies from zero at the ends to 4460 at the center of the diaphragm.) For No. 2 D. fir 2×6s, basic $F_t = 650$ psi.

$$\text{Req net } A = 4460/[650(1.33)] = 5.16 \text{ in.}^2$$

The chord will consist of two members. The one continuous piece must be able to carry the entire 4460-lb force at splice locations. Try a 2×6.

Assuming that the connection at the splice will use 1-in.-diameter bolts,

$$\text{Net } A = 8.25 - (17/16)(1.5)$$

$$= 6.66 \text{ in.}^2 > 5.16 \text{ in.}^2 \quad OK$$

Under compression (force C) the chord is laterally braced along its full length—vertically by the wall sheathing and horizontally by the roof sheathing. Thus $F_c' = F_c = 1.33(1050) > F_t$, so tension controls.

Since the chord force is less for east or west wind, the same chord may be used on the other two locations. (Design of these top plate members is not yet final; their size may actually be controlled by the

force in them when they are functioning as struts. See Section 12-6.)

Tentatively, use two No. 2 Douglas fir–larch 2 × 6s for chords.

Example 12-2

For north wind loads of Example 12-1, design the connection at splices in the south chord.

Bolt design at chord splices: Figure 12-5 shows that the bolts carry only half of the 4460-lb force through the splice when splices are large distances apart. The other half is carried directly by the continuous member.

For 1-in.-diameter bolts and length in main member = 1.5 in. allowable double shear value is

$$1890 \text{ lb/bolt}$$

Single shear value under wind equals

$$(1/2)(1890)(1.33) = 1257 \text{ lb/bolt}$$

Guessing that two bolts will be needed, no number-in-a-row modification is needed.

$$\text{Req } N = 2230/1257 = 2 \text{ bolts each side}$$

Even though the moment is smaller away from the center (where T is below 4460 lb), it is more practical to use two bolts each side at all splices.

Still tentatively, use chord of two 2 × 6s with two 1-in.-diameter bolts each side of each splice.

Alternative of nails at splices: Usually chord splices are only 4–6 ft apart when nails are used, rather than 10 ft or more when bolts are used. Therefore, in the nailed splice, the entire 4460-lb force would need to be transferred.

If 16d common nails were used to transfer load across this chord splice,

Nail value $= (1.5/1.78)(108)(1.33)$
$= 121$ lb/nail
Req $N = 4460/121 = 37$ nails per side of each
    splice of north or south chords.

Fig. 12-5. Chord splice for Example 12-2.

The UBC specifies that doubled top plates be nailed with 16d nails at 16 in. c/c and with two 16d nails at splices. Obviously, the nailing schedule given by the UBC is inadequate to cover a case such as this.

Example 12-3

Design nailed connections for the building of Example 12-1 as follows: (a) stud to top plate under north wind; (b) top plate to blocking under north wind; and (c) end joist to top plate under west wind.

a. Stud to top plate under north wind: Wind load perpendicular to the north wall at the eaves is

$$w = 228 \text{ lb/ft}$$

The studs transfer this load to the lower member of the upper plate (chord) by nails driven into end grain of the stud at location A of Fig. 12-3a.

Assuming 16d nails,

Nail value = 108(1.33)(2/3) = 95.8 lb
With studs at 24 in. c/c, force per stud
   = (228 lb/ft)(2 ft) = 456 lb
Nails req = 456/95.8 = 5 nails (16d) per stud

Because this is a large number of nails, framing anchors might be a better solution. One manufacturer gives a design load of 390 lb for one framing anchor. Thus two such anchors (or one anchor plus nails) would be required at each end of each stud.

An equal nailed connection is required to transfer the stud reaction from the lower member to the upper member of the top plate.

b. Top plate to blocking under north wind: Directly behind point B in Fig. 12-3a are toe nails (not shown) through full-depth blocking and into the top plate. (Note that if this were an east or west wall, the toe nails would be through the *joist* and into the top plate; i.e., whether it is a joist or blocking depends on which way the roof framing runs.) These toe nails transfer the perpendicular-to-wall wind load from the top plate into the full-depth blocking.

For 16d nails,

Nail value = 108(1.33)(5/6) = 120 lb
With blocking at 24 in. c/c, load per block =
   (228 lb/ft)(2 ft) = 456 lb
Nails required = 456/120 = 4 nails (16d) per
   block (or use framing anchors)

c. End joist to top plate under west wind: Nails at locations B and C of Fig. 12-3b transfer the 65-lb/ft unit shear in the roof plywood (Example 12-1) from the end joist into the top plate.

For 10d nails,

Nail value = 94(1.33)(1.5/1.63) = 115 lb
Required spacing = (115 lb)/(65 lb/ft)
   = 1.77 ft = 21 in. (at both locations B and C)

This is a large spacing. When making these same calculations for the east or west wall, the roof unit shear will be 200 lb/ft, which requires a spacing of 6.9 in. c/c. Therefore, for practicality, a 6-in. spacing will be used around the entire periphery.

Example 12-4

For the building of the previous three examples (see Fig. 12-4), calculate the maximum unit shear in the roof diaphragm for zone 4 seismic loads in the N-S direction knowing that wall dead load = 18 psf and roof dead load = 13 psf.

For N-S seismic loads, W = dead load to diaphragm.

From roof = (88 ft)(50 ft)(13 psf)
   = 57,200 lb
From N & S walls = (2 walls)(6.5 ft)(88 ft)(18
   psf) = 20,600 lb
W = 57,200 + 20,600 = 77,800 lb

(Weight of end walls is not included; their inertial forces affect only the end walls.)

Z = 1 (zone 4; see Section 4-7)
I = 1.0 (Section 4-7)
K = 1.0 (Appendix Table A-7)
CS = 0.14 (Section 4-7)
$V = ZIKCSW = 0.14W = 0.14(77,800)$
   = 10,900 lb

The uniform (seismic) load at the roof level is

$$w = V/88 = 10,900/88 = 124 \text{ lb/ft}$$

Maximum roof shear (near ends) is

$$w(88)/2 = 5460 \text{ lb}$$

Maximum unit roof shear equals

$$5460/50 = 109 \text{ lb/ft}$$

(This is smaller than that caused by wind in Example 12-1, so wind controls.)

Example 12-5

Without using Table 12-1, calculate the allowable unit shear for a blocked diaphragm that has $\frac{3}{8}$-in. C-C plywood, 2-in. nominal Douglas fir framing, and $8d$ nails at 6 in. c/c along all panel edges.

From Table 11-3, the shear-through-the-thickness allowable stress (Grade Stress Level S-1 and species Group 4) = 130 psi and effective thickness for shear = 0.278 in.

Calculating with factors in the same order as presented previously (Section 12-2),

$$V_{cp} = (130)(1.33)(12)(0.278) = 577 \text{ lb/ft}$$

$$V_{np} = (78)(1.3)(1.33)(2.25)(0.89)(0.90)$$

$$= 243 \text{ lb/ft}$$

The latter, 243 lb/ft, controls.

This compares well with the value of 240 lb/ft in Table 12-1. This table was derived from allowable lateral load nail values which have changed since the table was compiled. Therefore tabular and calculated values will not agree exactly, especially for $6d$ nails (which had the largest change in allowable load).

Example 12-6

For the two-story structure in Fig. 12-6 determine loads per foot, w_1 and w_2, transferred to each horizontal diaphragm for a 30-psf wind pressure acting on the long wall.

Analyzing the wall as two simple spans rather than one continuous one, the tributary width to the second floor extends 4.5 ft below second-floor level and 4.5 ft above second-floor level; therefore,

$$w_1 = (30 \text{ psf})(9 \text{ ft}) = 270 \text{ lb/ft}$$

Using Method 2 (Section 4-5), the tributary width to the eaves line is the vertical projection of the roof plus 4.5 ft.

$$w_2 = (30 \text{ psf})(2 \text{ ft} + 4.5 \text{ ft}) = 195 \text{ lb/ft}$$

Even though the roof is sloping, chord force at the eaves due to bending moment is found by Eq. 12-2 as before.

12-4. SHEARWALL DESIGN

The load from the horizontal diaphragm is transferred through the shearwalls (vertical diaphragms) to the ground. If a wood shearwall has no openings, it is said to consist of a single shear panel. If there are window or door openings, the shearwall will consist of multiple shear panels as shown in Fig. 12-7a. The height of a shearwall in a multistory building is generally accepted to be the distance between horizontal diaphragms (although a few challenge this).

For wood shearwalls, unlike concrete ones, the shear per unit width in every panel is assumed to be identical. This is because shear deformation predominates in wood walls. The total force in a panel will be in proportion to its length (horizontal dimension). The unit shear can be determined by dividing the total shear from the roof by the sum of the lengths of all shear panels.

For each panel to function as a separate element, the filler panels (Fig. 12-7a) may be nailed less heavily than the shear panels. Double full-height studs at the sides of openings separate the full-height shear panels from the filler panel. Panels are constrained by the drag struts to deflect together.

Chords of Shear Panels

The full-height studs at each side of a shear panel are the chords that resist all the bending

Fig. 12-6. Structure for Example 12-6.

Fig. 12-7. Two-panel wood shearwall.

moment for that panel. Figure 12-7b shows the cantilever bending moment in panel 1 resolved into a couple of a tension force, T_1, in the left chord and a compression force, C_1, in the right chord. From the free body of Fig. 12-7b it can be seen that

$$T_1 = C_1 = vh \qquad (12\text{-}3)$$

Design of these chords is similar to design of a horizontal-diaphragm chord. Wall, floor, or roof dead load tributary to a chord can be neglected in designing the tension chord (a conservative approach), but must be considered when designing the compression chord. Because we do not know the direction of the wind, all chords are designed for both tension and compression.

Covering Materials for Shearwalls

Plywood is probably the most common sheathing material in the wood shearwall, although particleboard is becoming more popular as more efficient use of wood is being made. For plywood and particleboard shear panels, the UBC (reference 5) specifies a maximum height-length ratio of 3.5 : 1 (blocked) or 2 : 1 (unblocked). Keeping the panels from being too tall and slender helps ensure that shear (deep beam) action will predominate and also limits deflection.

Table 12-2 shows allowable shears for shearwalls in which the plywood is applied directly to the framing or is applied over $\frac{1}{2}$-in. gypsum sheathing. The table also shows allowable shears for plywood panel siding (grooved panels that serve as exterior finish). Large nail heads appear unsightly on plywood siding, so Table 12-2 lists allowable shears for siding nailed with smaller-headed galvanized casing nails. Allowable shears are less when casing nails are used, because a failure due to nail head pull-through is more likely.

Table 12-2 for shearwalls was developed by taking ultimate loads from test results, applying a load factor, and placing the result in the table. Therefore, the shearwall table (Table 12-2) is not identical to the one for horizontal diaphragms (Table 12-1). Eventually the two tables may be combined into one; or, possibly, a design equation may replace the tables.

For small unit shears, lath and plaster or gypsum wallboard may be used as shear-resisting materials in a wood shearwall. The maximum height-to-length ratio for such panels is 2 : 1 (blocked) or 1.5 : 1 (unblocked). Other requirements for these panels can be found in Table 47-I of the UBC (reference 5). Fiberboard-sheathed shear panels are covered in UBC Table 25-P. Reference 4 covers the use of diagonally placed lumber as a shearwall covering.

The UBC will not allow addition of the shear resistance of two different materials in a composite shearwall (a wall with plywood attached to one side of the studs and gypsum sheathing

Table 12-2. Allowable Shear for Wind or Seismic Forces in Pounds per Foot for Plywood Shearwalls with Framing of Douglas Fir–Larch or Southern Pine[a,d].

Plywood Grade	Minimum Nominal Plywood Thickness (in.)	Minimum Nail Penetration in Framing (in.)	Nail-Size (common or galvanized box)	Plywood Applied Direct to Framing — Nail Spacing at Plywood Panel Edges				Nail Size (common or galvanized box)	Plywood Applied over $\frac{1}{2}$-in. Gypsum Sheathing — Nail Spacing at Plywood Panel Edges			
				6	4	3	2[b]		6	4	3	2[b]
Structural I	$\frac{5}{16}$	$1\frac{1}{4}$	6d	200	300	390	510	8d	200	300	390	510
	$\frac{3}{8}$	$1\frac{1}{2}$	8d	230[c]	360[c]	460[c]	610[c]	10d	280	430	550[b]	730[b]
	$\frac{15}{32}$	$1\frac{1}{2}$	8d	280	430	550	730	10d	280	430	550[b]	730
	$\frac{15}{32}$	$1\frac{5}{8}$	10d	340		665[b]	870	—	—	—	—	—
C-D, C-C Structural II and other grades covered in UBC Standard No. 25-9	$\frac{5}{16}$	$1\frac{1}{4}$	6d	180	270	350	450	8d	180	270	350	450
	$\frac{3}{8}$	$1\frac{1}{4}$	6d	200	300	390	510	8d	200	300	390	510
	$\frac{3}{8}$	$1\frac{1}{2}$	8d	220[c]	320[c]	410[c]	530[c]	10d	260	380	490[b]	640
	$\frac{15}{32}$	$1\frac{1}{2}$	8d	260	380	490	640	10d	260	380	490[b]	640
	$\frac{15}{32}$	$1\frac{5}{8}$	10d	310	460	600[b]	770	—	—	—	—	—
	$\frac{19}{32}$	$1\frac{5}{8}$	10d	340	510	665[b]	870	—	—	—	—	—
			Nail Size (galvanized casing)					Nail Size (galvanized casing)				
Plywood panel siding in grades covered in UBC Standard No. 25-9	$\frac{5}{16}$	$1\frac{1}{4}$	6d	140	210	275	360	8d	140	210	275	360
	$\frac{3}{8}$	$1\frac{1}{2}$	8d	130[e]	200[e]	260[e]	340[e]	10d	160	240	310[b]	410

[a] All panel edges backed with 2-in. nominal or wider framing. Plywood installed either horizontally or vertically. Space nails at 6 in. on center along intermediate framing members for $\frac{3}{8}$-in. plywood installed with face grain parallel to studs spaced 24 in. on center and 12 in. on center for other conditions and plywood thicknesses. These values are for short-time loads due to wind or earthquake and must be reduced 25% for normal loading.
Allowable shear values for nails in framing members of other species set forth in Table No. 25-17-J of UBC Standards shall be calculated for all grades by multiplying the values for common and galvanized box nails in Structural I and galvanized casing nails in other grades by the following factors: Group III, 0.82 and Group IV, 0.65.
[b] Framing shall be 3 in. nominal or wider and nails shall be staggered where nails are spaced 2 in. on center, and where 10d nails having penetration into framing of more than $1\frac{5}{8}$ in. are spaced 3 in. on center.
[c] The values for $\frac{3}{8}$-in. thick plywood applied direct to framing may be increased 20%, provided studs are spaced a maximum of 16 in. on center or plywood is applied with face grain across studs.
[d] Where plywood is applied on both faces of a wall and nail spacing is less than 6 in. on center on either side, panel joints shall be offset to fall on different framing members or framing shall be 3-in. nominal or thicker and nails on each side shall be staggered.
Source: Reproduced from the 1985 edition of the Uniform Building Code, copyright © 1985, with permission of the publishers, the International Conference of Building Officials.

to the other). The code says the gypsum must be disregarded, although recent tests (11) show that the contributions to strength and stiffness of the plywood and gypsum sheathings are additive.

Because light-frame residential structures have proved their performance over the years, building codes such as the UBC consider them to be "conventional construction," not requiring designed shearwalls. However, the UBC requires bracing by some method such as diagonal braces across the studs (let-in braces) $\frac{1}{2}$-in. fiberboard, $\frac{1}{2}$-in. gypsum board, and so on. Full details are given in UBC section 2517.

Deformations of Shearwalls

Shear deformations (racking) of shearwalls do not vary linearly with load, because load-slip nail behavior is highly nonlinear. When a wood shearwall is subjected to racking loads, the stud frame distorts as a parallelogram, the sheathing distorts to a different parallelogram, and the nails must deform. Therefore, it is nail slip that is the major factor affecting racking deformation. Stiffness of studs and sheathing plays only a secondary role. Reference 12 presents a convenient method for determining wall deformations at service loads. Gupta and Kuo (13) present an iterative model for determining racking load versus deformation (shear strain) that does not require finite-element analysis.

Example 12-7

Figure 12-8 shows the elevation of the west wall of the building considered in Example 12-1. Studs are 2×6 No. 2 D. fir at 24 in. c/c. Wall dead load = 18 psf and roof dead load = 13 psf. Using wind load as given in Example 12-1, choose chord sizes and sheathing for this shearwall.

The maximum height-width ratio allowed is 3.5:1. All panels meet this requirement. Noting from Fig. 12-2 that R is the reaction of the roof diaphragm which, in turn, loads the end shearwall,

$$R = w\ell/2 = 228(88)/2 = 10{,}000 \text{ lb}$$

Dividing this by the sum of the length of all shear panels,

$$\text{Unit } v = 10{,}000/(13.5 + 10 + 14) = 267 \text{ lb/ft}$$

From Table 12-2, *use 5/16-in. C-D or C-C plywood with 6d nails at 4 in. c/c around panel edges. Use nails at 12 in. c/c at intermediate members. Blocking is required.*
   Chord design for tension: Critical tension chords will be at the small opening because these chords carry the least dead load from the roof. Due to lateral load only,

$$T = vh = (267)(13) = 3470 \text{ lb}$$

The width of roof tributary to the west wall is 8 ft because the joists (Fig. 12-4a) are 16 ft long.

Roof dead load = $(13 \text{ psf})(8 \text{ ft}) = 104 \text{ lb/ft}$
Wall dead load when no openings
   = $(18 \text{ psf})(13 \text{ ft}) = 234 \text{ lb/ft}$
Wall dead load above door = $(18 \text{ psf})(5 \text{ ft})$
   = 90 lb/ft

With studs 2 ft c/c, there will be a 1-ft width of full-height wall tributary to the chord and 1.25 ft of tributary wall width above the door. The total chord tension is

$$T = 3470 - (104)(2.25) - (234)(1)$$
$$- (90)(1.25) = 2890 \text{ lb}$$

Assuming a 7/8-in. bolt (15/16-in. hole) for connecting a metal anchor bracket to the base of the 2 × 6 vertical chord member,

$$\text{Net area} = 1.5(5.5 - 15/16) = 6.84 \text{ in.}^2$$
$$f_t = 2890/6.84 = 423 \text{ psi}$$
$$F_t = 1.33 (650) = 864 \text{ psi}$$
$$\textit{One } 2 \times 6 \textit{ OK for tension}$$

Chord design for compression: Critical compression chords will be at the large opening since these carry more roof and wall dead load than those next to the small opening. Dead loads per foot of wall width are the same as used to design a tension chord. For studs at 2 ft c/c, there will be a 1-ft width of full-height wall tributary to the chord and a 5-ft width of wall above the door tributary to the chord.

$$C = 3470 \text{ lb} + (104 \text{ lb/ft})(6\text{ft})$$
$$+ 234(1) + 90(5) = 4780 \text{ lb}$$

Fig. 12-8. Elevation of wall for Example 12-7.

Nailing of the plywood sheathing to the narrow face of the 2 × 6 chord will prevent buckling about the weak axis; therefore, the controlling ratio

$\ell/d = 13(12)/5.5 = 28.4$
$K = 0.671\sqrt{1,700,000/(1.33 \times 1050)} = 23.4$
$28.4 > 23.4$, so long column
$F'_c = 0.30(1,700,000)/28.4^2 = 632$ psi

For a chord of two 2 × 6s,

$f_c = 4780/(2 \times 8.25) = 290$ psi *OK*

(Actually only one 2 × 6 is required for column strength, but a chord of two members is used to provide stiffness at the door openings.) Checking compressive stress in the chord at the bolt-hole location near the base,

$f_c = 4780/(2 \times 6.84) = 349$ psi

$F_c = 1.33(1050) = 1396$ psi *OK*

Checking bearing perpendicular to the grain on bottom plate,

$f_{c\perp} = 290$ psi $< F_{c\perp} = 625$ psi *OK*

Use two 2 × 6 studs for all shearwall chords.

Example 12-8

Consider again the same building as that in Fig. 12-6 and Example 12-6, recalling that the lateral load at the eaves line is 195 lb/ft and at second-floor level is 270 lb/ft. The building has two end shearwalls and a central shearwall as shown in Fig. 12-9. Calculate the lateral loads, R_1 and R_2, transferred to the central and end shearwalls. If the interior shearwall has no openings, what unit shear must the wall sheathing resist? Show the shear diagram for the floor diaphragm and for the roof diaphragm.

Calculation of lateral forces: Assuming two simple spans rather than continuous action, the tributary width to the central shearwall is 14 ft + 15 ft = 29 ft.

Central $R_1 = (195 \text{ lb/ft})(29 \text{ ft}) = 5655$ lb

Central $R_2 = (270 \text{ lb/ft})(29 \text{ ft}) = 7830$ lb

For the west wall, tributary width = 14 ft

West $R_1 = 195(14) = 2730$ lb

West $R_2 = 270(14) = 3780$ lb

(a) PLAN VIEW

(b) ELEVATION OF ANY OF THE THREE TRANSVERSE WALLS

Fig. 12-9. Building for Example 12-8.

For the east wall, tributary width = 15 ft

East $R_1 = 195(15) = 2925$ lb

East $R_2 = 270(15) = 4050$ lb

Unit shears in central wall: Central wall is 32 ft long. In upper story of central wall,

Unit shear = $5655/32 = 177$ lb/ft

In lower story of central wall,

Unit shear = $(5655 + 7830)/32 = 421$ lb/ft

(Now that these unit shears are known, thickness of plywood wall sheathing and spacing of nails can be easily chosen.)

Shear diagrams for roof and floor diaphragms: The shear diagrams are shown in Fig. 12-10. Because the maximum shear in the floor diaphragm is 4050 lb, the maximum unit shear in the floor is $4050/32 = 127$ lb/ft. Required plywood thickness can now be chosen for this unit shear.

12-5. ANCHORAGE OF SHEAR PANELS TO FOUNDATION

Shear panels must be anchored to the foundation so that they will not overturn because of

Fig. 12-10. Shear diagrams for horizontal diaphragms of Example 12-8.

Fig. 12-11. Anchor bracket.

the moment created by the lateral loads. (See also Section 4-5.) The tension chord force caused by this moment tends to lift the chord from the foundation. The typical connection used to resist this uplift is an anchor bracket or tie-down anchor as shown in Fig. 12-11. Bolts (in single shear) connect the bracket to the chord, and an anchor bolt (in tension) connects the bracket to the concrete foundation.

The UBC specifies a factor of safety of 1.5 against overturning due to wind. This applies to the entire structure and also to any one of its individual shear panels (see Examples 4-9 and 4-10). Tie-down anchors will not be necessary if the overturning moment is less than two-thirds the dead-load resisting moment. Anchorage for overturning will nearly always be more critical for the individual shear panel than for the entire wall.

For ordinary situations, Table 12-3 is often used to find the allowable tension on the anchor bolts that are resisting uplift. For more complicated situations, analysis will be necessary. It is quite common that anchor brackets have allowable loads given by the manufacturer, assuming a minimum embedment and minimum bolt diameter.

Anchor bolts in tension will fail due to yielding followed by fracture of the steel (long embedments), yielding and straightening of a hook, or pullout of a cone of concrete (shorter embedments). For the last cause, the failure surface is usually assumed to be a 45° cone (see Fig. 12-12) with uniform tensile stress distribution of $4\sqrt{f'_c}$ on the sides of the cone (uniform stress being used for simplicity). By statics this can be shown to be equivalent to applying this tensile stress to the cone's projected net area, A_0, as shown shaded in Fig. 12-12. Most current design procedures were developed for headed anchor studs which have a disk rather than a hook at the end. The 45° cone is assumed to branch out from the periphery of the disk, rather than the surface of the bolt, making A_0 a larger size. At present, the equivalent for a hooked bar is not known.

The projected area effective in resisting tension will be limited if the cone intersects a nearby vertical face of the concrete (too small a side-cover distance). If there were several anchor bolts in a close group, then overlapping

Table 12-3. Allowable Shear and Tension on Bolts (in pounds)[a,b].

Diameter (in.)	Minimum[c] Embedment (in.)	Shear[d] 2000	Shear[d] 3000	Tension[e] 2000 to 5000
		Minimum Concrete Strength (psi)		
$\frac{1}{4}$	$2\frac{1}{2}$	500	500	200
$\frac{3}{8}$	3	1100	1100	500
$\frac{1}{2}$	4	2000	2000	950
$\frac{5}{8}$	4	2750	3000	1500
$\frac{3}{4}$	5	2940	3560	2250
$\frac{7}{8}$	6	3580	4150	3200
1	7	3580	4150	3200
$1\frac{1}{8}$	8	3580	4500	3200
$1\frac{1}{4}$	9	3580	5300	3200

[a] Values are for natural stone aggregate concrete and bolts of at least A307 quality. Bolts shall have a standard bolt head or an equal deformity in the embedded portion.
[b] Values are based upon a bolt spacing of 12 diameters with a minimum edge distance of 6 diameters. Such spacing and edge distance may be reduced 50% with an equal reduction in value. Use linear interpolation for intermediate spacings and edge margins.
[c] An additional 2 in. of embedment shall be provided for anchor bolts located in the top of columns for buildings located in seismic zones nos. 2, 3, and 4.
[d] Values shown are for work with or without special inspection.
[e] Values shown are for work without special inspection. Where special inspection is provided values may be increased 100%.
Source: Reproduced from the 1985 edition of the Uniform Building Code, copyright © 1985, with permission of the publishers, the International Conference of Building Officials.

stress cones would also limit the effective projected area.

When headed anchor bolts in tension are designed, both the limit state of concrete failure and the limit state of steel failure must be checked. The design tensile (pullout) strength of the concrete equals

$$\phi P_n = \phi_c P_c = \phi_c A_0 4\sqrt{f'_c} \quad (12\text{-}4)$$

CONICAL TENSILE FAILURE SURFACE

HEAD DIAMETER

Fig. 12-12. Concrete failure cone around headed anchor bolt.

where $\phi_c = 0.65$ and f'_c is the concrete compressive strength in psi. The design tensile strength of the steel equals

$$\phi P_n = \phi_s P_s = \phi_s A_b f_y \quad (12\text{-}5)$$

where $\phi_s = 0.9$, A_b = nominal area of anchor bolt, and f_y = yield stress of the steel. If the grade of steel does not exhibit a definite yield point, then $f_y = 0.9 F_{ult}$ should be used.

The designer may wish to make the embedded length sufficiently large so that a concrete failure is prevented altogether. To find the embedment length required to prevent concrete pullout, the upper-bound strength of the steel will be calculated without reduction by ϕ; that is,

$$P_s \text{ upper bound} = A_b F_{ult} \quad (12\text{-}6)$$

Then the tensile capacity of the concrete is ensured to be greater than or equal to the upper-bound tensile capacity of the steel, as follows

$$\phi_c A_0 4\sqrt{f'_c} \geq A_b F_{ult}$$
$$\phi_c \pi l_e^2 4\sqrt{f'_c} \geq A_b F_{ult} \quad (12\text{-}7)$$

Equation 12-7 then can be solved for the

embedment length, l_e, which ensures that a concrete failure will not occur.

All of the above concerns the *strength* of the anchorage. The strength required is equal to the calculated force under service loads multiplied by a load factor (factor of safety). A load factor of 1.3–1.7 is usually used. References 14–17 contain further information on design of anchor bolts for tension and shear.

In addition to transferring tension uplift to the foundation, anchor bolts must transfer the horizontal shear in the shearwall to the foundation. This is accomplished by anchor bolts (in addition to those provided for uplift) that are embedded in the concrete foundation at points along the base of the shearwall. This series of anchor bolts passes through holes in the horizontal lumber member (sill plate) at the base of the stud wall. The UBC specifies a maximum spacing of 6 ft and minimum diameter of $\frac{1}{2}$ in. for these bolts. The strength of these bolts both in concrete and in wood parallel to the grain will need to be considered.

The same series of anchor bolts must also resist base reactions for perpendicular-to-wall wind load. In this case the sill bolts will bear on the wood perpendicular to the grain. Of course the bolts will be designed for whichever is the most severe condition—unit shear in the plane of the shearwall using parallel-to-grain bolt values, perpendicular-to-wall base shear using perpendicular-to-grain bolt values, or concrete failure.

For proper shear transfer, the plywood sheathing should extend down far enough that it can be securely nailed to the sill plate. The shear is transferred from wall sheathing to sill plate and then, by means of the anchor bolts, to the foundation. Experimental and analytical work (18) shows that increasing the number of nails through the plywood into the sill plate and into the adjacent header can greatly increase wall stiffness. The result is a reduction of deflection and an increase in resistance to uplift and shear loads.

Example 12-9

Design tie-down anchors for the center shear panel of Example 12-7 and Fig. 12-8.

Design of embedded anchor bolt for uplift: Unit $v = 267/\text{ft}$ (previously determined). Since the panel is 10 ft long, the total horizontal force at the top is

$$R = 267(10) = 2670 \text{ lb}$$

Total dead load equals roof dead load plus wall dead load, which equals

$$(104 \text{ lb/ft})(10 \text{ ft}) + (234)(10)$$
$$= 3380 \text{ lb (applied at panel centerline)}$$

Using a factor of safety of 1.5 against overturning,

$$1.5M = 1.5Rh = 1.5(2670)(13) = 52,100 \text{ ft-lb}$$

If the panel overturns, it will rotate on one lower corner. Dead load tends to resist this rotation with a moment equal to the dead load times the distance from the corner, or

Dead load resisting $M = (3380)(5) = 16,900$ ft-lb
Req M to be provided by anchors
   $= 52,100 - 16,900 = 35,200$ ft-lb

Assuming a lever arm of slightly less than 10 ft (say 9.7 ft), the required anchor bolt force is $35,200/9.7 = 3630$ lb. An embedded anchor bolt (like bolt A of Fig. 12-11) that has an allowable tension of 3630 lb must be provided for the connection of bracket and foundation. There will be one anchor bolt at the left chord of the panel and another 9.7 ft away at the right chord of the panel. From Table 12-3 (assuming special inspection), *use one $\frac{3}{4}$-in. bolt with at least 5 in. embedment for the bracket-to-foundation connection.*

Design of bolts connecting bracket and chord: For one 7/8-in. bolt in single shear due to wind in 3-in.-thick D. fir,

$$P = (1/2)(1.33)(3220) = 2140 \text{ lb/bolt}$$
$$N \text{ req} = 3630/2140 = 2 \text{ bolts}$$

Use two 7/8-in. bolts (bolts B in Fig. 12-11) for bracket-to-stud (chord) connection.

To minimize chances for construction errors, the same bolts should be used at the ends of the other two shear panels even though the uplift force is not as great for these panels.

Example 12-10

Design anchor bolts (sill bolts) to be distributed along the sill plate to prevent the wall of Fig. 12-8 from sliding.

These bolts must resist racking shear of 267 lb/ft (from Example 12-7) and (but not simultaneously) perpendicular - to - wall shear of $228\,(50/37.5) = 304$ lb/ft (from Example 12-1). Both the strength of the wood and the strength of the concrete must be checked. Try 5/8-in. bolts and D. fir sill.

For racking shear of 267 lb/ft: For a $1\frac{1}{2}$-in. wood sill anchored to concrete, the NDS specifies using a 3-in. thickness in Table C-12 to determine value parallel to the grain.

$$P = 1900 \text{ lb (double- shear value)}$$

For single shear and wind loading, the design value is

$$(1/2)\,(1.33)\,(1900) = 1264 \text{ lb}$$

Req pitch ≤ 1264 lb / 267 lb per ft = 4.7 ft

For shear strength of the bolt in concrete: Use Table 12-3 to find an allowable shear of 2000 lb.

$$\text{Req pitch} \leq 2000\,/\,304 = 6.6 \text{ ft}$$

For perpendicular-to-wall wind: Bolt load is perpendicular to the grain of the sill. Again, use the tabulated value for a 3-in. thickness.

$$Q = 970 \text{ lb (double shear)}$$

The bolt value for wind and single shear is

$$(1/2)\,(970)\,(1.33) = 645 \text{ lb}$$

Req pitch $\leq 645\,/\,304 = 2.1$ ft. *Controls*

Use $\frac{1}{2}$-in.-diameter anchor bolts at 2 ft c/c along sill plate.

Example 12-11

A $\frac{3}{4}$-in.-diameter headed anchor bolt with a head of about $1\frac{3}{16}$ in. diameter has a yield strength of 36 ksi. It is embedded 9 in. in concrete of compressive strength 3000 psi. What is the ultimate tensile load and the working load?

Tensile capacity as governed by concrete: The diameter of the projected area is $(2)(9) + 1.19 = 19.19$ in. The projected net area is

$$A_0 = \pi 19.19^2/4 - \pi 1.19^2/4 = 288 \text{ in.}^2$$

$$P_u = \phi_c P_c = \phi_c A_0\, 4 \sqrt{f_c'}$$

$$= (0.65)(288)(4)\sqrt{3000} = 41,000 \text{ lb}$$

Tensile capacity as governed by steel:

$$P_u = \phi_s A_b f_y$$

$$= (0.9)(0.442)(36,000) = 14,320 \text{ lb}$$

The ultimate capacity, controlled by steel, is 14,320 lb. The allowable load is less—14,320 divided by the applicable load factor.

12-6. STRUT DESIGN

If a shearwall has no openings, the strut (top plate parallel to lateral load) can transfer the load directly to the shearwall. However, a strut above an opening ("drag" strut) must transfer the unit shear in the diaphragm near the opening to the shear panels on either side of the opening. The force in a strut will vary along the length.

Figure 12-13 shows a shearwall with a single opening. The unit shear in the roof is

$$v_R = R/(b_1 + b_2 + b_3)$$

where R is the wind load reaction of the horizontal diaphragm. The unit shear in the shearwall panels is

$$v_S = R/(b_1 + b_3)$$

Figure 12-13b shows a portion of the building isolated as a free body. For the sum of the horizontal forces to equal zero, a strut force, S, is needed, because unit shears v_R and v_S are not of the same magnitude. Each force shown on the free body is a function of distance x.

Figure 12-13c shows a similar free body, but cut from the building at a location through the opening. In this case, the force at the bottom of the shearwall depends on dimension b_1, but not x. Force S can be determined from these free bodies, and will vary along the length of the strut.

Splices in struts will have connections that

(a) SHEARWALL

(b) STRUT FORCE IN LEFT PANEL (c) STRUT FORCE ABOVE OPENING

Fig. 12-13. Strut force, S.

are designed for either the strut force or the chord force (when the member serves as a chord), whichever is larger.

Struts may have the combined stresses of bending due to vertical loads and tension or compression due to lateral loads. This is particularly true of a drag strut (over an opening) which may be supporting roof or floor load. Sometimes the drag strut is not continuous with the rest of the strut (top plate) because it has been made out of a larger section (or a glulam) in order to support both gravity and lateral loads. In this case, the drag force at the edge of the opening becomes important, because connections have to be designed to transfer this force from the drag strut to the other parts of the chord of the horizontal diaphragm.

Example 12-12

Recalling that $R = 10,000$ lb when wind is from the north, find the strut force at all points along the strut for the shearwall in Fig. 12-8. What is the maximum force in this same top plate when wind is from the west and the top plate serves as a chord?

$$v_R = 10,000/50 = 200 \text{ lb/ft}$$

$$v_S = 10,000/[13.5 + 10 + 14] = 267 \text{ lb/ft}$$

For distance x varying from zero to 13.5 ft, summing forces in the horizontal direction for a free body similar to Fig. 12-13b gives

$$S + 200x - 267x = 0$$

$$S = 67x$$

So S will vary from 0 to 904.5 lb (tension) in the first 13.5 ft of strut. Now considering the drag strut over the first opening (for $13.5 < x < 16$), summing horizontal forces gives

$$S + 200x - 267(13.5) = 0$$

$$S = 3604.5 - 200x$$

So, at $x = 16$ ft,

$$S = 404.5 \text{ lb (tension)}$$

Making similar calculations for the other segments of the strut, the variation of strut force will be that of Fig. 12-14. The sign of the forces would be reversed if the wind reversed direction. The maximum force at any point in the strut is 1075 lb. The whole strut can be conservatively designed for the larger of this 1075-lb force or the maximum force obtained when it is serving as a chord (wind perpendicular to this wall).

When serving as a chord,

$$M = w\ell^2/8 = 228(50^2)/8 = 71,250 \text{ ft-lb}$$

Assuming the chords to be 88 ft $- 6/12 = 87.5$ ft apart,

$$T = C = 71,250/87.5 = 814 \text{ lb} < 1075 \text{ lb}$$

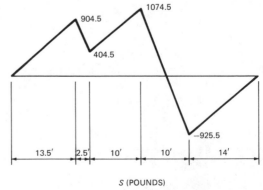

Fig. 12-14. Variation of strut force.

Maximum force in the west wall top plate is due to strut force (under north wind) rather than chord force (under west wind). The procedure for choosing a top plate member for this 1075-lb force is identical to choosing the top plate member in Example 12-1 (assuming that the vertical load from the roof over the opening is carried by a separate header). The procedure for designing the connection at splices in the strut is identical to that shown in Example 12-2.

12-7. WOOD DIAPHRAGM WITH MASONRY WALLS

When wood floor and roof diaphragms are combined with masonry or concrete walls, additional design constraints exist. The UBC (reference 5) requires that concrete or masonry walls be designed to resist bending between diaphragm anchors where anchor spacing is greater than 4 ft. Therefore, anchors (such as steel straps) which transfer perpendicular-to-wall forces from the masonry (or concrete) wall into the body of the wood diaphragm are often spaced at 4 ft c/c. These anchors will extend from the wall some distance inward, where the force will be transferred into larger framing members of the diaphragm. The anchors must be designed for a minimum seismic force of 200 lb/ft.

The UBC requires continuous ties between opposite diaphragm chords to distribute the anchorage forces into the diaphragm. If the framing members of the horizontal diaphragm were continuous in the two perpendicular directions, the requirement would be satisfied. However, even if framing is continuous in one direction, it is highly unlikely that it will be in the other. An expensive way of providing continuity would be to splice the purlins at crossings of the girders.

Another method to provide the necessary continuity, called the subdiaphragm method (references 2 and 19), has been developed to avoid the tie splice at every beam crossing. Certain portions of the total diaphragm are chosen as *subdiaphragms* and are designed as independent diaphragms. The normal-to-wall loads are transferred from wall to subdiaphragm and then to a member that is fully continuous across the building (such as a girder).

Another requirement of the UBC is that (in seismic zones 2, 3, or 4) wood diaphragms

cannot be anchored to concrete or masonry walls by using toe nails or nails in withdrawal.

A ledger (steel angle or lumber member) can be bolted with embedded anchor bolts to the concrete or masonry wall. The ledger can provide an attachment for wood roof trusses, floor joists, or other framing members to transfer diaphragm forces from the walls. Figure 12-15 shows a ledger-wall joint that has been carefully detailed to avoid cross-grain tension.

Wood can resist very little tensile stress perpendicular to the grain. The embedded strap in Fig. 12-15 prevents this tensile bending stress from occurring. Figure 12-16 shows how the ledger would be bent across the grain if the strap were not present. This failure mode was brought to light by cross-grain tension failures of ledgers during the San Fernando earthquake (20).

Rather than the wood ledger, continuous horizontal reinforcing bars in the concrete wall will serve as the tension chord of the diaphragm. Once the flange force, T (Eq. 12-2), has been determined and multiplied by a load factor of 1.3, the required area of reinforcing steel, A_s, can be found by

$$\text{Req } A_s = 1.3 \, T/(\phi f_y) \qquad (12\text{-}8)$$

where $\phi = 0.9$ for tension and f_y is the yield stress of reinforcing steel.

12-8. PLYWOOD-SHEATHED, METAL-STUD SHEARWALLS

Since the early 1970s, the use of steel studs for wall framing has increased for both commer-

Fig. 12-15. Concrete wall–diaphragm joint.

Fig. 12-16. Cross-grain tension in ledger.

cial and residential construction. Steel studs, cold-formed into C-shaped sections, are replacing wood studs in walls sheathed with plywood or other materials.

The structural response of both types of stud walls under combined vertical and lateral loads was tested at Oregon State University (21). Douglas fir studs of Stud grade and 20-gauge steel studs were compared, because they are used under similar conditions. The steel stud wall had its 3/8-in. plywood siding attached with screws, whereas the plywood for the wood stud wall was connected with nails. The ultimate load causing bending was 12% greater for the steel stud wall. However, all steel walls tested failed suddenly by buckling. Wood walls failed gradually and sustained additional load after the first stud failed.

Decreasing slip by applying a rigid adhesive between plywood coverings and studs increased ultimate load considerably, especially for the wood walls. The ultimate lateral (pressure) load increased 2.7 times for the wood stud wall and 1.8 times for the steel stud wall.

Steel stud walls conforming to certain specifications may be used for shearwalls as described in ICBO Report No. 2392P (22). Table 12-4 shows allowable unit shears for steel stud walls sheathed with plywood. The table is for walls with 25-gauge or thicker steel studs at 16 in. c/c maximum. Reference 22 also gives allowable loads for bearing walls and for shear walls of steel studs sheathed with gypsum wallboard.

12-9. TORSION IN DIAPHRAGMS

The design examples for the previous building dealt with nearly symmetrically placed shear walls. This is the ideal situation; otherwise, torsion of the entire structure will induce increased shearing forces on certain shearwalls. Torsion is caused when the location of the resultant lateral force above any level does not coincide with the center of rigidity of the shearwalls at that level.

If there is a lack of symmetry, then torsion must be considered whether due to wind or earthquake. For seismic design, an accidental torsion due to eccentricity of 5% of the maximum building dimension should be considered (reference 5). For wind design, accidental torsion need not be considered.

Table 12-4. Allowable Unit Racking Shear for Wind or Seismic Forces for Steel-Stud/Plywood Vertical Diaphragms.

Wall Construction	Screw Spacing (in.)	Shear Value (lb/ft)
One side: 4-ft-wide plywood	6	200[a]
dimension vertical; joints on studs;	6	115[b]
blocking provided at horizontal	4	300[a]
joints; studs at 16 in. c/c max.	4	175[b]
One side: 8-ft-long plywood dimension	6	170[a]
vertical; joints on studs; studs at 16	6	133[b]
in. c/c max.	4	250[a]
	4	200[b]

[a] Values when $\frac{1}{2}$-in. gypsum wallboard (screwed at 12 in. c/c) is applied to opposite side of the wall.
[b] Where plywood is applied to both sides of wall, these values may be doubled.
Note: Minimum stud to be used: 25-gauge. Minimum plywood to be used: $\frac{3}{8}$-in. Structural I. Screws shall be 1 in. long, coated, double-lead thread drywall screws, spaced as indicated with $\frac{3}{8}$-in. edge distance, in all studs at joints, and panel blocking (where blocking is indicated), bottom track and top track or wood top plate. Screw spacing to all other studs not at panel edges shall be 12 in. c/c. Values above are as reported in reference 22.

12-10. OTHER LOADS ON WALL SHEATHING

Discussion of wall sheathing in earlier parts of this chapter have centered on the sheathing's ability to resist unit shear due to lateral load. In addition, wall sheathing (1) transmits wind loads to wall studs and (2) keeps studs from buckling in their weak direction. Choosing wall plywood or other sheathing for these latter structural functions is based mainly on accepted practice rather than hard design criteria. The recommendations given in the UBC for stud and nail spacing for wall sheathing are usually sufficient for residential construction. The most critical structural function of the sheathing is diaphragm (shearwall) action, so choice will be based on shearwall performance.

REFERENCES

1. Peterson, J., "Bibliography on Lumber and Wood Panel Diaphragms," *Journal of Structural Engineering, ASCE*, **109**(12), Dec. 1983.
2. Applied Technology Council, *Guidelines for the Design of Horizontal Wood Diaphragms*, ATC, Berkeley, CA, 1981.
3. *National Design Specification for Wood Construction*, National Forest Products Association, Washington, DC, 1986.
4. *Western Woods Use Book*, 2nd ed., Western Wood Products Association, Portland, OR, 1979.
5. *Uniform Building Code*, International Conference of Building Officials, Whittier, CA, 1985.
6. Tarpy, T. S., D. J. Thomas, and L. A. Soltis, "Continuous Timber Diaphragms," *Journal of Structural Engineering, ASCE*, **111**(5), May 1985, pp. 992–1002.
7. Tissell, J. R., and J. R. Elliott, Research Report 138, American Plywood Association, Tacoma, WA, 1983.
8. *Plywood Design Specification*, American Plywood Association, Tacoma, WA, 1985.
9. *Plywood Diaphragm Construction*, American Plywood Association, Tacoma, WA, 1976.
10. Bower, W. H., "Lateral Analysis of Plywood Diaphragms," *ASCE, Journal of the Structural Division*, **100**(ST4), Apr. 1974, pp. 759–772.
11. Patton-Mallory, M., R. W. Wolfe, L. A. Soltis and R. M. Gutkowski, "Light-Frame Shear Wall Length and Opening Effects," *Journal of Structural Engineering, ASCE*, **111**(10), Oct. 1985, pp. 2227–2239.
12. McCutcheon, W. J., "Racking Deformations in Wood Shear Walls," *Journal of Structural Engineering, ASCE*, **111**(2), Feb. 1985, pp. 257–269.
13. Gupta, A. K., and G. P. Kuo, "Behavior of Wood-Framed Shear Walls," *Journal of Structural Engineering, ASCE*, **111**(8), Aug. 1985, pp. 1722–1733.
14. Marsh, M. L., and E. G. Burdette, "Anchorage of Steel Building Components to Concrete," *Engineering Journal, AISC*, **22**(1), 1st Quarter, 1985.
15. Shipp, J. G., and E. R. Haninger, "Design of Headed Anchor Bolts," *Engineering Journal, AISC*, **20**(2), 2nd Quarter, 1983.
16. ACI Committee 349, *Code Requirements for Nuclear Safety Related Concrete Structures* (ACI 349-80), Appendix B—Steel Embedments, American Concrete Institute, Detroit MI, 1980.
17. *PCI Design Handbook*, 2nd ed., Prestressed Concrete Institute, Chicago, IL, 1978.
18. Polensek, Anton, and Boyd Schimel, "Rotational Restraint of Wood-Stud Wall Supports," *Journal of Structural Engineering, ASCE*, **112**(6), June 1986, pp. 1247–1262.
19. Breyer, Donald E., *Design of Wood Structures*, McGraw-Hill, New York, 1980.
20. Diekmann, E. F., "Design Details for the Transfer of Forces in Wood Diaphragms to Vertical Elements," *Proceedings of Workshop on Design of Horizontal Wood Diaphragms*, Applied Technology Council, Berkeley, CA, 1980.
21. Polensek, Anton, "Strength and Stiffness of Walls with Wood and Steel Studs," *Forest Products Journal*, **27**(2), 1977, pp. 45–53.
22. Report No. 2392P, International Conference of Building Officials, Whittier, CA, June, 1982.

PROBLEMS

12-1. For the north wall of Example 12-1, recall that there is a perpendicular wind load of 228 lb/ft at the eaves. For the blocking perpendicular to this wall, how many 8*d* nails per blocking member (at location *D* of Fig. 12-3a) will be needed to transfer the eaves load out of the block and into the plywood?

12-2. For the roof and floor diaphragms of Example 12-6, choose plywood and nailing.

12-3. What plywood thickness and nail spacing are required for the central and end shear walls of Example 12-8?

12-4. Table 12-1 gives an allowable shear of 185 lb/ft for a blocked horizontal diaphragm with 5/16-in. Structural I plywood, 2-in. nominal D. fir framing, and 6*d* nails at 6 in. c/c along all panel edges. Show how this allowable shear might be derived from an equation.

12-5. A blocked horizontal diaphragm consists of 5/8-in. C-D plywood nailed to 3-in. nominal D. fir framing members with 10*d* common nails. There are two lines of nails, each with spacing of 4 in. at boundaries and 6 in. at other panel edges. Calculate the allowable unit shear.

12-6. Using No. 3 D. fir–larch, choose top plate

members for the shearwall of Example 12-12. How many nails will be required at splices in the top plate?

12-7. The warehouse of Fig. 12-17, constructed with No. 1 D. fir–larch wall and roof framing, is to be designed for a 70-mph wind load. The roof diaphragm and chords are at the 23-ft level and there is a 2-ft parapet. (a) Using UBC Method 2, find the maximum unit shear in the roof diaphragm (north wind); (b) assuming 2 × framing at 24 in. c/c E-W for the roof, select plywood thickness, layout (case no.), blocking requirements, and nail spacing.

12-8. Assuming that the lateral north wind load to the roof diaphragm of Problem 12-7 is 300 lb/ft, select a top plate (chord) size based on maximum chord force and design a bolted connection at splices.

12-9. Assuming that the lateral north wind load to the roof diaphragm of Problem 12-7 is 300 lb/ft, (a) determine the unit shear in the shear panels of the east wall; (b) select plywood and nail spacing for the wall; (c) sketch a diagram of the variation of strut force in the east wall top plate; (d) find required top plate size based on maximum strut force.

12-10. Assuming that the lateral north wind load to the roof diaphragm of Problem 12-7 is 300 lb/ft, what is the required anchor bolt force at each end of each shear panel of the east wall? Use a factor of safety of 1.5 against

(a) PLAN VIEW

(b) ELEVATION OF EAST WALL

Fig. 12-17. Warehouse for Problem 12-7.

Fig. 12-18. Continuity tie for Problem 12-14.

overturning. Make reasonable estimates of dead load from Table A-1.

12-11. Example 12-9 said that uplift force for chords of the wider panels is not as great as uplift for chords of the narrower panel. Prove it.

12-12. Review Example 4-10, which calculated the necessary anchor bolt force to give a 1.5 factor of safety against overturning. The solution was based on a wall with only small openings acting as a single panel. Design anchor bolts to be located at corners for this 7.09-k tie-down force. Choose bolts for connecting anchor bracket to chord; $f_c' = 3000$ psi and $f_y = 36,000$ psi.

12-13. For the anchor bolt designed in Problem 12-12, what embedment length is required to ensure that failure will be in the bolt rather than the concrete?

12-14. In a two-story building, the second-story shearwall chord is not continuous with the first-story chord. A threaded rod has been chosen for the continuity tie. The second-story (vertical) chord force of 6100 lb must be transferred to the first-story chord. Will the detail shown in Fig. 12-18 be satisfactory to maintain continuity? What diameter threaded A36 rod is needed? (Use an unmodified allowable tensile stress in A36 steel of 19 ksi, based on gross area of the rod.)

13

Built-Up and Composite Members

To make efficient use of materials, built-up and composite members of wood are common. Plywood and lumber can be used in an all wood assembly to form a plywood-lumber box beam or a stressed-skin panel. Wood can also be used in combination with other materials, such as steel or honeycombed paper, to form structural components.

Many of these built-up or composite members, such as joists with I-shaped cross sections, are proprietary in nature. These members are frequently used in larger residences, apartments, and commercial buildings.

13-1. PLYWOOD-LUMBER BEAMS

Plywood-lumber beams have plywood webs glued to dimension lumber flanges. As shown in Fig. 13-1, these beams may be box- or I-shaped. The cross section may be varied along the length of the beam by adding extra webs in areas of high shear near the supports. In a similar manner to a steel wide flange, the flanges of the plywood-lumber beam resist most of the bending load, while the web resists the shear.

Allowable Stresses for Design

Because each flange has nearly uniform compression or nearly uniform tension, the allowable stress in bending for the flange lumber will be the smaller of the table values for F_c or F_t, rather than F_b. The Plywood Design Specification (1) allows for an increase in the allowable horizontal (shear-through-the-thickness)

shear stress if the plywood is rigidly glued to continuous framing members. If the framing is glued to only two panel edges parallel to the face grain, the increase is 19%. For gluing to framing around all panel edges, the increase is 33%. Because of stress concentrations, the allowable rolling-shear stress at the flange-web joint is reduced by 50% (2).

The moment of inertia used to check bending stress is the net I, consisting of (in a two-web beam) contributions from the flanges and one web. Because of staggered web splices, only one of the two webs is effective in bending at any one section.

Deflection

Deflection of a plywood-lumber beam can be approximated by multiplying the bending deflection by a factor to account for shear deflection. As a function of span/depth ratio, the factors (reference 2) are

Span/Depth	Factor
10	1.5
15	1.2
20	1.0

Intermediate factors may be obtained by interpolation. Use the elastic modulus of the flanges in the calculations. Deflection limits for floor beams are $\ell/360$ for live load and $\ell/240$ for total load. For roof beams the limits are $\ell/240$ for live load and $\ell/180$ for total load (reference 2).

Fig. 13-1. Plywood-lumber beam sections.

Stiffeners

Bearing stiffeners are required at the beam ends and also at points of other concentrated loads or reactions. The cross-sectional dimension of the bearing stiffener parallel to the span will be determined based on the rolling-shear strength of the plywood web and the bearing strength perpendicular to the grain for the flange that the stiffener bears on. Intermediate stiffeners of nominal 2-in. lumber are spaced at a maximum of 48-in. c/c to provide stability.

Example 13-1

Design plywood-lumber box beams to span between two masonry walls. The span is 30 ft and beams will be at 8 ft c/c. Use Douglas fir Select Structural flanges and Structural I Interior plywood webs. The beams support a roof system (mild slope) having the following loads:

Suspended ceiling	8 psf
Roofing	7 psf
Insulation	2 psf
Decking and beams	9 psf
Snow	30 psf
	56 psf

Total $w = 56$ (8 ft) = 448 lb/ft; maximum M = $(448)(30^2)/8 = 50,400$ ft-lb. Choose a trial section:

Use depth approx = span/10 = $30(12)/10$ = 36 in.
Basic F_t for flanges is 1200 psi

Assuming 2×8 flanges, the allowable bending stress (using footnote 15 of Table B-3) is

$$F_b = (1.15)(0.90)(1200) = 1240 \text{ psi}$$

For 2×8s and total depth of 36 in., the moment arm between the centroid of the flanges is $36 - 7.25 = 28.75$ in. Equating internal and external moments,

$$1240 A_f(28.75) = 50,400(12)$$
Req $A_f = 17.0$ in.^{2}
Area provided by two 2×8s = $2(10.875)$ = 21.8 in.^{2}
Try two 2×8s for each flange.

Choose plywood based on the thickness effective for shear, t_s. The following empirical equation can be used to estimate the required thickness effective for shear.

$$\text{Req } t_s = 5V/(4hF_v)$$

where h is overall depth. For species Group 1 and Grade Stress Level S-2,

$$F_v = 1.15(190) = 218 \text{ psi}$$
$$V = 448(30)/2 = 6720 \text{ lb}$$
Req $t_s = 5(6720)/[4(36)(218)] = 1.07$ in.
t_s provided by two $\frac{3}{4}$-in. panels = $2(0.739)$ = 1.5 in.

Try $\frac{3}{4}$-in. Structural I plywood for each outside web. Figure 13-2 shows the trial section. The flanges have been surfaced so that a pair of 2×8s is 7 in. deep by 2.75 in. wide. Surfacing has decreased the net overall depth to 35.5 in.

Bending check:

$$I_{\text{flanges}} = 2[(1/12)(2.75)(7)^3 + 2.75(7)(14.25^2)]$$
$$= 7975 \text{ in.}^4$$

Specifying that the plywood face grain is horizontal, the thickness of plywood effective in resisting bending is $A_{\parallel}/12 = 4.219/12 = 0.352$ in.

$$I_{\text{one web}} = (1/12)(0.352)(35.5)^3 = 1312 \text{ in.}^4$$
$$I_{\text{total}} = 7975 + 2(1312) = 10,600 \text{ in.}^4$$
$$I_{\text{net}} = 7975 + 1312 = 9290 \text{ in.}^4$$

For the plywood, allowable bending stress, F_c is $1.15(1540) = 1770$ psi. (The tabular value F_b is not used, because it is for flatwise bending only.)

Fig. 13-2. Plywood-lumber beam for Example 13-1.

The allowable bending stress for the flange lumber is 1240 psi. (See above.) The 1240-psi value controls.

$$I \text{ req} = Mc/F_b = 50,400(12)(17.75)/1240$$
$$= 8657 \text{ in.}^4$$
$$I \text{ provided} = I_{net} = 9290 \text{ in.}^4 \quad OK$$

Shear check:

$$Q_{flanges} = bd(h/2 - d/2)$$
$$= (2.75)(7)(35.5/2 - 7/2) = 274.3 \text{ in.}^3$$
$$Q_{webs} = 2t(h/2)(h/4)$$
$$= 2(4.219/12)(17.75)(8.875)$$
$$= 110.8 \text{ in.}^3$$
$$Q_{total} = 274.3 + 110.8 = 385.1 \text{ in.}^3$$

Horizontal shear: For continuous glue along framing parallel to face grain, allowable horizontal shear stress may be increased 19%. The allowable shear force is

$$V_h = F_v I_t t_s / Q$$
$$= (1.19 \times 1.15 \times 190)(10,600)$$
$$\times (2 \times 0.739)/385.1$$
$$= 10,580 \text{ lb}$$

Actual $V = 448(15) = 6720$ lb   *OK*

Flange-web (rolling) shear: The basic rolling-shear stress is reduced 50% for flange-web shear (sect. 3.8.2, ref 1). The allowable shear force is

$$V_s = 2F_s \, dI_t / Q_{flanges}$$
$$= 2(0.50 \times 1.15 \times 75)(7)(10,600)/274.3$$
$$= 23,300 \text{ lb} > 6720 \text{ lb} \quad OK$$

Deflection: Span/depth ratio = 10:1, so factor = 1.5

$$\text{Total-load } \Delta = 1.5(5) \, w\ell^4/(384 \, EI_t)$$
$$= \frac{1.5(5)(448)(30^4)(1728)}{384(1,800,000)(10,600)}$$
$$= 0.64 \text{ in.}$$

Total-load Δ limit $= \ell/180 = 30(12)/180$
$$= 2.0 \text{ in.} \quad OK$$

Live-load $\Delta = (30/56)(0.64) = 0.34$ in.
$$< \ell/240 = 1.5 \text{ in.} \quad OK$$

Bearing stiffeners: Stiffener width required for bearing of stiffener perpendicular to grain of flange:

End reaction $= R = 448(15) = 6720$ lb

$F_{c\perp}$ (of flange lumber) = 625 psi

Req bearing area $= R/F_{c\perp} = 6720/625$
$$= 10.75 \text{ in.}^2$$

Using a pair of 2× stiffeners (surfaced to $1\frac{3}{8}$ in. each), the required stiffener width is

$$10.75/(2 \times 1.375) = 3.91 \text{ in.}^2$$

Double 2 × 6s are OK for bearing perpendicular to grain.

Stiffener width required for rolling shear at each of the two stiffener-web interfaces:

Req surface area $= R/F_s = 6720/(75 \times 1.15)$
$$= 77.9 \text{ in.}^2$$

Req stiffener width $= 77.9/(2 \times 35.5)$
$$= 1.10 \text{ in.}$$

Use double 2 × 6s at each end for bearing stiffeners.

13-2. STRESSED-SKIN PANELS

Stressed-skin panels are wood assemblies consisting of a plywood skin securely glued to $2 \times$ stringers (webs) so that the unit acts compositely. With composite action, no slippage occurs between the skin and stringers. Horizontal shears are developed along the glue line as the parts are constrained to deflect together. Stressed-skin panels are structurally efficient, resulting in less overall weight for the roof or floor system.

Typical panel sections are shown in Fig. 13-3. Panels may have two skins, a single skin, or a single skin with 1- or 2-in lumber for the opposite flange (T-flange panel). One-skin and T-flange panels are useful where access to space for piping and ductwork is necessary. Skins are usually a minimum of $5/16$ in. thick. Thinner plywood may have adequate strength, but may also have an undesirable bowed appearance (3).

In addition to skins and stringers, panels have headers at their ends and usually have blocking to give edge support to the plywood and lateral support to the stringers. The blocking also backs up splice plates at butt joints in the plywood and helps distribute concentrated loads.

Because the stringer lumber can be obtained in relatively long lengths, a single stressed-skin panel may be made of two or more sheets of plywood (4 ft wide by a multiple of 8 ft long), which cuts on handling and erection costs. Often stressed-skin panels are continuous over three or more supports.

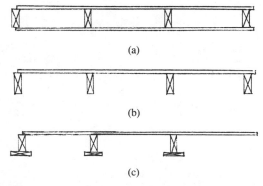

Fig. 13-3. Stressed-skin panels: (a) two-skin, (b) one-skin, (c) T-flange.

Bending

Because the plywood and stringers usually have different moduli of elasticity, the net effective moment of inertia for bending is calculated by the transformed-section method. The properties of the whole 4-ft-wide assembly are used. Only those plywood plies whose grain is parallel to the span of the assembly are used in calculating section properties. The larger cross-sectional dimension of the stringers will be $1/8$ in. less than the sawn lumber width because of smoothing of the gluing surfaces.

If the spacing between stringers is large, not all of the skin is effective in resisting bending. Basic spacings, b, (of plywood effective in bending) are listed in Table 13-1. The effective width of the flange (skin) cannot exceed $t + b$, where t is stringer thickness; that is, the effective part of the skin extends no further than $b/2$ out from either face of the stringer. Only the effective portions of the skins are used in determining the net moment of inertia for bending-stress calculations.

Because the skins have nearly uniform tension or compression, allowable extreme fiber stress in bending is either F_c or F_t. Repetitive-member allowable bending stresses may be used for stringers (if closer than 24 in. apart). Allowable bending stress in the plywood is subject to a modification, C, to provide a proper factor of safety against buckling of the skin (reference 3). Where c is the clear spacing between stringers,

For $c/b \leq 0.5$, $C = 1.0$
For $0.5 < c/b < 1.0$, $C = 0.67 \ (2 - c/b)$
For $c/b \geq 1.0$, $C = 0.67$

Table 13-1. Basic Spacing, b, of Unsanded Plywood Effective in Bending for Stressed-Skin Panels.

Thickness (in.)	Basic Spacing, b (in.)	
	Face Grain Parallel to Stringers	Face Grain $\perp$ to Stringers
$\frac{5}{16}$	12	13
$\frac{3}{8}$	14	17
$\frac{15}{32}, \frac{1}{2}$	18	21
$\frac{19}{32}, \frac{5}{8}$	23	22
$\frac{23}{32}, \frac{3}{4}$	31	29

Deflection

Deflection (of either the whole assembly or of the skin) often controls the design of stressed-skin panels. In calculations for deflection, using the published values for E of plywood and stringers (which have an increase to account for shear deflection) results in overly conservative assembly deflection values. To obtain a bending-only modulus when designing stressed-skin panels, the published value of E for plywood should be increased 10%, and for lumber increased 3%. The gross moment of inertia for deflection calculations is also found by the transformed-section method.

For checking horizontal shear, rolling shear, and deflection, the section properties should be based on the gross section of all material having its grain parallel with the direction of principal stress.

Example 13-2

A trial section has been chosen for a stressed-skin roof panel (two-skin) for a 12-ft simple span with loads of $D = 11$ psf and $S = 30$ psf. The deflection limit under live load is $\ell/240$ for both the assembly as a whole and for the top skin alone. The assembly will have overall dimensions of 4 ft by 12 ft. The trial section is composed of the following:

Top skin: 15/32-in. Rated Sheathing (C-D), Exp 1, 32/16
Stringers: Four 2 × 6 stringers running parallel to face grain (No. 1 D. fir–larch)
Bottom skin: 5/16-in. Rated Sheathing (C-D), Exp 1, 20/0

Because the assembly has a length of 12 ft, the plywood will have to be spliced once. Clear distance between stringers calculates as 13.9 in. Check the trial section for bending and deflection.

Properties of a 4-ft width of top skin are

$$A_\parallel = 2.292(4) = 9.17 \text{ in.}^2$$
$$I_\parallel = 0.067(4) = 0.268 \text{ in.}^4$$
$$I_\perp = 0.004(4) = 0.016 \text{ in.}^4$$
$$E = 1,800,000(1.10) = 1,980,000 \text{ psi}$$

Properties of four stringers after resurfacing are

$$A = 8.06(4) = 32.2 \text{ in.}^2$$
$$I = 19.4(4) = 77.6 \text{ in.}^4$$
$$E = 1,800,000(1.03) = 1,850,000 \text{ psi}$$

Properties of a 4-ft width of bottom skin are

$$A_\parallel = 1.491(4) = 5.96 \text{ in.}^2$$

$$I_\parallel = 0.022(4) = 0.088 \text{ in.}^4$$

$$E = 1,980,000 \text{ psi}$$

$$n = E \text{ stringer}/E \text{ plywood} = 1.85/1.98 = 0.934$$

Find N.A. of the gross transformed section (see Fig. 13-4a): Because the bending stress in the assembly runs parallel to the plywood face grain, only the areas of those plies whose grain is parallel to the face grain are used. The stringer material is transformed into material having the stiffness of plywood. Therefore the transformed area of the stringers is 32.2(0.934) = 30.07.

	A	y	Ay
Top skin	9.17	5.922	54.30
Stringers	30.07	3.000	90.21
Bottom skin	5.96	0.156	0.93
	45.20		145.44

To the N.A., $\bar{y} = 145.44/45.20 = 3.22$ in.

Using this neutral axis, the moment of inertia of the gross transformed section, I_g, can be found.

I_0	A	d	d^2	Ad^2	$I_0 + Ad^2$
0.268	9.17	2.702	7.30	66.9	67.2
77.6(0.934)	30.07	0.22	0.05	1.5	74.0
0.088	5.96	3.064	9.39	56.0	56.1
				$I_g =$	197.3

$$EI_g = (1,980,000)(197.3)$$
$$= 3.91 \times 10^8 \text{ lb-in.}^2 \text{ per 4-ft width}$$

Check deflection of the assembly: Bending deflection (using a 4' width) is

$$\Delta_b = 5(4w)\ell^4/(384EI_g)$$

where w = live load in psf

$$= \frac{5(4 \times 30)(12^4)(1728)}{384(391,000,000)}$$

$$= 0.14 \text{ in.}$$

(a) GROSS SECTION FOR DEFLECTION

(b) NET EFFECTIVE SECTION FOR BENDING

Fig. 13-4. Stressed-skin panel for Example 13-2.

For shear deflection, the properties of the stringers are used assuming $G = 0.06E$. The shear deflection is given by

$$\Delta_s = 0.15(4w)\ell^2/AG$$
$$= \frac{0.15(4 \times 30)(12^2)(12)}{(32.2)(0.06 \times 1,850,000)}$$
$$= 0.01 \text{ in.}$$

Total deflection = 0.15 in.

$$\ell/240 = 0.60 \text{ in.} \quad OK$$

Check deflection of the top skin: Considering bending between stringers of the top skin acting alone, stress is perpendicular to the face grain. Therefore the correct moment of inertia to calculate deflection of the skin is $I_\perp = 0.016$. For a 4-ft width, assuming the plywood acts like a beam with fixed ends, the live load deflection is

$$\Delta = (4w)(\ell^4)/(384EI)$$
$$= \frac{(4 \times 30)(13.9/12)^4(1728)}{384(1,800,000)(0.016)}$$
$$= 0.03 \text{ in.} < 13.9/240 = 0.06 \text{ in.} \quad OK$$

Determine skin effective in bending: For unsanded plywood with face grain parallel to stringers,

$b = 18$ in. for top skin
$b = 12$ in. for bottom skin (Table 13-1)

For the top skin, the effective width beyond the face of the stringer on either side = $18/2 = 9$ in. However this effective distance cannot be more than half the clear span = $13.9/2 = 6.95$ in. So the entire top skin is effective in bending. For the bottom skin, only $12/2 = 6$ in. beyond each stringer edge is effective, so there will be three ineffective strips of width $2(0.95) = 1.9$ in. (see Fig. 13.4b). This leaves $48 - 1.9(3) = 42.3$ in. effective; that is, $42.3/48$ of the bottom skin is effective.

Find N.A. of the section effective in bending: The effective area of the bottom flange is $5.96(42.3/48) = 5.25$.

A	y	Ay
9.17	5.922	54.30
30.07	3.000	90.21
5.25	0.156	0.82
44.49		145.33

$\bar{} = 145.33/44.49 = 3.27$ in.

Calculate net effective (in bending) moment of inertia:

I_0	A	d	d^2	Ad^2	$I_0 + Ad^2$
0.268	9.17	2.652	7.03	64.5	64.8
77.6(0.934)	30.07	0.27	0.07	2.1	74.6
0.088(42.3/48)	5.25	3.114	9.70	50.9	51.0
				$I_n =$	190.4

Find allowable bending stress:

For top skin, $c/b = 13.9/18 = 0.77$, so reduction $C = 0.82$

For bottom skin, $c/b = 13.9/12 = 1.16$, so $C = 0.67$

Top skin (in compression): $F_c = (0.82)(1540) = 1260$ psi (plywood species 1; Grade Stress Level S-2)

Bottom skin (in tension): $F_t = (0.67)(1650) = 1100$ psi

Find actual bending stress: For a 4-ft width,

$$M = (4w)(\ell^2)/8$$
$$= (4 \times 41)(12)^2/8 = 2950 \text{ ft-lb}$$

At top,

$$f_b = Mc/I_n = (2950)(12)(2.89)/190.4$$
$$= 537 \text{ psi} < 1260 \text{ psi} \quad OK$$

At bottom,

$$f_b = (2950)(12)(3.27)/190.4$$
$$= 608 \text{ psi} < 1100 \quad OK$$

Trial assembly is OK for deflection and bending. (For a complete design, splice plates would need to be designed and checks on rolling shear and horizontal shear would be required.)

13-3. SANDWICH PANELS

A sandwich panel is made up of a plywood face and back bonded to a thicker central core. Often the core is paper honeycomb, but may be a variety of other materials such as plastic foam or formed metal. The facing is not necessarily plywood, but may be particleboard, hardboard, or sheet aluminum.

The face and back provide the bending strength and rigidity, the core providing shear resistance. It is an economical assembly because the core material is cheaper than the facing material and the faces are advantageously positioned apart from one another. Sandwich construction results in a lightweight member which, by proper selection of materials, can be moisture resistant, heat resistant, and fire resistant.

The sandwich panel must be designed to resist bending loads and shear. The panel also must be checked for buckling, wrinkling of the faces, and shear crimping. Design information is given in references 4 and 5.

13-4. WOOD BEAMS REINFORCED WITH METAL

Wood beams occasionally have to be reinforced with metal plates, either to improve strength or to reduce deflection. Some of the ways in which this may be done are shown by Fig. 13-5. The vertically laminated (flitch) beam may be used with original construction to make possible a larger span than would be possible with an all-wood beam of equal depth. The wood beam with reinforcing plates top and bottom is not so common, although its strength and stiffness can be equal to that of a steel beam of equal depth. Ordinarily, it would be simpler and more effective to use a steel beam rather than reinforcing a wood beam in this manner. The wood beam with a steel plate on the bottom only, however, could be used for adding to the strength or stiffness of a wood beam already installed in a structure.

To design any of these types, the transformed-section method must be used. The transformed section is an imaginary section of stiffness equal to that of the actual reinforced beam. Its dimensions and properties are found by imagining the steel area to be replaced by an elastically equivalent area of wood. The elastically equivalent area of wood is the area of steel multiplied by the ratio of the modulus of elasticity of steel to that of wood.

Designers should be cautioned that the vari-

Fig. 13-5. Steel-reinforced beams.

ation in modulus of elasticity of wood affects the assumptions made. The coefficient of variation of E in wood is around 25–30%. For example, extreme values in a batch of wood with a nominal E of 1,800,000 psi may range from 1,000,000 psi to 3,000,000 psi.

The action of the reinforced beams is based on the fact that the deflected shape of the wood and steel portions are compatible. For the flitch beam, for example, both the wood and the steel plate bend to the same deflected shape. If the loads include long-term loads, a lesser E-value should be used for the wood.

Example 13-3

Two 3 × 12 No. 2 hem-fir beams are connected to a $\frac{1}{4}$-in. × 11-in. steel plate, as shown by Fig. 13-6, to form a flitch beam. Assuming full lateral support, what is the allowable normal-duration bending moment?

Modulus of elasticity, E, for the steel is 29,000,000 psi, and E for the wood is 1,400,000 psi. The modular ratio is $29/1.4 = 20.7$. The transformed section is found by replacing the $\frac{1}{4}$-in. thickness of steel with wood 20.7 times this thick, or 5.175-in. thick, as shown by Fig. 13-6. (Note that the imaginary wood that is substituted for the steel must be at the same location relative to the neutral axis as the actual steel; thus its total height normal to that axis is still 11 in.) The transformed section has a moment of inertia of

$$I_t = 2(2.5)(11.25^3)/12 + 5.175(11^3)/12 = 1167 \text{ in.}^4$$

Bending stress on the transformed section is $f_b = My/I_t$. The allowable bending stress in the wood is 1000 psi. In the steel it is 22,000 psi for A36 steel. The allowable bending moment, M, must be calculated two ways: first assuming that the extreme fiber

of the wood will be stressed to 1000 psi; and second, assuming that the extreme fiber of the steel will be stressed to 22,000 psi. Based on the wood,

$$M = F_b(I/c) = 1000(1167/5.625)$$
$$= 207,000 \text{ in.-lb, or } 17.3 \text{ ft-k}$$

Based on the steel, assume a stress at the top of the transformed steel to be $22,000/n$, or 1063 psi at 5.5 in. from the neutral axis. Thus

$$M = 1063(1167)/5.5$$
$$= 225,500 \text{ in.-lb, or } 18.8 \text{ ft-k}$$

The lower value, 17.3 ft-k, controls. The wood will be fully stressed when $M = 17.3$ ft-k, but the steel stress will be less than 22,000 psi.

What to do for connections between the wood and the steel depends on how the vertical loads are applied to the beam and whether or not the steel plate bears on the supports at the end. The need for connections is minimized if both the steel and the wood bear on the end supports. If they do, and if the load also bears on all three pieces, there should be no shearing load on the bolts holding the parts together.

Example 13-4

A 6 × 12 beam, hem-fir No. 2, in place in a building, needs to be strengthened to carry increased load. Decking for the floor above is attached to the top surface, but the bottom surface is accessible for adding a reinforcing plate. The beam spans 15 ft, is laterally supported, and carries a controlling load of normal duration. How much increased capacity can be had by shoring up the beam to remove all load and then attaching a steel plate 3/16 in. × 4 in. to the bottom only?

In this case, the transformed section is unsymmetrical. The plate is replaced by imaginary wood of total width $4(20.7) = 82.8$ in., as shown by Fig. 13-7. The centroid of the transformed section is at distance y below the top of the wood beam, and is computed as follows:

Item	Area	y	Ay
Wood 6 × 12	63.25 in.²	5.75	363.7
Imaginary wood 3/16 × 82.8 in.	15.52	11.59	179.9
Totals	78.77 in.²		543.6

Distance of centroid below top
$$= 543.6/78.77 = 6.90 \text{ in.}$$

ℝ $\frac{1}{4}$ × 11

3 × 12

N.A.

11" × 5.175"

Fig. 13-6. Flitch beam for Example 13-3.

Fig. 13-7. Reinforced beam for Example 13-4.

This centroid of the transformed area is the neutral axis for bending of the combined section. The moment of inertia, I_t, for the transformed section is the sum of

$$(1/12)(5.5)(11.5)^3 \quad = \quad 697.1$$
$$5.5(11.5)(5.75 - 6.90)^2 = \quad 83.6$$
$$(1/12)(82.8)(3/16)^3 \quad = \quad 0.0$$
$$15.52(11.59 - 6.90)^2 \quad = \quad \underline{341.4}$$
$$I_t = 1122 \text{ in.}^4$$

Flexural stress in the wood will be highest at the top, where distance $y = 6.90$ in. Flexural stress in the steel will be largest at $y = 4.788$ in. and will be n times the theoretical stress in the transformed section at the same location. Or, the stress in the imaginary material will be $1/20.7$ times the actual steel stress.

Based on wood,

$$\text{Allowable } M = 1000(1122)/6.90$$
$$= 162,600 \text{ in.-lb, or } 13.6 \text{ ft-k}$$

Based on steel,

$$\text{Allowable } M = (22,000/20.7)(1122)/4.788$$
$$= 249,100 \text{ in.-lb, or } 20.8 \text{ ft-k}$$

Stress in the wood controls, and the allowable M for the reinforced section is 13.6 ft-k.

Without the added steel plate, the section modulus of the beam would be 121.2 in.^{3}, and

$$\text{Allowable } M = 1000(121.2)/12,000 = 10.1 \text{ ft-k}$$

Adding the plate caused only a 35% increase in bending-moment capacity. A single plate is seldom very effective, but is often the only way available to strengthen a beam.

Connecting the plate to the wood beam is another part of the total design problem. To determine the required spacing of lag screws, bolts, or nails, the designer must compute the longitudinal shear stress *for the transformed section* at the junction of the steel and the wood, and then choose the proper size and spacing of connectors to resist this shear stress. The maximum shear stress in the wood member is at the centroid of the transformed section, and this too must be checked.

REFERENCES

1. *Plywood Design Specification*, American Plywood Association, Tacoma, WA, 1986.
2. *Design and Fabrication of Plywood-Lumber Beams*, American Plywood Association, Tacoma, WA, 1983.
3. *Design and Fabrication of Plywood Stressed-Skin Panels*, American Plywood Association, Tacoma, WA, 1982.
4. Uniform Building Code Standard No. 25-18, International Conference of Building Officials, Whittier, Calif., 1985.
5. *Wood Handbook*, U.S. Forest Products Laboratory, U.S. Department of Agriculture, Washington, D.C., 1974.

PROBLEMS

13-1. Design plywood-lumber roof beams to span 22 ft. Use flanges of No. 1 hem-fir. These box beams are spaced at 7 ft c/c. The applied dead loads total 22 psf, and snow load is 40 psf.

13-2. Design plywood-lumber I-shaped beams for floor loading of 40-psf live load and 11-psf dead load. Beams span 14 ft and are placed at 4-ft centers. Use No. 2 Douglas fir–larch flanges.

13-3. A stressed-skin panel is 4 ft wide and spans 16 ft. The top skin is $\frac{1}{2}$-in. Structural I Exterior plywood; the bottom skin is similar, except 3/8 in. thick. The four stringers (which run parallel to the plywood face grain) are 2

× 6 D. fir–larch No. 2. Clear distance between stringers is 13.9 in. Dead load is 15 psf and snow live load 35 psf. Using a deflection limit of $\ell/240$, check the assembly for bending and deflection.

13-4. What is the maximum allowable total load for the stressed-skin panel of Problem 13-3 based on bending? Based on deflection?

13-5. Check the panel of Problem 13-3 for horizontal shear at the neutral axis of the gross section and for rolling shear at the junction of the inner-face ply and the adjoining crossband.

13-6. Will the stressed-skin panel of Problem 13-3 be OK for bending and deflection if the bottom skin is replaced by 1 × 6s as in Fig. 13-3c? (*Hint*: Distance between stringers is no longer 13.9 in.)

13-7. Determine a size and spacing of lag screws to connect the 3/16- × 4-in. plate to the 6 × 12 in Example 13-4.

13-8. Assume that is is found possible to add another similar plate to the top of the beam of Example 13-4. By what percentage will this increase the bending capacity of the 6 × 12? For a span of 15 ft, what is the allowable load per foot, based on bending? Will shear now control, and if so, what is the allowable load?

13-9. Two 2 × 12s of D. fir–larch No. 1 are used on a 15-ft span to carry a load of 350 lb/ft. Design a flitch beam to do the job.

14

Formwork for Concrete

14-1. INTRODUCTION

One of the important uses of lumber and structural plywood is one in which the wood does not even remain as a part of the completed structure. Plywood and lumber used as concrete formwork are, of course, stripped (removed) after the concrete hardens. Wood formwork can be used to make a simple concrete wall or column or a complex shell roof.

Economy is a major consideration in the design of concrete formwork, since the formwork may represent about half the total cost of a concrete structure. Formwork economy depends to a large degree on the concrete designer. Keeping beam and column sizes the same for several floors will permit reuse of the forms from floor to floor. Columns can be kept the same size over several floors by merely changing the amount of steel reinforcing and concrete compressive strength. Planning for reuse of the forms is an important money-saving measure. With careful removal and proper cleaning and oiling after each use, a plywood form can be reused 30 times or more. Before forms are used they are coated with a form-release agent (or oil) that keeps the concrete from sticking to the forms and reduces the absorption of water into the plywood.

Years ago formwork consisted of many lumber planks, the evidence of this still remaining in the form of horizontal stripes visible on old retaining and foundation walls. (See Fig. 14-1.) Today most of the forms use special types of plywood that are reusable and labor saving. The plywood forms also result in large joint-free, smooth concrete surfaces and can be curved for forming curved grade beams and walls. Steel, plastic, and paper-fiber forms are also used in building concrete structures. Wood forms, however, are the most common.

14-2. PLYWOOD FOR FORMWORK

The most common type of plywood for formwork is B-B Plyform, which has both face and back of grade B veneers. Grade B is a better grade than the C or D grade veneer in most structural plywood. It has the solid surface necessary for wet concrete. Plyform comes in Class I or Class II, both of which are exterior type. The more commonly available Plyform Class I has Group 1 species faces, Group 1 or 2 crossbands, and Group 1, 2, 3, or 4 centers. Class II is somewhat lower quality. Plyform is sanded on both sides, mill-oiled, and edge-sealed to provide protection against moisture. Because it is sanded, it will have different section properties than other structural plywoods of the same thickness.

The most common thicknesses are $\frac{5}{8}$ in. and $\frac{3}{4}$ in. Table 11-1 shows that Plyform is Grade Stress Level S-2. For determining allowable stress, Plyform Class I is designated as Group 1 species and Class II as Group 3 species. Other exterior types of plywood may also be used for formwork, especially when surface evenness and form reuse are not of prime concern.

A coated plywood with a very smooth surface, called high-density overlay (HDO), is also used. The overlay consists of a cellulose fiber sheet, having no less than 45% resin solids, which is fused under heat and pressure. Overlaid plywood is sometimes referred to as plastic coated, but should be distinguished from plywood that is user-coated with a plastic com-

Fig. 14-1. Imprint left on concrete retaining wall by use of board forms instead of plywood. (Photograph by authors.)

pound. HDO is used where a smooth, grainless concrete surface is desired, because the resin blocks out the grain and knot pattern. HDO is often reused 200 times.

Allowable stresses for plywood used in formwork are generally modified with a duration of load factor of 1.25. The designer may need to adjust this for forms that will be reused. The American Plywood Association also recommends modifying plywood allowable stresses with an "experience factor" of 1.30 (1). Stresses for "wet" plywood should be used.

For dimension lumber members that support the plywood sheathing, allowable bending stresses for single-member use should be used (2). The repetitive-member stresses should be used only for assemblies such as crane-handled panels carefully constructed for reuse.

14-3. DESIGN FORCES

Formwork must be designed to support both vertical and lateral loads. Forms for slabs or beams must support the dead load of the concrete and reinforcing steel in addition to the vertical live loads of workers, materials, and equipment on freshly hardened slabs. Live loads during construction can often be more severe than those imposed after occupancy of the building. The form designer must be alert to unusual loads that might be applied while the structure is under construction. A *minimum* live load of 50 psf is recommended by ACI Committee 347, Formwork for Concrete.

For walls and columns it is the lateral force (lateral pressure) that constitutes the design load. The amount of this pressure due to the fresh concrete depends on pour rate, job temperature, slump, cement type, concrete density, method of vibration, and height of form. A fast pour rate will result in a higher pressure. Concrete that is placed in cold weather will remain unhardened for a longer time and to a greater depth; thus, forms for cold-weather concrete should be designed for higher pressures.

The ACI publishes recommended equations (experimentally determined) for calculating lateral concrete pressures. In the equations that follow, p is the maximum equivalent liquid pressure, in pounds per square foot, at any elevation in the form; R is the rate of concrete placement, in feet per hour; T is the temperature, in degrees Fahrenheit, of the concrete in the forms; and h is the maximum height, in feet, of fresh concrete above the point being considered. The equations are for normal weight, internally vibrated concrete, having slump no greater than 4 in., of Type I cement, with no admixtures.

For columns:

$$p = 150 + 9000R/T \qquad (14\text{-}1)$$

but not over 3000 psf or 150h

For walls with $R < 7\,ft/hr$:

$$p = 150 + 9000R/T \qquad (14\text{-}2)$$

but not over 2000 psf or 150h

For walls with R of 7–$10\,ft/hr$:

$$p = 150 + 43{,}400/T + 2800\,R/T \qquad (14\text{-}3)$$

but not over 2000 psf or 150h

For walls with $R > 10\,ft/hr$:

$$p = 150h \qquad (14\text{-}4)$$

(Adjustments to these equations for external vibration when consolidating, superplasticizing admixtures, and others are too varied to discuss here in this basic treatment.)

14-4. TERMINOLOGY

Figure 14-2 shows the location of the various formwork members in a wall form, and Fig. 14-3 the parts of a slab form. The designer needs to be familiar with names of the parts of the typical formwork assembly. These terms are defined below.

Sheathing: The plywood that forms the surface against which the wet concrete is placed.

Studs: The vertical members (usually $2 \times$'s) that support the plywood sheathing in wall forms.

Wales: The horizontal members that support the studs. Each wale usually consists of two timbers, so that holes do not have to be drilled

Fig. 14-2. Typical wall form. (Courtesy of American Concrete Institute.)

Fig. 14-3. Typical slab formwork. (Courtesy of American Concrete Institute.)

in a wale member to accept the ties (see *ties* below).

Ties: Metal hardware that ties together the opposite sides of the form, preventing them from separating under lateral pressure of the concrete. The form ties pass through the entire assembly and may (or may not) also spread the sheathing the correct distance apart. A particular type of tie called a *snap tie* is commonly used. A snap tie is twisted off just below the concrete surface after the concrete has hardened, the resulting hole then being filled with mortar.

Tie wedge or tie holder: The part of the tie assembly (often wedge-shaped) that is external to the wall form and presses against the wales. It is driven to put tension into the ties.

Joists: The horizontal beams that support the plywood sheathing for floor or roof slab formwork.

Stringers: Beams running perpendicular to and supporting the joists of floor or roof slab formwork.

Shores: The vertical posts or struts that support the roof or floor formwork assembly.

14-5. WALL FORMWORK

The maximum lateral pressures on wall and column forms are given by Eqs. 14-1 to 14-4. These pressures occur in the lower part of the forms, the pressure near the top tapering to zero. However, it is usually more practical to assume this maximum pressure to be uniform over the entire height of the form. The wale

and tie spacings could be increased near the top of the wall forms if the designer desired.

Steps in the design procedure for wall forms are as follows:

1. Calculate the design lateral pressure.
2. Check plywood sheathing for bending, deflection, and shear. Either the sheathing thickness or the spacing of its supports (i.e., the stud spacing) will be predetermined based on economy and what materials are available. If the plywood thickness is fixed, the bending check will consist of finding the maximum allowable span of the sheathing. This, of course, will be the stud spacing. But, if the stud spacing is fixed, the bending check will consist of finding the required section modulus of plywood based on bending stresses. The deflection limit is $\ell/360$ for the plywood and also for the lumber members.
3. Design studs, considering bending, shear, and deflection.
4. Design wales, considering bending, shear, and deflection.
5. Choose ties and spacing.
6. Check bearing stresses (studs on wales and tie holders on wales).
7. Design lateral bracing.

Example 14-1

Design wall forms for a 12-ft-high wall, concreted at 3 ft/hr. The concrete has a temperature of 60°F and is internally vibrated; that is, a concrete vibrator is immersed in the wet concrete but not applied to the form itself. Class I, $\frac{3}{4}$-in. Plyform is available and will not be reused. Ties with an allowable tensile force of 3000 lb are on hand. Use construction grade Douglas fir for the studs and wales.

From Eq. 14-2,

$$\text{Max pressure} = p = 150 + 9000(3)/60$$
$$= 600 \text{ psf}$$

The lateral pressure varies (as a straight line) with height from the top of the concrete surface. Pressure is γh, where γ is the specific weight of concrete = 150 pcf. At 1 ft from the top of the form the pressure will be 150 psf, at 2 ft it will be 300 psf, and so on. The maximum pressure of 600 psf is reached at only 4 ft from the top of the form. At greater depth, the maximum is used, so it is logical to design the entire height of form for a uniform pressure of 600 psf.

Check sheathing for bending: For species Group 1, Grade Stress Level S-2, the basic (wet) $F_b = 1190$ psi. This is modified by an "experience factor" of 1.3 and by a duration of load factor of 1.25. Therefore,

$$F_b = 1190(1.3)(1.25) = 1930 \text{ psi}$$

For sanded $\frac{3}{4}$-in. plywood, Table 11-4A shows

$$KS = 0.412 \text{ in.}^3/\text{ft}$$

(for a panel with 8-ft dimension horizontal; i.e. with bending stress parallel to the face grain).

Assuming continuous over three spans (Table A-9),

$$M = F_b(KS) = w\ell^2/10$$

$$w = (600 \text{ psf})(1 \text{ ft}) = 600 \text{ lb/ft} = 50 \text{ lb/in.}$$

$$\ell = \sqrt{10(1930)(0.412)/50} = 12.6 \text{ in.}$$

(maximum allowable span of plywood based on bending)

Check sheathing for deflection: Deflection should be limited to $\ell/360$; I (from Table 11-4A) is 0.197 in.^{4}/ft.

$$\Delta\text{max} = w\ell^4/145EI = \ell/360$$

$$\ell = \sqrt[3]{145EI/360w}$$

$$= \sqrt[3]{145(1.5 \times 10^6)(0.197)/[360(50)]}$$

$$= 13.4 \text{ in.}$$

(maximum allowable span of plywood based on a deflection limit of $\ell/360$)

Check sheathing for rolling shear: From Table 11-4A, $Ib/Q = 6.762$

$$V = 0.6w\ell$$
$$f_s = VQ/Ib$$
$$F_s = 44(1.3)(1.25) = 71.5 \text{ psi}$$

Set $F_s = f_s$

$$71.5 = 0.6(50)(\ell)/6.762$$
$$\ell = 16.1 \text{ in.} \quad \text{(max span based on shear)}$$

So bending controls and the maximum allowable span of the plywood (maximum spacing of the studs) is 12.6 in. The stud spacing should be an even frac-

tional part of the 96-in. horizontal dimension of the plywood. *Use studs at 12 in. c/c.*

Stud design for bending: With studs spaced at 12 in. c/c, the tributary width to a single stud is 1 ft, so $w = 600 \text{ psf} \times 1 \text{ ft} = 600 \text{ lb/ft} = 50 \text{ lb/in.}$ Assuming 2×4s and single-member use,

$$F_b = 1050(1.25) = 1313 \text{ psi and } S_x = 3.06 \text{ in.}^3$$

$$\ell = \sqrt{10(1313)(3.06)/50} = 28.3 \text{ in.}$$

(max span based on bending)

Stud design for deflection:

$$I = 5.36 \text{ in.}^4 \quad \text{and} \quad E = 1,500,000 \text{ psi}$$

Using a deflection limit of $\ell/360$,

$$\ell = 40.1 \text{ in.} \quad \text{(max span based on deflection)}$$

Stud design for shear:
Hurd (2) recommends multiplying the allowable shear stress by a load-duration factor of 1.25 and by a "two-beam" factor of 1.5. Therefore,

$$F_v = 95(1.25)(1.5) = 178 \text{ psi}$$

Shear at a distance d from the support is

$$V = 0.6w\ell_c - wd$$
$$f_v = 1.5V/A$$
$$= \frac{1.5(0.6w\ell_c - wd)}{bd}$$

With $d = 3.5$ in., $b = 1.5$ in., and $w = 50 \text{ lb/in.}$, and setting the allowable stress equal to the actual stress, the maximum stud clear span based on shear is

$$\ell_c = 26.6 \text{ in.}$$

With doubled wales of $2\times$'s, the approximate maximum span is

$$\ell = 26.6 + 2(1.5)$$
$$= 29.6 \text{ in.} \quad \text{(max span based on shear)}$$

Bending controls and the maximum allowable span of the studs (spacing of the wales) is 28.3 in. *Place top and bottom wales one foot from the top and bottom of the form (a common construction practice) and space other wales at 24 in. c/c.*

Design wales for bending: The tributary width to a single wale is 2 ft, so

$w = (600 \text{ psf})(2 \text{ ft}) = 1200 \text{ lb/ft} = 100 \text{ lb/in.}$

Assuming ties are spaced at 24 in. c/c,

$$\text{Req } S = 0.1w\ell^2/F_b$$
$$= 0.1(100)(24^2)/1313 = 4.39 \text{ in.}^3$$

S provided by doubled 2 × 4s = 6.12 in.^{3}

Design wales for shear:

$$F_v = 178 \text{ psi}$$
$$f_v = \frac{1.5(\, 0.6(100)\,(24) - (100)\,(3.5)\,)}{(2 \times 1.5)(3.5)}$$
$$= 156 \text{ psi}$$

Wales of doubled 2 × 4s are OK if ties are spaced at 24 in. c/c.
Check ties:

Req spacing of ties = tie capacity/wale load
= 3000 lb/1200 lb per ft
= 2.5 ft > 2 ft   *OK*

Check bearing stresses: Two locations need to be investigated for bearing (compression perpendicular to grain): (1) where the tie wedges bear on the wales and (2) where the studs bear on the wales. This check is left as a student problem.

Lateral bracing: The forms will require lateral bracing for the local code-specified wind loads. These braces could be wood struts on a single side, in which case they must resist either compression or tension. Guy wire bracing on both sides of the wall could also be used.

14-6. FORMWORK FOR ROOF AND FLOOR SLABS

There is no accepted sequence for design of formwork for slabs. The design sequence may depend upon the fact that a particular sheathing is available, that a particular set of shores is available, or that a certain construction module is used. In the design, however, the plywood, joists, and stringers must each be checked for bending, deflection, and shear. Also, the shores must be designed (including lateral braces) and bearing stresses checked.

The shores (posts) that support the stringers are designed exactly as any other wood column (Chapter 7). Each shore supports the load on the tributary formwork area extending halfway to the adjacent shore on all sides. Because shores are usually reused repeatedly, no increase in allowable stress for load duration should be used. Bearing stress in the member the shore supports can often be the governing design criterion. The bearing area may be less than the full area of the shore if a narrow member rests on the shore. Metal or hardwood plates attached to the top of the shores may be used to extend the bearing area and reduce the bearing stress.

Example 14-2

A 9-in.-thick floor slab of normal weight concrete is to be formed using $\frac{3}{4}$-in. Class II plyform. Bays are 15 ft × 15 ft. Joists are construction grade D. fir–larch. Determine the spacing of the joists and the stringers. Forms will be used only once.

Determine loads:

Concrete 150(9/12)	=	113 psf
Recommended live load	=	50 psf
Est. form wt.	=	4 psf
Total	=	167 psf

Plywood for bending: For species Group 3, basic $F_b = 820$ psi (wet)
$w = 167$ lb/ft = 13.9 lb/in. per ft width
$F_b = 820(1.3)(1.25) = 1332$ psi
$F_b(KS) = w\ell^2/10$

Effective section modulus is $KS = 0.412$ in.^{3}/ft width

$$\ell = \sqrt{10(1332)(0.412)/13.9}$$
$$= 19.9 \text{ in.}$$

Plywood for deflection: For stress parallel to face grain,

$$I = 0.197 \text{ in.}^4/\text{ft}$$
$$w\ell^4/145EI = \ell/360$$
$$\ell = \sqrt[3]{145(1,100,000)(0.197)/[360(13.9)]}$$
$$= 18.4 \text{ in.}$$

Plywood for rolling shear:

$$V = 0.6w\ell$$
$$f_s = VQ/Ib = 0.6w\ell/6.762$$
$$F_s = 44(1.3)(1.25) = 71.5 \text{ psi}$$
$$71.5 = 0.6(13.9)\ell/6.762$$
$$\ell = 58.0 \text{ in.}$$

Deflection governs and joist spacing can be no more than 18.4 in. *Use joists at 16 in. c/c* in order to make joist spacing an even fractional part of the 96 in. plywood dimension.

Joist for bending: Assume 2 × 4 joists. The tributary width to a joist is 16 in.

$$w = 167(16/12) = 223 \text{ lb/ft}$$
$$= 18.6 \text{ lb/in. on a 1-ft width}$$
$$F_b = (1050)(1.25) = 1313 \text{ psi} \quad \text{and}$$
$$S = 3.06 \text{ in.}^3$$
$$\ell = \sqrt{10(1313)(3.06)/18.6}$$
$$= 46.5 \text{ in.}$$

Joist for deflection:

$$w\ell^4/145EI = \ell/360$$
$$\ell = \sqrt[3]{145(1,500,000)(5.36)/[360(18.6)]}$$
$$= 55.8 \text{ in.}$$

Joist for shear:

$$F_v = 178 \text{ psi (see Example 14-1)}$$
$$V = 0.6w\ell_c - wd$$
$$f_v = 1.5V/bd = 178$$
$$\ell_c = 61.7 \text{ in.}$$
$$\ell = 61.7 + 1.5 = 63.2 \text{ in.}$$

Bending controls and maximum span of joists (spacing of stringers) is 46.5 in.

For uniform spacing across the bay, *use stringer spacing of approximately 38.4 in.* for five equal spaces. Based on strength and stiffness of stringers, the spacing between shores (maximum stringer span) can be determined in a similar manner.

REFERENCES

1. *Concrete Forming*, American Plywood Association, Tacoma, WA, 1986.
2. Hurd, M. K., *Formwork for Concrete*, 4th ed., American Concrete Institute, Detroit, MI, 1979.

Fig. 14-4. Bearing area of tie wedge for Problem 14-3.

PROBLEMS

14-1. Design the forms for a 12-ft-high concrete wall for which the concrete is to be placed at a rate of 4 ft/hr at a temperature of 50°F. Class I $\frac{5}{8}$-in. Plyform is on hand. Use No. 2 D. fir–larch for studs and wales. Ties can carry 4000 lb. Contact area of tie wedge is 3 in.^{2}.

14-2. Design the wall forms for Problem 14-1, except that the wall is concreted at the rate of 7 ft/hr and Plyform is $\frac{7}{8}$ in.

14-3. If the contact area between tie wedge and doubled wale of Example 14-1 is as shaded in Fig. 14-4, compare actual bearing stress with the allowable stress.

14-4. Is the junction of stud and wale for Example 14-1 satisfactory in bearing?

14-5. Design forms to support an 8-in. concrete floor slab using $\frac{23}{32}$-in. Class I Plyform. Use Construction grade Douglas fir framing members and shoring. Ceiling height is 9 ft, and bays are 22 ft × 22 ft. Forms will be used only once.

14-6. Find the maximum shore spacing for Example 14-2 assuming stringers are 2 × 6s of No. 2 grade Douglas fir.

14-7. Design shores for Example 14-2 using No. 2 D. fir–larch. Clear height between floors is 9.5 ft. Assume shores are spaced every 10 ft along the length of the stringer. What is the minimum bearing area between shore and the No. 2 grade D. fir stringer?

15

Miscellaneous Structure Types

Light-frame construction, pole buildings, and timber bridges were introduced in Chapter 1 but essentially ignored thereafter. Each of these deserves further attention, so this chapter covers them in greater detail. It also introduces the subject of wood foundations. Design features of the other principal building types (heavy timber, and post and beam) have already been discussed.

Timber bridges are not covered here in great detail. To do so would double or triple the length of this book. However, some attention should be given to this important use of wood, so the authors have elected to discuss the common timber bridge types and to give a few design examples to illustrate design features that differ somewhat from those for wood buildings.

15-1. LIGHT-FRAME CONSTRUCTION

Figures 15-1 and 15-2 show two types of light-frame construction. The platform frame of Fig. 15-1 is by far the more common. It is the simpler to construct, therefore usually less expensive. It can be built one story at a time. As one story is completed, wall framing for the story above can be assembled using the completed story below as a work platform.

A disadvantage of platform framing is that its total vertical shrinkage is the sum of the longitudinal shrinkage of the studs plus the transverse shrinkage of the horizontal members. The latter, shrinkage (or swelling) in width of the horizontal members, can be much larger than the change of length of the studs. Thus, the total vertical dimension change is fairly concentrated at the horizontal members. This can be

troublesome, leading to cracking in outside finish materials, such as brick veneer or stucco. Also, in buildings of more than two stories, it may cause difficulties with rigid plumbing and mechanical systems. Problems of vertical shrinkage in platform framing can be lessened by use of I-shaped joists having plywood webs.

Vertical shrinkage is better controlled by the balloon system of Fig. 15-2. It includes only the longitudinal shrinkage of the vertical members, which extend over the entire two stories. Transverse shrinkage of the horizontal members does not add to the total. The balloon frame is the preferred type where the exterior finish surface is to be a brittle material without expansion or contraction joints at story levels (such as masonry veneer or stucco). A disadvantage is that it requires longer studs, and a two-story or three-story preassembled wall is much harder to lift into position than a one-story wall assembly of the platform type.

Frequently, for small buildings of either type of light-frame construction, little engineering analysis is done. However, with increasing material scarcity and material costs, builders are encouraged to find ways of using material more efficiently. As a result, engineering analysis is becoming more important, and the future may see more sophisticated designs in light-frame construction. Examples already evident are the substitution of parallel-chord trusses or of shop-fabricated wood beams for sawn-lumber joists, and substitution of prefabricated wall panels for the conventional studs and sheathing.

If traditional light-frame structures are examined using modern structural-design criteria, they may be discovered in some respects to be overdesigned; that is, they are stronger than

RAFTER (OR TRUSS T.C.)

ROOF PLYWOOD

BLOCKING

TOP PLATE
(2 – 2 X'S)

CEILING JOIST (OR
TRUSS B.C.)

LET-IN BRACING
(WOOD OR METAL)

2ND-LEVEL STUD

SOLE PLATE

HEADER

TOP PLATE
(2 – 2 X'S)

X-BRIDGING

2ND-FLOOR JOISTS

SIDING

LET-IN BRACING
(WOOD OR METAL)

1ST-LEVEL STUD

SOLE PLATE

FLOOR PLYWOOD

1ST-FLOOR JOISTS

HEADER

SILL, BEDDED ON
CEMENT GROUT

X-BRIDGING

FOUNDATION

LET-IN BRACING IS OFTEN OMMITED IF OUTSIDE
SHEATHING CAN ACT AS A RIGID DIAPHRAGM.

Fig. 15-1. Platform framing: cross section through exterior wall.

RAFTER (OR
TRUSS T.C.)

ROOF PLYWOOD

BLOCKING

TOP PLATE
(2 – 2 X'S)

CEILING JOIST
(OR TRUSS B.C.)

FLOOR PLYWOOD
(NOTCHED TO PASS
AROUND STUD)

STUD CONTINUOUS

X-BRIDGING

RIBBON, SUPPORTS
FLOOR JOISTS

2ND-FLOOR JOISTS

SIDING

FLOOR PLYWOOD (NOTCHED
TO PASS AROUND STUD)

SILL, BEDDED IN
CEMENT GROUT

X-BRIDGING

FOUNDATION

1ST-FLOOR JOISTS

Fig. 15-2. Balloon framing: cross section through exterior wall.

necessary for the expected loads. In other respects, however, they may be found to be not strong enough. One example of the latter is the plate or sill, the 2-in. nominal horizontal member on which rafters, joists, or ends of the wall studs bear. Bearing stresses normal to the grain of the plate frequently exceed the allowables for compression perpendicular to the grain. Yet we use these 2-in. nominal plates without problems. Why? The answer is that a 0.04-in. de-

formation (bearing failure, as it is defined by the code) probably occurs, but collapse does not. A deformation this small is normally undetectable and is considered harmless in the finished structure.

Most design decisions for light-frame structures can be made using tables. Some engineering design knowledge went into preparing these tables, and so did experience. Thus, the designer may usually use the tables with confidence, but should be alert to any limitations placed on their use. The UBC (1) allows design using tables alone for light-frame structures up to two stories high. For higher buildings it requires engineered design—not for structural purposes, but to ensure that vertical shrinkage does not raise trouble for mechanical, electrical, or plumbing systems.

Much of light-frame design (and design of other types, too) is limited by module sizes—length and width of available plywood sheets, for example. Thus, the builder is encouraged to choose stud, joist, and truss spacings that are even fractions of those dimensions.

Nailing schedules tell how to connect the members of light-frame buildings together. These schedules give connections that are satisfactory in most cases. However, the designer should give special attention to connections wherever unusual conditions occur. Under these conditions, better-than-normal connections may be required.

As one example of the latter, consider buildings with roof overhangs. Standard nailing schedules are based on the assumption that roof uplift forces are relatively small and are transferred by nails from one member to the next: from roof sheathing to rafter, to top plate, to wall sheathing, to bottom plate, to anchor bolt, and finally to the foundation. In localities with high winds, this system of nailing may prove inadequate for roof members with large overhangs. For such cases, metal timber connectors may be necessary. For extreme cases, a better solution is to provide full-height steel anchor rods, running all the way from the roof member to the foundation anchor bolts. If this is done, uplift forces are transferred directly from the rafters to the foundation, bypassing the sequence of nailed members occurring between the roof and the foundation.

Substructures

Light-frame buildings are usually built on foundations of either masonry or concrete. If the foundation is a wall, the first wood member is a sill plate (or sill), not engineered, placed on a leveling bed of grout (cement mortar). The sill plate is usually of 2-in. nominal thickness

and 4-in. (or more) nominal width. The UBC (reference 1) requires these plates to be either redwood or preservative-treated lumber. Anchor bolts connect the plate to the foundation.

If the light-frame building is supported on isolated footings, then stub columns or pedestals will support a beam—either a glulam, a wood beam built by nailing together (side by side) the required number of 2-in. nominal members, or a steel or concrete beam. If the beam is concrete, a properly anchored sill plate is placed on top of the beam. If the beam is steel, a wood sill plate is bolted to the top flange of the steel beam. The first-floor structure is mounted on the sill plate or on the beam as shown by Fig. 15-3.

Wall Structure

Wall studs are connected to a sole plate, usually the same width as the studs. A double plate, made of two 2-in. nominal pieces is attached to the top end of the studs. The wall assembly must have bracing to resist shearing loads parallel to the length of the wall. This bracing may be *let-in* bracing (as shown by Fig. 15-1), which may be either metal or wood. If the bracing is wood, it is of 1×4 nominal size. It should be set into recesses cut into the studs and preferably also into both the top and bottom plates. (Figure 15-1 shows it let into the studs only.) As an alternate to let-in bracing, certain types of rigid sheathing, nailed to the studs and plates, may be used.

In the United States, plywood and other sheathing materials are used almost to the exclusion of let-in bracing. Also, in the United States, the wall studs, plates, and at least the end panels of plywood sheathing are assembled on a flat working surface, and then the entire assembly is tilted to the desired position on the structure. If let-in bracing is used, however, installation of sheathing is frequently left until the wall framing is raised into its vertical position.

Solid blocking, horizontal pieces of 2×4 nailed between studs (usually at mid-height) was formerly considered to have structural value, reducing the weak-axis ℓ/d ratio of the studs. Attached sheathing materials are now realized to be much more effective for this pur-

Fig. 15-3. Connection of light-frame building to substructure. (Courtesy of American Society of Civil Engineers, D. H. Percival and S. K. Suddarth, "Light-Frame Construction," *Wood Structures, a Design and Commentary*, 1975.)

pose than blocking. Thus, blocking between studs is usually omitted, unless it is needed in a particular location for edge nailing of sheathing panels, as fire blocking, or as support for nonstructural elements such as electrical boxes.

15-2. POLE BUILDINGS

Pole buildings have been used since ancient times. Early American Indians in the Missouri valley used living tree trunks as poles for buildings of up to 40-ft diameter, with low side walls and a center pole about 14 ft high. In the Fiji Islands, deeply buried poles were used to help buildings resist the forces of frequent hurricanes, the structures acting as rigid frames but with considerable flexibility. In Japan, pole buildings with relatively light walls were found to be a good answer to the lateral forces of earthquakes.

Modern pole construction, dating from the 1950s, became feasible when preservative treatment materials and methods were developed to prevent rotting of wood in contact with the ground. Recent pole construction was orig-

inally for agricultural buildings but now often finds use for residences, commercial and industrial buildings, churches, and schools. It is claimed that there are some 600,000 pole buildings in the United States and Canada (2).

The typical pole building has vertical poles, embedded in the ground, serving both as columns and as members to resist lateral forces. The poles are usually round members, but might instead be rectangular sawn timbers or glulams. The floor and roof structures attached to and supported by the poles may be strictly of light-frame construction or may include glulam members or trusses.

The poles, being set into the ground, *must* be treated with preservative. The American Wood Preservers' Association standards (3) specify appropriate preservatives and methods of treatment.

There are two main types of pole building. In the first type, poles extend only to the bottom of the first-floor framing, as in Fig. 15-4, and serve merely as foundations for the platform-type frame structure above. In the second type, the poles extend all the way to the roof, as in Fig. 15-5, in which case they serve as

Fig. 15-4. Platform framing for pole building.

building columns also. The connection of the top of the poles to the roof structure may be either pinned (nonrigid) or made rigid by means of bracing. In either type of pole building, bents as shown are normally in the range from 10 to 20 ft apart (i.e., the bays are 10–20 ft long). Spans of up to 100 ft have been built where roof trusses are used, but 60 ft is the usual practical limit (4).

Advantages of Pole Construction

1. Less costly excavation may be needed than for buildings with concrete foundations or foundation walls. Normally the only needs are round drilled holes to receive the poles, plus the usual digging for water lines and sewers.

2. After the poles are placed, the roof can be installed immediately, providing shelter under which construction may continue.

3. If the poles go all the way to the roof, the walls need to carry only their own weight. Thus, walls might be of lighter construction. For example, there would be no need to provide the usual 16-in. stud spacing (local building code permitting, of course).

4. Pole construction is good for "difficult" locations, such as hillsides, swampy ground, permafrost areas, and seaside areas where high wind and waves may be present.

Fig. 15-5. Two methods of framing a pole building with poles extending to roof.

Arrangements of Poles and Walls

Figure 15-6 shows three ways of locating walls relative to the poles. In the first, the pole actually becomes a part of the wall structure. This system appears simplest, but it has disadvantages. Round timbers used for poles are seldom straight, and sealing the joint between a crooked pole and the wall material is difficult. With the poles inside, as in Fig. 15-6b, otherwise useful space inside the building is consumed. If the appearance of the round pole inside the building is objectionable, inside wall finish material (drywall, for example) can be placed around the pole to enclose it. Sealing against weather is simplest of all with the arrangement in Fig. 15-6c, in which the poles are outside the building. This scheme also keeps the possibly odorous and toxic preservative chemicals outside the building. Exposed ends of floor girders may require treatment.

Connections

Horizontal structural members may be connected to the poles using bolts or timber connectors. If possible, the poles should be *dapped*, that is, notched as shown by Fig. 15-7. If the dap is carefully prepared and the horizontal cut deep enough, the vertical reaction of the beam is resisted by direct bearing of the beam on the pole, so that the bolts merely ensure that the beam does not shift sidewise and slide off its seat. In this case the bolts carry no load. If large enough daps are not possible, then the bolts (or timber connectors) would be in vertical shear, carrying the entire reaction of the beam into the pole.

Spikes might be used instead of bolts. If spikes are used, however, it will be preferable to transfer the vertical reaction by beam bearing on the dapped pole, using the spikes only to hold the parts in contact.

Allowable Design Stresses for Poles

Table 15-1 shows allowable stresses for poles that are treated with preservative. The stresses are for single-member use. Increase of F_b for repetitive-member use is not allowed. Also, the form factor for round members has already been considered in formulating the table; it should

(a) POLE EMBEDDED IN WALL

(b) POLE INSIDE

ENDS OF FLOOR GIRDERS

(c) POLE OUTSIDE

Fig. 15-6. Three positions of pole relative to wall.

FLAT SURFACE

Fig. 15-7. Pole dapped to provide seat and flat faces for connecting girders (corner pole shown).

Table 15-1. Design Values for Preservatively Treated Poles,[a, b] Wet and Dry Conditions of Use.

Species	Bending F_b (psi)	Compression Parallel to Grain F_c (psi)	Modulus of Elasticity E (psi)	Compression Perpendicular to Grain $F_{c\perp}$ (psi)	Horizontal Shear F_v (psi)	End Grain in Bearing[c] F_g (psi)
Douglas fir, coast	1,850	1,000[d]	1,500,000	375	115	1,200
Jack pine	1,500	800	1,100,000	280	95	870
Lodgepole pine	1,350	700	1,100,000	240	85	870
Northern white cedar	1,050	525	600,000	225	80	670
Ponderosa pine	1,300	650	1,000,000	320	90	820
Red or Norway pine	1,450	725	1,300,000	265	85	790
Southern pine	1,700	900[d]	1,500,000	320	105	1,120
Western red cedar	1,350	750	900,000	255	95	940
Western hemlock	1,650	900	1,300,000	245	115	1,120
Western larch	2,050	1,075	1,500,000	375	120	1,210

[a]The design values are based on ASTM D 2899-74, *Method for Establishing Design Stresses for Round Timber Piles*. The values are for single-member uses of poles and assume that the conditioning prior to teatment was in accordance with ASTM D 1760-76. If the poles are conditioned by air or kiln drying only prior to treatment, the design values, with the exception of modulus of elasticity, may be increased by application of the modifying factor for seasoning conditioning, C_{co}, which is 1.18 for southern pine and 1.11 for other species.
[b]The compression perpendicular to grain design values $F_{c\perp}$ are based on the design values for sawn lumber under wet conditions of use, which have been reduced for conditioning. For dry conditions of use, multiply by 1.5.
[c]End grain in bearing design values, F_g, are based on the design values for sawn lumber used in a wet location and reduced for conditioning. When used in a dry location, they may be increased by 10% ($C_M = 1.10$).
[d]Design values for compression parallel to grain, F_c, in Douglas fir and southern pine may be increased 0.2% for each foot of length from the tip of the pole to the critical section. This increase shall not exceed 10%. The modification factor for location of critical section, C_{cs} = 1 + 0.002L, where L = distance (ft) from the tip to the critical section; C_{cs} shall not exceed 1.10.
Source: Copyright ASTM. Reprinted with permission.

not be used again to modify the tabulated value of F_b. The allowables are to be modified for load duration, however.

Distribution of Horizontal Load Among Poles

This part of the design is merely structural analysis. For many configurations of pole building, the analysis may be aided by simplifying assumptions, such as those shown by Patterson (6). The shear in each pole depends on pole stiffness and the location of horizontal loads on the structure. Poles resist horizontal loads, acting as embedded cantilevers. The roof system is usually sufficiently rigid that the tops of the poles all deflect by the same amount. The same is true for floor systems and for the pole deflections where the floors are attached. Stating these conditions mathematically permits solving for the amount of horizontal load resisted by each pole and for the horizontal forces transmitted by the floor and roof systems.

Poles may act as simple cantilevers, having the superstructure above pin-connected to the poles (nonrigid connection), or they may have some degree of top fixity by being connected more rigidly to the superstructure. Means of providing a degree of top fixity would include connecting the pole to fairly deep trusses, glulams, or deep wall panels, or using bracing.

If the pole tops are free to rotate with respect to the roof structure, the analysis of wind load distribution to poles that support only a roof and walls is fairly simple, as shown below.

Figure 15-8a shows wind loads on a simple two-pole bent, with the poles pin-connected to the roof structure above. Girts (horizontal beams spanning from pole to pole) support the wall and transfer its wind loads to the poles. These loads on the poles may be treated as uniformly distributed loads. A redundant horizontal force, F, in the roof system causes both poles to deflect equally at the top. Figure 15-8 shows free bodies of the left and right poles, with the loads acting directly on each pole, and

(a) CROSS SECTION (SCHEMATIC) AND WIND PRESSURES
ON WALLS (WIND UPLIFT ON ROOF NOT SHOWN)

(b) LOADS ON POLES AND POLE DEFLECTIONS

Fig. 15-8. Distribution of wind load to poles.

the unknown redundant force F. In this case, with uniform lateral load over the full pole height, the lateral deflections of each pole are easily computed using superposition of expressions for deflection of a cantilever. Thus, *for this case*, the lateral deflections of the left and right poles are:

$$\Delta_1 = w_1 h^4/8EI - Fh^3/3EI$$
$$\Delta_2 = w_2 h^4/8EI + Fh^3/3EI$$

Equating the two deflections and solving for F gives

$$F = (w_1 - w_2)(3h/16) \qquad (15\text{-}1)$$

Knowing the value of the redundant force, F, we can now solve for the base shear and bending moment for each pole.

Figure 15-9 shows another simple case, a building with outside poles and supporting one floor as well as a roof. Assuming that wind

forces on the exposed poles are negligible compared to those on the broad expanse of the light-frame walls, wind loads are transferred to the poles at roof and floor levels only. Since both poles deflect equally at each level and there is negligible distributed load applied along the pole length, *each pole is bent to the same shape*. Thus, the base shears are the same for each pole, and analysis is even simpler than in the first example above.

For most other conditions (such as poles of unequal stiffness or length to the ground line, or more than two poles, or a frame in which the poles are braced to the roof system) the analysis is based on similar principles, but is more complex (reference 6).

Foundation

Space around the embedded pole must be filled with well-compacted soil, gravel, or sand. This backfill is important for three reasons. Skin

(a) BUILDING CROSS SECTION

Same shape
$\Delta_1 = \Delta_2$ AND
$\Delta_3 = \Delta_4$
$\therefore P_1 = P_2$ AND
$P_3 = P_4$

(b) DEFLECTIONS AND POLE LOADS

Fig. 15-9. Distribution of shear to poles with concentrated loads only.

friction (friction between the pole cylindrical surface and the soil) increases the ability of the pole to transfer vertical force to the soil. The backfill also raises the resistance to lateral loads, and (by reducing the lever arm of lateral loads) reduces the bending moment for which the pole must be designed. Finally, the skin friction also resists uplift forces on the pole.

Greatest economy usually results if the pole itself can transfer the entire vertical load to the soil through end bearing and side friction. However, if this requires a pole that is either too long or too large in diameter, then other means may be used. The simplest solution in this case is to drill a hole larger than required for the pole itself, casting concrete in the bottom of the hole to serve as a footing, or *punch pad* as shown by Fig. 15-10. The footing must be large enough in diameter to transfer the load to the soil by bearing. It must be thick enough

Fig. 15-10. Pole with concrete footing.

that the end of the pole will not punch through the footing (punching shear failure).

Another alternative is to drill an oversize hole, filling the space around the pole with concrete, all the way to the ground surface, or to the original ground surface, whichever is less. When this is done, it provides an end bearing area larger than that of the pole, and also a larger perimeter for transferring load to the soil by skin friction.

Pole Embedment

This is largely a soil mechanics problem rather than one of timber design. To determine necessary embedment requires knowing allowable soil properties. In the absence of field tests, codes such as the UBC (reference 1) often give allowable pressure for various types of soil. Table 15-2, taken from the UBC, is one of these.

Derivations of equations for required embedment depth may be found in texts on soil mechanics. Examples of such equations (from reference 5) are shown below: Eq. 15-2 for cases in which there is lateral restraint at the ground line (such as would be provided by a concrete slab surrounding the pole), and Eq. 15-3 for

cases in which there is no such restraint. Figure 15-11 shows dimensions referred to in these equations.

$$d = 1.63 \sqrt[3]{\frac{Ph}{BS_4}} \qquad (15\text{-}2)$$

$$d = 1.97 \sqrt[3]{\frac{P(h + 0.93\,d)}{BS_4}} \qquad (15\text{-}3)$$

Definitions of other terms in the two equations are:

P = lateral force applied to pole (lb)

S_4 = modified allowable lateral soil pressure as adjusted for load duration and allowable movement (footnote c on Table 15-2) (psf per ft of depth)

B = bottom end diameter of round pole, or diameter of concrete casing (if used), or diagonal dimension of a square pole (ft).

In Eq. 15-3, each side of the equation includes the unknown whose value is sought. Thus, it is necessary to solve by successive approximation (trial and error), assuming a value for the un-

Table 15-2. Allowable Foundation and Lateral Pressure.

Class of Materials[a]	Allowable Foundation Pressure S_B (psf)[b]	Lateral Bearing of Depth below Natural Grade S_0 (psf)[c]
1. Massive crystalline bedrock	4000	1200
2. Sedimentary and foliated rock	2000	400
3. Sandy gravel and/or gravel (GW and GP)	2000	200
4. Sand, silty sand, clayey sand, silty gravel, and clayey gravel, (SW, SP, SM, SC, GM, and GC)	1500	150
5. Clay, sandy clay, silty clay, and clayey silt (CM, ML, MH, and CH)	1000[d]	100

[a]For soil classifications such as organic clays and peat, a foundation investigation shall be required.
[b]All values of allowable foundation pressure are for footings having a minimum width of 12 in. and a minimum depth of 12 in. into natural grade. Except as in footnote d below, an increase of 20% is allowed for each additional foot of width and/or depth to a maximum value of 3 times the designated value.
[c]May be increased by the amount of the designated value for each additional foot of depth to a maximum of 15 times the designated value. Isolated poles for uses such as flagpoles or signs and poles used to support buildings that are not adversely affected by a $\frac{1}{2}$-in. motion at ground surface due to short-term lateral loads may be designed using lateral bearing values equal to 2 times the tabulated values.
[d]No increase for width is allowed.
Source: Reproduced from the 1985 edition of the Uniform Building Code, copyright © 1985, with permission of the publishers, the International Conference of Building Officials.

Fig. 15-11. Depth of embedment.

known d before each trial. This will be shown by Example 15-1.

Pole Standards

Table 15-3 shows published standard dimensions of poles of several classes. Notice that pole circumferences, rather than diameters, are given by the table. This is because it is easy to measure the circumference of a somewhat round pole, but harder to measure its diameter.

Suggested Design Procedure

Reference 4 laments with respect to pole buildings, "Unfortunately, because of their simplicity, they are often built without adequate design." Often standard plans are followed, without special attention to details that raise unique problems at a particular site. It is preferable to have all parts of the structural design, especially the foundation, checked by a qualified designer.

Design procedure may vary according to number of poles in a bent, whether the poles are all of the same size, whether the terrain is level or sloping, the relative stiffness of poles and superstructure, and the rigidity of the connection of floor and roof structures to the poles.

The procedure suggested here should be satisfactory for most cases, though.

1. Compute design loads on the structure, separately for dead, live, snow, and wind load. (Usually, a combination that includes wind will control the pole design.) Consider both wind pressure on the windward side of the building and suction on the leeward side. Also compute wind forces (if they are significant) on exposed parts of the poles. This requires estimating the pole diameters.

2. Compute distribution of the total wind forces among the poles. Each pole will have shear due to a portion of the total wind force.

3. Estimate the pole circumference and determine the required pole embedment. This requires knowing the soil conditions and allowable values for both direct vertical bearing and lateral passive pressure. These values should preferably be determined by a soils engineer, but often the local building code suggests suitable values.

4. Design the poles for whichever load combination is critical. For poles near the outside wall lines, this is usually a combination that includes wind, in which case the 1.33 load-duration factor applies to the allowables for both bending and compression (or tension) parallel to grain. (In buildings having an interior row of poles as well as rows along each wall line, the controlling load combination for the interior poles is often the one that causes the greatest vertical load on the pole, such as full snow load but no wind.) In any case, *interaction of bending and axial load must be considered.* Maximum bending moment in the pole occurs at the point of zero horizontal shear. For a pole embedded in earth, this will be at a point below the ground surface; often it is assumed to be at one-fourth of the embedment depth below the surface. If the pole is embedded in concrete or is firmly supported by a concrete slab cast around the pole at surface level, the maximum bending moment may be assumed to occur at the top of the concrete.

5. If a diameter different from the assumed diameter is required, recycle to step 3, above.

6. Design the concrete footing, if one is needed.

Table 15-3. Dimensions of Douglas Fir (Both Types) and Southern Pine Poles (Based on a Fiber Stress of 8000 psi).

Class		1	2	3	4	5	6	7	9	10
Minimum Circumference at Top (in.)		27	25	23	21	19	17	15	15	12
Length of Pole (ft)	Groundline Distance from Butt[a] (ft)	Minimum Circumferences at 6 Feet from Butt (in.)								
20	4	31.0	29.0	27.0	25.0	23.0	**21.0**	**19.5**	17.5	**14.0**
25	5	33.5	31.5	29.5	27.5	25.5	**23.0**	**21.5**	19.5	**15.0**
30	$5\frac{1}{2}$	36.5	**34.0**	**32.0**	**29.5**	27.5	25.0	23.5	20.5	
35	6	39.0	**36.5**	**34.0**	**31.5**	**29.0**	27.0	25.0		
40	6	**41.0**	**38.5**	**36.0**	**33.5**	**31.0**	**28.5**	26.5		
45	$6\frac{1}{2}$	**43.0**	**40.5**	**37.5**	**35.0**	**32.5**	30.0	28.0		
50	7	**45.0**	**42.0**	**39.0**	36.5	34.0	31.5	29.0		
55	$7\frac{1}{2}$	**46.5**	**43.5**	**40.5**	38.0	35.0	32.5			
60	8	**48.0**	**45.0**	**42.0**	39.0	36.0	33.5			
65	$8\frac{1}{2}$	**49.5**	**46.5**	**43.5**	40.5	37.5				
70	9	**51.0**	**48.0**	**45.0**	41.5	38.5				
75	$9\frac{1}{2}$	**52.5**	**49.0**	**46.0**	43.0					
80	10	**54.0**	**50.5**	47.0	44.0					
85	$10\frac{1}{2}$	**55.0**	**51.5**	48.0						
90	11	**56.0**	**53.0**	49.0						
95	11	**57.0**	**54.0**	50.0						
100	11	**58.5**	**55.0**	51.0						
105	12	**59.5**	**56.0**	52.0						
110	12	**60.5**	**57.0**	53.0						
115	12	**61.5**	**58.0**							
120	12	**62.5**	**59.0**							
125	12	**63.5**	**59.5**							

[a]The figures in this column are intended for use only when a definition of groundline is necessary in order to apply requirements relating to scars, straightness, etc.

Note: Classes and lengths for which circumferences at 6 ft from the butt are listed in boldface type are the preferred standard sizes. Those shown in light type are included for engineering purposes only.

Source: Courtesy American Wood Preservers Institute.

Example 15-1

Choose poles for the building shown by Fig. 15-12a. Use the UBC (reference 1) to compute wind pressures. The basic wind speed is 80 mph and the area is defined as open. Pole bents are 12 ft apart; dead load plus roof live load applies 2000 lb to each pole. No bracing is used at the tops of the poles, so the roof structure can be assumed pin-connected to each pole; that is, roof members can transfer only vertical and horizontal loads (not moment) to the poles. Treated southern pine poles are available. Soil is classified as sandy gravel.

Wind pressures for this condition were solved in Example 4-9. They were 16.3 psf inward for the windward wall and 11.0 psf outward (suction) for the lee wall. Multiplying pressures by the tributary wall width (12 ft) gives the distributed wind load per foot for each pole. Starting with this information, the solution is shown in Fig. 15-13 as a designer's computation sheet.

15-3. WOOD FOUNDATIONS

Light-frame buildings are usually supported by either concrete or masonry walls which, in turn, rest on concrete wall-type footings (Figs. 15-1 and 15-2). An alternative, now feasible using preservative treatment, is to use wood for both

(a) CROSS SECTION

(b) POLE LATERAL LOADS

Fig. 15-12. Pole building for Example 15-1.

foundation wall and footing. This has been done since the early 1960s and the practice is recognized by many building codes. Wood foundations are of several varieties: (1) poles of pole-type buildings, (2) box-beam foundation, (3) wall foundation system, and (4) bent foundation system.

All four of these types are discussed in reference 4, but only the third type, also known as the *permanent wood foundation system*, is covered here. At the time of this writing, about 50,000 buildings have been built in the United States using this type of foundation.

Untreated wood may be used in foundations, but only if it will remain well below the groundwater table at all times. Perhaps the best known of many examples is New York's famous Brooklyn Bridge. The massive timber caissons of this bridge form permanent components of the supporting piers. Being always

covered with water, they are protected against all agents of deterioration. Another, but much smaller, example familiar to the authors is a truss bridge in Denver. Its piers are of sandstone blocks, extending to about 2 ft below the water line. The stone piers rest on a grillage consisting of two layers of 12 × 12 timbers (in alternating directions) capped and tied together by a diagonal layer of 3 × 12s. In England, timber piles that had been below the water line for over 2000 years were pulled up, and the oak wood was found to be well preserved and reusable for other purposes.

The modern building foundation using treated lumber differs from these in that it is built entirely above the water line and must be protected so that the wood will remain as dry as possible. Most advantages claimed for using wood foundations for buildings relate to economy: foundations as well as superstructure are

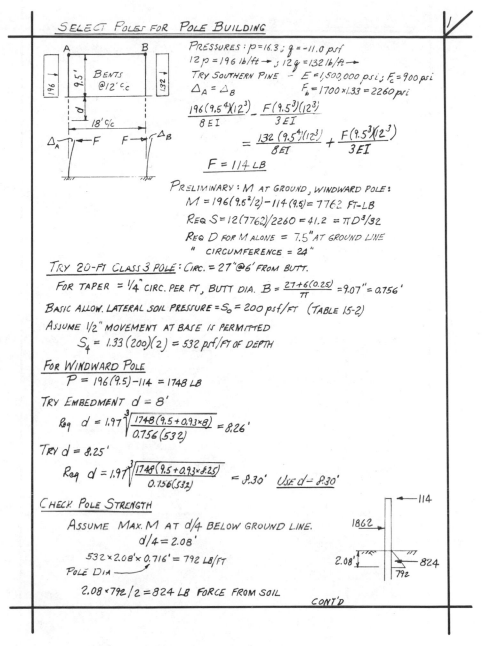

Fig. 15-13. Designer's computations for Example 15-1.

installed by carpenters, so that there is no need to schedule a masonry crew; work on the foundation can continue in freezing weather; and construction is faster.

The permanent wood foundation system uses pressure-treated standard 2-in. nominal lumber and pressure-treated exterior plywood. Figure 15-14 shows a typical example of a basement wall foundation. Reference 7 shows configurations for several other applications and gives specifications for the materials and their installation.

The plywood sheathing on the outside of the foundation wall must be pressure treated with preservative. Similarly, the studs, footing plate, bottom plate, and top plate must be pressure

<u>POLE SELECTION - CONT'D</u>   2

$M = 1862(9.5/2 + 2.08) - 114(9.5 + 2.08) - 824(2.08/3) = 10,830$ FT-LB

CIRCUMF = 27" @ 2.08' BELOW GROUND

$$\therefore S = 62.3 \text{ in}^3$$

$$f_b = M/S = 12 \times 10,830/63.2 = 2086 \text{ psi} < F_b = 2260$$

AXIAL LOAD = 2000 - 822 WIND UPLIFT = 1178 LB

AT 2.08' BELOW, POLE CANNOT BUCKLE

$$\therefore F_c = 900(1.33) = 1200 \text{ psi}$$

$$f_c = 1178/58.1 = 20 \text{ psi}$$

$$\frac{20}{1200} + \frac{2086}{2260} = 0.94 < 1.00 \quad \underline{OK}$$

<u>CHECK COLUMN BUCKLING</u>

AT 1/3 FROM TOP (AFTER CUTTING TO 17.8' LONG) CIRCUMF = 25.53"

AND DIA = 8.13" USE 7.2" EQUIV. SQUARE

$$\frac{K_e L}{D} = \frac{2.1 \times 12(11.58)}{7.2} = 40.5 \quad LONG$$

$$F_c' = 0.3(1,500,000)/40.5^2 = 274 \text{ psi}$$

$$f_c = \frac{2000 + 822}{51.8} = 54.5 \text{ psi} < F_c'$$

LEEWARD COL. OK FOR AXIAL LD ALONE

"   " HAS SMALLER M @ 2.08'

$$\therefore \underline{\text{USE 20' CLASS 3 POLE - SOUTHERN PINE}}$$

Fig. 15-13. Designer's computations for Example 15-1, continued.

treated. The plywood sheathing, studs, top plate, and bottom plate are preassembled, moved into place, and connected to the wood footing plate. A field-applied top plate and building elements above it need not be pressure treated.

With this type of foundation, it is essential to drain water away from the area and to pre-vent moisture from being absorbed by the footing plate or the wall foundation assembly. Nevertheless, the footing and the sheathing are designed using allowable stresses for wet use. Nails and other fasteners used in the pressure-treated wood parts must be of silicon bronze, copper, or stainless steel, but galvanized nails are permitted in certain locations. Reference 7

Fig. 15-14. Wood foundation wall for basement. (Courtesy of National Forest Products Association, Washington, DC.)

should be used to determine the proper types of fasteners for each application and the proper measures to protect against moisture.

Outside backfill material applies a horizontal force to the wall. Thus, the wall must act as a vertical beam as well as carrying the vertical load from the structure above. The base of the wall is free to rotate, so the wall cannot act effectively as a cantilever. Therefore, it acts as a simple beam, subject to horizontal distributed load from the backfill and having horizontal reactions at the base and at the first-floor struc-

ture. The reaction at the first floor causes a compressive force in the floor joists, and the connection of these joists to the top plate must be able to transfer this force. At the bottom, a concrete basement floor slab (or a wood basement floor system) provides the necessary reaction, preventing the bottom of the wall from sliding inward under the backfill pressure.

If the horizontal loads acting on opposite sides of the building are not equal and opposite, then the end walls of the building receive racking loads; that is, they must act as vertical diaphragms to carry the unbalanced load to the foundation. An unbalance of lateral loads will always be present when wind is considered or when the depth of earth fill on opposite walls is not the same.

Reference 7 gives rules, procedures, and helpful tables for structural design. With one small exception, however, these rules are identical with those of the NDS. That one exception is that the allowable compressive stress $F_{c\perp}$ where the bottom end of the foundation wall studs bear on the bottom plate *is* modified for load duration. This disagreement with the NDS is generally on the safe side, since permanent load will probably be the controlling load condition.

Suggested Design Sequence

1. Choose a size and spacing of wall studs. This is a trial-and-error procedure. In the United States, the horizontal length of a panel will usually be 8 ft, the length of a sheet of plywood, so 12 in., 16 in., or 24 in., would be suitable stud spacings. The interaction of axial load and bending moment must be considered. Stud bending is caused by the horizontal pressure of the earth backfill against the plywood sheathing. The lateral soil pressure should preferably be determined by a geotechnical engineer, but again, local codes may suggest suitable values. The equivalent fluid pressure method is usually used to compute the lateral design pressures.

Figure 15-15 shows the pressure variation in pounds per square foot. Since the stud spacing may not yet be known, it is best at this stage to determine all wall loads, shears, and bending moments *per foot of wall length*. Top and bottom reactions to the wall pressures must be found next. Maximum bending moment occurs where the shear in the stud is zero. (Finding the location of zero shear and then the maximum bending moment is illustrated by Example

$$\text{SHEAR} = 0 \text{ AT DEPTH } y$$
$$0 = R_T - wy^2/2$$

$$\text{MAX. } M = R_T(y+a) - \frac{wy^2}{2}(y/3)$$

Fig. 15-15. Lateral pressure on foundation wall.

15-2.) Reference 7 uses dimension H for the span of the vertical beam, but the authors believe that to use the dimension shown as HH in Fig. 15-15 is more realistic. This is the distance between supporting members, from the center of the basement slab to the center of the first-floor structure.

2. Check the adequacy of the stud size and spacing using the interaction equation for members with combined axial compression and bending moment (Chapter 7). It is not easy to know in advance which load combination will control, so all reasonable ones should be checked. This may sound like a lot of work, but it is really quite easy if the designer will tabulate the terms going into the interaction equation. Again, Example 15-2 will show how. Notice that the lateral earth pressure is assumed to be a permanent load. Notice also that the stud design is based on dry use, even though the footing and sheathing designs are based on wet use.

Having found a size and spacing that is OK for bending, check to see whether the studs are satisfactory in shear. The critical section for shear will be at a distance equal to stud depth from the bottom (same as for a beam under the NDS).

3. Check the selected studs to be sure that they are not overloaded in compression perpendicular to the grain where they push horizontally against the concrete basement slab (or wood floor system) to transfer the necessary bottom reaction. The allowable stress *is* modified for permanent load duration. (This differs from NDS requirements.)

4. Check to determine whether or not the selected studs will overstress the bottom plate in compression perpendicular to the grain. Again, the allowable for the bottom plate *is* modified for load duration and for wet use.

5. Knowing the stud size, and knowing that the bottom plate will probably be of the same width, determine an allowable soil bearing pressure and find a required width of footing plate. The allowable pressure should preferably be specified by a geotechnical engineer, but in many cases local building codes will suggest values that are adequate for the area where the building will be located. Allowable soil bear-

ing pressures are *not* adjusted for load duration. The footing width required (inches)

$$= \frac{12 \text{ (load per ft of wall)}}{\text{allowable soil pressure}}$$

For the footing plate, select a standard width of lumber that is equal to or larger than the required width. In further calculations use the *actual* soil pressure rather than the allowable.

If the footing is wider than the bottom plate, it will have transverse bending moment due to the short cantilever that projects beyond the edge of the plate. This bending moment will cause flexural stresses perpendicular to the grain, and the stress on the bottom of the projecting cantilever will be tension. The NDS gives no allowable stress for tension due to bending in this direction (perpendicular to the grain), but reference 8 suggests allowing an amount equal to one-third of the allowable shear stress.

If the footing selected is found to be overstressed, it can be made thicker, or it can be reinforced with plywood. (*Reinforced* is a poor word, however, since the recommended design procedure is to assume that the plywood alone resists the entire bending moment.)

6. Choose plywood for the foundation sheathing, considering bending due to the active soil pressure, and racking. The sheathing acts as a beam, spanning from stud to stud, and is subject to a uniform load (from soil pressure) in that distance. The horizontal bending moment is largest near the bottom (where soil pressure is the highest) and reduces to zero at the level of the top of the soil. The lower edge of the sheathing overlaps the bottom plate, so the plywood there is in two-way bending. Horizontal bending is assumed to be maximum at about $\frac{1}{2}$ ft above the bottom edge, so bending moment (per foot of width) at that depth is used to compute the required plywood thickness. The horizontal bending moment per foot width of plywood at $\frac{1}{2}$ ft above the bottom edge is

$$M = w(h - 0.5)s^2/8 \qquad (15\text{-}4)$$

for single spans or for two-span panels, and

$$M = w(h - 0.5)s^2/10 \qquad (15\text{-}5)$$

for panels that are continuous over three or more spans.

The required effective section modulus per foot width (vertical) of plywood is

$$KS = M/F_b \qquad (15\text{-}6)$$

In the above equations

F_b = allowable bending stress adjusted for wet conditions and permanent duration
w = equivalent-fluid weight of soil
s = stud spacing
h = height of backfill (ft)
All with compatible units

The racking force is caused by unbalanced earth loads and by wind (or seismic) loads. Unless the building has diaphragm-type walls crossing the building between the ends, the entire racking force must be resisted by the two end walls alone. If the end walls have openings, that fact must be considered in determining the amount of racking force (shear through the thickness) per linear foot of wall (Section 12-4). The thickness of plywood selected for lateral soil pressure should now be checked using Table 15-4 to determine whether it is satisfactory for racking, and if so, what size and spacing of nails are required at panel edges (7).

7. Provide adequate connections, as follows:

a. Wall studs to top plate. Nails or framing anchors must be sufficient to transmit the inward horizontal reaction at the top of the studs. (Consider duration factor.)
b. Field-applied plate to top plate. Nails must transmit force equal to the top reaction of the studs.
c. Floor framing to field-applied plate. Nails or framing anchors connecting the floor system to the field-applied top plate must be able to transfer load at least equal to the top reaction of the studs. (As an alternative, the top end of the studs may transfer their load directly to the floor system by bearing against the ends of the joists. In this case, the studs must be checked for compression perpendicular to the grain.)

Table 15-4. Recommended Allowable Design Shear for Lumber-Framed Foundation Walls Sheathed With $\frac{1}{2}$, $\frac{5}{8}$, and $\frac{3}{4}$ Inch Plywood (Pounds per Foot of Wall).[a,b,c]

| | Plywood[a] | | | Nail Spacing at Panel Edges (in.)[e] | | | | |
| | | | | All Panel Edges Supported | | | | One or More Panel Edges Not Supported |
Loading	Minimum Thickness (in.)	Grade	Nail Size[d]	6	4	$2\frac{1}{2}$	2	6
Wind or	$\frac{1}{2}$	All grades	8d	280	430	640	730	215
wind plus	$\frac{1}{2}$	All grades	10d	310	460	640	730	230
unequal	$\frac{1}{2}$	Structural I grades	10d	340	510	690	780	255
backfill	$\frac{5}{8}, \frac{3}{4}$	All grades	10d	340	510	690	780	255
	$\frac{1}{2}$	All grades	8d	190	290	430	490	145
	$\frac{1}{2}$	All grades	10d	210	310	430	490	155
Unequal backfill	$\frac{1}{2}$	Structural I grades	10d	230	345	465	530	175
	$\frac{5}{8}, \frac{3}{4}$	All grades	10d	230	345	465	530	175

[a]Values apply to all plywood conforming to Article 2.2.1 except plywood made with Group 5 species (basswood, poplar, and balsam).
[b]Values shown are for studs of species/grade combinations B-1, B-2, B-3 from Table 3 (Group II per National Design Specification 8.1A). Use 81% of these design shears for combinations C-1, C-2, C-3, D-1, D-2, D-3 in Table 3 (NDS Group III). For NDS Group IV studs with 10d nails use 65% of these values, or 62% for Group IV with 8d nails.
[c]Design shears for $\frac{3}{8}$-in. plywood used in crawl space construction shall be 90% of the values for 8d nails. This adjustment is cumulative with the adjustment per footnote b.
[d]Nails to be hot-dipped zinc-coated steel; or copper, silicon bronze, or stainless steel Types 304 or 316 having equivalent lateral load capacity. Values apply to standard common wire nails or box nails of the following minimum sizes: 6d—2.5″ × 0.113″ (0.281″ head); 10d—3″ × 0.128″ (0.312″ head).
[e]All panel edges backed with 2-in. nominal or wider framing, except where noted. Space nails at 12 in. c/c along intermediate framing members.
Source: Courtesy National Forest Products Association, Washington, DC.

Many other details and connection requirements are shown by reference 7.

Example 15-2

Design a wood foundation and basement wall structure for the building shown by Fig. 15-16. Roof trusses (not shown) are parallel to the short walls; thus, the long walls are bearing walls and support all of the roof load.

Loads in pounds per foot of each long wall are:

From roof, 220 dead (includes ceiling weight) and 270 snow
From first floor, 120 dead and 360 live
Wall, first floor to roof, 160 dead

Wind load from building above = 200 lb/ft, with resultant at 15 ft above footing.

There is active lateral soil pressure; assume it is equal to pressure from a 30-pcf equivalent fluid. Use NDS allowable stresses for No. 2 D. fir–larch. The geotechnical engineer has specified an allowable soil bearing pressure of 2000 psf.

Figure 15-17 shows the designer's solution. Footing loads are calculated first. Load combinations considered by the designer were only: D alone, $D + L$; $D + L + S$; and $D + L + W$. The designer considered simultaneous full snow load and full wind load to be improbable. (However, some local building codes might require considering a combination including wind and a reduced snow load, or an unsymmetrical snow load.)

After finding the vertical loads per foot of wall, the designer next considered the lateral active pressure of the earth backfill against the stud wall. For this, he considered only the wall with the greater

(a) BASEMENT WALL PLAN

(b) SECTION *A-A*

Fig. 15-16. Building for Example 15-2.

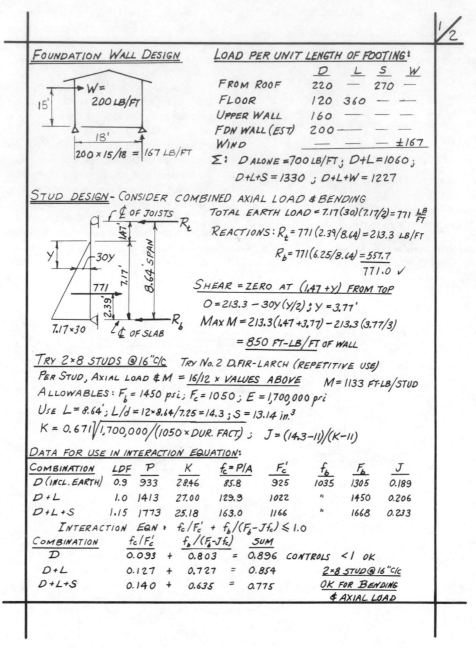

Fig. 15-17. Designer's calculations for Example 15-2.

backfill depth. (Bending moments for the other wall will be smaller.)

Up to this point the design calculations are all in terms of force per unit length (ft) of wall. For studs at 16 in. c/c, all of these values are 16/12 times as large. Composite action of the studs and the plywood sheathing is not considered, so the studs alone must resist the entire vertical load and bending moment. The studs can buckle with respect to their major axis, but the plywood sheathing prevents them from buckling in their weak direction. After finding a stud size and spacing that were satisfactory for combined axial load and bending, the designer verified that they were also OK in longitudinal shear.

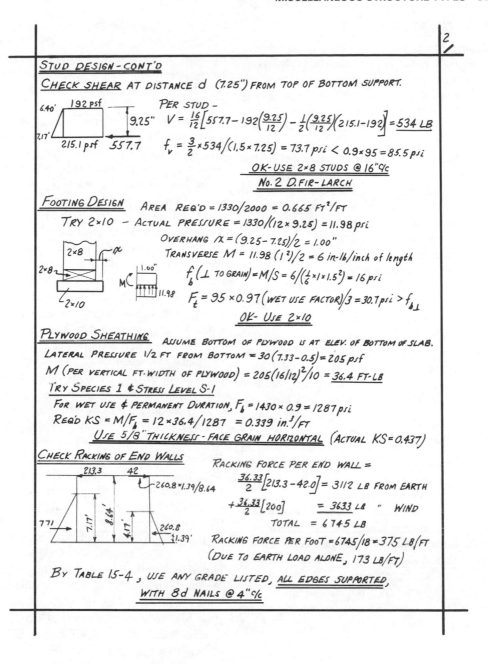

Fig. 15-17. Designer's calculations for Example 15-2, continued.

(The designer should also have considered end bearing and lateral bearing, but decided instead to inflict this task on the reader, by means of Problem 15-5.)

Finally, the designer selected a footing size and a thickness and grade of plywood sheathing. The table referred to for checking racking strength of the end walls shows that *all* edges of the sheathing must be nailed either to studs or to blocking. The sheathing will have its face grain horizontal. The plywood sheet is only 4 ft 0 in. wide, so this will require that solid 2 × 4 (or larger) blocking be nailed between the studs, 48 in. from the bottom, so that the horizontal edges of the plywood sheets can be supported.

15-4. TIMBER BRIDGES

Timber bridges have been used for centuries, yet wood's potential for modern bridge construction is not generally appreciated. Concern about deterioration is probably the main reason for selecting steel or reinforced concrete rather than wood for bridge construction. But, if the wood is properly treated with preservative and the bridge is properly designed, a timber bridge will give trouble-free service for many years. Given the rate at which steel bridges corrode and concrete bridges crack and spall, there is every reason to consider wood for new bridges of short or moderate spans.

Advantages of wood for bridges are many. Wood is not adversely affected by deicing agents such as salt. It is usually more economical than other materials. Construction is easier and can be done in weather that would not be suitable for concrete work, for steel painting, or for welding. Fairly large prefabricated wood assemblies can be installed easily without long delays to traffic. That wood bridges are durable when properly designed and properly protected from the elements is proved by experience with covered bridges. There are still some 850 of these in use in the United States, most of them over 100 years old (8), and all built before the days of preservative treatment.

Timber bridge design is not covered in a comprehensive manner here; to do that would require much more space, perhaps several volumes. Rather, the subject of timber bridge design is merely introduced, with examples to illustrate a few of the main ways in which wood bridge design differs from that for buildings. Most of the principles for wood building design given earlier in this book apply also to bridge design.

Only those parts of the American Association of State Highway and Transportation Officials (ASSHTO) specifications (9) needed to accompany the design examples are given here. A designer intending to do wood bridge design should consult the complete AASHTO specification and the many references on timber bridges given at the end of this chapter.

Types of Wood Bridges

Timber bridges serve many purposes; they carry pedestrian, roadway, or railway traffic. Special-purpose bridges include those carrying pipelines, flumes, or conveyors. Similarly, timber bridges are of many structural types. Even suspension bridges (except for the main cables, of course) are built of wood.

Trestles. Wood trestles are used for both highways and railways. They are the most common type of wood bridge for railway use. Trestles consist of longitudinal beams (stringers) and cross ties, supported by closely spaced timber bents. (See Fig. 15-18.) Each bent has four or more vertical members, either wood posts resting on concrete footings or wood piling driven into the ground to serve as both foundation and vertical member of the bent. Usually, the outer post (or pile) on each side is battered (slightly sloped) so that the bent will be able to resist lateral load. The posts (or piles) are capped with a heavy transverse member, 12×12 or larger, and the stringers rest on this cap. Solid blocking placed between the stringers at their ends holds them in line, prevents them from rotating, and serves as a fire stop. Similar blocking at the one-third points of the stringer spans restrains the stringers against lateral buckling.

Girder Bridges. The type of timber bridge most widely used for highways is the bridge with a transverse deck supported on several longitudinal stringers, which may reach across the entire span of the bridge. Girder bridges may also have the deck spanning between transverse floor beams, which in turn are supported by longitudinal girders. The girders may be either sawn-timber members or glulams. The choice will depend on the span. Thirty feet is probably the maximum practical span for sawn timbers, but with glulam girders, clear spans of up to 130 ft have been possible.

Truss Bridges. Wood truss bridges are practical for spans as high as 250 ft (reference 4) and may be either through-truss (see Fig. 15-19) or deck-truss bridges. Deck-truss bridges are usually the more economical.

Truss design involves no principles not already covered in Chapter 10. Perhaps the main difference between trusses for bridges and those for buildings is that sturdier cross frames are required between bridge trusses to prevent top-chord buckling. Also, horizontal-plane bracing

Fig. 15-18. Typical bent for railway trestle. (Courtesy of TECO Products & Testing Corp.)

to resist lateral loads is necessary. This lateral bracing system usually consists of a truss in a horizontal plane at the level of the top chord and another at bottom-chord level. The lateral-system trusses usually share chord members with the main, vertical trusses. (See Fig. 15-20.) The transverse floor beams act as members of one of the lateral-system trusses.

Top-chord bracing for a deck-truss bridge is easy to provide, since it will lie just below the bridge deck. If the deck is rigid enough to act as a horizontal diaphragm, it could possibly serve to brace the top chord against lateral buckling and also to resist lateral loads from traffic, wind, and other causes. This does not justify omitting a top-chord lateral system,

Fig. 15-19. Pedestrian and pipeline bridge of 170-ft clear span. Bowstring trusses are made entirely of glulams. Top chord is braced by outrigged diagonals attached to ends of floor beams. (Courtesy of Western Wood Structures, Inc.)

however, since one will be needed to hold the assembly of trusses, floor beams, and stringers in line during construction. For a through-truss bridge, however, the top-chord bracing must be high enough above the roadway to provide necessary clearance for vehicles. A compromise is the half-through or "pony truss" bridge, in which the top chord is laterally braced by outrigged diagonal members, located on the outer side of the truss and connecting the top chord to projecting ends of the transverse floor beams. Panel lengths for bridge trusses are determined

Fig. 15-20. Deck-truss bridge—schematic only, showing "anatomy."

by what is a practical and economical length for the longitudinal stringers, that is, the spacing of the transverse floorbeams.

Wood trusses have declined in popularity, largely because of the difficulty of protecting the members and connections from the weather. However, using glulams makes possible longer chord members, with fewer splices. When the glulam chords are used with web members of the same thickness and with steel gusset plates on each side (as discussed in Section 10-8 for buildings), problems of weathering should be reduced. This type of construction and perhaps the increasing use of exotic, durable wood species may bring about a return in the popularity of truss bridges for new construction.

Other Structural Types. Arch bridges are possible and practical using glued laminated members. Such bridges may use either two-hinged or three-hinged arches, depending on the spans.

For short spans, bridges may consist of merely a deck, with no other supporting members. The longitudinal deck bridge would be most effective if adjacent elements could be connected to each other well enough to ensure that the deck acts as a structural plate. Though this ideal situation probably cannot be reached, progress toward it is being made. Devices being considered to make the performance of such decks more effective are transverse distributor beams, doweling of adjacent panels, and transverse post-tensioning.

Wood may be used for bridges in combination with other materials; for example, a wood deck may be used with a system of steel beams and girders or with steel trusses.

15-5. BRIDGE DECK TYPES

Wood bridge decks include plank decks, nailed laminated decks, glued laminated decks, and composite decks. Which type to select depends on traffic density, expected life of the bridge, and economics.

Plank Decks

The simplest deck is the plank deck, shown by Fig. 15-21a. Unfortunately, the plank deck is suitable only for light traffic or for temporary

(a) PLANK DECK

(b) NAILED LAMINATED DECK

Fig. 15-21. Two types of wood deck.

structures. It consists of transverse planks, set flatwise and spiked to the stringers. When unseasoned planks are used, they are placed in side-by-side contact. As the planks shrink, gaps form between adjacent planks. Where seasoned planks are used, they should be placed with a gap between planks, so as to allow for swelling that will occur in wet weather.

To protect the planks from abrasion, longitudinal running planks are usually placed above the transverse ones, along the line the vehicle wheels will follow. Occasionally, the longitudinal planks are placed all across the width of the bridge to create a two-layer deck.

The problem with the plank deck is that its open spaces allow water to accumulate on the stringers below. If the stringers are not treated with preservative, rotting is a problem. An asphaltic wearing surface may be added to the plank deck but does little to protect the wood from moisture. Since the adjacent planks deflect independently as the load rolls across, the

asphalt surface cracks and moisture penetrates easily.

Nailed Laminated Decks

Nailed laminated decks are made using 2-in. nominal planks placed with their larger cross-sectional dimension vertical, as shown by Fig. 15-21b. The planks are nailed to each other with 30d nails, long enough to penetrate three planks. Each plank is also toe nailed to alternate stringers. Compared to the plank deck, the nailed laminated deck gives a little better protection to the stringers below and can be both stronger and stiffer.

Nailed laminated decks have the disadvantage of the nails loosening with repeated loading by traffic and with alternate wetting and drying. Transverse post-tensioning has been used with nailed laminated decks to make them perform better (10). Figure 15-22 shows two methods of post-tensioning: one for new work,

(a) Prestress Detail for New Designs

(b) Prestress Detail for Rehabilitation

Fig. 15-22. Post-tensioning laminated decks. (Courtesy of Ontario Ministry of Transportation and Communications.)

and one for repairing old nailed laminated decks. In the latter, the upper steel rods are embedded in the pavement material, and as long as the pavement is intact, they are protected from damage by bridge traffic.

Glued Laminated Decks

The glued laminated deck is also made of 2-in. nominal lumber placed with its larger cross-sectional dimension vertical; that is, it is vertically laminated. The depth of the deck varies from 4- to 16-in. nominal according to the width of the lumber used. As for glulam beams, material is planed from both surfaces after gluing, so as to remove irregularities. Thus, the deck depths available are the same as the standard widths available for glulam beams (Chapter 8).

The deck is preassembled into panels 3–4 ft wide and of length equal to the width of the bridge. Panels can be fully shop fabricated, with all necessary holes drilled, and the panel can then be treated with preservative before delivery to the bridge site.

The glued laminated deck is stiffer than either of the other deck types discussed above. Being stiffer, it may allow the use of a wider spacing between stringers or beams. Since it provides an almost solid surface, without gaps, it offers the best protection to the supporting wood members below. Tests show (11) that glued laminated decks are as much as 45% stronger than nailed laminated decks of the same thickness.

Tests show that the glued laminated decks are more effective than nailed laminated ones for distributing concentrated wheel loads, deflection directly under the load being much less

than it is with an equivalent thickness of nailed laminated deck. When the laminations are merely nailed to each other, there is slip between adjacent laminations. There is no slip when they are glued together.

Glued laminated panels may be doweled to each other in the bridge deck, as shown in Fig. 15-23, or they may be undoweled. The purpose of the steel dowels is to force each panel to act together with the adjacent ones rather than deflecting independently. Design procedures for doweled glued laminated decks were developed at the Forest Products Laboratory (11) and have become the basis of the AASHTO design method (reference 9).

Glued laminated decks find application not only for new bridges, but also for repair and upgrading of existing bridges (12). Nondoweled construction is often used. The laminated panels in this type construction are connected by a clip system (reference 5). Clips show on the underside of a pedestrian bridge in Fig. 15-24.

15-6. BRIDGE DESIGN CRITERIA

The AASHTO specifications (reference 9) give design criteria for highway bridges.

Highway Bridge Loads

Truck wheel loads, as defined by AASHTO, are a specified percentage of the gross weight of the truck and its contents. For example, each front wheel carries 10% of the gross weight and each rear wheel carries 40%. The width of traffic lane required is 12 ft. If a bridge has more than two lanes, the AASHTO specification allows reducing the design load used so as

Fig. 15-23. Glued laminated panel deck.

Fig. 15-24. Connection detail. Attachment of nailed laminated deck to glulam crossbeams of pedestrian bridge. (Photograph by authors.)

to consider the lesser probability that several lanes will be fully loaded at the same time. If there are three lanes, only 90% of full load need be considered; if there are four or more lanes, only 75% need be used in design.

Wheel Load Distribution to Deck Members

The first step in designing a highway bridge deck is to determine how the wheel load is distributed; that is, over what area is the wheel load actually applied, and what width of deck participates in resisting this load?

Figure 15-25a shows two views of a wheel load and gives the length and width of its area of contact with the bridge deck. Assuming that the tire air pressure is 100 psi, the area required (square inches) to carry the wheel load is $0.01P$ (P is the wheel load in pounds). AASHTO specifies that the width of the tire "footprint" is to be 2.5 times its length. Figure 15-25b shows this wheel footprint as the loaded area on a deck whose members run transverse to the traffic and rest on longitudinal stringers.

The wheel load deforms the deck, causing a dishlike depression around the loaded area. Contour lines drawn on the deck surface in Fig.

15-25b indicate the dished shape. The beam strip directly under the load deflects and, being connected to adjacent strips, causes them to bend also, but to lesser degree. The total necessary resisting moment is provided by combined action of the strip immediately beneath the load and strips alongside. Computing the resistance is facilitated by assuming the wheel load to be distributed over an effective width of strip that is defined by the AASHTO specifications (reference 9).

Just what width may be assumed to be effective depends on the manner of construction. The effective width is largest for decks whose adjacent elements are most firmly connected to each other. It is the largest for glued laminated decks and least for decks that consist of merely adjacent planks that are not connected to each other.

AASHTO gives two different sets of requirements for distribution of wheel loads to an effective width of deck; one is for a deck structure that is transverse to the direction of traffic movement; the other is for a deck structure parallel to traffic movement. The rules are summarized here.

For transverse flooring: In the direction of

DECK WIDTH TO CONSIDER
EFFECTIVE DEPENDS ON HOW
DECK STRIPS ARE CONNECTED
TO EACH OTHER

P

TIRE WIDTH b

CONTACT AREA LENGTH = WIDTH/2.5

(a) LENGTH AND WIDTH OF TIRE IN CONTACT WITH DECK

WOOD DECK
(TRANSVERSE)

TIRE CONTACT AREA
(SHADED)
$= 0.01P = b^2/2.5$

LONGITUDINAL SUPPORT
(STRINGER)

DEPRESSION CONTOURS SHOW
BENDING OF ADJACENT STRIPS
OF DECK

(b) DEFORMATION OF DECK BY WHEEL LOAD

Fig. 15-25. Distribution of wheel load on transverse wood deck.

the flooring span, the wheel load is distributed over the width of the tire. Normal to the span (i.e., in the direction to traffic) the load shall be distributed over a length as shown below.

For plank floors, distribution is over the width of plank only. For nailed laminated floors with laminated panels not interconnected, distribution is over a 15-in. width, but not more than the width of a laminated panel. For noninterconnected glued laminated panels, distribution is over a width equal to 15 in. plus the floor thickness, but not more than the panel width. For interconnected glued laminated panels the load may be distributed over 15 in. plus twice the floor thickness, but not more than the panel width.

For longitudinal flooring: In the flooring span direction, no distribution is allowed; the wheel load is considered as a point load.

Normal to the flooring span (i.e., transverse to the traffic direction) the load is distributed over the width of a plank for a plank floor. For noninterconnected nailed laminated or glued laminated panels, use the width of tire plus floor thickness, but not more than the panel width. For adequately interconnected nailed laminated or glued laminated panels, use the wheel width plus twice the floor thickness.

Deck Spans

In either direction, the span of the deck should be taken as the smaller of:

Clear span plus one-half the width of the supporting beam
Clear span plus the floor thickness

Wheel Load Distribution to Longitudinal Stringers

Figure 15-26 shows several bridge stringers and one wheel load near the center of the stringer span. The stringer immediately beneath the

ABUTMENT

DEFLECTED STRINGERS

δ

WHEEL FOOTPRINT

δ

δ

DEPENDING ON TRANSVERSE DECK
STIFFNESS, ADJACENT STRINGERS
SHARE IN RESISTING WHEEL LOAD

STRINGERS
(DECK NOT SHOWN)

Fig. 15-26. Distribution of wheel load among longitudinal stringers.

wheel load deflects, and the stiffness of the deck members (not shown) causes adjacent stringers to deflect also. Thus, the stringer beneath the load is getting help from the adjacent stringers; they are participating in resisting the bending moment due to the load shown.

AASHTO defines the effect of this participation by a table showing what fraction of the wheel load is to be assumed as resisted by a single beam. Table 15-5 shows those portions of the AASHTO table that apply to the design of timber bridges.

Table 15-6 summarizes the AASHTO load distributions for transverse glued laminated timber decks. Distributions for design of longitudinal glued laminated decks are recommended by reference 13 and may be included in the AASHTO specification shortly.

Example 15-3

A rural bridge consisting of a simple plank deck on wood stringers is to carry occasional H-15-44 trucks. A one-lane bridge will be sufficient. Not over 20 truck passages per day are expected, and it is intended that the bridge will have only an eight-year useful life. Use concrete abutments (end supports for the stringers). Provide a running strip of longitudinal planks and an 8 × 8 rough wood curb. The clear span between abutments will be 18 ft 0 in.

Load-duration factors for timber design are the same under AASHTO as under NDS. However, AASHTO does not suggest what duration to assume for the actual load on the bridge. So the designer must analyze the situation and decide what the controlling duration will be. For this particular bridge, the expected life is only eight years. Realizing that temporary structures rarely are as short-lived as originally planned, the designer of this bridge feels that it would be wise to double the 8-year figure. For a 16-year life, there would be 16 × 365 × 20 = 116,800 passages of full live load. If the average time for each passage were $\frac{1}{4}$ minute, this would amount to 29,200-minute total duration of full load, or about 20 days. This is almost one month, so the designer selects a duration factor of 1.15, as for snow load.

Deck plank design: The bridge cross section will be as shown by Fig. 15-27a. As a first step in designing the deck, the designer must decide how many stringers to use and must estimate the stringer width. If either guess proves wrong, the designer must start over with the correct assumptions. For this bridge, the designer has decided to try eight stringers (seven spaces between stringers) and estimates a stringer width of $9\frac{1}{2}$ in.

Assuming 1.79 ft c/c of stringers, the clear distance between stringers is 1.79 − 9.5/12 = 1.00 ft

Table 15-5.   Distribution of Wheel Loads in Longitudinal Beams.

Kind of Floor	Bridge Designed for One Traffic Lane	Bridge Designed for Two or More Traffic Lanes
Timber:[a]		
Plank[b]	$S/4.0$	$S/3.75$
Nail laminated[c] 4″ thick or multiple layer[d] floors over 5″ thick	$S/4.5$	$S/4.0$
Nail laminated[c] 6″ or more thick	$S/5.0$	$S/4.25$
	If S exceeds 5′ use footnote f	If S exceeds 6.5′ use footnote f
Glued laminated[e] panels on glued laminated stringers		
4″ thick	$S/4.5$	$S/4.0$
6″ or more thick	$S/6.0$	$S/5.0$
	If S exceeds 6′ use footnote f	If S exceeds 7.5′ use footnote f
On steel stringers		
4″ thick	$S/4.5$	$S/4.0$
6″ or more thick	$S/5.25$	$S/4.5$
	If S exceeds 5.5′ use footnote f	If S exceeds 7′ use footnote f
Concrete:		
On timber stringers	$S/6.0$	$S/5.0$
	If S exceeds 6′ use footnote f	If S exceeds 10′ use footnote f

S = average stringer spacing in feet.
[a]Timber dimensions shown are for nominal thickness.
[b]Plank floors consist of pieces of lumber laid edge to edge with the wide faces bearing on the supports (see Article 20.17—Division II).
[c]Nail laminated floors consist of pieces of lumber laid face to face with the narrow edges bearing on the supports, each piece being nailed to the preceding piece (see Article 20.18—Division II).
[d]Muliple-layer floors consist of two or more layers of planks, each layer being laid at an angle to the other (see Article 20.17—Division II).
[e]Glued laminated panel floors consist of vertically glued laminated members with the narrow edges of the laminations bearing on the supports (see Article 20.1.1—Division II).
[f]In this case the load on each stringer shall be the reaction of the wheel loads, assuming the flooring between the stringers to act as a simple beam.
Source: Standard Specifications for Highway Bridges, Washington, DC: The American Association of State Highway and Transportation Officials, copyright 1984. Used by permission.

Assuming that 3-in. rough planks will be used, the deck span will be the smaller of

$$1.00 + 0.5(9.5/12) = 1.40 \text{ ft} \text{ or}$$
$$1.00 + 3/12 = 1.25 \text{ ft} \textit{Controls}$$

For an H-15 truck (15 tons gross wt), one rear wheel load is $0.4(30,000) = 12,000$ lb. The 12,000-lb wheel load is to be distributed over an area of $0.01 \times 12,000$, or 120 in.^{2} Setting this equal to tire width, b, times footprint length, $b/2.5$, width $b = 17.3$ in. or 1.44 ft. As shown by Fig. 15-27b, the 12,000-lb wheel load is distributed over a distance larger than the deck span. The live load per unit length is $12,000/1.44 = 8333$ lb/ft. The deck is continuous over several supports, so the maximum bending moment is not the same as for a simple beam. To relieve the designer of having to consider each such case individually, AASHTO gives the rules shown in Table 15-7. These permit modifying the simple bending moment to obtain reasonably safe values for the positive and negative bending moments in continuous deck spans. (The values shown are intended for use with composite wood/concrete spans, but should be reasonably satisfactory for plank spans also.)

$$\text{Simple moment} = w\ell^2/8 = 8333(1.25^2)/8$$
$$= 1628 \text{ ft-lb due to live load alone}$$

AASHTO specifies that timber (either treated or untreated) be assumed to weigh 50 pcf. Thus, the design weight for the planks is about $50 \times 3/12 = 13$ lb/ft. Simple dead-load bending moment is $13(1.25^2)/8 = 3$ ft-lb.

Correcting for continuity, positive M for the end span equals

$$(0.7 \times 3) + (0.85 \times 1628) = 1386 \text{ ft-lb}$$

Try southern pine, No. 2 dense, for the planks. The basic allowable must be adjusted for load duration, flatwise use, and use under moist conditions. (The AASHTO correction for MC > 19% is almost the same as the NDS correction.) Assuming that 3-in. thickness will be used,

$$F_b = 1400(1.15)(1.16)(0.80)$$
$$= 1494 \text{ psi}$$
$$\text{Required } S = M/F_b$$
$$= 12 \times 1386/1484$$
$$= 11.13 \text{ in.}^3 \text{ per plank}$$

Actual S for a 3×12 rough is 18.0 in.^{3}. *OK for bending.* (Narrower planks could be used, but would not save lumber. Use 12 in.)

Check 3 × 12 planks for shear: AASHTO specifies that the critical section and starting position for the edge of the wheel load shall be at whichever distance from the face of the support is smaller: three times the plank thickness; or one-fourth of the span.

Table 15-6.　AASHTO Method for Interconnected Glued Laminated Decks.

3.25.1.3　One design method for interconnected glued laminated panel floors is as follows: For glued laminated panel decks using vertically laminated lumber with the panel placed in a transverse direction to the stringers and with panels interconnected using steel dowels, the determination of the deck thickness shall be based on the following equations for maximum unit primary moment and shear.[a] The maximum shear is for a wheel position assumed to be 15 in. or less from the centerline of the support. The maximum moment is for a wheel position assumed to be centered between the supports.

$$M_x = P(.51 \log_{10}s - K) \qquad (3\text{-}23)$$

$$R_x = .034P \qquad (3\text{-}24)$$

Thus

$$t = \sqrt{\frac{6M_x}{F_b}} \qquad (3\text{-}25)$$

or

$$t = \frac{3R_x}{2F_v} \text{ whichever is greater} \qquad (3\text{-}26)$$

where

M_x = primary bending moment in inch-pounds per inch
R_x = primary shear in pounds per inch
x = denotes direction perpendicular to longitudinal stringers
P = design wheel load in pounds
s = effective deck span in inches
t = deck thickness, in inches, based on moment or shear, whichever controls
K = design constant depending on design load as follows:
　　　H 15　　$K = 0.47$
　　　H 20　　$K = 0.51$
F_b = allowable bending stress, in pounds per square inch, based on load applied parallel to the wide face of the laminations
F_v = allowable shear stress, in pounds per square inch, based on load applied parallel to the wide face of the laminations

3.25.1.4　The determination of the minimum size and spacing required of the steel dowels required to transfer the load between panels shall be based on the following equation:

$$n = \frac{1000}{\sigma_{PL}} \times \left[\frac{\overline{R}_y}{R_D} + \frac{\overline{M}_y}{M_D} \right] \qquad (3\text{-}27)$$

where

n = number of steel dowels required for the given spans
σ_{PL} = proportional limit stress perpendicular to grain (for Douglas fir or southern pine, use 1000 psi)
$\overline{R}_y$ = total secondary shear transferred, in pounds, determined by the relationship

$$\overline{R}_y = 6Ps/1,000 \text{ for } s \leq 50 \text{ in.} \qquad (3\text{-}28)$$

or

$$\overline{R}_y = \frac{P}{2s}(s - 20) \text{ for } s > 50 \text{ in.} \qquad (3\text{-}29)$$

$\overline{M}_y$ = total secondary moment transferred, in inch-pound, determined by the relationship

$$\overline{M}_y = \frac{Ps}{1600}(s - 10) \text{ for } s \leq 50 \text{ in.} \qquad (3\text{-}30)$$

$$\overline{M}_y = \frac{Ps(s - 30)}{20(s - 10)} \text{ for } s > 50 \text{ in.} \qquad (3\text{-}31)$$

R_D and M_D = shear and moment capacities, respectively, as given in the following table:

Diameter of Dowel (in.)	Shear Capacity R_D (lb.)	Moment Capacity M_D (in.-lb.)	Steel Stress Coefficients C_R (1/in.²)	Steel Stress Coefficients C_M (1/in.³)	Total Dowel Length Required (in.)
0.5	600	850	36.9	81.5	8.50
0.625	800	1340	22.3	41.7	10.00
0.75	1020	1960	14.8	24.1	11.50
0.875	1260	2720	10.5	15.2	13.00
1.0	1520	3630	7.75	10.2	14.50
1.125	1790	4680	5.94	7.15	15.50
1.25	2100	5950	4.69	5.22	17.00
1.375	2420	7360	3.78	3.92	18.00
1.5	2770	8990	3.11	3.02	19.50

3.25.1.5　In addition, the dowels shall be checked to ensure that the allowable stress of the steel is not exceeded using the following equation:

$$\sigma = \frac{1}{n}(C_R\overline{R}_y + C_M\overline{M}_y) \qquad (3\text{-}32)$$

where

σ = minimum yield point of steel pins in pounds per square inch (see Table 10.32.1A);
$n, \overline{R}_y, \overline{M}_y$ = as previously defined;
C_R, C_M = steel stress coefficients as given in preceding table.

[a]The equations are developed for deck panel spans equal to or greater than the width of the tire (as specified in Article 3.30), but not greater than 200 in.
Source: Standard Specifications for Highway Bridges, Washington, DC: The American Association of State Highway and Transportation Officials, copyright 1984. Used by permission.

(a) CROSS SECTION OF BRIDGE

(b) WHEEL IN POSITION FOR MAXIMUM POSITIVE
BENDING MOMENT IN SIMPLE-SPAN DECK

(c) WHEEL POSITIONED TO CAUSE
MAXIMUM SHEAR IN DECK

(d) LOADS POSITIONED TO CAUSE
MAXIMUM SHEAR IN A STRINGER

Fig. 15-27. Plank-deck bridge for Example 15-3.

For this case, these distances are 9 in. and 3 in., respectively. The latter controls.

Clear distance between stringer faces is 1.0 ft. Figure 15-27c shows the wheel located to cause the largest possible shear at that distance. With the wheel load in that position, the load actually on the span is $8333(9/12) = 6250$ lb, and the shear (equal to the left end reaction) is $6250(4.5/12) = 2344$ lb. Dead-

Table 15-7. Maximum Bending Moments—Percent of Simple-Span Moment.[a]

	Maximum Uniform Dead Load Moments				Maximum Live Load Moments			
	Wood Subdeck		Composite Slab		Concentrated Load		Uniform Load	
Span	Pos.	Neg.	Pos.	Neg.	Pos.	Neg.	Pos.	Neg.
Interior	50	50	55	45	75	25	75	55
End	70	60	70	60	85	30	85	65
2-span[b]	65	70	60	75	85	30	80	75

[a]Shown by AASHTO for composite wood-concrete members.
[b]Continuous beam of two equal spans.
Source: Standard Specifications for Highway Bridges, Washington, DC: The American Association of State Highway and Transportation Officials, copyright 1984. Used by permission.

load shear adds about 3 lb, for a total $V = 2347$ lb. The unit shear stress is

$$f_v = 1.5(2347)/(3 \times 12) = 98 \text{ psi}$$

The modified allowable shear stress (using AASHTO's 0.95 for MC > 19%) is

$$F_v = 90(1.15)(0.95) = 98 \text{ psi}$$

Use 3 × 12 rough sawn planks.

Stringer design: Assume that the steel bearing plates at each end will be set back 2 in. from the face of the concrete abutments, and that the required bearing length will be 3 in. at each end. If so, the span for bending of the stringer will be

$$18 + 2(2/12) + 2(3/2)/12 = 18.58 \text{ ft}$$

and the clear span for calculating shear, 18.33 ft.

Table 15-5 shows $S/4.0$ to be the fraction of a wheel load to be resisted (in bending) by one stringer. Spacing S is 1.79 ft, so the fraction to be used is $1.79/4 = 0.45$. The bridge span is so short that the largest bending moment will occur when only the rear wheels are on the bridge and located at midspan. The live-load bending moment for one stringer will be

$$M = 0.45(12,000)(18.58)/4 = 25,080 \text{ ft-lb}$$

Dead load carried by one stringer is:

Deck $50(3/12)(1.79)$	$= 22.4$
Running strip (same)	$= 22.4$
Stringer (est)	$= \underline{50.0}$
Total w_d	$= 95 \text{ lb/ft}$

Dead-load $M = 95(18.58^2)/8 = 4100$ ft-lb. Total bending moment is

$$M = 29,200 \text{ ft-lb}$$

Try No. 2 SR grade southern pine. The basic allowable stresses must be modified for load duration and wet use.

$$F_b = 1100(1.15)(1.00)C_F = 1265 C_F$$

Assuming $C_F = 1.0$,

$$\text{Required } S = 12(29,200)/1265 = 277 \text{ in.}^3$$

Try 10 × 14 (S4S)

$$C_F = (12/13.5)^{1/9} = 0.987$$
$$F_b = 0.987(1265) = 1250 \text{ psi}$$

Actual $f_b = 12(29,200)/288.6 = 1214$ psi   *OK*

Check for shear: Place the entire 12,000-lb wheel load above a stringer, as shown by Fig. 15-27d. To cause maximum shear in the stringer, this load must be placed at distance x from one end, where x is the smaller of three times the stringer depth or one-fourth of the span. Assuming the the stringer depth will be 13.5 in., the two values are $18.58/4 = 4.64$ ft and $3 \times 13.5/12 = 3.38$ ft. The latter value controls.

For shear, no distribution to other stringers is allowed for the heavy rear wheel close to the end of the span, but the smaller wheel load (from the front axle) may be distributed among stringers. The load on this stringer, coming from the smaller wheel load of $0.1(30,000) = 3000$ lb, will be (Table 15-5)

$$3000(1.79)/4 = 1342 \text{ lb}$$

and the computed live-load shear will be 9857 lb as shown in Fig. 15-25d.

Shear at distance x from the left end due to dead load will be

$$95(18.33/2 - 3.38) = 550 \text{ lb}$$

and the total shear is the sum, or 10,410 lb.

$$F_v = 95(1.15)(1.00) = 109 \text{ psi}$$
$$\text{Actual } f_v = 1.5(10,410)/(9.5 \times 13.5)$$
$$= 122 \text{ psi} > F_v \quad NG$$

If the stringer size is increased to 10 × 16, it will be satisfactory for both shear and flexure, the actual shear stress f_v reducing to 102 psi. An alternate solution would be to change the grade to No. 1 SR, for which the modified F_v would be 126 psi.

Example 15-4

Consider the same bridge as in Example 15-3, but design a transverse nail-laminated deck with a 2-in. asphaltic wearing surface. (AASHTO specifies 150 pcf for the weight of this material.)

Figure 15-28a shows a cross section for this bridge. Again, to design the deck it is necessary to estimate the stringer width and the number of stringers. In this case, the designer assumed that there would be five stringers, equally spaced. The distance between stringers is 3.14 ft c/c or 2.35 ft clear. The thickness of the deck is not yet known, but it must be used to determine the deck span. Assuming that the deck will consist of 6-in. nominal lumber (5.5 in thick) and that the stringer width will be 9.5 in., the span will be the smaller of

$$2.35 + \tfrac{1}{2}(9.5/12) = 2.75 \text{ ft} \quad \text{or} \quad \textit{Controls}$$
$$2.35 + 5.5/12 = 2.81 \text{ ft}$$

For noninterconnected transverse nail-laminated decks, the wheel load may be assumed distributed over a 15-in. width of deck. Figure 15-28b shows one span of the deck, with the entire 12,000-lb wheel load centered on the deck span. Dead load on the 15-in. width is estimated to be

Deck $(15/12)(50)(5.5/12)$ $= 29$ lb/ft
Wearing surface $(15/12)(150)2/12) = \underline{31}$ lb/ft
    Total dead load per 15-in. width $= 60$ lb/ft

The simple bending moments are

$$6000(2.75/2 - 1.44/4)$$
$$= 6090 \text{ ft-lb from live load}$$
$$60(2.75^2)/8 = 57 \text{ ft-lb from dead load}$$

Correcting for continuity, the maximum positive bending moment is

$$0.85(6090) + 0.7(57)$$
$$= 5220 \text{ ft-lb per 15-in. width}$$

Assuming that the asphalt surface will keep the deck dry, the allowable stresses need be adjusted only for

(a) CROSS SECTION

(b) LOADS FOR MAXIMUM *M* ON
    SIMPLE-SPAN DECK

(c) LOADS FOR MAXIMUM SHEAR
    IN DECK

Fig. 15-28. Nailed laminated deck for Example 15-4.

load duration. For No. 2 southern pine, the modified allowables will be

$$F_b = 1200(1.15) = 1380 \text{ psi}$$
$$F_v = 90(1.15) = 104 \text{ psi}$$

Required section modulus is

$$S = 12(5220)/1380 = 45.4 \text{ in.}^3 \text{ per 15-in. width}$$

The required thickness is

$$\sqrt{6(45.4)/15} = 4.26 \text{ in.}$$
Try 2 × 6 laminations

Check shear: The critical section for shear is at whichever distance from the support is less,

$3 \times$ deck thickness of 5.5 in. = 16.5 in. or
Clear span$/4 = 12(2.35)/4 = 7.05$ in.
$= 0.59$ ft *Controls*

Figure 15-28c shows the 12,000-lb wheel load in this position. The shear is

$$V = 12,000(0.88)/2.35 + 60(2.35/2 - 0.59)$$
$$= 4530 \text{ lb}$$

Shear stress is

$$f_v = 1.5(4530)/(15 \times 5.5) = 82 \text{ psi} < F_v$$
OK

Example 15-5

A two-lane bridge to carry H-15 truck loads spans 30 ft and has a 26-ft clear roadway width. The designer is considering using a 3-in. asphalt wearing surface on a transverse deck of interconnected glued laminated panels, supported by glulam stringers. One of the designer's trial designs assumed the bridge cross section shown by Fig. 15-29. Design the glued laminated deck, assuming interconnected panels. This is not a temporary structure, so use normal-duration allowable stresses.

If four stringers are used, they will be 8 ft 4 in. c/c. Assume that the deck will be 6.75 in. thick and the stringers 10.75 in. wide. (These are both standard widths for glulams; see Chapter 8.) With these assumptions, the clear distance between stringers will be 7.44 ft. The deck span will be the lesser of

$7.44 + \frac{1}{2}(10.75/12) = 7.89$ ft or   *Controls*
$7.44 + 6.75/12 = 8.00$ ft

Fig. 15-29. Bridge for Example 15-5.

From this point on, bending moments and shears must be computed using the empirical equations of the AASHTO specification. These are shown by Table 15-6. Using the equations shown, the live-load bending moment is

$$M_x = P(0.51 \log_{10} s - K)$$

in which $s = 12(7.89) = 94.7$ in. and K (defined by AASHTO) is 0.47 for H-15 loading.

$$M_x = 12,000(0.51 \times 1.976 - 0.47)$$
$$= 6453 \text{ ft-lb/ft}$$

Deck dead load is

Estimated deck weight $50(6.75)/12 = 28$ psf
Asphalt $150(3/12) =$ $\underline{38}$ psf
  Total $=$ 66 psf

Simple dead-load bending moment is

$$66(7.89^2)/8 = 514 \text{ ft-lb/ft}$$

(*Note:* AASHTO is written to imply that dead-load bending moment may be ignored. The authors think it should be considered.) Using the same fraction of simple dead-load moment as for composite slabs, the total bending moment is

$$(0.7)(514) + 6453 = 6810 \text{ ft-lb/ft}$$

Allowable stresses are given by Appendix Table B-5 for glulams loaded parallel to the wide face of the laminations. Here, the UBC allowable stresses are difficult to correlate with those of AASHTO. The

UBC allowables appear to be more conservative (safer), so this example will proceed using the UBC values. Assuming southern pine combination 20F-V1, the basic allowables are

$$F_b = 1450 \text{ psi} \quad \text{and} \quad F_v = 175 \text{ psi}$$

The required section modulus is

$$S \text{ req} = 12(6810)/1450 = 56.4 \text{ in.}^3/\text{ft width}$$

and the thickness required is

$$t = \sqrt{6(56.2)/12} = 5.31 \text{ in.}$$

$6\frac{3}{4}$-in. deck OK for bending

Check shear: The AASHTO empirical method uses shear equal to $R_x = 0.034P$, which figures in this case to $0.034(12,000) = 408$ lb *per inch* of width. (For this check, AASHTO permits ignoring dead-load shear. The authors agree, since including it would make but a trivial difference.) Setting the computed actual shear stress per inch of width equal to the allowable, the thickness required for shear is

$$t \text{ req} = 1.5 \, R_x/F_v = 1.5(408)/175 = 3.50 \text{ in.}$$

Use $6\frac{3}{4}$-in. deck

Determine required size and spacing of dowels: The method of determining these, shown by Table 15-6, is based on analysis by the method of beam on an elastic foundation (14) and has been confirmed by experiment.

Secondary shear occurs between adjacent panels due to their interaction with one another. For a span of 94.7 in., this secondary shear is

$$R_y = P(s - 20)/2s$$
$$= 12,000(74.7)/(2 \times 94.7) = 4730 \text{ lb}$$

The secondary moment transferred is

$$M_y = \frac{Ps(s - 30)}{20(s - 10)} = \frac{12,000(94.7)(64.7)}{20(84.7)}$$
$$= 43,400 \text{ in.-lb}$$

Try 1.0-in.-diameter dowels. Table 15-6 shows moment and shear capacities for these dowels as $M_D = 3630$ in.-lb and $R_D = 1520$ lb, respectively. All values needed to determine the required number of dowels are now known. Substituting in the equation for n,

$$\text{Req } n = [1000/1000] \times [(4730/1520) + (43,400/3630)]$$
$$= 15.0$$

With this deck thickness and dowel diameter, the dowel spacing will be $12(8.33 \text{ ft})/15 = 6\frac{1}{2}$ in., which is not practical. If the dowel size is increased to $1\frac{1}{2}$-in. diameter, the dowels can be spaced 15 in. c/c, which is better. The problem of dowel spacing could be alleviated also by using one more stringer, reducing the spacing of stringers to 6 ft 3 in. c/c.

15-7. COMPOSITE DECKS

In a composite deck, a layer of concrete is placed above the wood layer of the deck. The concrete layer resists the flexural compression and the wood the tension. An advantage of the composite deck is that the structural concrete may serve also as the wearing surface.

There are two types of composite deck, the composite T-beam deck and the composite slab deck. The two are shown by Fig. 15-30.

T-Beam Deck

In the T-beam type, shown by Fig. 15-30a, wood stringers form the stems of the T-beams and the concrete slab forms the flanges. Horizontal shear is transferred from stringer to slab through notches cut into the top of the stringers and through mechanical fasteners driven into the wood stringers. The slab, in addition to participating as the main compression element of the longitudinal composite T-beam, acts as a continuous beam spanning transversely from stringer to stringer. To construct this type of deck requires building concrete forms between the stringers.

Composite Slab Deck

Figure 15-30b shows the composite slab bridge deck. (This deck is useful also for floors of buildings carrying heavy live loads.) It consists of 2-in. nominal lumber in the vertical position and nail-laminated, with a concrete slab cast above. The top surface of the wood portion has 2-in.-deep grooves, formed either by using two different widths of lumber for the laminations or by staggering the position of laminations to

(a) COMPOSITE T-BEAM DECK

(b) COMPOSITE SLAB DECK

Fig. 15-30. Types of composite deck.

create a 2-in. offset. Longitudinal shear is transferred from timber to concrete by slots cut into the projecting laminations or by metal shear developers driven into the top of the wood, or by both.

An advantage of the composite slab deck is that the wood portion serves as the form on which to cast the concrete; no additional support is needed. If a long (parallel to the laminations) deck is needed, individual laminations can be cut to differing lengths, so that the pattern of staggered overlapping members is like a large-scale finger joint.

Since the concrete slab serves merely as the compression part of a composite member, the concrete requires reinforcement only for (a) resisting negative bending (tension on the top side) of the deck, as at locations over intermediate supports; and (b) resisting temperature and shrinkage stresses.

Design of Composite Slab Decks

In the composite slab deck, as shown by Fig. 15-30b, the timber portion alone must support its own weight plus the weight of the concrete. It may also have to support some additional load caused by construction equipment. Only after the concrete hardens does composite action occur. Composite action is effective only

Fig. 15-31. Transformed sections and stresses in composite slab deck.

for carrying live load or such additional dead load that is applied to the bridge after the concrete hardens.

The effective sections and the stresses at various stages are shown graphically by Fig. 15-31. Under positive bending, the concrete is in compression and is fully effective. The concrete and the timber will have different values for modulus of elasticity, so one of the materials is imagined to be replaced with an elastically equivalent area of the other to create the transformed section, as shown by Fig. 15-31b. In this section, the concrete has been replaced by imaginary timber, having area equal to that of the concrete times the modular ratio, n (ratio $n = E_c/E_w$).

Stresses caused by the permanent loads present while the concrete is curing are shown by Fig. 15-31c. These are merely the flexural stresses, My/I, on the timber deck alone. The concrete at this stage is unstressed. When live load is added as traffic crosses the bridge, the entire composite section resists additional bending moment, and the stresses added are as shown by Fig. 15-31d. The position of the neu-

tral axis for these added stresses may be within the wood, at the wood-concrete junction, or in the concrete, depending on the thicknesses of the two portions and upon the modular ratio, $n = E_c/E_w$. The total stresses, shown by Fig. 15-31c, are merely the sum of the stresses shown for dead load alone and for live load alone.

In regions where bending causes tension on the top, reinforcing must be placed in the slab and the concrete is ineffective. The transformed section for this case is shown by Fig. 15-31f. Here, the steel is imagined to be replaced by equivalent wood having n times as much area and using a modular ratio $n = E_s/E_w$.

For a much more comprehensive coverage of composite deck design, see reference 15.

15-8. CONCLUSION

The authors have attempted to cover adequately many types of wood structure and to consider various wood-based construction materials. We realize, however, that many more subjects could have been addressed were it not for limited space and limited time to complete this manuscript. For example, topics intentionally omitted include lamella roof structures, shells, domes, and folded plates.

As long as structural designers are innovative, the list of possible structural forms will continue to grow. And as long as researchers remain interested in conserving resources and using them to best advantage, the varieties of new structural materials—many of them wood-based—will also grow. Future editions of this book may include information on those structural forms and new wood-based construction materials that prove important.

REFERENCES

1. *Uniform Building Code*, International Conference of Building Officials, Whittier, CA, 1982.
2. Wolfe, R., *Low Cost Pole Construction*, Garden Way Publishing Co., VT, 1980.
3. *Book of Standards*, American Wood Preservers' Association, Stevensville, MD, 1984.
4. *Wood Structures, A Design Guide and Commentary*, American Society of Civil Engineers, New York, 1975.
5. American Institute of Timber Construction, *Timber Construction Manual*, 3rd ed., Wiley, New York, 1986.
6. Patterson, Donald, *Pole Building Design*, 6th ed., American Wood Preservers Institute, McLean, VA, 1981.
7. *Permanent Wood Foundation System*, National Forest Products Association, Washington, DC, 1988.
8. *Evaluation, Maintenance and Upgrading of Wood Structures, A Guide and Commentary*, American Society of Civil Engineers, New York, 1982.
9. *Standard Specifications for Highway Bridges*, American Association of State Highway and Transportation Officials, Washington, DC, 1983.
10. Taylor, R. J., B. Batchelor, and K. VanDalen, *Prestressed Wood Bridges*, Ontario Ministry of Transportation and Communications, Downsview, Ontario, Canada, 1983.
11. McCutcheon, W. J., and R. L. Tuomi, "Design Procedure for Glued-Laminated Bridge Decks," *Forest Products Journal*, **23**(6), June 1973, pp. 36–42.
12. Barnhardt, J., "Ohio's Experiences with Treated Timber for Bridge Construction," Transportation Research Record 1053, Transportation Research Board, National Research Council, Washington, DC, 1986.
13. Wipe, C., F. W. Klaiber, and W. W. Sanders, "Load Distribution Criteria for Glued Laminated Longitudinal Timber Deck Highway Bridges," Transportation Research Record 1053, Transportation Research Board, National Research Council, Washington, DC, 1986.
14. Hetenyi, M., *Beams on Elastic Foundation*, University of Michigan Press, Ann Arbor, 1946.
15. Gurfinkel, G., *Wood Engineering*, Southern Forest Products Association, New Orleans, 1973.

PROBLEMS

15-1. Choose pole sizes and depths of embedment for the pole building shown by Fig. 15-32. Assume loads as follows: roof dead load applied to pole = 1200 lb; snow load = 25 psf; wind load = 15 psf pressure on windward wall and 7 psf suction on leeward wall. Soil is classed as silty gravel (see Table 15-2). Use D. fir poles.

15-2. Choose pole size and embedment for the building shown by Fig. 15-33. Roof dead load is 12 psf and snow load is 20 psf. Wind pressure is as shown for Problem 15-1. Soil is silty gravel. Consider the combination of W + S/2. Note that bending may control for the outside poles, but design of the inside ones *may* be controlled by the maximum vertical load. Use southern pine.

15-3. Check the pole of Example 15-1 for end bearing. The allowable pressure is given in Table 15-2. Design concrete footing if required.

15-4. For the wood foundation of Example 15-2 (and Figs. 15-16 and 15-17), determine the size and spacing of nails required to connect: (a) upper end of the studs to the top plate; (b) top plate to field-installed plate; and (c) field-installed plate to floor system.

Fig. 15-32. Pole building for Problem 15-1.

Fig. 15-33. Interior bent. Pole building for Problem 15-2.

15-5. For the wood foundation of Example 15-2, determine whether or not the studs chosen are satisfactory: (a) in bearing (compression perpendicular to the grain) where they press against the edge of the concrete slab; and (b) in end bearing against the bottom plate.

15-6. Design a wood foundation system for the building shown by Figure 15-16, but with the clear height of the basement increased to 9 ft 0 in.

15-7. Example 15-2 did not consider the load combination involving full wind along with live load and snow load on only one side of the peaked roof. Check the solution shown by Fig. 15-17 to determine whether or not the foundation designed will be satisfactory. Assume that the unsymmetrical snow load causes a downward force of 210 lb per foot of wall on the lee side.

15-8. Check the stringers of Example 15-3 for required bearing length and for deflection. Note that bearing length must be checked using the maximum *reaction*, not maximum shear. Note also that load-duration factors do not apply to allowable compression stress perpendicular to the grain.

15-9. Redesign the plank bridge of Example 15-3, but using S4S Douglas fir planks and using glulams for the stringers. Change the number of stringers if so directed by your instructor.

15-10. Repeat Example 15-5 using five equally spaced stringers to see if a practical spacing of 1-in. dowels results.

15-11. Design stringers for the bridge of Problem 15-10. After selecting for flexure and shear, determine the required bearing length at each end, and check the deflection.

16

Wood Durability, Protection, and Preservation

Wood has three shortcomings that tend to limit its structural use: (1) it is subject to deterioration by weathering and by attack of various organisms, (2) it is combustible, and (3) it is subject to swelling and shrinking.

Yet, in spite of these shortcomings, properly designed wood structures can be as long-lived as either steel-frame or reinforced-concrete structures, can resist fire better than unprotected steel structures, and can accommodate the swelling and shrinkage without distress. As is true for each of the materials (wood, structural steel, or reinforced concrete), to ensure successful performance and long life of the structure requires that the designer understand the behavior of the material and plan so as to avoid problems that its properties may raise. The builder also has an important role in that success, and so does the final user.

The first requirement in learning to design for best permanence and durability in any material is to identify the problems: In the case of wood, this means to identify the various agents of destruction.

16-1. CAUSES OF DETERIORATION

Borrowing terms from the science of physical geology, wood deterioration might be divided into two classes: decomposition, in which the chemical makeup is affected; and disintegration, in which the wood is merely broken apart physically but not changed chemically. Of course, both types can occur together.

Decomposition of wood is brought about by the action of bacteria, fungi, molds, chemicals, or fire. These may affect the wood while still in the tree, during lumber production and storage, or after it is installed in the structure. Decomposition may result in changes of appearance only or may result in structural weakening of the wood. The common causes of disintegration are insects, marine borers, interaction of treated wood with metals, and weathering. Disintegration almost always weakens the wood. Mechanical damage by abrasion during normal use might be considered to be a human-caused type of disintegration.

Microorganisms Affecting Wood

Four classes of microorganism affecting wood are wood-destroying fungi, wood-staining fungi, molds, and bacteria. The first class is the most important from a structural-strength standpoint. The wood-destroying fungi feed on materials in the cell walls. Thus, they destroy the cell walls and drastically reduce all strength properties of the wood. The resulting damage, called *decay*, can occur at any time: in the live tree, in the forest after the tree is cut, during processing and storage of the lumber, or in the completed structure.

It is reasonable to assume that all wood at some time or another has been infected with the spores of these fungi. All that is necessary to cause the wood to decay is to provide the conditions necessary for these spores to grow. Conversely, all that is necessary to prevent these spores from growing into fungi that will destroy the wood is to eliminate one or more of the things they need in order to grow—a food supply, air, moisture, and suitable temperature.

Decay can also occur as wood-destroying fungi spread directly from the soil or from infected wood into undamaged wood. This occurs where the wood parts of a structure are too close to the ground. It occurs also in lumber storage where the lumber is separated by infected stickers or where logs are stored on the ground.

Decay is serious even though it may affect only a small portion of the total volume of wood in a structural member. For example, a long timber beam may be rendered completely useless when only a small volume of wood has decayed near the location of maximum shearing or maximum bearing stress. A wood column that is decayed near its lower end is completely useless, even though most of its length consists of high-strength, undeteriorated wood. The bolted joint in a wood truss cannot transfer load effectively from one member to another if the wood bearing against one of the bolts is rotten; thus, the entire truss may be worthless even though the volume of defective wood is quite small.

After wood is infected by airborne spores, microscopic, threadlike structures grow in the wood. These structures, known as hyphae, can form on the wood surface or in openings such as surface checks. They can also penetrate the cell walls, passing through the wood from one cell to another. In the early stages of decay, the presence of the hyphae may not be obvious. At more advanced stages, however, the wood may become soft and will have lost almost entirely all of its structural value.

Three recognized general types of decay are white rot, brown rot, and soft rot (1). Fungi causing white rot consume both carbohydrates and lignin from the cell walls, leaving the wood a lighter color. Fungi causing brown rot remove the cellulose and hemicellulose, and the crumbly brown residue consists largely of the lignin from the cell walls. The bending strength is reduced markedly even in the early stages of decay by brown rot fungi. In soft rot, the surface of the wood becomes softened, but the depth of damaged wood is not so great. The rotted surface can be scraped away from the undamaged wood beneath.

The above classification relates only to the general nature of the rotted wood. Each type of decay can be caused by many species of fungus. Some can destroy only particular species of wood, while others can affect a wide variety of wood species.

A term frequently heard is *dry rot*. This term is normally used to describe the final stages of brown rot. It is also a misnomer, as wood will not decay if it is kept dry, regardless of whether it is infected or not. However, wood in which the growth of fungi has been arrested because of dry conditions may return to active growth again once the level of moisture increases. Further, these fungi have the ability to transport moisture from one location to another in the wood. Thus, fungi that started to develop in wood in damp locations (such as in basements or where the wood is in contact with soil) may transport moisture as far as 20 ft into wood in other parts of a structure. They may also form a coating of porous, rootlike strands over masonry to reach wood that is in drier locations.

Effect of Decay on Properties

Table 16-1 (from reference 2) shows expected strength losses in wood that has been partially destroyed by decay. The measure used to define the degree of decay is the percent loss of weight. The table is for softwoods only and for

Table 16-1. Probable Strength of Wood in Early Stage of Decay (5 to 10% weight loss).

Strength Property	Probable Remaining Strength (% of Original Strength)
Static bending	30
Impact bending	20
Modulus of elasticity	30
Compression parallel to grain	55
Tension parallel to grain	40
Compression perpendicular to grain	40
Shear	80

Based on information in Reference 16-2.

just one type of decay. Much additional research would be needed to allow reliable determination of the strength remaining in actual structural members. The striking feature of the table is that it shows that bending strength, and compression and tension strengths parallel to the grain—the identical strength properties that most often control our design of structural members—are the ones most severely affected by decay.

Resistance to Decay

The natural resistance to decay varies among woods of different species. As with most other wood properties, decay resistance also varies between trees of the same species, and even among pieces from the same tree. For example, lumber from the sapwood of most trees is easily decayed, while the resistance of lumber made from the heartwood is much higher. This is due to toxic substances called *extractives* which are deposited in the wood as it changes from sapwood to heartwood. These substances might be regarded as natural preservatives, protecting the wood by poisoning the food supply of the attacking microorganism. In the living tree, however, the heartwood is more susceptible to decay than the sapwood. This is probably because the sapwood contains so much water that the amount of air present is insufficient for growth of the fungi.

Some species of wood commonly used for lumber, notably California redwood, junipers, and cedars, are generally considered to be very durable. Thus redwood is often specified as the species to use for sill plates, the first piece of wood above the foundation in light-frame construction, and cedar is often preferred for exterior siding. Table 16-2 shows a listing of some of the common species according to approximate average resistance to decay. Certain dense tropical hardwoods (bongassi, basrolocus, and bilinga) are believed to have superior resistance to deterioration when used in climates such as those of North America.

Table 16-2. Grouping of Some Domestic Woods According to Heartwood Decay.

Resistant or Very Resistant	Moderately Resistant	Slightly or Nonresistant
Baldcypress (old growth)[a]	Baldcypress (young growth)[a]	Alder
Catalpa	Douglas fir	Ashes
Cedars	Honey locust	Aspens
Cherry, black	Larch, western	Basswood
Chestnut	Oak, swamp chestnut	Beech
Cypress, Arizona	Pine, eastern white[a]	Birches
Junipers	Southern pine:	Buckeye
Locust, black[b]	Longleaf[a]	Butternut
Mesquite	Slash[a]	Cottonwood
Mulberry, red[b]	Tamarack	Elms
Oaks:		Hackberry
Bur		Hemlocks
Chestnut		Hickories
Gambel		Magnolia
Oregon white		Maples
Post		Oak (red and black species)
White		Pines (other than longleaf, slash, and eastern white)
Osage orange[b]		Poplars
Redwood		Spruces
Sassafras		Sweetgum
Walnut, black		True firs (western and eastern)
Yew, Pacific[b]		Willows
		Yellow poplar

[a] The southern and eastern pines and baldcypress are now largely second growth, with a large proportion of sapwood. Consequently, substantial quantities of heartwood lumber of these species are not available.
[b] These woods have exceptionally high decay resistance.
Source: Courtesy Van Nostrand Reinhold Co.

Necessary Elements for Decay

For decay-causing fungi to grow four things are necessary: a food supply, suitable temperature, sufficient moisture, and oxygen.

Food Supply. The food supply is the wood itself. Different species of microorganism show preferences for different species of wood and for different compounds or structural parts of the wood.

Suitable Temperature. Temperatures under which wood-destroying fungi may develop vary over a fairly wide range. Normally, the fungus grows most vigorously during the hotter and more moist parts of the year. Optimum growth temperatures for most species of wood-destroying fungi are between 68°F and 97°F (reference 1). As the temperatures drop to low levels, the fungus becomes dormant but is not killed. Growth resumes when the temperature rises to an acceptable level. Temperatures higher than those suitable for fungal growth will retard growth significantly and will even kill the fungus. This is fortunate, as the temperatures reached in normal kiln drying of lumber are sufficient to kill the fungi or spores already present.

However, using high temperatures to kill fungi is of little use in infected completed structures. To raise the temperature of an entire structure to a suitable degree and to maintain that temperature long enough to kill the fungi is impractical. Furthermore, even if the infected wood parts were to be sterilized by heat, there would be no assurance that reinfection would not occur if the wood were exposed again to fungal spores.

Sufficient Moisture. For most wood-destroying fungi to grow, the moisture content of the wood must be at or above the fiber saturation point, but a few species will cause decay with less moisture. Water in excess of that causing fiber saturation is present as free water in the cell cavities. This free water carries enzymes of the fungi and promotes the deterioration of the wood. Without this free water, the enzymes cannot enter the cell walls and growth of the fungus is arrested. Consequently, wood that has

been properly dried or air seasoned will not decay provided its moisture content is kept below the fiber saturation point (FSP) at every location within the piece. If any one location has water in excess of the FSP, however, decay may occur at that particular location and the entire piece may become useless.

Notice, however, that the dry-rot fungi are an exception to the above. These fungi can *carry* water to drier parts of the wood, moistening it and causing it to decay.

Oxygen. A small amount of air is all that is needed to promote the growth of wood-destroying fungi. Conditions in normal stockpiles or in a completed wood structure are sufficient. Even the air present in the cell cavities of wood is sometimes sufficient for growth of the fungi.

It is probably because air is excluded by the large amount of water present in sapwood of a living tree that its sapwood is more resistant to decay than its heartwood. Wood piling that is driven into the ground but remains below the water table may last for hundreds of years, even though thoroughly saturated. It is the exclusion of air that prevents these piles from decaying. Should the top of the pile be above the water table or should the water table become lowered, decay will probably commence.

Insects

Insects that damage wood include termites, powder-post beetles, and carpenter ants. Of these, the termites are by far the most important.

Termites. Termites are frequently called "white ants," but are not ants and generally are not white. Worldwide, there are more than 2100 species of termite. The 41 species found in North America can be divided according to their living habits into three groups: subterranean termites, dry-wood termites, and damp-wood termites.

Subterranean termites are the most troublesome in wood structures. They live in moist ground, entering wood that is either near the ground or in contact with it. They will enter either sound or decaying wood. They construct tunnels of mud and partially digested wood fibers upward from the ground to reach the wood. Once they enter the wood, they can eat their way upwards to as high as the second or third story of a building. Using the cellulose of the cell walls as their food, and generally eating parallel to the grain, they create interconnected galleries of burrows in the wood, hollowing out the interior but rarely penetrating the surface of the lumber. Thus, the damage they do is often not detected until serious weakening has occurred. Subterranean termites are found in almost every part of the United States and cause considerable damage to embedded poles, bridge timbers, and foundation members of buildings.

Dry-wood termites seldom enter the ground, but live almost entirely in the wood, from which they obtain sufficient moisture. The winged form of reproductive adult can attack sound wood directly from the outside. Once they have entered, they plug the entrance with chewed wood and proceed inside to develop a new colony of termites. Since they occasionally open the wood again to the outside to discharge pellets of waste material, it is sometimes possible to detect their presence before the wood member in question has been severely damaged. Since dry-wood termites can live in wood with as low as 10–12% MC, they may inhabit well-seasoned or dried lumber. Dry-wood termites are found in the United States only along a narrow strip along the southern border and along the Atlantic coast as far north as Virginia.

Damp-wood termites also enter wood from the air. They require considerable moisture to thrive. Consequently, they usually are active in only decaying wood or very moist wood. They can penetrate sound wood from adjoining decayed wood, however.

Powder-post and Other Beetles.
There are many species of powder-post beetle, some of which attack wood in dying trees or freshly cut green wood. Others can attack wood with moisture content as low as 10% (3). Beetles cause most damage in hardwoods, although occasional population explosions of the pine-bark beetle cause considerable damage in western softwoods.

Beetles deposit eggs in open pores of cut lumber or through the bark of standing trees. The larvae that hatch burrow through the wood, using certain wood components as their food. After metamorphosis, the emerging adult beetles proceed to reproduce, infecting new wood at some distance away or perhaps even laying their eggs in the same timber from which they emerged. The damage by beetle attack is a series of irregular burrows, sometimes just following the surface under the bark, but in most cases penetrating the solid wood in random directions. The pine-bark beetle not only leaves burrows, but also infects the tree with a wood-staining fungus. Even though this does not seriously affect the structural properties of the wood from infected trees, it does severely reduce its value and salability.

Ants.
Ants use wood particles for shelter rather than food. Normally, they confine their activities to wood that has already been softened by decay.

Marine Borers

Marine borers are of two classes—wood-boring mollusks (related to clams and oysters) and crustaceans (related to crabs and lobsters). Damage by marine borers is common throughout salt or brackish water regions of the world, especially so in warm regions where the water temperature is 50°F or above (4). In North America, it is common along the Pacific, Gulf, and Atlantic coasts. Marine borers can completely destroy untreated timber piles in less than a year.

Mollusks.
Three important genera of mollusks that damage wood on coasts of the United States are *Teredo*, *Bankia*, and *Martesia*. There are a number of species of each genus. Species of the first two are wormlike in form and are commonly called shipworms. Members of the third genus are more like clams in general appearance.

Free-swimming larvae of shipworms have small bivalve shells that may be used for protection. After the larva attaches itself to a piece of wood, it uses these to eat its way inside. Once inside the piece of wood, the shipworm remains there for life, growing rapidly in both diameter and length. Using the wood as part of its food supply, it extends its burrows, often completely riddling the interior of the wood. The wood pile, for example, may become practically a hollow shell with only the small entrance hole visible on the outside.

Wood-boring crustaceans do not remain in the wood for life, so the outside of a timber attacked by them will show numerous holes. The damage is sometimes superficial but may be so thorough that the entire surface can be removed by mechanical action of the water. This exposes new wood, and action of the crustacean borers continues. In this manner, the structural value of the wood member can be completely destroyed. Of the three species of wood-boring crustacean found in United States coastal water, *Limnoria* is the most destructive.

Weathering

Several types of defect that can be caused by exposure to the weather are raised grain, checks, splits, large cracks, warping, twisting, loosening of nails, roughened surface, and change of color. The wood surface often becomes friable, with splinters and other fragments falling away. These defects can occur if the wood is not protected from wetting and drying and from exposure to wind or to sunlight. (See Fig. 16-1.)

Weathering is not the same as decay, although the defects caused by weathering may provide conditions conducive to decay. The splits and raised grain, for example, provide easy entrance into the wood by fungal spores or insects.

Changes due to weathering may be classified as those due to chemical, mechanical, and light energies (5). The principal cause of weathering changes is repeated wetting and drying. Wood is hygroscopic and wood fibers near the surface respond quickly to changes in atmospheric humidity, as well as to wetting by rain, dew, or snow, whereas those away from the surface respond slowly, if at all. The result is that wood fibers near the exposed surface have varying moisture content (MC), swelling and shrinking, while those below the surface remain at more nearly constant size. This causes a steep gradient in MC and thus a steep gradient in stresses. Swelling and shrinkage are greatest perpendicular to the grain, and strength is the least. Thus, the alternating tensile and compressive stresses that result normal to the grain

Fig. 16-1. Badly weathered redwood. Note splits and eroded surface. (Photograph by authors.)

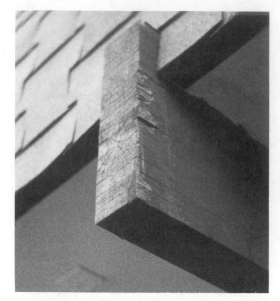

Fig. 16-2. Splits and checks in end of glulam exposed to weather. (Photograph by authors.)

cause checks and splitting parallel to the grain. With prolonged exposure, the small surface checks become larger, even culminating in splits that go through the entire thickness of the piece. (See Fig. 16-2.)

Raising of the grain, often the first sign of weathering, is due to differential swelling and shrinking of springwood and summerwood.

Erosion due to blowing grit or sand (sometimes referred to as natural sandblasting) often occurs in drier localities having high winds, such as the mountain areas of the western United States. The moving sand abrades grooves where the softer springwood is at the surface, leaving ridges where the harder summerwood is exposed. (Where this appearance is considered desirable, wood is sometimes intentionally sandblasted to create the impression of weathered wood.)

Sunlight. Both ultraviolet and visible parts of sunlight cause darkening of wood and chemical degradation of the lignin. Some of the products of lignin degradation are soluble, and when these products are washed away by rainwater, the wood becomes more gray in color (5). Grayed wood is sometimes considered attractive, and chemicals are sometimes used inten-

tionally to accelerate the color change. For example, redwood can be given a surface treatment that quickly gives it the silvery grey appearance of old, weathered redwood.

Effect of Weathering on Strength. The few studies that have been done indicate that weathering has little effect on modulus of rupture, strength in compression parallel to the grain, and modulus of elasticity. Only toughness, abrasion resistance, and (for very old wood) splitting and impact strengths decrease.

Weathering of wood products such as plywood and particleboard involves weathering of the adhesives as well as the wood veneers or particles, and as this is written, little information is available regarding the effect of weathering on strength.

Preventing Deterioration

Damage from the various agents of deterioration described above can be prevented by either (1) preservative treatment or (2) preventative construction and maintenance.

16-2. PRESERVATIVE TREATMENT

The science and practice of wood preservation is so involved and complex that the most that can be done in this short chapter is to present a few of the fundamentals and describe a few of the many methods used to treat wood.

As pointed out earlier, four essential conditions are required for growth of wood-destroying fungus. Most wood preservatives poison the food supply, thus eliminating one of these essentials. This type of treatment can also be effective as protection against attack by certain insects.

To a degree, some preservative treatments also protect by waterproofing the wood. If the interior of the piece of wood has less moisture than is needed for a damaging fungus to grow, then waterproofing will prevent the moisture content from rising to the danger level. Stains and paints can also provide this kind of protection, but often they are of short-term value and must be replaced at frequent intervals.

To be an effective wood preservative a chemical must be:

1. Toxic to the organisms that can destroy wood
2. Soluble to some degree in the body fluids of the organism it is intended to repel or kill
3. Capable of penetrating the wood to considerable depth, so that wood at the bottom of checks and splits will be protected
4. Permanent, not leaching out of the wood during the years it is expected to remain in service.

There is no perfect solution, however, since different organisms have different body fluids and their reactions to various chemicals differ.

Preservative Chemicals

Generally, water-soluble preservatives soak into the cell walls, but creosote and similar ones only fill (or partially fill) the cell cavities. Reference 1 lists the principal chemicals used today in preservative treatment and gives criteria for selecting which chemical to use for various applications. The structural designer should probably not make the selection without seeking the advice of experts in wood preservation. However, the designer should be aware that wood preservatives are poisonous, because special precautions may be needed where persons, animals, or foodstuffs may come in contact with the treated wood. Some (creosote, for example) have an odor that makes them unsuited for use inside of most buildings. They may also ooze out of the timber, so that personal or animal contact with treated exterior timbers may be undesirable. Some preservatives can be used for wood that will be stained or painted, while others cannot.

Treatment Methods

Treatments include pressure methods and nonpressure methods, and there are several varieties of each. The nonpressure methods generally are both less expensive and less effective than pressure methods.

16-3. NONPRESSURE TREATMENT METHODS

There are dozens of these methods, some of them proprietary.

1. Surface treatments include brushing or spraying. The preservative is drawn into the wood only by capillary attraction, and the degree of penetration is slight. Coal tar creosote or oil-borne solutions of toxic chemicals are generally used. The advantage of brushing or spraying is that it is cheap and requires no extensive physical plant or equipment. The disadvantage is that the degree of protection is minimal. Brushing, however, is a necessary additional treatment to be used where wood treated by better means is cut during fabrication or installation. This occurs particularly at cut ends, where, fortunately, capillary attraction may cause slightly better penetration.

2. Dipping is merely immersing in a bath of the preservative for a few minutes, hoping that the chemical will have sufficient penetration. It is more costly than brushing or spraying since it requires a large vessel and considerably more preservative. However, it gives better assurance that all parts of the surface are treated. Dipping is useful with finished products such as window frames, sash, and doors. Penetration is about the same as for brushing or spraying, poor in side grain but better in end grain.

3. Steeping is merely dipping for longer periods of time, for several days to a few weeks. If steeping is used for seasoned lumber, both water and dissolved preservative are absorbed by the wood. When steeping is used for green wood, the liquid in which the preservative is dissolved cannot enter the already saturated wood. In this case the preservative enters the wood by diffusion, the preservative salt leaving the liquid in the vat and entering the water already present in the green wood. Again, penetration is not very good. To counter this, the solution used is more concentrated than is used in the pressure processes. One would think that penetration would be better for seasoned wood than for green wood, but this is not the case. The chemical entering the green wood by dif-

fusion continues to spread within the wood after the wood is removed from the vat, as long as the moisture content of the wood remains fairly high.

Reference 1 lists 24 other nonpressure processes.

16-4. PRESSURE TREATMENT METHODS

There are numerous pressure processes, many of them proprietary. Broadly, they may be classified as empty-cell processes and full-cell processes. In each type of process, the wood is treated in a large, cylindrical tank into which the preservative may be entered and placed under pressure to force it into the wood.

Empty-Cell Processes

In empty-cell processes the preservative tends to coat the walls of the wood cells and to be absorbed by them, but after the wood is removed from the treatment tank the cells themselves are essentially empty, little preservative remaining in the cell cavities. Usually creosote or preservative oils are used, although the process may also be used with aqueous solutions of preservatives. The processes are used where good penetration is needed but maximum absorption is not.

A typical empty-cell process has the following steps:

1. After placing the wood in the tank and closing the end covers, the tank is filled with preservative at atmospheric pressure.
2. Pressure is applied to the fluid in the tank, forcing the preservative into the cell cavities. Pressures as high as 250 psi, but normally not over 150 psi, are used. These pressures are held for a few hours.
3. The pressure is released and the tank drained.
4. A partial vacuum is applied and held for about 30 minutes to remove free preservative.

In one of the empty-cell processes, air pressure of 25–75 psi is entered into the tank first and then the preservative is entered under pressure. Air in the cell cavities is initially compressed and it helps to increase the percent of preservative recovered during the vacuum step.

The steps described above are applied to seasoned (air-dried) wood. If green wood is to be treated, it must be conditioned first. One way of doing this is by steaming followed by partial vacuum; another is by boiling under a partial vacuum.

Full-Cell Processes

Full-cell treatment is used to treat wood with creosote or either oil or aqueous solutions of various salts. The amount of chemical retained by the wood is greater than with empty-cell processes. Thus, one would expect the full-cell process to be both more expensive and more effective than the empty-cell process, provided the depth of penetration is the same for each.

In full-cell processes, the preservative coats and is absorbed by the cell walls, just as in the empty-cell process, but also the cell cavities themselves contain a significant amount of free preservative. The difference between empty-cell and full-cell processes is that the full-cell process begins (after the wood is enclosed in the tank) with application of a partial vacuum to remove as much as possible of the air present in the wood cell cavities. The partial vacuum created in the cell cavities helps to draw the preservative into the wood. Also, less preservative is pushed out of the cell cavities during the final vacuum step.

Reference 1 describes numerous variations to the full-cell process.

16-5. EFFECTIVENESS OF TREATMENT

Penetration and retention are the two measures by which the effectiveness of treatment with a particular preservative may be measured. Penetration is the more important of the two measures, but it is hard to verify without damaging the wood.

The average retention (preservative weight per unit volume of wood) is easily measured by weighing the timber before and after treatment.

Though this shows how much preservative is retained, it does not show how it is distributed. For an entire batch being treated, retention can be measured by the difference between the volume of preservative introduced into the vessel and the volume remaining after treatment.

For a given treatment, both penetration and retention vary with species, between sapwood and heartwood, and with density, condition, and moisture content. Retention also varies with size of cross section of the wood, so, when possible, different sizes of timber are treated separately.

Depth of penetration cannot be observed on the outside surface. It can be measured only by boring, crosscutting, or splitting the treated piece of lumber. Whichever method is used, the wood is damaged.

A small-diameter increment borer (a tool, not an insect) can be used to remove a small cylindrical portion for observation. Since the wood is variable, however, it would be necessary to make such borings at several locations to determine the least penetration. As an alternative, holes may be made using ordinary wood bits, measuring the penetration by observing the color of the chips removed at various depths. For wood treated with creosote or similar materials, however, this is not a good method; the preservative material may be worked deeper into the wood by the bit, giving a false impression of penetration.

In either method, the hole opens up the interior of the wood for infection. Consequently, the holes must be filled with tight-fitting, treated wood plugs.

In treated lumber that is longer than necessary, penetration depth is shown at sections where the piece is cut to proper length before installation. This shows the depth only at the location of the cut. The depth shown at a cut near the end of the treated length piece is normally greater than at locations farther from the end. Again, such a cut exposes untreated wood, and infection can occur in spite of the preservative treatment. If the preservative treatment is to be effective, it is essential to brush or spray all such areas with preservative before completing the installation. This applies to end cuts, bolt holes, grooves for timber connectors, and cuts or notches made for end bearing or other

reasons. It is especially important in wood for foundations.

The most accurate method of determining penetration is to split the treated member along its entire length. This gives a better view of penetration, but unfortunately destroys the piece of wood. Performed on a sufficient number of pieces from a treated batch, however, it could serve as an effective means of quality control.

Amount of Retention Needed

The amounts of retention specified vary with species, size of timber, and use. Higher retention (in pounds per cubic foot) is usually specified for small timbers than for large ones. The amount needed depends on what the wood is to be used for. For piling above the water table and for piling in salt water, the retention must be fairly high. For wood for most buildings, it can be less.

The ideal situation would be to have all cell walls completely saturated with the preservative. This is difficult (or impossible) to do, so we settle for lesser practical retentions. Reference 6 gives a more nearly complete list of recommendations for various applications, wood species, and chemicals.

Penetration Requirements

In glued laminated members, it is unsafe to assume that the preservative penetrates (normal to the lamination) beyond the first glue line, although it might penetrate laterally somewhat further.

For plywood and particleboard, treatment of the *edges* (including the edges that are cut during installation) is important in preventing deterioration by weathering.

For sawn timbers, specifications usually require that all (or a high percentage) of the sapwood be impregnated, as well as some of the heartwood. A designer preparing specifications should consult either reference 6 or 7 for guidance and should be careful that the specifications are feasible for the species and sizes involved.

16-6. USING PROPER DESIGN DETAILS TO PREVENT DECAY

The presence and continued use of so many early U.S. colonial buildings and old timber bridges, all constructed without preservative treatment, suggest that properly detailed wood structures can have long life, even in moist environments that one would expect to encourage decay. The secret? Merely design so as to shed water away from the wood, to prevent moisture-laden vapors from condensing in the wood, and to drain or ventilate spaces where water may accumulate.

"There is, of course, a right way and a wrong way to do everything with wood, and pursuing the wrong way will eventually lead to trouble" (reference 2). The main agents of wood deterioration are decay and termite damage, and good construction details can go a long way toward protecting from either agent.

It is reasonable to assume that all wood as it is received for construction has already been infected, so decay can proceed if conditions are right. The wood substance itself is the food supply. The three additional things needed are air, proper temperature, and water. To eliminate any one of these is to prevent decay. In most situations where wood is used, air is available and the temperatures are suitable for decay to occur. The one necessity that we can control is water; eliminate water and decay cannot occur.

What Are Some of the Improper Details?

The list of improper details is probably not complete, and never will be. As new types of construction are attempted, new errors will be made. What is important is to learn from those errors, as has been done for the ones listed here.

1. Untreated wood in contact with the ground. In most cases, the MC of wood touching the ground rises sufficiently to allow fungal spores to grow. It also allows easy access for termites. Most building codes specify the minimum distance to be provided between the ground and untreated wood.

2. Details that can trap water from plumbing leaks, seepage, or condensation.

3. Contact between wood and a porous material that can draw water from the ground up to the wood. (A too-low concrete foundation wall, for example.)

4. Inadequate ventilation of enclosed spaces that may accumulate moisture. Examples would be enclosed soffits (overhanging roofs with the underside covered), crawl spaces beneath wood floors next to the ground, and heavily insulated but enclosed attics.

5. Inadequate flashing. Flashing is a membrane, usually metal, placed at edges and at changes in direction of roofing material so as to direct the flow of water from one surface to the next and prevent it from getting under the roofing. A common error is to have the flashing too narrow.

6. Using no vapor barrier where one is needed, or having the vapor barrier in the wrong location. A vapor barrier can be explained using Fig. 16-3. Assume that the interior is heated and the air outside is cold. The interior air has a higher relative humidity than the colder outside air. Vapor tends to move, through the wall, from inside toward outside. When it contacts material whose temperature is below the dew point, the vapor condenses. The water that forms is then absorbed by whatever permeable material is present: insulation, wood studs, or plywood siding. The vapor barrier should stop the movement of the vapor at a point where the temperature of the wall material is above the dew point. This means that the vapor barrier should be located at the warm side of the wall, as shown by Fig. 16-3. A frequent error is to place it near the outside, in which case it does more harm than good.

7. Broken membranes, such as building paper (a pervious layer used on the outside of the structural sheathing of a wall), flashing, or vapor barrier. These are frequently damaged during installation or by workers doing subsequent work.

8. Use of an impervious membrane where a pervious one should be used. An example is in some of the modern combination insulation and sheathing boards used instead of plywood for exterior wall sheathing. Often these have a

Fig. 16-3. Vapor barrier location.

layer of aluminum foil on the outside face. This is fine for reflecting heat, but the foil may form an impervious barrier to the outward passage of water vapor. If there is a break in the inside vapor barrier, or if there is no such barrier, then the exterior surface acts as an incorrectly placed vapor barrier. Water accumulates and decay follows. Realizing this potential difficulty, some builders deliberately knock holes in the combination sheathing to provide ventilation to the wall (reference 2).

9. Unprotected timbers projecting beyond a building roof so that they are exposed to weathering. (See Fig. 16-4.) The situation is even worse if the timbers are horizontal, in which case rain falling on the top surface of the timber can be carried by capillary attraction into the contact between the timber and the roof decking, encouraging decay of both members.

What Details Would Be Better?

1. The sequence of wood pieces should shed water completely at the first opportunity, rather than passing it along from one piece to another. One way of doing this is to make sure that the lower edge of an exposed piece is below that of the adjacent part, as in Fig. 16-5a. Alternatives would be to provide a groove beyond which water cannot move (Fig. 16-5b) or flashing to shed the water directly.

2. Use a gutter system to prevent the water from touching any wood at the roof eaves.

3. Wherever the average January temperature is 35°F or less, use a vapor barrier *on the warm side* of exterior walls of heated buildings (reference 3).

4. Ventilate closed spaces so that water that does enter can evaporate rather than accumulate. This rule applies especially to crawl spaces where wood floor systems are close to the

Fig. 16-4. Improper detail caused all of these projecting beams to split badly. (Photograph by authors.)

(a) KEEP ADJACENT MEMBER HIGHER
THAN EXPOSED MEMBER

(b) PROVIDE WATER-SHEDDING GROOVE
ON BOTTOM OF EXPOSED MEMBER

Fig. 16-5. Details to avoid wetting of adjacent member by rainwater.

ground beneath the building. It seems logical to enclose the area all around to conserve heat, yet to do so is just inviting trouble. Adequate ventilation must be provided and the area must be kept open. The ground provides water vapor, which condenses on the underside of wood members near the perimeter of the area. In addition, an impervious plastic cover may be placed over the soil to retard movement of water from the soil into the crawl space air.

5. Protect wood beams that extend beyond the cover of the roof. Flashing placed over the projecting timbers should allow air space between the wood and the protecting metal. The flashing should be so formed that water will be shed and will not pass by capillary attraction along the length or across the top of the beam.

6. Bases of unprotected wood posts or columns should be elevated above surfaces on which water may collect. Columns bearing directly on concrete floors, for example, are very prone to rotting at the base. Even though the column is an interior one that is protected from rain water, the concrete below will transmit soil moisture into the end grain of the wood. Figure 9-10 shows a suitable column base detail. References 2 and 3 give additional recommendations.

16-7. TERMITE PROTECTION

Measures that may be used to prevent termite damage are:

1. Remove all tree stumps, roots, wood scraps, wood stakes, and concrete formwork from the building site. Do not bury any of these in the backfill. (The reason for this precaution appears at first to be merely cosmetic, but the real purpose is to prevent these scraps from decaying. Wood with brown rot is an attractive food for termites.)

2. Use only treated lumber within 18 in. of the top of the soil or highest backfill (8). In some localities, more extensive use of pressure-treated lumber is advised. Such an area is Hawaii, where all framing materials in light-frame stud walls should be treated to protect against the Formosan termite. In Puerto Rico pressure treatment is advised for all structural members to prevent attack by dry-wood termites (reference 3).

3. Use solid concrete foundation walls, or masonry walls with portland cement mortar (to avoid cracks along which the termites can travel). Extend these foundation walls to at least 18 in. above the soil or highest backfill.

4. Poison the soil along each face of the foundation to kill termites before they can build tunnels to the wood above. An advisory committee has recommended to the Department of Housing and Urban Development (HUD) that soil treatment be *required* in the southeastern United States.

5. Use poison bait blocks around the foundation, both inside and outside. The poison used with this method has been banned for environmental reasons, but poison may return as a viable method when a satisfactory alternate chemical has been found.

6. Use termite shields. These are barriers of sheet metal to prevent the termites from reaching the wood by building tunnels on masonry or concrete foundations, or along sewer and

water pipes or electrical conduits that extend from the ground to the wood above. A 2-in. width of the sheet metal barrier should slope downward at 45° on all sides of the wood being protected and should be completely sealed at all corners by soldering or crimping, so that there is no way the termite can pass the barrier except by crawling down the bottom of the 2-in. sloping width and back up on the top surface.

Termite shields are now considered to be not as effective as using treated lumber (reference 3). They are accepted by some building codes but not by the U.S. Minimum Property Standard.

16-8. FIRE DAMAGE

Untreated wood is flammable, yet wood structures are not necessarily more easily destroyed by fire than buildings of other materials. Further, the maximum fire hazard is usually from burning of building contents rather than the building itself. Even though a fire begins in the building contents, a prolonged fire will certainly damage the building also.

Two distinct problems in materials used for buildings are fire resistance and surface flammability. Fire resistance is the ability of the material to restrict being penetrated by the fire and to retain structural strength. Surface flammability measures how fast the flame will spread along the material surface. Ignition speed for the wood depends on temperature, whether the heat source is a flame or merely hot air, and the length of time of exposure to the elevated temperature.

The spread of fire in a wood structure depends on flammability, on size and arrangement of the wood members, and on details that allow or restrict air movement through and around the wood members. Generally, the larger the cross section of a wood member, the longer it will take to destroy the member. (Thus, five 2 × 10s would be destroyed more quickly than a single 8 × 10 of the same strength.) The burning wood develops a layer of "char" (black charcoal) on the burning surfaces, and this insulates the wood beneath to retard it from burning. The char layer proceeds through the burning wood at an average rate of about $1\frac{1}{2}$ in. per hour (reference 4).

For fire considerations, wood structures are classified as (1) heavy-timber construction, (2) ordinary construction, and (3) wood frame construction. Heavy-timber construction was defined in Chapter 1. Ordinary construction has masonry exterior walls and wood framing members smaller than those meeting the limitations for heavy-timber construction. Wood frame construction is the same, but has wood-framed exterior walls.

In heavy-timber construction buildings, the char protects the main structural members so that, after the fire is extinguished, sufficient strength is usually present to carry at least all of the dead load. Thus the building does not collapse. Depending on actual design loads and member sizes, it may have sufficient remaining strength to allow continued use. Or it may be possible to reinforce, replace, or strengthen members selectively, as needed. In calculating the remaining strength in large timbers, however, the designer should realize the char depth is greater at the corners of the member than on the flat surfaces.

Subjected to a similar fire, however, a building with an unprotected steel frame would probably collapse completely, the yield strength and modulus of elasticity of its members being reduced by the high temperatures. The steel could be protected, of course, by fire-protective coatings or by enclosure in masonry, concrete, or other fire-resistant material. Similarly, wood parts can be treated to make them harder to ignite and to retard their burning.

16-9. DESIGN TO MINIMIZE FIRE DANGER

The most important step in minimizing fire danger is, of course, to protect the building *contents* against becoming ignited. Beyond this, there are two main categories of protection for the building itself: (1) appropriate design details to prevent ignition and the spread of fire from one part of the building to another and (2) protection or treatment of the wood to make ignition harder and retard burning.

Details

Reference 4 gives a good summary list of the detail design deficiencies responsible for spread of fire, heat, and smoke in buildings. Most of these deficiencies are addressed by building codes such as the UBC (9). The deficiencies in general consist of openings that allow heat, flame, or smoke to travel from one part of the building to another. Examples include unprotected wall openings for dumbwaiters, stairways without doors at floor levels, framed walls without firestops at concealed spaces below floors or ceilings, and improper anchorage of structural framing members in masonry walls. Figure 6-14 shows ways of framing floor or roof joists into masonry walls. The first and second of these details allow the floor of a burning building to collapse without bringing down the masonry wall at the same time.

Other design deficiencies include the lack of fire-resistant walls between adjoining buildings, the lack of protective doors in passageways between adjacent buildings, and the lack of separation between living areas and areas such as shops or boiler rooms. These requirements apply to buildings of any material—steel, concrete, masonry, or wood.

For some building uses, codes require fire doors that close automatically in case of fire. Automatic sprinklers may be required, but their function is mainly to extinguish fire in the building contents; thus their need is more dictated by building use than building material.

The designer should observe carefully all requirements of building codes to avoid most of these deficiencies.

Protection or Treatment

Using formulas published by the Council of American Building Officials, the designer may calculate the required size of members necessary for fire resistance. The formulas give the fire-resistance rating in minutes for timber beams and columns with smaller dimensions not less than 6 in. nominal. The formulas are:

For beams—
Exposed to fire on all four surfaces

$$2.54Zb(4 - 2b/d) \qquad (16\text{-}1)$$

Exposed to fire on only three surfaces

$$2.54Zb(4 - b/d) \qquad (16\text{-}2)$$

For columns—
Exposed to fire on all four surfaces

$$2.54Zd(3 - d/b) \qquad (16\text{-}3)$$

Exposed to fire on only three surfaces

$$2.54Zd(3 - b/2d) \qquad (16\text{-}4)$$

In each of the above

$b =$ the horizontal dimension of a beam cross section, or the *larger* side dimension of a column

$d =$ the vertical depth of a beam, or the *small* side dimension of a column

$Z =$ a "load" factor that depends on the percent of allowable load actually carried by the member; Figure 16-6 shows how to determine Z

Using the above expressions, the designer may estimate the member dimensions needed for unprotected heavy-timber construction.

Fig. 16-6. Factor Z for fire-resistance equations. (Courtesy of American Institute of Timber Construction.)

For ordinary and wood frame construction, the steps to protect against fire include, first, following all building code requirements and avoiding construction detail deficiencies. The next step, if it is judged necessary, is to reduce flammability of the wood. This can be done in three ways:

1. Encase the wood members in protective material such as gypsum plaster or plasterboard.
2. Apply an intumescent coating, painted onto the exposed surfaces, which expands when heated to produce an insulating layer. This treatment is better reserved for upgrading existing buildings. (For new buildings, treatment before the wood is installed is better.)
3. Use fire-retardant treatment, either for all wood parts or for those parts whose size or exposure makes them the most likely to ignite and fuel a fire.

Fire-retardant treatment has been in use since the fourth century B.C. (reference 1). Modern fire-retardant treatment (it should not be called fireproofing) consists of pressure treatment with aqueous solutions of various salts, followed by drying to remove the water and to lower the MC to the desired level (19% or less for lumber not over 2 in. thick, and 15% or less for plywood). Chemicals used most widely are combinations including zinc chloride, ammonium sulfate, borax, or boric acid, with lesser amounts of sodium dichromate or diammonium phosphate (reference 6). The effectiveness of fire-retardant treatment depends on the chemicals used and on the amount of chemical retained by the wood.

Some of the salts used in treatment are hygroscopic, so retardant-treated wood should not be used where the relative humidity will be 80% or more for long periods. To do so would cause the MC of the wood to rise to an undesirable level. (For localities where the average relative humidity is 70% or more, the designer should specify an exterior nonhygroscopic chemical.) Also, if water contacts the treated members, there can be leaching of the salt. This can be retarded by surface painting. However, if treated wood is to be stained, painted, or varnished, the designer should investigate carefully before specifying treated wood to be certain that the intended finish can be applied successfully.

Fire-retardant treatment does two things that are important to the structural designer. It increases the weight of the member, and it may decrease the allowable stresses. Weight of the member depends on the weight of chemical retained per unit volume of wood. The NDS (10) requires that the allowable stresses be reduced by:

20% for tension parallel to the grain
15% for bending
10% for fastener design loads, all other stresses, and modulus of elasticity

16-10. EVALUATION AND REPAIR OF WOOD STRUCTURES

Structural designers are often called on to evaluate existing structures and either to design repair measures to correct structural deficiencies detected or to determine the capability of the structure to serve some new purpose, perhaps with increased loading. In the case of wood structures, this occurs frequently as architects and developers attempt to preserve existing structures, adapting them to new uses.

What things should the designer look for in evaluating an existing wood structure? The following lists many of them:

1. Evidence of past changes in use. Revised framing might indicate this. Revised use might also be detected by comparing the present structure to that illustrated by the drawings, both the originals and those prepared during past remodeling.
2. Evidence of past repairs. These might be detected by the presence of obviously newer lumber in some members or might be evidenced by "beefed-up" connections. Any such repair (unless it was a step made to increase the load capacity beyond

that intended by the original designer) probably indicates some past structural distress.

3. Evidence of foundation problems. Unequal settlement of foundations can cause cracks in walls, poor fit of doors and windows, and sloping floors.

4. Roof surface irregularities. These might be the result of foundation problems, but more often indicate a weakness in certain roof members.

5. Abnormal deflections of beam members. This could be the result of creep. If the deflections are continuing to increase without increase in load, the possibility of collapse due to tertiary creep should be considered. (See Chapter 2.)

6. Fire damage. Should the wood show evidence of fire damage, the metal parts should also be examined.

7. Evidence of chemical deterioration or rust of metal fittings.

8. Loosened connectors. Loosening often occurs because of wood shrinkage. (If shrinkage is expected, an obviously wise maintenance step is to tighten the connectors after wood shrinkage has occurred.)

9. Splits in the wood at bolted connections.

10. Water stains on the wood, or evidence of chemical deterioration.

11. Ends of beams that rest on masonry or concrete walls and are totally enclosed to prevent adequate ventilation. Rotting may be detected at such locations.

12. Moist wood at boxed-in connection details. (Fig. 9-10e, for example, especially if the weep holes are missing or blocked.)

13. Decay at the ends of posts or columns that rest directly upon concrete floors.

14. Longitudinal separations at the glue line of glued laminated members. Separation in the glue itself, rather than the wood, could be serious and might be a warning of more serious failure.

The above and several others are discussed in greater detail in reference 11.

Repairs to correct structural distress range from simple maintenance to prevent further deterioration to complete replacement of the distressed component. Repair techniques are not discussed here, but are covered in some detail by reference 11.

REFERENCES

1. Hunt, G. M., and G. A. Garratt, *Wood Preservation*, 3rd ed., McGraw-Hill, New York, 1967.

2. Wilcox, W. W., and A. F. Rosenberg, "Architectural and Construction Deficiencies Contributing to Decay of Wood in Buildings," in Meyer, R. W., and R. M. Kellogg, *Structural Uses of Wood in Adverse Environments*, Van Nostrand Reinhold, New York, 1982.

3. DeGroot, R. C., and G. R. Esenther, "Microbiological and Entomological Effects," in Meyer, R. W., and R. M. Kellogg, *Structural Uses of Wood in Adverse Environments*, Van Nostrand Reinhold, New York, 1982.

4. *Wood Structures, A Design Guide and Commentary*, American Society of Civil Engineers, New York, 1975.

5. Feist, W. C., "Weathering of Wood in Structural Uses," in Meyer, R. W., and R. M. Kellogg, *Structural Uses of Wood in Adverse Environments*, Van Nostrand Reinhold, New York, 1982.

6. *AWPA Standards*, American Wood Preservers' Association, Philadelphia, PA.

7. *Federal Specification—Wood Preservation Treating Practices*, Federal Specification TT-W-571, General Services Administration, Washington, DC.

8. American Institute of Timber Construction, *Timber Construction Manual*, 3rd ed., Wiley, New York, 1986.

9. *Uniform Building Code*, International Conference of Building Officials, Whittier, CA, 1985.

10. *National Design Specification for Wood Construction*, National Forest Products Association, Washington, DC, 1986.

11. *Evaluation, Maintenance and Upgrading of Wood Structures, A Guide and Commentary*, American Society of Civil Engineers, New York, 1982.

PROBLEMS

16-1. Number 2, 8 × 8 D. fir–larch timbers are specified to be pressure treated with creosote, with 10-pcf specified retention. What will be the probable weight per linear foot of the treated timbers?

16-2. Re-solve Problem 6-14, assuming that all members will have fire-retardant treatment.

16-3. From the standpoint of subjects of this chapter, tell what is wrong with each of the design details shown by Fig. 16-7. For each, show a better detail.

Fig. 16-7. Defective details for Problem 16-3.

16-4. A residential building used a nailed laminated floor system (such as shown by Fig. 15-21b for a bridge) of 20-ft span. In one location, the planks forming this nailed laminated floor were extended 5 ft to the outside to form a cantilever balcony. A membrane placed to protect the balcony structure developed cracks, the wood rotted, and the balcony collapsed. Explain what went wrong and suggest a better detail. (Could alternate laminations have been eliminated from the balcony without jeopardizing strength?)

Appendix A

Table A-1. Weights of Building Materials.

Materials	Weight (psf)	Materials	Weight (psf)
Ceilings		Partitions	
Channel suspended system	1	Clay Tile	
Lathing and plastering	See Partitions	3 in.	17
Acoustical fiber tile	1	4 in.	18
		6 in.	28
Floors		8 in.	34
Steel Deck	See Manufacturer	10 in.	40
Concrete–reinforced 1 in.		Gypsum block	
Stone	$12\frac{1}{2}$	2 in.	$9\frac{1}{2}$
Slag	$11\frac{1}{2}$	3 in.	$10\frac{1}{2}$
Lightweight	6 to 10	4 in.	$12\frac{1}{2}$
		5 in.	14
Concrete plain 1 in.		6 in.	$18\frac{1}{2}$
Stone	12	Wood studs 2 × 4	
Slag	11	12–16 in. o.c.	2
Lightweight	3 to 9	Steel partitions	4
		Plaster 1 in.	
Fills 1 in.		Cement	10
Gypsum	6	Gypsum	5
Sand	8	Lathing	
Cinders	4	Metal	$\frac{1}{2}$
		Gypsum board $\frac{1}{2}$ in.	2
Finishes			
Terrazzo 1 in.	13	Walls	
Ceramic or quarry tile $\frac{3}{4}$ in.	10	Brick	
		4 in.	40
Linoleum $\frac{1}{4}$ in.	1	8 in.	80
Mastic $\frac{3}{4}$ in.	9	12 in.	120
Hardwood $\frac{7}{8}$ in.	4	Hollow concrete block (heavy	
Softwood $\frac{3}{4}$ in.	$2\frac{1}{2}$	aggregate)	
		4 in.	30
Roofs		6 in.	43
Copper or tin	1	8 in.	55
3-ply ready roofing	1	$12\frac{1}{2}$ in.	80
3-ply felt and gravel	$5\frac{1}{2}$	Hollow concrete block (light	
5-ply felt and gravel	6	aggregate)	
		4 in.	21
Shingles		6 in.	30
Wood	2	8 in.	38
Asphalt	3	12 in.	55
Clay tile	9 to 14	Clay tile (load bearing)	
Slate $\frac{1}{4}$	10	4 in.	25
		6 in.	30

Table A-1. (*Continued*)

Materials	Weight (psf)	Materials	Weight (psf)
Sheathing		8 in.	33
Wood $\frac{3}{4}$ in.	3	12 in.	45
Gypsum 1 in.	4	Stone 4 in.	55
		Glass block 4 in.	18
Insulation 1 in.		Windows, glass, frame & sash	8
Loose	$\frac{1}{2}$	Curtain walls	See Manufacturer
Poured in place	2	Structural glass 1 in.	15
Rigid	$1\frac{1}{2}$	Corrugated cement asbestos $\frac{1}{4}$ in.	3

Source: Courtesy of American Institute of Steel Construction.

Table A-2   (UBC Table 23-A). Uniform and Concentrated Loads.

Use or Occupancy		Uniform Load[1]	Concentrated Load
Category	Description		
1. Access floor systems	Office Use	50	2000[2]
	Computer Use	100	2000[2]
2. Armories		150	0
3. Assembly areas[3] and auditoriums and balconies therewith	Fixed seating areas	50	0
	Movable seating and other areas	100	0
	Stage areas and enclosed platforms	125	0
4. Cornices, marquees, and residential balconies		60	0
5. Exit facilities[4]		100	0[5]
6. Garages	General storage and/or repair	100	6
	Private or pleasure-type motor vehicle storage	50	6
7. Hospitals	Wards and rooms	40	1000[2]
8. Libraries	Reading rooms	60	1000[2]
	Stack rooms	125	1500[2]
9. Manufacturing	Light	75	2000[2]
	Heavy	125	3000[2]
10. Offices		50	2000[2]
11. Printing plants	Press rooms	150	2500[2]
	Composing and linotype rooms	100	2000[2]
12. Residential[7]		40	0[5]
13. Rest rooms[8]			
14. Reviewing stands, grandstands, and bleachers		100	0
15. Roof deck	Same as area served or for the type of occupancy accommodated		
16. Schools	Classrooms	40	1000[2]
17. Sidewalks and driveways	Public access	250	6
18. Storage	Light	125	
	Heavy	250	
19. Stores	Retail	75	2000[2]
	Wholesale	100	3000[2]

[1]See Section 2306 for live-load reductions.
[2]See Section 2304 (c), first paragraph, for area of load application.
[3]Assembly areas include such occupancies as dance halls, drill rooms, gymnasiums, playgrounds, plazas, terraces, and similar occupancies which are generally accessible to the public.
[4]Exit facilities shall include such uses as corridors serving an occupant load of 10 or more persons, exterior exit balconies, stairways, fire escapes, and similar uses.
[5]Individual stair treads shall be designed to support a 300-lb concentrated load placed in a position which would cause maximum stress. Stair stringers may be designed for the uniform load set forth in the table.
[6]See Section 2304(c), second paragraph, for concentrated loads.
[7]Residential occupancies include private dwellings, apartments, and hotel guest rooms.
[8]Rest room loads shall be not less than the load for the occupancy with which they are associated, but need not exceed 50 psf.
Source: Reproduced from the 1985 edition of the Uniform Building Code, copyright c 1985, with permission of the publishers, the International Conference of Building Officials.

Table A-3 (UBC Table 23-C).   Minimum Roof Live Loads,[1] (psf)

	Method 1			Method 2		
	Tributary Loaded Area in Square Feet for any Structural Member			Uniform Load[2]	Rate of Reduction r (%)	Maximum Reduction R (%)
Roof Slope	0 to 200	201 to 600	Over 600			
1. Flat or rise less than 4 in./ft. Arch or dome with rise less than one-eighth of span	20	16	12	20	.08	40
2. Rise 4 in/ft to less than 12 in./ft. Arch or dome with rise one-eighth of span to less than three-eighths of span	16	14	12	16	.06	25
3. Rise 12 in./ft and greater. Arch or dome with rise three-eighths of span or greater	12	12	12	12	No reductions permitted	
4. Awnings except cloth covered[3]	5	5	5	5	No reductions permitted	
5. Greenhouses, lath houses, and agricultural buildings[4]	10	10	10	10	No reductions permitted	

[1]Where snow loads occur, the roof structure shall be designed for such loads as determined by the building official. See Section 2305 (d). For special purpose roofs, see Section 2305 (e).
[2]See Section 2306 for live load reductions. The rate of reduction r in Section 2306 Formula (6-1) shall be as indicated in the table. The maximum reduction R shall not exceed the value indicated in the table.
[3]As defined in Section 4506.
[4]See Section 2305 (e) for concentrated load requirements for greenhouse roof members.
Source: Reproduced from the 1985 edition of the Uniform Building Code, copyright c 1985, with permission of the publishers, the International Conference of Building Officials.

Table A-4 (UBC Table 23G).
Combined Height, Exposure and Gust Factor Coefficient, C_e.

Height Above Average Level of Adjoining Ground (ft)	Exposure C	Exposure B
0–20	1.2	0.7
20–40	1.3	0.8
40–60	1.5	1.0
60–100	1.6	1.1
100–150	1.8	1.3
150–200	1.9	1.4
200–300	2.1	1.6
300–400	2.2	1.8

Source: Reproduced from the 1985 edition of the Uniform Building Code, copyright c 1985, with permission of the publishers, the International Conference of Building Officials.

Table A-5 (UBC Table 23-H).   Pressure Coefficients, C_q.

Structure or part thereof	Description	C_q Factor
1. Primary frames and systems	**Method 1** (Normal Force Method)	
	Windward wall	0.8 inward
	Leeward wall	0.5 outward
	Leeward roof or flat roof	0.7 outward
	Windward roof	
	Slope < 9 : 12	0.7 outward
	Slope 9 : 12 to 12 : 12	0.4 inward
	Slope > 12 : 12	0.7 inward
	Wind parallel to ridge	
	Enclosed structures	0.7 outward
	Open structures[1]	1.2 outward
	Method 2 (Projected Area Method)	
	On vertical projected area	
	Structures 40 ft or less in height	1.3 horiz. any direction
	Structures over 40 ft in height	1.4 horiz. any direction
	On horizontal projected area	
	Enclosed structure	0.7 upward
	Open structure[1]	1.2 upward
2. Elements and components	Wall elements	
	All structures	1.2 inward
	Enclosed structures	1.1 outward
	Open structures	1.6 outward
	Parapets	1.3 inward or outward
	Roof elements	
	Enclosed structures	
	Slope < 9 : 12	1.1 outward
	Slope 9 : 12 to 12 : 12	1.1 outward or 0.8 inward
	Slope > 12 : 12	1.1 outward or inward
	Open structures[1]	
	Slope < 9 : 12	1.6 outward
	Slope 9 : 12 to 12 : 12	1.6 outward or 0.8 inward
	Slope > 12 : 12	1.6 outward or 1.1 inward
3. Local areas at discontinuities[2]	Wall corners	2.0 outward
	Canopies or overhangs at eaves or rakes	2.8 upward
	Roof ridges at ends of buildings or eaves and roof edges at building corners	3.0 upward
	Eaves or rakes without overhangs away from building corners and ridges away from ends of building	2.0 upward
	Cladding connections	
	Add 0.5 to outward or upward	
	C_q for appropriate location	
4. Chimneys, tanks, and solid towers	Square or rectangular	1.4 any direction
	Hexagonal or octagonal	1.1 any direction
	Round or elliptical	0.8 any direction
5. Open-frame towers[3]		
Rectangular		2.0 any direction
Triangular		1.8 any direction
6. Signs, flagpoles, lightpoles, minor structures		1.4 any direction

[1]A structure with more than 30% of any one side open shall be considered an open structure. Nonimpact-resistant glazing shall be considered as an opening.

[2]Local pressures shall apply over a distance from the discontinuity of 10 ft or 0.1 times the least width of the structure, whichever is smaller.

[3]The design wind forces shall be calculated based on the area of all exposed members and elements projected on a vertical plane normal to the wind direction. The forces shall be assumed to act parallel to the wind direction.

Source: Reproduced from the 1985 edition of the Uniform Building Code, copyright c 1985, with permission of the publishers, the International Conference of Building Officials.

Table A-6 (UBC Table 23-F).   Wind Stagnation Pressure, q_s, at Standard Height of 30 Feet.

Basic wind speed (mph)[1]	70	80	90	100	110	120	130
Pressure q_s (psf)	13	17	21	26	31	37	44

[1]Wind speed from Section 2311 (b).
Source: Reproduced from the 1985 edition of the Uniform Building Code, copyright c 1985, with permission of the publishers, the International Conference of Building Officials.

Table A-7 (UBC Table 23-I).   Horizontal Seismic Force Factor, K, for Buildings or Other Structures.[1]

Type or Arrangement of Resisting Elements	Value[2] of K
1. All building framing systems except as hereinafter classified	1.00
2. Buildings with a box system as specified in Section 2312 (b) EXCEPTION: Buildings not more than three stories in height with stud wall framing and using plywood horizontal diaphragms and plywood vertical shear panels for the lateral force system may use $K = 1.0$.	1.33
3. Buildings with a dual bracing system consisting of a ductile moment-resisting space frame and shear walls or braced frames using the following design criteria: a. The frames and shear walls or braced frames shall resist the total lateral force in accordance with their relative rigidities considering the interaction of the shear walls and frames b. The shear walls or braced frames acting independently of the ductile moment-resisting portions of the space frame shall resist the total required lateral forces c. The ductile moment-resisting space frame shall have the capacity to resist not less than 25% of the required lateral force	0.80
4. Buildings with a ductile moment-resisting space frame designed in accordance with the following criteria: The ductile moment-resisting space frame shall have the capacity to resist the total required lateral force	0.67
5. Elevated tanks plus full contents, on four or more cross-braced legs and not supported by a building	2.5[3]
6. Structures other than buildings and other than those set forth in Table No. 23-J	2.00

[1]Where wind load as specified in Section 2311 would produce higher stresses, this load shall be used in lieu of the loads resulting from earthquake forces.
[2]See Figures Nos. 1, 2 and 3 in this chapter and definition of Z as specified in Section 2312 (c).
[3]The minimum value of KC shall be 0.12 and the maximum value of KC need not exceed 0.25.
   The tower shall be designed for an accidental torsion of 5 percent as specified in Section 2312 (e) 4. Elevated tanks which are supported by buildings or do not conform to type or arrangement of supporting elements as described above shall be designed in accordance with section 2312 (g) using $C_p = .3$.
Source: Reproduced from the 1985 edition of the Uniform Building Code, copyright c 1985, with permission of the publishers, the International Conference of Building Officials.

Table A-8 (UBC Table 23-J).   Horizontal Seismic Force Factor, C_p, for Elements of Structures and Nonstructural Components.

Part or Portion of Buildings	Direction of Horizontal Force	Value of C_p[1]
1. Exterior bearing and nonbearing walls, interior bearing walls and partitions, interior nonbearing walls and partitions—see also Section 2312 (j) 3 C. Masonry or concrete fences over 6 ft high	Normal to flat surface	0.3[6]
2. Cantilever elements:	Normal to flat surfaces	
a. Parapets		0.8
b. Chimneys or stacks	Any direction	0.8
3. Exterior and interior ornamentations and appendages	Any direction	0.8
4. When connected to, part of, or housed within a building:		
a. Penthouse, anchorage and supports for chimneys, stacks and tanks, including contents[3]		0.3
b. Storage racks with upper storage level at more than 8 ft in height, plus contents[2]	Any direction	0.3
c. All equipment or machinery[3]		0.3
d. Fire sprinkler system		0.3
e. Supports and bracing, equipment racks and piping for hazardous production material		0.45
5. Suspended ceiling framing systems (applies to Seismic-Zones Nos. 2, 3, and 4 only)—see also Section 4701 (e)	Any direction	0.3[4,7]
6. Connections for prefabricated structural elements other than walls, with force applied at center of gravity of assembly	Any direction	0.3[5]
7. Access floor systems	Any direction	0.3[8]

[1]C_p for elements laterally self-supported only at the ground level may be two-thirds of value shown.

[2]W_p for storage racks shall be the weight of the racks plus contents. The value of C_p for racks over two storage support levels in height shall be 0.24 for the levels below the top two levels. In lieu of the tabulated values, steel storage racks may be designed in accordance with UBC Standard No. 27-11.

   Where a number of storage rack units are interconnected so there are a minimum of four vertical elements in each direction on each column line designed to resist horizontal forces, the design coefficients may be as for a building with K values from Table No. 23-I, CS = 0.2 for use in the formula $V = ZIKCSW$ and W equal to the total dead load plus 50% of the rack-rated capacity. Where the design and rack configurations are in accordance with this paragraph, the design provisions in UBC Standard No. 27-11 do not apply.

[3]For flexible and flexibly mounted equipment and machinery, the appropriate values of C_p shall be determined with consideration given to both the dynamic properties of the equipment and machinery and to the building or structure in which it is placed but shall be not less than the listed values. The design of the equipment and machinery and their anchorage is an integral part of the design and specification of such equipment and machinery.

   For essential facilities and life safety systems, the design and detailing of equipment which must remain in place and be functional following a major earthquake shall consider drifts in accordance with Section 2312 (k).

[4]Ceiling weight shall include all light fixtures and other equipment which is laterally supported by the ceiling. For purposes of determining the lateral force, a ceiling weight of not less than 4 psf shall be used.

[5]The force shall be resisted by positive anchorage and not by friction.

[6]See also Section 2309 (b) for minimum load and deflection criteria for interior partitions.

[7]Does not apply to ceilings constructed of lath and plaster or gypsum board screw or nail attached to suspended members that support a ceiling at one level extending from wall to wall.

[8]W_p for access floor systems shall be the dead load of the access floor system plus 25% of the floor live load and a 10 psf partition load.

Source: Reproduced from the 1985 edition of the Uniform Building Code, copyright c 1985, with permission of the publishers, the International Conference of Building Officials.

Table A-9.   Maximum Beam Bending Moments, Shears, and Deflections.

	Maximum Moment	Maximum Shear	Maximum Deflection
	$0.125w\ell^2$	$0.5w\ell$	$5w\ell^4/(384EI)$
	$0.125w\ell^2$	$0.625w\ell$	$w\ell^4/(185EI)$
	$0.1w\ell^2$	$0.6w\ell$	$w\ell^4/(145EI)$
	$0.107w\ell^2$	$0.607w\ell$	$w\ell^4/(154EI)$

Appendix B

Table B-1. Properties of Sawn-Timber Sections.

Nominal Size $b \times d$ (in.)	Standard Dressed Size (S4S) $b \times d$ (in.)	Area of Section A (in.²)	X-X Axis Moment of Inertia I (in.⁴)	S_x (in.³)	Effective $S_x = C_F \times S$ (in.³)	Y-Y Axis Moment of Inertia I (in.⁴)	S_y (in.³)	Effective $S_y = C_F \times S$ (in.³)	Board Measure per Lineal Foot	Weight (lb/ft) 30 pcf	40 pcf	50 pcf
1 × 3	$\frac{3}{4} \times 2\frac{1}{2}$	1.875	0.977	0.78	n.a.	0.088	0.234	n.a.	$\frac{1}{4}$	0.391	0.521	0.651
1 × 4	$\frac{3}{4} \times 3\frac{1}{2}$	2.625	2.680	1.53	n.a.	0.123	0.328	n.a.	$\frac{1}{3}$	0.547	0.729	0.911
1 × 6	$\frac{3}{4} \times 5\frac{1}{2}$	4.125	10.40	3.78	n.a.	0.193	0.516	n.a.	$\frac{1}{2}$	0.859	1.146	1.432
1 × 8	$\frac{3}{4} \times 7\frac{1}{4}$	5.438	23.82	6.57	n.a.	0.255	0.680	n.a.	$\frac{2}{3}$	1.133	1.510	1.888
1 × 10	$\frac{3}{4} \times 9\frac{1}{4}$	6.938	49.47	10.70	n.a.	0.325	0.867	n.a.	$\frac{5}{6}$	1.445	1.927	2.409
1 × 12	$\frac{3}{4} \times 11\frac{1}{4}$	8.438	88.99	15.82	n.a.	0.396	1.055	n.a.	1	1.758	2.344	2.930
2 × 3	$1\frac{1}{2} \times 2\frac{1}{2}$	3.750	1.953	1.56	n.a.	0.703	0.938	n.a.	$\frac{1}{2}$	0.781	1.042	1.302
2 × 4	$1\frac{1}{2} \times 3\frac{1}{2}$	5.250	5.359	3.06	n.a.	0.984	1.313	n.a.	$\frac{2}{3}$	1.094	1.458	1.823
2 × 6	$1\frac{1}{2} \times 5\frac{1}{2}$	8.250	20.80	7.56	n.a.	1.547	2.063	n.a.	1	1.719	2.292	2.865
2 × 8	$1\frac{1}{2} \times 7\frac{1}{4}$	10.88	47.64	13.14	n.a.	2.039	2.719	n.a.	$1\frac{1}{3}$	2.266	3.021	3.776
2 × 10	$1\frac{1}{2} \times 9\frac{1}{4}$	13.88	98.93	21.39	n.a.	2.602	3.469	n.a.	$1\frac{2}{3}$	2.891	3.854	4.818
2 × 12	$1\frac{1}{2} \times 11\frac{1}{4}$	16.88	178.0	31.64	n.a.	3.164	4.219	n.a.	2	3.516	4.688	5.859
2 × 14	$1\frac{1}{2} \times 13\frac{1}{4}$	19.88	290.8	43.89	n.a.	3.727	4.969	n.a.	$2\frac{1}{3}$	4.141	5.521	6.901
3 × 4	$2\frac{1}{2} \times 3\frac{1}{2}$	8.750	8.932	5.10	n.a.	4.557	3.646	n.a.	1	1.823	2.431	3.038
3 × 6	$2\frac{1}{2} \times 5\frac{1}{2}$	13.75	34.66	12.60	n.a.	7.161	5.729	n.a.	$1\frac{1}{2}$	2.865	3.819	4.774
3 × 8	$2\frac{1}{2} \times 7\frac{1}{4}$	18.12	79.39	21.90	n.a.	9.440	7.552	n.a.	2	3.776	5.035	6.293
3 × 10	$2\frac{1}{2} \times 9\frac{1}{4}$	23.12	164.9	35.65	n.a.	12.04	9.635	n.a.	$2\frac{1}{2}$	4.818	6.424	8.030
3 × 12	$2\frac{1}{2} \times 11\frac{1}{4}$	28.12	296.6	52.73	n.a.	14.65	11.72	n.a.	3	5.859	7.813	9.766
3 × 14	$2\frac{1}{2} \times 13\frac{1}{4}$	33.12	484.6	73.15	n.a.	17.25	13.80	n.a.	$3\frac{1}{2}$	6.901	9.201	11.50
3 × 16	$2\frac{1}{2} \times 15\frac{1}{4}$	38.12	738.9	96.90	n.a.	19.86	15.88	n.a.	4	7.943	10.590	13.24
4 × 4	$3\frac{1}{2} \times 3\frac{1}{2}$	12.25	12.50	7.15	n.a.	12.50	7.15	n.a.	$1\frac{1}{3}$	2.552	3.403	4.253
4 × 6	$3\frac{1}{2} \times 5\frac{1}{2}$	19.25	48.53	17.65	n.a.	19.65	11.23	n.a.	2	4.010	5.347	6.684
4 × 8	$3\frac{1}{2} \times 7\frac{1}{4}$	25.38	111.1	30.66	n.a.	25.90	14.80	n.a.	$2\frac{2}{3}$	5.286	7.049	8.811
4 × 10	$3\frac{1}{2} \times 9\frac{1}{4}$	32.38	230.8	49.91	n.a.	33.05	18.88	n.a.	$3\frac{1}{3}$	6.745	8.933	11.24
4 × 12	$3\frac{1}{2} \times 11\frac{1}{4}$	39.38	415.3	73.83	n.a.	40.20	22.97	n.a.	4	8.203	10.94	13.67
4 × 14	$3\frac{1}{2} \times 13\frac{1}{4}$	46.38	678.5	102.4	n.a.	47.34	27.05	n.a.	$4\frac{2}{3}$	9.657	12.88	16.09
4 × 16	$3\frac{1}{2} \times 15\frac{1}{4}$	53.38	1,034	135.7	n.a.	54.49	31.14	n.a.	$5\frac{1}{3}$	11.12	14.83	18.54
6 × 6	$5\frac{1}{2} \times 5\frac{1}{2}$	30.25	76.26	27.73	n.a.	76.26	27.73	n.a.	3	6.302	8.403	10.50
6 × 8	$5\frac{1}{2} \times 7\frac{1}{2}$	41.25	193.4	51.56	n.a.	104.0	37.81	n.a.	4	8.594	11.46	14.32
6 × 10	$5\frac{1}{2} \times 9\frac{1}{2}$	52.25	393.0	82.73	n.a.	131.7	47.90	n.a.	5	10.88	14.51	18.14
6 × 12	$5\frac{1}{2} \times 11\frac{1}{2}$	63.25	697.1	121.3	n.a.	159.4	57.98	n.a.	6	13.18	17.57	21.96
6 × 14	$5\frac{1}{2} \times 13\frac{1}{2}$	74.25	1,128	167.1	164.9	187.2	68.06	n.a.	7	15.47	20.62	25.78
6 × 16	$5\frac{1}{2} \times 15\frac{1}{2}$	85.25	1,707	220.2	214.1	214.9	78.15	n.a.	8	17.76	23.68	29.60
6 × 18	$5\frac{1}{2} \times 17\frac{1}{2}$	96.25	2,456	280.7	269.2	242.6	88.23	n.a.	9	20.05	26.74	33.42
6 × 20	$5\frac{1}{2} \times 19\frac{1}{2}$	107.2	3,398	348.6	330.3	270.4	98.31	n.a.	10	22.34	29.79	37.24
6 × 22	$5\frac{1}{2} \times 21\frac{1}{2}$	118.2	4,555	423.7	397.1	298.1	103.4	n.a.	11	24.64	32.85	41.06
6 × 24	$5\frac{1}{2} \times 23\frac{1}{2}$	129.2	5,948	506.2	469.8	325.8	118.5	n.a.	12	26.93	35.90	44.88
8 × 8	$7\frac{1}{2} \times 7\frac{1}{2}$	56.25	263.7	70.31	n.a.	263.7	70.31	n.a.	$5\frac{1}{3}$	11.72	15.62	19.53
8 × 10	$7\frac{1}{2} \times 9\frac{1}{2}$	71.25	535.9	112.8	n.a.	334.0	89.06	n.a.	$6\frac{2}{3}$	14.84	19.79	24.74

Table B-1. (Continued)

Nominal Size b × d (in.)	Standard Dressed Size (S4S) b × d (in.)	Area of Section A (in.²)	X-X Axis Moment of Inertia I (in.⁴)	S_x (in.³)	Effective $S_x = C_F \times S$ (in.³)	Y-Y Axis Moment of Inertia I (in.⁴)	S_y (in.³)	Effective $S_y = C_F \times S$ (in.³)	Board Measure per Lineal Foot	Weight (lb/ft) 30 pcf	40 pcf	50 pcf
8 × 12	$7\frac{1}{2} \times 11\frac{1}{2}$	86.25	950.5	165.3	n.a.	404.3	107.8	n.a.	8	17.97	23.96	29.95
8 × 14	$7\frac{1}{2} \times 13\frac{1}{2}$	101.2	1,538	227.8	224.9	474.6	126.6	n.a.	$9\frac{1}{3}$	21.09	28.12	35.16
8 × 16	$7\frac{1}{2} \times 15\frac{1}{2}$	116.2	2,327	300.3	291.9	544.9	145.3	n.a.	$10\frac{2}{3}$	24.22	32.29	40.36
8 × 18	$7\frac{1}{2} \times 17\frac{1}{2}$	131.2	3,350	382.8	367.1	615.2	164.1	n.a.	12	27.34	36.46	45.57
8 × 20	$7\frac{1}{2} \times 19\frac{1}{2}$	146.2	4,634	475.3	450.4	685.5	182.8	n.a.	$13\frac{1}{3}$	30.47	40.62	50.78
8 × 22	$7\frac{1}{2} \times 21\frac{1}{2}$	161.2	6,211	577.8	541.6	755.9	201.6	n.a.	$14\frac{2}{3}$	33.59	44.79	55.99
8 × 24	$7\frac{1}{2} \times 23\frac{1}{2}$	176.2	8,111	690.3	640.6	826.2	220.3	n.a.	16	36.72	48.96	61.20
10 × 10	$9\frac{1}{2} \times 9\frac{1}{2}$	90.25	678.8	142.9	n.a.	678.8	142.9	n.a.	$8\frac{1}{3}$	18.80	25.07	31.34
10 × 12	$9\frac{1}{2} \times 11\frac{1}{2}$	109.2	1,204	209.4	n.a.	821.7	173.0	n.a.	10	22.76	30.35	37.93
10 × 14	$9\frac{1}{2} \times 13\frac{1}{2}$	128.2	1,948	288.6	284.8	964.6	203.1	n.a.	$11\frac{2}{3}$	26.72	35.62	44.53
10 × 16	$9\frac{1}{2} \times 15\frac{1}{2}$	147.2	2,948	380.4	369.7	1,107	233.1	n.a.	$13\frac{1}{3}$	30.68	40.90	51.13
10 × 18	$9\frac{1}{2} \times 17\frac{1}{2}$	166.2	4,243	484.9	465.0	1,250	263.2	n.a.	15	34.64	46.18	57.73
10 × 20	$9\frac{1}{2} \times 19\frac{1}{2}$	185.2	5,870	602.1	570.4	1,393	293.3	n.a.	$16\frac{2}{3}$	38.59	51.46	64.32
10 × 22	$9\frac{1}{2} \times 21\frac{1}{2}$	204.2	7,868	731.9	686.0	1,536	323.4	n.a.	$18\frac{1}{3}$	42.55	56.74	70.92
10 × 24	$9\frac{1}{2} \times 23\frac{1}{2}$	223.2	10,274	874.4	811.5	1,679	353.5	n.a.	20	46.51	62.01	77.52
12 × 12	$11\frac{1}{2} \times 11\frac{1}{2}$	132.2	1,458	253.5	n.a.	1,458	253.5	n.a.	12	27.55	36.74	45.92
12 × 14	$11\frac{1}{2} \times 13\frac{1}{2}$	155.2	2,358	349.3	344.8	1,711	297.6	n.a.	14	32.34	43.12	53.91
12 × 16	$11\frac{1}{2} \times 15\frac{1}{2}$	178.2	3,569	460.5	447.6	1,964	341.6	n.a.	16	37.14	49.51	61.89
12 × 18	$11\frac{1}{2} \times 17\frac{1}{2}$	201.2	5,136	587.0	562.9	2,218	385.7	n.a.	18	41.93	55.90	69.88
12 × 20	$11\frac{1}{2} \times 19\frac{1}{2}$	224.2	7,106	728.8	690.5	2,471	429.8	n.a.	20	46.72	62.29	77.86
12 × 22	$11\frac{1}{2} \times 21\frac{1}{2}$	247.2	9,524	886.0	830.4	2,725	473.9	n.a.	22	51.51	68.68	85.85
12 × 24	$11\frac{1}{2} \times 23\frac{1}{2}$	270.2	12,437	1,058	982.3	2,978	518.0	n.a.	24	56.30	75.07	93.84
14 × 16	$13\frac{1}{2} \times 15\frac{1}{2}$	209.2	4,189	540.6	525.4	3,178	470.8	464.7	$18\frac{2}{3}$	43.59	58.12	72.66
14 × 18	$13\frac{1}{2} \times 17\frac{1}{2}$	236.2	6,029	689.1	660.8	3,588	531.6	524.7	21	49.22	65.62	82.03
14 × 20	$13\frac{1}{2} \times 19\frac{1}{2}$	263.2	8,342	855.6	810.6	3,998	592.3	584.6	$23\frac{1}{3}$	58.84	73.12	91.41
14 × 22	$13\frac{1}{2} \times 21\frac{1}{2}$	290.2	11,181	1,040	974.8	4,408	653.1	644.6	$25\frac{2}{3}$	60.47	80.62	100.8
14 × 24	$13\frac{1}{2} \times 23\frac{1}{2}$	317.2	14,600	1,243	1,153	4,818	713.8	704.5	28	66.09	88.12	110.2

Table B-2. Properties of Glued Laminated Timber Sections.

For glulams with $1\frac{1}{2}''$ laminations only. Units are: section modulus, S, and effective section modulus, SC_F, in.³; moment of inertia, I, in.⁴.

Glulam Width			$2\frac{1}{8}''$			$3\frac{1}{8}''$			$5\frac{1}{8}''$		
Depth (in.)	No. of Lams	Size Factor C_F	S	SC_F	I	S	SC_F	I	S	SC_F	I
3	2	1.00	3.2	3.2	4.8	4.7	4.7	7.0			
$4\frac{1}{2}$	3	1.00	7.2	7.2	16.1	10.5	10.5	23.7	17.3	17.3	38.9
6	4	1.00	12.8	12.8	38.2	18.8	18.8	56.2	30.8	30.8	92.2
$7\frac{1}{2}$	5	1.00	19.9	19.9	74.7	29.3	29.3	110	48.0	48.0	180
9	6	1.00	28.7	28.7	129	42.2	42.2	190	69.2	69.2	311
$10\frac{1}{2}$	7	1.00	39.0	39.0	205	57.4	57.4	301	94.2	94.2	494
12	8	1.00	51.0	51.0	306	75.0	75.0	450	123	123	738
$13\frac{1}{2}$	9	0.99	64.6	63.7	436	94.9	93.7	641	156	154	1,051
15	10	0.98	79.7	77.7	598	117	114	879	192	188	1,441
$16\frac{1}{2}$	11	0.97				142	137	1,170	232	224	1,919
18	12	0.96				169	161	1,519	277	265	2,491
$19\frac{1}{2}$	13	0.95				198	188	1,931	325	308	3,167
21	14	0.94				230	216	2,412	377	354	3,955
$22\frac{1}{2}$	15	0.93				264	246	2,966	432	403	4,865

Table B-2.   (*Continued*)

Glulam Width			$2\frac{1}{8}''$			$3\frac{1}{8}''$			$5\frac{1}{8}''$		
Depth (in.)	No. of Lams	Size Factor C_F	S	SC_F	I	S	SC_F	I	S	SC_F	I
24	16	0.93				300	278	3,600	492	456	5,904
$25\frac{1}{2}$	17	0.92				339	311	4,318	555	511	7,082
27	18	0.91				380	347	5,126	623	569	8,406
$28\frac{1}{2}$	19	0.91				423	384	6,028	694	630	9,887
30	20	0.90				469	423	7,031	769	694	11,530
$31\frac{1}{2}$	21	0.90							848	761	13,350
33	22	0.89							930	831	15,350
$34\frac{1}{2}$	23	0.89							1,017	904	17,540
36	24	0.89							1,107	980	19,930

Glulam Width			$6\frac{3}{4}''$			$8\frac{3}{4}''$			$10\frac{3}{4}''$		
Depth (in.)	No. of Lams	Size Factor C_F	S	SC_F	I	S	SC_F	I	S	SC_F	I
6	4	1.00	40.5	40.5	122						
$7\frac{1}{2}$	5	1.00	63.3	63.3	237						
9	6	1.00	91.1	91.1	410	118	118	532			
$10\frac{1}{2}$	7	1.00	124	124	651	161	161	844	198	198	1,037
12	8	1.00	162	162	972	210	210	1,260	258	258	1,548
$13\frac{1}{2}$	9	0.99	205	202	1,384	266	262	1,794	327	322	2,204
15	10	0.98	253	247	1,898	328	320	2,461	403	393	3,023
$16\frac{1}{2}$	11	0.97	306	296	2,527	397	383	3,276	488	471	4,024
18	12	0.96	364	348	3,280	472	452	4,252	580	555	5,224
$19\frac{1}{2}$	13	0.95	428	405	4,171	554	525	5,407	681	646	6,642
21	14	0.94	496	466	5,209	643	604	6,753	790	742	8,296
$22\frac{1}{2}$	15	0.93	570	531	6,407	738	688	8,306	907	846	10,200
24	16	0.93	648	600	7,776	840	778	10,080	1,032	956	12,380
$25\frac{1}{2}$	17	0.92	732	673	9,327	948	872	12,090	1,165	1,071	14,850
27	18	0.91	820	749	11,070	1,063	972	14,350	1,306	1,194	17,630
$28\frac{1}{2}$	19	0.91	914	830	13,020	1,185	1,076	16,880	1,455	1,322	20,740
30	20	0.90	1,012	914	15,190	1,312	1,185	19,690	1,612	1,456	24,190
$31\frac{1}{2}$	21	0.90	1,116	1,003	17,580	1,447	1,300	22,790	1,778	1,597	28,000
33	22	0.89	1,225	1,095	20,210	1,588	1,419	26,200	1,951	1,744	32,190
$34\frac{1}{2}$	23	0.89	1,339	1,191	23,100	1,736	1,544	29,940	2,132	1,896	36,790
36	24	0.89	1,458	1,290	26,240	1,890	1,673	34,020	2,322	2,055	41,800
$37\frac{1}{2}$	25	0.88	1,582	1,394	29,660	2,051	1,807	38,450	2,520	2,220	47,240
39	26	0.88	1,711	1,501	33,370	2,218	1,946	43,250	2,725	2,391	53,140
$40\frac{1}{2}$	27	0.87	1,845	1,612	37,370	2,392	2,090	48,440	2,939	2,567	59,510
42	28	0.87	1,984	1,727	41,670	2,572	2,238	54,020	3,160	2,750	66,370
$43\frac{1}{2}$	29	0.87	2,129	1,845	46,300	2,760	2,392	60,020	3,390	2,938	73,740
45	30	0.86	2,278	1,967	51,260	2,953	2,550	66,450	3,628	3,133	81,630
$46\frac{1}{2}$	31	0.86	2,433	2,093	56,560	3,153	2,713	73,310	3,874	3,333	90,070
48	32	0.86	2,592	2,222	62,200	3,360	2,880	80,640	4,128	3,539	99,070
$49\frac{1}{2}$	33	0.85				3,573	3,053	88,440	4,390	3,750	108,650
51	34	0.85				3,793	3,230	96,720	4,660	3,968	118,800
$52\frac{1}{2}$	35	0.85				4,020	3,412	105,500	4,938	4,191	129,600
54	36	0.85				4,252	3,598	114,800	5,224	4,420	141,100
$55\frac{1}{2}$	37	0.84				4,492	3,789	124,700	5,519	4,655	153,100
57	38	0.84				4,738	3,985	135,000	5,821	4,896	165,900
$58\frac{1}{2}$	39	0.84				4,991	4,185	146,000	6,132	5,142	179,300
60	40	0.84				5,250	4,390	157,500	6,450	5,394	193,500
$61\frac{1}{2}$	41	0.83				5,516	4,600	169,600	6,777	5,651	208,400
63	42	0.83				5,788	4,814	182,300	7,111	5,915	224,000

Table B-2. (Continued)

Glulam Width			$6\frac{3}{4}''$			$8\frac{3}{4}''$			$10\frac{3}{4}''$		
Depth (in.)	No. of Lams	Size Factor C_F	S	SC_F	I	S	SC_F	I	S	SC_F	I
$64\frac{1}{2}$	43	0.83							7,454	6,183	240,400
66	44	0.83							7,804	6,478	257,500
$67\frac{1}{2}$	45	0.83							8,163	6,738	275,500
69	46	0.82							8,530	7,023	294,300
$70\frac{1}{2}$	47	0.82							8,905	7,315	313,900
72	48	0.82							9,288	7,611	334,400
$73\frac{1}{2}$	49	0.82							9,679	7,914	355,700
75	50	0.82							10,078	8,221	377,900

Glulam Width			$12\frac{1}{4}''$			$14\frac{1}{4}''$		
Depth (in.)	No. of Lams	Size Factor C_F	S	SC_F	I	S	SC_F	I
12	8	1.00	294	294	1,764			
$13\frac{1}{2}$	9	0.99	372	367	2,512	433	427	2,922
15	10	0.98	459	448	3,445	534	521	4,008
$16\frac{1}{2}$	11	0.97	556	537	4,586	647	624	5,334
18	12	0.96	662	632	5,954	770	736	6,926
$19\frac{1}{2}$	13	0.95	776	736	7,569	903	856	8,805
21	14	0.94	900	846	9,454	1,047	984	11,000
$22\frac{1}{2}$	15	0.93	1,034	964	11,630	1,202	1,121	13,530
24	16	0.93	1,176	1,089	14,110	1,368	1,267	16,420
$25\frac{1}{2}$	17	0.92	1,328	1,221	16,930	1,544	1,420	19,690
27	18	0.91	1,488	1,360	20,090	1,731	1,582	23,370
$28\frac{1}{2}$	19	0.91	1,658	1,506	23,630	1,929	1,752	27,490
30	20	0.90	1,838	1,660	27,560	2,138	1,931	32,060
$31\frac{1}{2}$	21	0.90	2,026	1,820	31,910	2,357	2,117	37,120
33	22	0.89	2,223	1,987	36,690	2,586	2,311	42,680
$34\frac{1}{2}$	23	0.89	2,430	2,161	41,920	2,827	2,514	48,760
36	24	0.89	2,646	2,342	47,630	3,078	2,724	55,400
$37\frac{1}{2}$	25	0.88	2,871	2,530	53,830	3,340	2,943	62,620
39	26	0.88	3,105	2,724	60,550	3,612	3,169	70,440
$40\frac{1}{2}$	27	0.87	3,349	2,925	67,810	3,896	3,403	78,890
42	28	0.87	3,601	3,134	75,630	4,190	3,645	87,980
$43\frac{1}{2}$	29	0.87	3,863	3,348	84,030	4,494	3,895	97,750
45	30	0.86	4,134	3,570	93,020	4,809	4,152	108,200
$46\frac{1}{2}$	31	0.86	4,415	3,798	102,600	5,135	4,418	119,400
48	32	0.86	4,704	4,032	112,900	5,472	4,691	131,300
$49\frac{1}{2}$	33	0.85	5,003	4,274	123,800	5,819	4,972	144,000
51	34	0.85	5,310	4,522	135,400	6,177	5,260	157,500
$52\frac{1}{2}$	35	0.85	5,627	4,776	147,700	6,546	5,556	171,800
54	36	0.85	5,954	5,037	160,700	6,926	5,860	187,000
$55\frac{1}{2}$	37	0.84	6,289	5,305	174,500	7,316	6,171	203,000
57	38	0.84	6,633	5,579	189,100	7,716	6,490	219,900
$58\frac{1}{2}$	39	0.84	6,987	5,859	204,400	8,128	6,816	237,700
60	40	0.84	7,350	6,146	220,500	8,550	7,150	256,500
$61\frac{1}{2}$	41	0.83	7,722	6,440	237,500	8,983	7,491	276,200
63	42	0.83	8,103	6,740	255,300	9,426	7,840	296,900
$64\frac{1}{2}$	43	0.83	8,494	7,046	273,900	9,881	8,197	318,600
66	44	0.83	8,893	7,359	293,500	10,350	8,560	341,400
$67\frac{1}{2}$	45	0.83	9,302	7,678	314,000	10,820	8,931	365,200
69	46	0.82	9,720	8,003	335,400	11,310	9,310	390,100
$70\frac{1}{2}$	47	0.82	10,150	8,335	357,700	11,800	9,696	416,100

Table B-2.   (*Continued*)

Depth (in.)	No. of Lams	Size Factor C_F	$12\frac{1}{4}''$			$14\frac{1}{4}''$		
			S	SC_F	I	S	SC_F	I
72	48	0.82	10,580	8,673	381,000	12,310	10,090	443,200
$73\frac{1}{2}$	49	0.82	11,030	9,018	405,300	12,830	10,490	471,500
75	50	0.82	11,480	9,369	430,700	13,360	10,900	501,000
$76\frac{1}{2}$	51	0.81	11,950	9,726	457,000	13,900	11,310	531,600
78	52	0.81	12,420	10,090	484,400	14,450	11,740	563,500
$79\frac{1}{2}$	53	0.81	12,900	10,460	512,900	15,010	12,170	596,700
81	54	0.81	13,400	10,830	542,500	15,580	12,600	631,100
$82\frac{1}{2}$	55	0.81	13,900	11,220	573,200	16,160	13,050	666,800
84	56	0.81	14,410	11,600	605,100	16,760	13,500	703,800
$85\frac{1}{2}$	57	0.80				17,360	13,960	742,200
87	58	0.80				17,980	14,420	782,000
$88\frac{1}{2}$	59	0.80				18,600	14,900	823,100
90	60	0.80				19,240	15,380	865,700
$91\frac{1}{2}$	61	0.80				19,880	15,870	909,700
93	62	0.80				20,540	16,360	955,200
$94\frac{1}{2}$	63	0.80				21,210	16,860	1,002,000
96	64	0.79				21,890	17,370	1,051,000

Note: "Glulam Width" spans the Depth, No. of Lams, and Size Factor columns in the header.

Table B-3. Design Allowable Stresses and Modulus of Elasticity for Visually Graded Structural Lumber Under Normal-Duration Loading.

Species and Commercial Grade	Size Classification	Extreme Fiber in Bending F_b		Tension Parallel to Grain F_t	Horizontal Shear F_v	Compression perpendicular to Grain $F_c \perp^{21}$	Compression Parallel to Grain F_c	Modulus of Elasticity E^{21}	U.B.C. Stds Under Which Graded
		Single-member Uses	Repetitive-member Uses						
CALIFORNIA REDWOOD (Surfaced dry or surfaced green. Used at 19% max. m.c.)									
Clear Heart Structural	4" & less thick, any width	2300	2650	1500	145	650	2150	1,400,000	
Clear Structural		2300	2650	1500	145	650	2150	1,400,000	
Select Structural	2" to 4" thick	2050	2350	1200	80	650	1750	1,400,000	
Select Structural, Open grain	2" to 4" wide	1600	1850	950	80	425	1300	1,100,000	
No. 1		1700	1950	975	80	650	1400	1,400,000	
No. 1, Open grain	2" to 4" thick	1350	1550	775	80	425	1050	1,100,000	
No. 2		1400	1600	800	80	650	1100	1,250,000	
No. 2, Open grain	2" to 4" wide	1100	1250	625	80	425	825	1,000,000	
No. 3		800	900	475	80	650	675	1,100,000	
No. 3, Open grain		625	725	375	80	425	500	900,000	
Stud		625	725	375	80	425	500	900,000	
Construction	2" to 4" thick	825	950	475	80	425	925	900,000	
Standard		450	525	250	80	425	775	900,000	
Utility	4" wide	225	250	125	80	425	500	900,000	RIS 25-7 (See footnotes 2 through 7, 9, 13, 15 and 16)
Select Structural		1750	2000	1150	80	650	1550	1,400,000	
Select Structural, Open grain		1400	1600	925	80	425	1150	1,100,000	
No. 1	2" to 4" thick	1500	1700	975	80	650	1400	1,400,000	
No. 1, Open grain		1150	1350	775	80	425	1050	1,100,000	
No. 2	5" and wider	1200	1400	650	80	650	1200	1,250,000	
No. 2, Open grain		950	1100	500	80	425	875	1,000,000	
No. 3		700	800	375	80	650	725	1,100,000	
No. 3, Open grain		550	650	350	80	425	525	900,000	
Clear Heart Structural		1850	—	1250	135	650	1650	1,300,000	
Clear Structural		1850	—	1250	135	650	1650	1,300,000	
Select Structural	5" by 5" and larger	1400	—	950	95	650	1200	1,300,000	
No. 1		1200	—	800	95	650	1050	1,300,000	
No. 2		975	—	650	95	650	900	1,100,000	
No. 3		550	—	375	95	650	550	1,000,000	

Table B-3. (Continued)

Species and Commercial Grade	Size Classification	Extreme Fiber in Bending F_b Single-member Uses	Extreme Fiber in Bending F_b Repetitive-member Uses	Tension Parallel to Grain F_t	Horizontal Shear F_v	Compression perpendicular to Grain $F_c\perp^{21}$	Compression Parallel to Grain F_c	Modulus of Elasticity E^{21}	U.B.C. Stds Under Which Graded
Select Decking, Close grain	Decking 2" thick	1850	2150	—	—	—	—	1,400,000	25-7
Select Decking	thick	1450	1700	—	—	—	—	1,100,000	(See footnotes
Commercial Decking	6" & wider	1200	1350	—	—	—	—	1,000,000	2, 5 and 6)
DOUGLAS FIR-LARCH (Surfaced dry or surfaced green. Used at 19% max. m.c.)									
DOUGLAS FIR-LARCH (North)									
Dense Select Structural	2" to 4" thick	2450	2800	1400	95	730	1850	1,900,000	
Select Structural		2100	2400	1200	95	625	1600	1,800,000	
Dense No. 1		2050	2400	1200	95	730	1450	1,900,000	NLGA
No. 1	2" to 4" thick	1750	2050	1050	95	625	1250	1,800,000	WCLIB
Dense No. 2		1700	1950	1000	95	730	1150	1,700,000	WNPA
No. 2	2" to 4" wide	1450	1650	850	95	625	1000	1,700,000	25-2
No. 3		800	925	475	95	625	600	1,500,000	25-3
Appearance		1750	2050	1050	95	625	1500	1,800,000	and
Stud		800	925	475	95	625	600	1,500,000	25-4
Construction	2" to 4" thick	1050	1200	625	95	625	1150	1,500,000	(See footnotes
Standard		600	675	350	95	625	925	1,500,000	2 through 9,
Utility	4" wide	275	325	175	95	625	600	1,500,000	11, 13, 15 and
Dense Select Structural		2100	2400	1400	95	730	1650	1,900,000	16)
Select Structural		1800	2050	1200	95	625	1400	1,800,000	
No. 1	2" to 4" thick	1800	2050	1200	95	730	1450	1,900,000	
Dense No. 2	5" and wider	1500	1750	1000	95	625	1250	1,800,000	
No. 2		1450	1700	775	95	730	1250	1,700,000	
No. 3 and Stud		1250	1450	650	95	625	1050	1,700,000	
Appearance		725	850	375	95	625	675	1,500,000	
		1500	1750	1000	95	625	1500	1,800,000	
Dense Select Structural	Beams and Stringers[12]	1900	—	1100	85	730	1300	1,700,000	
Select Structural		1600	—	950	85	625	1100	1,600,000	WCLIB
Dense No. 1		1550	—	775	85	730	1100	1,700,000	25-3
No. 1		1300	—	675	85	625	925	1,600,000	
Dense Select Structural	Posts and Timbers[12]	1750	—	1150	85	730	1350	1,700,000	(See footnotes
Select Structural		1500	—	1000	85	625	1150	1,600,000	2 through 9)
Dense No. 1		1400	—	950	85	730	1200	1,700,000	
No. 1		1200	—	825	85	625	1000	1,600,000	
Select Dex	Decking	1750	2000	—	—	625	—	1,800,000	
Commercial Dex		1450	1650	—	—	625	—	1,700,000	

Allowable Unit Stresses in Pounds per Square Inch

Table B-3. (Continued)

Species and Commercial Grade	Size Classification	Extreme Fiber in Bending F_b Single-member Uses	Extreme Fiber in Bending F_b Repetitive-member Uses	Tension Parallel to Grain F_t	Horizontal Shear F_v	Compression perpendicular to Grain $F_c \perp^{21}$	Compression Parallel to Grain F_c	Modulus of Elasticity E^{21}	U.B.C. Stds Under Which Graded
				Allowable Unit Stresses in Pounds per Square Inch					
Dense Select Structural	Beams and Stringers[12]	1850	—	1100	85	730	1300	1,700,000	
Select Structural		1600	—	950	85	625	1100	1,600,000	
Dense No. 1		1550	—	775	85	730	1100	1,700,000	WWPA 25-4
No. 1		1350	—	675	85	625	925	1,600,000	(See footnotes 2 through 10)
Dense No. 2		1000	—	500	85	730	700	1,400,000	
No. 2		875	—	425	85	625	600	1,300,000	
Dense Select Structural	Posts and Timbers[12]	1750	—	1150	85	730	1350	1,700,000	
Select Structural		1500	—	1000	85	625	1150	1,600,000	
Dense No. 1		1400	—	950	85	730	1200	1,700,000	
No. 1		1200	—	825	85	625	1000	1,600,000	
Dense No. 2		800	—	550	85	730	550	1,400,000	
No. 2		700	—	475	85	625	475	1,300,000	
Selected Decking	Decking	—	2000	—	—	—	—	1,800,000	
Commercial Decking		—	1650	—	—	—	—	1,700,000	
Selected Decking	Decking	—	2150	—	—	—	—	1,900,000	
Commercial Decking		—	1800	(Surfaced at 15% max. m.c. and used at 15% max. m.c.)				1,700,000	
Select Structural	Beams and Stringers	1600	—	950	85	625	1100	1,600,000	NLGA 25-2
No. 1		1300	—	675	85	625	925	1,600,000	
Select Structural	Posts and Timbers	1500	—	1000	85	625	1150	1,600,000	(See footnotes 2 through 9 and 11)
No. 1		1200	—	825	85	625	1000	1,600,000	
Select	Decking	2000	1750	—	—	625	—	1,800,000	
Commercial		1650	1450	—	—	625	—	1,700,000	

Table B-3.   (Continued)

Allowable Unit Stresses in Pounds per Square Inch

Species and Commercial Grade	Size Classification	Extreme Fiber in Bending F_b Single-member Uses	Extreme Fiber in Bending F_b Repetitive-member Uses	Tension Parallel to Grain F_t	Horizontal Shear F_v	Compression perpendicular to Grain $F_c \perp$ [21]	Compression Parallel to Grain F_c	Modulus of Elasticity E [21]	U.B.C. Stds Under Which Graded
HEM-FIR (Surfaced dry or surfaced green. Used at 19% max. m.c.)									
Select Structural	2" to 4" thick 2" to 4" wide	1650	1900	975	75	405	1300	1,500,000	WCLIB
No. 1		1400	1600	825	75	405	1050	1,500,000	WWPA
No. 2		1150	1350	675	75	405	825	1,400,000	25-3 and 25-4
No. 3		650	725	375	75	405	500	1,200,000	(See footnotes
Appearance		1400	1600	825	75	405	1250	1,500,000	2 through 9,
Stud		650	725	375	75	405	500	1,200,000	11, 12, 13, 15
Construction	2" to 4" thick 4" wide	825	975	500	75	405	925	1,200,000	and 16)
Standard		475	550	275	75	405	775	1,200,000	
Utility		225	250	125	75	405	500	1,200,000	
Select Structural	2" to 4" thick 5" and wider	1400	1650	950	75	405	1150	1,500,000	
No. 1		1200	1400	800	75	405	1050	1,500,000	
No. 2		1000	1150	525	75	405	875	1,400,000	
No. 3 and Stud		575	675	300	75	405	550	1,200,000	
Appearance		1200	1400	800	75	405	1250	1,500,000	
Select Structural	Beams and Stringers	1300	—	750	70	405	925	1,300,000	WCLIB 25-3
No.1		1050	—	525	70	405	750	1,300,000	
Select Structural	Posts and Timbers	1200	—	800	70	405	975	1,300,000	(See footnotes 2 through 9, 15 and 16)
No. 1		975	—	650	70	405	850	1,300,000	
Select Dex	Decking	1400	1600	—	—	405	—	1,500,000	
Commercial Dex		1150	1350	—	—	405	—	1,400,000	
Select Structural	Beams and Stringers	1250	—	725	70	405	925	1,300,000	WWPA 25-4
No. 1		1050	—	525	70	405	775	1,300,000	
No. 2		675	—	325	70	405	475	1,100,000	
Select Structural	Posts and Timbers	1200	—	800	70	405	975	1,300,000	(See footnotes 2 through 9, 15 and 16)
No. 1		950	—	650	70	405	850	1,300,000	
No. 2		525	—	350	70	405	375	1,100,000	

Table B-3. (Continued)

Allowable Unit Stresses in Pounds per Square Inch

Species and Commercial Grade	Size Classification	Extreme Fiber in Bending F_b Single-member Uses	Extreme Fiber in Bending F_b Repetitive-member Uses	Tension Parallel to Grain F_t	Horizontal Shear F_v	Compression perpendicular to Grain $F_c\perp^{21}$	Compression Parallel to Grain F_c	Modulus of Elasticity E^{21}	U.B.C. Stds Under Which Graded
Select Structural	Beams and Stringers[12]	1250	–	850	70	405	925	1,300,000	NLGA 25-2
No. 1		1050	–	725	70	405	775	1,300,000	(See footnotes 2 through 12)
Select Structural	Posts and Timbers	1200	–	800	70	405	975	1,300,000	
No. 1		950	–	650	70	405	850	1,300,000	(See footnotes 2 through 12)
Selected Decking	Decking	–	1600	–	–	–	–	1,500,000	NLGA WWPA
Commercial Decking		–	1350	–	–	–	–	1,400,000	25-2 and 25-4
Selected Decking	Decking	–	1700	(Surfaced at 15% max. m.c. and used at 15% max. m.c.)				1,600,000	(See footnotes 2 through 12)
Commercial Decking		–	1450					1,400,000	
Select	Decking	1350	1550	–	–	405	–	1,500,000	
Commercial		1150	1300	–	–	405	–	1,400,000	
SOUTHERN PINE (Surfaced at 15% moisture content, K.D. Used at 15% max. m.c.)									
Select Structural	2" to 4" thick 2" to 4" wide	2150	2500	1250	105	565	1800	1,800,000	
Dense Select Structural		2500	2900	1500	105	660	2100	1,900,000	
No. 1		1850	2100	1050	105	565	1450	1,800,000	
No. 1 Dense		2150	2450	1250	105	660	1700	1,900,000	
No. 2		1550	1750	900	95	565	1150	1,600,000	
No. 2 Dense		1800	2050	1050	95	660	1350	1,700,000	
No. 3		850	975	500	95	565	675	1,500,000	
No. 3 Dense		1000	1150	575	95	660	800	1,500,000	
Stud		850	975	500	95	565	675	1,500,000	
Construction	2" to 4" thick 4" wide	1100	1250	650	105	565	1300	1,500,000	SPIB 25-6
Standard		625	725	375	95	565	1050	1,500,000	(See footnotes 3, 4, 9, 13, 15, 16, 18 and 19)
Utility		275	300	175	95	565	675	1,500,000	
Select Structural	2" to 4" thick 5" and wider	1850	2150	1200	95	565	1600	1,800,000	
Dense Select Structural		2200	2500	1450	95	660	1850	1,900,000	
No. 1		1600	1850	1050	95	565	1450	1,800,000	
No. 1 Dense		1850	2150	1250	95	660	1700	1,900,000	
No. 2		1300	1500	675	95	565	1200	1,600,000	
No. 2 Dense		1550	1750	800	95	660	1400	1,700,000	
No. 3		750	875	400	95	565	725	1,500,000	
No. 3 Dense		875	1000	450	95	660	850	1,500,000	
Stud		800	900	400	95	565	725	1,500,000	

Table B-3. (Continued)

Allowable Unit Stresses in Pounds per Square Inch

Species and Commercial Grade	Size Classification	Extreme Fiber in Bending F_b		Tension Parallel to Grain F_t	Horizontal Shear F_v	Compression perpendicular to Grain $F_c \perp^{21}$	Compression Parallel to Grain F_c	Modulus of Elasticity E^{21}	U.B.C. Stds Under Which Graded
		Single-member Uses	Repetitive-member Uses						
Dense Standard Decking	2" to 4" thick	2150	2450	1250	105	660	1700	1,900,000	SP1B 25-6 (See footnotes 4, 9, 15, 16, 18 and 19)
Select Decking		1550	1750	900	95	565	1150	1,600,000	
Dense Select Decking	2" and wider Decking	1800	2050	1050	95	660	1350	1,700,000	
Commercial Decking		1550	1750	900	95	565	1150	1,600,000	
Dense Commercial Decking		1800	2050	1050	95	660	1350	1,700,000	
Dense Structural 86	2" to 4" thick	2800	3250	1900	165	660	2300	1,900,000	
Dense Structural 72		2400	2750	1600	135	660	1950	1,900,000	
Dense Structural 65		2150	2450	1450	125	660	1750	1,900,000	
SOUTHERN PINE (Surfaced dry. Used at 19% max. m.c.)									
Select Structural	2" to 4" thick	2000	2300	1150	100	565	1550	1,700,000	SPIB 25-6 (See footnotes 3, 4, 9, 13, 15, 16, 18 and 19)
Dense Select Structural		2350	2700	1350	100	660	1800	1,800,000	
No. 1		1700	1950	1000	100	565	1250	1,700,000	
No. 1 Dense		2000	2300	1150	100	660	1450	1,800,000	
No. 2	2" to 4" wide	1400	1650	825	90	565	975	1,600,000	
No. 2 Dense		1650	1900	975	90	660	1150	1,600,000	
No. 3		775	900	450	90	565	575	1,400,000	
No. 3 Dense		925	1050	525	90	660	675	1,500,000	
Stud		775	900	450	90	565	575	1,400,000	
Construction	2" to 4" thick	1000	1150	600	100	565	1100	1,400,000	
Standard		575	675	350	90	565	900	1,400,000	
Utility	4" wide	275	300	150	90	565	575	1,400,000	
Select Structural	2" to 4" thick	1750	2000	1150	90	565	1350	1,700,000	
Dense Select Structural		2050	2350	1300	90	660	1600	1,800,000	
No. 1		1450	1700	975	90	565	1250	1,700,000	
No. 1 Dense		1700	2000	1150	90	660	1450	1,800,000	
No. 2	5" and wider	1200	1400	625	90	565	1000	1,600,000	
No. 2 Dense		1400	1650	725	90	660	1200	1,600,000	
No. 3		700	800	350	90	565	625	1,400,000	
No. 3 Dense		825	925	425	90	660	725	1,500,000	
Stud		725	850	350	90	565	625	1,400,000	

Table B-3.   (Continued)

Species and Commercial Grade	Size Classification	Extreme Fiber in Bending F_b		Tension Parallel to Grain F_t	Horizontal Shear F_v	Compression perpendicular to Grain $F_{c\perp}$ [21]	Compression Parallel to Grain F_c	Modulus of Elasticity E [21]	U.B.C. Stds Under Which Graded
		Single-member Uses	Repetitive-member Uses						
Dense Standard Decking	2" to 4" thick	2000	2300	1150	100	660	1450	1,800,000	SPIB 25-6 (See footnotes 3, 4, 9, 13, 15, 16, 17, 18 and 19)
Select Decking		1400	1650	825	90	565	975	1,600,000	
Dense Select Decking	2" and wider Decking	1650	1900	975	90	660	1150	1,600,000	
Commercial Decking		1400	1650	825	90	565	975	1,600,000	
Dense Commercial Decking		1650	1900	975	90	660	1150	1,600,000	
Dense Structural 86	2" to 4" thick	2600	3000	1750	155	660	2000	1,800,000	
Dense Structural 72		2200	2550	1450	130	660	1650	1,800,000	
Dense Structural 65		2000	2800	1300	115	660	1500	1,800,000	
SOUTHERN PINE (Surfaced green. Used any condition)									
Select Structural	2½" to 4" thick	1600	1850	925	95	375	1050	1,500,000	
Dense Select Structural		1850	2150	1100	95	440	1200	1,600,000	
No. 1		1350	1550	800	95	375	825	1,500,000	
No. 1 Dense	2½" to 4" thick	1600	1800	925	95	440	950	1,600,000	
No. 2		1150	1300	675	85	375	650	1,400,000	
No. 2 Dense	2½" to 4" wide	1350	1500	775	85	440	750	1,400,000	
No. 3		625	725	375	85	375	400	1,200,000	
No. 3 Dense		725	850	425	85	440	450	1,300,000	
Stud		625	725	375	85	375	400	1,200,000	
Construction	2½" to 4" thick	825	925	475	95	375	725	1,200,000	
Standard		475	525	275	85	375	600	1,200,000	
Utility	4" wide	200	259	125	85	375	400	1,200,000	

Allowable Unit Stresses in Pounds per Square Inch

Table B-3.　(Continued)

Allowable Unit Stresses in Pounds per Square Inch

Species and Commercial Grade	Size Classification	Extreme Fiber in Bending F_b		Tension Parallel to Grain F_t	Horizontal Shear F_v	Compression perpendicular to Grain $F_c \perp$ [2]	Compression Parallel to Grain F_c	Modulus of Elasticity E [2]	U.B.C. Stds Under Which Graded
		Single-member Uses	Repetitive-member Uses						
Select Structural		1400	1600	900	85	375	900	1,500,000	SP1B
Dense Select Structural		1600	1850	1050	85	440	1050	1,600,000	25-6
No. 1		1200	1350	775	85	375	825	1,500,000	(See footnotes
No. 1 Dense	2½" to 4" thick	1400	1600	925	85	440	950	1,600,000	3, 4, 9, 13,
No. 2		975	1100	500	85	375	675	1,400,000	15, 16, 17, 18
No. 2 Dense	5" and wider	1150	1300	600	85	440	800	1,400,000	and 19)
No. 3		550	650	300	85	375	425	1,200,000	
No. 3 Dense		650	750	350	85	440	475	1,300,000	
Stud		575	675	300	85	375	425	1,200,000	
Dense Standard Decking	2½" to 4" thick	1600	1800	925	95	440	950	1,600,000	
Select Decking		1150	1300	675	85	375	650	1,400,000	
Dense Select Decking	2" and wider Decking	1350	1500	775	85	440	750	1,400,000	
Commercial Decking		1150	1300	675	85	375	650	1,400,000	
Dense Commercial Decking		1350	1500	775	85	440	750	1,400,000	
No. 1 SR		1350	—	875	110	375	775	1,500,000	
No. 1 Dense SR	5" and thicker	1550	—	1050	110	440	925	1,600,000	
No. 2 SR		1100	—	725	95	375	625	1,400,000	
No. 2 Dense SR		1250	—	850	95	440	725	1,400,000	
Dense Structural 86	2½" and thicker	2100	2400	1400	145	440	1300	1,600,000	
Dense Structural 72		1750	2050	1200	120	440	1100	1,600,000	
Dense Structural 65		1600	1800	1050	110	440	1000	1,600,000	

Footnotes for Table B-3 (Note: This Table is number 25–A–1 in the Uniform Building Code. Section numbers referred to are those of the 1985 edition of UBC.)

[1]Where eastern spruce and balsam fir are shipped in a combination, the tabulated values for balsam fir shall apply:

[2]The design values shown in Table No. 25–A–1 are applicable to lumber that will be used under dry conditions such as in most covered structures. For 2-inch-to 4-inch-thick lumber the DRY surfaced size shall be used. In calculating design values, the natural gain in strength and stiffness that occurs as lumber dries has been taken into consideration as well as the reduction in size that occurs when unseasoned lumber shrinks. The gain in load-carrying capacity due to increased strength and stiffness resulting from drying more than offsets the design effect of size reductions due to shrinkage. For 5-inch and thicker lumber, the surfaced sizes also may be used because design values have been adjusted to compensate for any loss in size by shrinkage which may occur.

[3]Values for F_b, F_t and F_c for the grades of Construction, Standard and Utility apply only to 4-inch widths.

Table B-3.   (Continued) Footnotes

[4]The values in Table No. 25-A-1 for dimension 2 inches to 4 inches are based on edgewise use. Where such lumber is used flatwise, the recommended design values for extreme fiber stress in bending may be multiplied by the following factors:

WIDTH	THICKNESS		
	2″	3″	4″
2 inches to 4 inches	1.10	1.04	1.00
5 inches and wider	1.22	1.16	1.11

Values for decking may be increased by 10 percent for 2-inch decking and 4 percent for 3-inch decking.

[5]When 2-inch- to 4-inch-thick lumber is manufactured at a maximum moisture content of 15 percent and used in a condition where the moisture content does not exceed 15 percent, the design values shown in Table No. 25-A-1 for surfaced dry and surfaced green may be multiplied by the following factors:

EXTREME FIBER IN BENDING F_b	TENSION PARALLEL TO GRAIN F_t	HORIZONTAL SHEAR F_v	COMPRESSION PERPENDICULAR TO GRAIN $F_c \perp$	COMPRESSION PARALLEL TO GRAIN F_c	MODULUS OF ELASTICITY E
1.08	1.08	1.05	1.00	1.17*	1.05*

*For redwood use 1.15 for F_c and 1.04 for E.

[6]When 2-inch to 4-inch-thick lumber is designed for use where the moisture content will exceed 19 percent for an extended period of time, the values shown in Table No. 25-A-1 shall be multiplied by the following factors:

EXTREME FIBER IN BENDING F_b	TENSION PARALLEL TO GRAIN F_t	HORIZONTAL SHEAR F_v	COMPRESSION PERPENDICULAR TO GRAIN $F_c \perp$	COMPRESSION PARALLEL TO GRAIN F_c	MODULUS OF ELASTICITY E
0.86	0.84	0.97	0.67	0.70	0.97

[7]When lumber 5 inches and thicker is designed for use where the moisture content will exceed 19 percent for an extended period of time, the values shown in Table No. 25-A-1 shall be multiplied by the following factors:

EXTREME FIBER IN BENDING F_b	TENSION PARALLEL TO GRAIN F_t	HORIZONTAL SHEAR F_v	COMPRESSION PERPENDICULAR TO GRAIN $F_c \perp$	COMPRESSION PARALLEL TO GRAIN F_c	MODULUS OF ELASTICITY E
1.00	1.00	1.00	0.67	0.91	1.00

[8]The tabulated horizontal shear values shown herein for lumber 4 inches and thinner shall be multiplied by a factor of 0.92 when such lumber is manufactured unseasoned. Specific horizontal shear values may be established by using the following tables when the length of split or check is known:

WHEN LENGTH OF SPLIT ON WIDE FACE IS:	MULTIPLY TABULATED F_v VALUE BY: (NOMINAL 2-INCH LUMBER)
No split	2.00
$\frac{1}{2}$ × wide face	1.67
$\frac{3}{4}$ × wide face	1.50
1 × wide face	1.33
$1\frac{1}{2}$ × wide face or more	1.00

Table B-3. (*Continued*) Footnotes

WHEN LENGTH OF SPLIT ON WIDE FACE IS:	MULTIPLY TABULATED F_v VALUE BY: (3-INCH AND THICKER LUMBER)
No split	2.00
$\frac{1}{2}$ × narrow face	1.67
1 × narrow face	1.33
$1\frac{1}{2}$ × narrow face or more	1.00

[9] Stress-rated boards of nominal 1-inch, $1\frac{1}{4}$-inch and $1\frac{1}{2}$-inch thickness, 2 inches and wider, are permitted the recommended design values shown for Select Structural, No. 1, No. 2, No. 3, Construction, Standard, Utility, Appearance, Clear Heart Structural and Clear Structural grades as shown in the 2-inch- to 4-inch-thick categories herein, where graded in accordance with the stress-rated board provisions in the applicable grading rules.

[10] When decking is used where the moisture content will exceed 15 percent for an extended period of time, the tabulated design values shall be multiplied by the following factors: Extreme Fiber in Bending F_b—0.79; Modulus of Elasticity E—0.92.

[11] Where lumber is graded under U.B.C. Standard No. 25-2 values shown for Select Structural, No. 1, No. 2, No. 3, and Stud grades are not applicable to 3-inch × 4-inch and 4-inch × 4-inch sizes.

[12] Lumber in the beam and stringer or post and timber size classification may be assigned different working stresses for the same grade name and species based upon the grading rules of the specific agency involved. It is therefore necessary that the grading rule agency be identified to properly correlate permitted design stresses with the grade mark.

[13] Utility grades of all species may be used only under conditions specifically approved by the building official.

[14] A horizontal shear F_v of 70 may be used for eastern white pine graded under U.B.C. Standards No. 25-5 and No. 25-8 (grading rules of Northern Hardwood and Pine Manufacturers, Inc. and Northeastern Lumber Manufacturers Association, Inc.)

[15] Tabulated tension parallel to grain values for species 5 inches and wider, 2 inches to 4 inches thick (and $2\frac{1}{2}$ inches to 4 inches thick) size classifications apply to 5-inch and 6-inch widths only, for grades of Select Structural, No. 1, No. 2, No. 3, Appearance and Stud (including dense grades). For lumber wider than 6 inches in these grades, the tabulated F_t values shall be multiplied by the following factors:

GRADE

(2 inches to 4 inches thick, 5 inches and wider)
($2\frac{1}{2}$ inches to 4 inches thick, 5 inches and wider)
(includes "Dense" grades)

	Multiply tabulated F_t values by		
	5 inches and 6 inches wide	8 inches wide	10 inches and wider
Select Structural	1.00	0.90	0.80
No. 1, No. 2, No. 3 and Appearance	1.00	0.80	0.60
Stud	1.00		

[16] Design values for all species of Stud grade in 5-inch and wider size classifications apply to 5-inch and 6-inch widths only.

[17] Repetitive member design values for extreme fiber in bending for southern pine grades of Dense Structural 86, 72 and 65 apply to 2-inch to 4-inch thicknesses only.

[18] When 2-inch- to 4-inch-thick southern pine lumber is surfaced dry or at 15 percent maximum moisture content (KD) and is designed for use where the moisture content will exceed 19 percent for an extended period of time, the design values in Table No. 25-A-1 for the corresponding grades of $2\frac{1}{2}$-inch- to 4-inch-thick surfaced green southern pine lumber shall be used. The net green size may be used in such designs.

[19] When 2-inch- to 4-inch-thick southern pine lumber is surfaced dry or at 15 percent maximum moisture content (KD) and is designed for use under dry conditions, such as in most covered structures, the net DRY size shall be used in design. For other sizes and conditions of use, the net green size may be used in design.

[20] Values apply only to ponderosa pine graded under U.B.C. Standard No. 25-2.

[21] The duration of load modification factors given in Section 2504 (c) 4 shall not apply.

Table B-3 and footnotes are reproduced from the 1985 edition of the Uniform Building Code (copyright © 1985) with permission of the publishers, the International Conference of Building Officials.

Table B-4.   Allowable Design Stresses and Modulus of Elasticity for Machine Stress Rated Lumber Under Normal-Duration Loading.

| Grade Designation[10] | Grading rules U.B.C. Std. No. (See footnotes 1,2,3,4) | Size Classification | Design values in pounds per square inch[9] | | Tension parallel to grain F_t | Compression parallel to grain F_c | Modulus of elasticity E |
| | | | Extreme fiber in bending F_b[7] | | | | |
			Single-member uses	Repetitive-member uses			
900f-1.0E	3,4		900	1050	350	725	1,000,000
1200f-1.2E	1,2,3,4		1200	1400	600	950	1,200,000
1350f-1.3E	2,4		1350	1550	750	1075[11]	1,300,000
1450f-1.3E	1,3,4		1450	1650	800	1150	1,300,000
1500f-1.3E	2		1500	1750	900	1200	1,300,000
1500f-1.4E	1,2,3,4		1500	1750	900	1200	1,400,000
1650f-1.4E	2		1650	1900	1020	1320	1,400,000
1650f-1.5E	1,2,3,4		1650	1900	1020	1320	1,500,000
1800f-1.6E	1,2,3,4		1800	2050	1175	1450	1,600,000
1950f-1.5E	2	Machine rated lumber 2" thick or less All widths	1950	2250	1375	1550	1,500,000
1950f-1.7E	1,2,4		1950	2250	1375	1550	1,700,000
2100f-1.8E	1,2,3,4		2100	2400	1575	1700	1,800,000
2250f-1.6E	2		2250	2600	1750	1800	1,600,000
2250f-1.9E	1,2,4		2250	2600	1750	1800	1,900,000
2400f-1.7E	2		2400	2750	1925	1925	1,700,000
2400f-2.0E	1,2,3,4		2400	2750	1925	1925	2,000,000
2550f-2.1E	1,2,4		2550	2950	2050	2050	2,100,000
2700f-2.2E	1,2,3,4		2700	3100	2150	2150	2,200,000
2850f-2.3E	2		2850	3300	2300	2300	2,300,000
3000f-2.4E	1,2		3000	3450	2400	2400	2,400,000
3150f-2.5E	2		3150	3600	2500	2500	2,500,000
3300f-2.6E	2		3300	3800	2650	2650	2,600,000
900f-1.0E	1,2,3	See footnote 5	900	1050	350	725	1,000,000
900f-1.2E	1,2,3		900	1050	350	725	1,200,000
1200f-1.5E	1,2,3		1200	1400	600	950	1,500,000
1350f-1.8E	1,2		1350	1550	750	1075	1,800,000
1500f-1.8E	3		1500	1750	900	1200	1,800,000
1800f-2.1E	1,2,3		1800	2050	1175	1450	2,100,000

Footnotes for Table B-4 (Note: This is Table 25-A-2 of the Uniform Building Code.)

[1] U.B.C. Standard No. 25-2.

[2] U.B.C. Standard No. 25-6.

[3] U.B.C. Standard No. 25-3.

[4] U.B.C. Standard No. 25-4.

[5] Size classifications for these grades are:

   U.B.C. Standard No. 25-2—Machine Rated Lumber; 2" thick or less, all widths.

   U.B.C. Standard No. 25-6—Machine Rated Lumber; 2" thick or less, all widths.

   U.B.C. Standard No. 25-3—Machine Rated Joists; 2" thick or less. 6" and wider.

[6] Stresses apply for lumber used at 19 percent maximum moisture content.

[7] Tabulated extreme fiber in bending values F_b are applicable to lumber loaded on edge. When loaded flatwise, these values may be increased by multiplying by the following factors:

Nominal width (in.)	3"	4"	5"	6"	8"	10"	12"	14"
Factor	1.06	1.10	1.12	1.15	1.19	1.22	1.25	1.28

[8] Footnotes 1,2,9,11 and 19 to Table B-3 apply also to machine stress rated lumber.

[9] Design values for horizontal shear F_v and compression perpendicular to grain $F_c\perp$ for lumber used under dry conditions are as follows:

	Douglas Fir-Larch	Douglas Fir-South	Hem-Fir	Western Hemlock	Lodgepole-Ponderosa-Sugar Pine	Englemann Spruce	Cedar	Southern Pine		White Woods-Mixed Species
								KD	S-DRY	
Horizontal Shear F_v	95	90	75	90	70	70	75	95	90	70
Compression Perpendicular to Grain $F_c\perp$	625	520	405	410	375	320	425	565	565	315

Reproduced from the 1985 edition of the Uniform Building Code (copyright © 1985) with permission of the publishers, the International Conference of Building Officials.

Table B-5. Design Allowable Stresses and Modulus of Elasticity for Structural Glued Laminated Lumber Under Normal-Duration Loading and Stressed Primarily in Bending with the Load Applied Perpendicular to the Wide Face of the Laminations.

COMBINATION SYMBOL[20]	SPECIES OUTER LAMINATIONS/CORE LAMINATIONS[5]	BENDING ABOUT X-X AXIS (Loaded Perpendicular to Wide Faces of Laminations)						BENDING ABOUT Y-Y AXIS (Loaded Parallel to Wide Faces of Laminations)					AXIALLY LOADED		
		Extreme Fiber in Bending F_{bxx}		Compression Perpendicular to Grain $F_{c\perp xx}$		Horizontal Shear F_{vxx} psi	Modulus of Elasticity E_{xx}[22] x10^6 psi	Extreme Fiber in Bending[6,18] F_{byy} psi	Compression Perpendicular to Grain $F_{c\perp yy}$ psi	Horizontal Shear F_{vyy} psi	Horizontal Shear F_{vyy} psi (For members with multiple-piece laminations which are not edge glued)[21]	Modulus of Elasticity E_{yy}[22] x10^6 psi	Tension Parallel to Grain F_t psi	Compression Parallel to Grain F_c psi	Modulus of Elasticity E[22] x10^6 psi
		Tension Zone Stressed in Tension[6] psi	Compression Zone Stressed in Tension[7] psi	Tension Face psi	Compression Face psi										
1	2	3	4	5	6	7	8	9	10	11	12	13	14	15	16
Visually Graded Western Species															
16F-V1	DF/WW	1600	800	385[8,9]	385[8]	140[19]	1.3	950	190	130	65	1.1	675	975	1.1
16F-V2	HF/HF			385[10]	245[10]	155	1.4	1250	245	135	70	1.3	875	1300	1.3
16F-V3	DF/DF			385[8,9]	385[8]	165	1.5	1450	385	145	75	1.5	950	1550	1.5
16F-V8	DFS/DFS			395	335	165	1.2	1200	335	145	75	1.1	825	1350	1.1
The following two combinations are intended for straight or slightly cambered members for dry use and industrial appearance.[11]															
16F-V4	DF/N3WW	1600	800	450	385[8]	90[10,19]	1.5	900	190	130	65	1.3	650	600	1.3
16F-V5	DF/M3DF			450	385[8]	90[13]	1.6	1000	270	135	70	1.5	750	875	1.5
The following two combinations are balanced and are intended for members continuous or cantilevered over supports and provide equal capacity in both positive and negative bending.															
16F-V6	DF/DF	1600	1600	385[8,9]	385[8]	165	1.5	1450	385	145	75	1.4	950	1550	1.5
16F-V7	HF/HF			245[10]	245[10]	155	1.4	1200	245	135	70	1.3	850	1350	1.3
20F-V1	DF/WW	2000	1000	450	385[8]	140[19]	1.4	1000	190	130	65	1.2	750	1000	1.2
20F-V2	HF/HF			385[10]	245[10]	155	1.5	1200	245	135	70	1.4	950	1350	1.4
20F-V3	DF/DF			450	385[10]	165	1.6	1450	385	145	75	1.5	1000	1550	1.5
20F-V4	DF/DF			410[8,9]	385[10]	165	1.6	1450	385	145	75	1.6	1000	1550	1.6
20F-V10	DF/HF			450	385	155	1.5	1300	245	135	70	1.4	950	1500	1.4
20F-V11	DFS/DFS			395	335	165	1.3	1400	335	145	75	1.1	900	1400	1.1

Table B-5. (Continued)

COMBINATION SYMBOL[20]	SPECIES OUTER LAMINATIONS/CORE LAMINATIONS[5]	BENDING ABOUT X-X AXIS — Loaded Perpendicular to Wide Faces of Laminations — Extreme Fiber in Bending F_{bxx} — Tension Zone Stressed in Tension[6] psi	Compression Zone Stressed in Tension[7] psi	Compression Perpendicular to Grain $F_{c\perp xx}$ — Tension Face psi	Compression Face psi	Horizontal Shear F_{vxx} psi	Modulus of Elasticity E_{xx}^{22} x10⁶ psi	BENDING ABOUT Y-Y AXIS — Loaded Parallel to Wide Faces of Laminations — Extreme Fiber in Bending[6,18] F_{byy} psi	Compression Perpendicular to Grain $F_{c\perp yy}$ psi	Horizontal Shear F_{vyy} psi	Horizontal Shear F_{vyy} psi (For members with multiple-piece laminations which are not edge glued)[21]	Modulus of Elasticity $E_{yy}^{22} \times 10^6$ psi	AXIALLY LOADED — Tension Parallel to Grain F_t psi	Compression Parallel to Grain F_c psi	Modulus of Elasticity E^{22} x10⁶ psi
1	2	3	4	5	6	7	8	9	10	11	12	13	14	15	16
20F-V5	DF/N3WW	2000	1000	450	385[10]	90[12,19]	1.6	1000	190	135	70	1.3	750	725	1.3
20F-V6	DF/M3DF	2000	1000	450	385[10]	90[13]	1.6	1000	270	135	70	1.5	775	900	1.5

The following two combinations are intended for straight or slightly cambered members for dry use and industrial appearance.[11]

The following three combinations are balanced and are intended for members continuous or cantilevered over supports and provide equal capacity in both positive and negative bending.

COMBINATION SYMBOL[20]	SPECIES	3	4	5	6	7	8	9	10	11	12	13	14	15	16
20F-V7	DF/DF	2000	2000	450	450	165	1.6	1450	385	145	75	1.6	1000	1600	1.6
20F-V8	DF/DF	2000	2000	410[8,9]	410[8,9]	165	1.7	1450	385	145	75	1.6	1000	1600	1.6
20F-V9	HF/HF	2000	2000	385[10]	385[10]	155	1.5	1430	245	135	70	1.4	975	1400	1.4
22F-V1	DF/WW	2200	1100	450	385[8]	140[19]	1.6	1050	190	130	65	1.3	850	1100	1.3
22F-V2	HF/HF	2200	1100	385[10]	385[10]	155	1.5	1250	245	135	70	1.4	950	1350	1.4
22F-V3	DF/DF	2200	1100	450	385[8]	165	1.7	1450	385	145	75	1.6	1050	1500	1.6
22F-V4	DF/DF	2200	1100	410[8,9]	385[8]	165	1.7	1450	385	145	75	1.6	1000	1550	1.6
22F-V10	DF/DFS	2200	1100	450	385	165	1.6	1600	335	145	75	1.3	1000	1400	1.3

The following two combinations are intended for straight or slightly cambered members for dry use and industrial appearance.[11]

COMBINATION SYMBOL[20]	SPECIES	3	4	5	6	7	8	9	10	11	12	13	14	15	16
22F-V5	DF/N3WW	2200	1100	450	385[8]	90[12,19]	1.6	1100	190	135	75	1.4	800	725	1.4
22F-V6	DF/M3DF	2200	1100	450	385[8]	90[13]	1.7	1250	270	135	75	1.6	900	925	1.6

Table B-5. (Continued)

		BENDING ABOUT X-X AXIS						BENDING ABOUT Y-Y AXIS					AXIALLY LOADED		
		Loaded Perpendicular to Wide Faces of Laminations						Loaded Parallel to Wide Faces of Laminations							
		Extreme Fiber in Bending F_{bxx}		Compression Perpendicular to Grain $F_{c\perp xx}$		Horizontal Shear F_{vxx} psi	Modulus of Elasticity E_{xx} x10⁶ psi	Extreme Fiber in Bending[6,18] F_{byy} psi	Compression Perpendicular to Grain $F_{c\perp yy}$ psi	Horizontal Shear F_{vyy} psi	Horizontal Shear F_{vyy} psi (For members with multiple-piece laminations which are not edge glued)[21]	Modulus of Elasticity E_{yy}[22] x10⁶ psi	Tension Parallel to Grain F_t psi	Compression Parallel to Grain F_c psi	Modulus of Elasticity E[22] x10⁶ psi
COMBINATION SYMBOL[20]	SPECIES OUTER LAMINATIONS/CORE LAMINATIONS[5]	Tension Zone Stressed in Tension[6] psi	Compression Zone Stressed in Tension[7] psi	Tension Face psi	Compression Face psi										
1	2	3	4	5	6	7	8	9	10	11	12	13	14	15	16

The following three combinations are balanced and are intended for members continuous or cantilevered over supports and provide equal capacity in both positive and negative bending.

22F-V7	DF/DF	2200	2200	450	450	165	1.8	1450	385	145	75	1.6	1100	1650	1.6
22F-V8	DF/DF			410[8,9]	410[8,9]	165	1.7	1450	385	145	75	1.6	1050	1650	1.6
22F-V9	HF/HF			385[10]	385[10]	155	1.5	1250	245	135	70	1.4	975	1400	1.4
24F-V1	DF/WW	2400	1200	450	450	140	1.7	1250	190	130	70	1.4	1000	1300	1.4
24F-V2	HF/HF			385[10]	385[10]	155	1.5	1250	245	135	70	1.4	950	1300	1.4
24F-V3	DF/DF			450	385[8]	165	1.7	1500	385	145	75	1.6	1100	1600	1.6
24F-V4	DF/DF			450	450	165	1.8	1500	385	145	75	1.6	1150	1650	1.6
24F-V5	DF/HF			450	450	155	1.7	1350	245	140	70	1.5	1100	1450	1.5
24F-V11	DF/DFS			450	385	165	1.7	1600	335	145	75	1.4	1150	1700	1.4

The following two combinations are intended for straight or slightly cambered members for dry use and industrial appearance.[11]

| 24F-V6 | DF/N3WW | 2400 | 1200 | 450 | 385[8] | 90[12,19] | 1.7 | 1200 | 190 | 140 | 70 | 1.5 | 950 | 800 | 1.5 |
| 24F-V7 | DF/M3DF | | | 450 | 385[8] | 90[13] | 1.7 | 1250 | 270 | 135 | 70 | 1.6 | 900 | 950 | 1.6 |

The following three combinations are balanced and are intended for members continuous or cantilevered over supports and provide equal capacity in both positive and negative bending.

24F-V8	DF/DF	2400	2400	450	450	165	1.8	1450	385	145	75	1.6	1100	1650	1.6
24F-V9	HF/HF			385[10]	385[10]	155	1.5	1500	245	135	70	1.4	1000	1450	1.4
24F-V10	DF/HF			450	450	155	1.8	1400	245	140	70	1.6	1150	1600	1.6
Wet-use factors[2]		0.8	0.8	0.667	0.667	0.875	0.833	0.8	0.667	0.875	0.875	0.833	0.8	0.73	0.833

Table B-5. (Continued)

		BENDING ABOUT X-X AXIS						BENDING ABOUT Y-Y AXIS					AXIALLY LOADED		
		Loaded Perpendicular to Wide Faces of Laminations						Loaded Parallel to Wide Faces of Laminations							
COMBINATION SYMBOL[20]	SPECIES OUTER LAMINATIONS/CORE LAMINATIONS[5]	Extreme Fiber in Bending F_{bxx}		Compression Perpendicular to Grain $F_{c\perp xx}$		Horizontal Shear F_{vxx} psi	Modulus of Elasticity E_{xx}[22] x10^6 psi	Extreme Fiber in Bending[6,18] F_{byy} psi	Compression Perpendicular to Grain $F_{c\perp yy}$ psi	Horizontal Shear F_{vyy} psi	Horizontal Shear F_{vyy} psi (For members with multiple-piece laminations which are not edge glued)[21]	Modulus of Elasticity E_{yy}[22] × 10^6 psi	Tension Parallel to Grain F_t psi	Compression Parallel to Grain F_c psi	Modulus of Elasticity E[22] × 10^6 psi
		Tension Zone Stressed in Tension[6] psi	Compression Zone Stressed in Tension[7] psi	Tension Face psi	Compression Face psi										
1	2	3	4	5	6	7	8	9	10	11	12	13	14	15	16
						E-Rated Western Species									
16F-E1	WW/WW	1600	800	190[14]	190[14]	140[19]	1.3	1050	190	125	65	1.2	725	925	1.2
16F-E2[16]	HF/HF			245[15]	245[15]	155	1.4	1250	245	135	70	1.3	825	1200	1.3
16F-E3	DF/DF			385[15]	385[15]	165	1.6	1450	385	145	75	1.5	975	1600	1.5
		The following two combinations are intended for straight or slightly cambered members for dry use and industrial appearance.[11]													
16F-E4	DF/N3WW	1600	800	385[15]	385[15]	90[12,19]	1.6	900	190	130	65	1.3	675	675	1.3
16F-E5	DF/M3DF			385[15]	385[15]	90[13]	1.6	1050	270	135	70	1.5	700	900	1.5
		The following two combinations are balanced and are intended for members continuous or cantilevered over supports and provide equal capacity in both positive and negative bending.													
16F-E6	DF/DF	1600	1600	385[15]	385[15]	165	1.6	1500	385	145	75	1.5	1000	1600	1.5
16F-E7[16]	HF/HF			245[15]	245[15]	155	1.4	1250	245	135	70	1.3	850	1150	1.3
20F-E1	WW/WW	2000	1000	190[14]	190[14]	140[14]	1.6	1100	190	125	65	1.3	800	1050	1.3
20F-E2[16]	HF/HF			385[15]	385[15]	155	1.6	1400	245	135	70	1.4	925	1550	1.4
20F-E3	DF/DF			385[15]	385[15]	165	1.7	1550	385	145	75	1.6	1050	1650	1.6
		The following two combinations are intended for straight or slightly cambered members for dry use and industrial appearance.[11]													
20F-E4	DF/N3WW	2000	1000	450	385[15]	90[12,19]	1.6	1100	190	130	65	1.4	800	700	1.4
20F-E5	DF/N3DF			385[15]	385[15]	90[13]	1.7	1300	270	135	70	1.6	825	975	1.6

Table B-5. (Continued)

COMBINATION SYMBOL[20]	SPECIES OUTER LAMINATIONS/ CORE LAMINATIONS[5]	BENDING ABOUT X-X AXIS — Loaded Perpendicular to Wide Faces of Laminations						BENDING ABOUT Y-Y AXIS — Loaded Parallel to Wide Faces of Laminations					AXIALLY LOADED		
		Extreme Fiber in Bending F_{bxx} — Tension Zone Stressed in Tension[6] psi	Extreme Fiber in Bending F_{bxx} — Compression Zone Stressed in Tension[7] psi	Compression Perpendicular to Grain $F_{c\perp xx}$ — Tension Face psi	Compression Perpendicular to Grain $F_{c\perp xx}$ — Compression Face psi	Horizontal Shear F_{vxx} psi	Modulus of Elasticity E_{xx}^{22} x10^6 psi	Extreme Fiber in Bending[6,18] F_{byy} psi	Compression Perpendicular to Grain $F_{c\perp yy}$ psi	Horizontal Shear F_{vyy} psi	Horizontal Shear F_{vyy} psi (For members with multiple-piece laminations which are not edge glued)[21]	Modulus of Elasticity E_{yy}^{22} x10^6 psi	Tension Parallel to Grain F_t psi	Compression Parallel to Grain F_c psi	Modulus of Elasticity E^{22} x10^6 psi
1	2	3	4	5	6	7	8	9	10	11	12	13	14	15	16
		The following two combinations are balanced and are intended for members continuous or cantilevered over supports and provide equal capacity in both positive and negative bending.													
20F-E6	DF/DF	2000	2000	385[15]	385[15]	165	1.7	1600	385	145	75	1.6	1150	1650	1.6
20F-E7[16]	HF/HF			385[15]	385[15]	155	1.6	1500	245	135	70	1.4	1050	1550	1.4
22F-E1	DF/DF	2200	1100	450	385[15]	165	1.7	1550	385	145	75	1.6	1050	1600	1.6
22F-E2[16]	HF/HF			385[15]	385[15]	155	1.6	1400	245	135	70	1.4	950	1400	1.4
		The following two combinations are intended for straight or slightly cambered members for dry use and industrial appearance.[11]													
22F-E3	DF/N3WW	2200	1100	450	385[15]	90[12,19]	1.7	1250	190	135	70	1.4	825	750	1.4
22F-E4	DF/N3DF			450	385[15]	90[13]	1.8	1350	270	135	70	1.6	950	950	1.6
		The following two combinations are balanced and are intended for members continuous or cantilevered over supports and provide equal capacity in both positive and negative bending.													
22F-E5	DF/DF	2200	2200	450	450	165	1.7	1650	385	145	75	1.6	1100	1650	1.6
22F-E6[16]	HF/HF			385[15]	385[15]	155	1.7	1550	245	135	70	1.5	1050	1500	1.5
24F-E1	DF/DF	2400	1200	450	450	165	1.8	1550	385	145	75	1.6	1100	1600	1.6
24F-E2[16]	HF/HF			385[15]	385[15]	155	1.7	1450	245	135	70	1.5	1000	1400	1.5
24F-E3	DF/HF			450	385[15]	155	1.8	1500	245	135	70	1.5	1050	1550	1.5
24F-E4	DF/DF			450	450	165	1.8	1650	385	145	75	1.7	1100	1700	1.7
24F-E5	DF/DF			450	385[15]	165	1.8	1650	385	145	75	1.6	1100	1550	1.6
24F-E6[16]	HF/WW			385[15]	385[15]	140[14]	1.8	1250	190	130	65	1.4	925	1350	1.4
		The following three combinations are intended for straight or slightly cambered members for dry use and industrial appearance.[11]													
24F-E7	DF/N3WW	2400	1200	450	450	90[12,19]	1.9	1400	190	135	70	1.6	975	875	1.6
24F-E8	DF/N3DF			450	450	90[13]	1.9	1400	270	135	70	1.7	1000	1050	1.7
24F-E9[16]	HF/N3HF			385[15]	385[15]	90[12]	1.8	1350	245	135	70	1.6	950	825	1.6

Table B-5. (Continued)

		BENDING ABOUT X-X AXIS						BENDING ABOUT Y-Y AXIS					AXIALLY LOADED		
		Loaded Perpendicular to Wide Faces of Laminations						Loaded Parallel to Wide Faces of Laminations							
		Extreme Fiber in Bending F_{bxx}		Compression Perpendicular to Grain $F_{c\perp xx}$		Horizontal Shear F_{rxx} psi	Modulus of Elasticity E_{xx}^{22} x10⁶ psi	Extreme Fiber in Bending[6,18] F_{byy} psi	Compression Perpendicular to Grain $F_{c\perp yy}$ psi	Horizontal Shear F_{ryy} psi	Horizontal Shear F_{ryy} psi (For members with multiple laminations which are not edge glued)[21]	Modulus of Elasticity E_{yy}^{22} x10⁶ psi	Tension Parallel to Grain F_t psi	Compression Parallel to Grain F_c psi	Modulus of Elasticity E^{22} x10⁶ psi
COMBINATION SYMBOL[20]	SPECIES OUTER LAMINATIONS/CORE LAMINATIONS[5]	Tension Zone Stressed in Tension[6] psi	Compression Zone Stressed in Tension[7] psi	Tension Face psi	Compression Face psi										
1	2	3	4	5	6	7	8	9	10	11	12	13	14	15	16

The following four combinations are balanced and are intended for members continuous or cantilevered over supports and provide equal capacity in both positive and negative bending.

Comb.	Species	3	4	5	6	7	8	9	10	11	12	13	14	15	16
24F-E10	DF/DF	2400	2400	450	450	165	1.9	1850	385	145	75	1.7	1300	1750	1.7
24F-E11[16]	HF/HF			385[15]	385[5]	155	1.8	1600	245	135	70	1.5	1150	1550	1.5
24F-E12	DF/HF			450	450	155	1.9	1750	245	135	70	1.6	1200	1600	1.6
24F-E13	DF/DF			450	450	165	1.8	1950	385	145	75	1.7	1250	1700	1.7
24F-E14	DF/DF	2400	1200	450	385	165	1.8	1600	385	145	75	1.6	1050	1600	1.6
Wet-use factors[2]		0.8	0.8	0.667	0.667	0.875	0.833	0.8	0.667	0.875	0.875	0.833	0.8	0.73	0.833

Visually Graded Southern Pine

16F-V1	SP/SP	1600	800	385[8,9]	385[8]	200	1.4	1450	385	175	90	1.3	950	1450	1.3
16F-V2	SP/SP			385[8,9]	385[8]	200	1.4	1600	385	175	90	1.4	1000	1550	1.4
16F-V3	SP/SP			450	450	200	1.4	1450	385	175	90	1.3	975	1450	1.3

The following combination is intended for straight or slightly cambered members for dry use and industrial appearance.[11]

| 16F-V4 | SP/SP | 1600 | 800 | 385[8,9] | 385[8] | 90[17] | 1.3 | 975 | 270 | 150 | 75 | 1.2 | 650 | 950 | 1.2 |

The following combination is balanced and intended for members continuous or cantilevered over supports and provides equal capacity in both positive and negative bending.

16F-V5	SP/SP	1600	1600	385[8,9]	385[8,9]	200	1.4	1600	385.	175	85	1.4	1000	1550	1.4
20F-V1	SP/SP	2000	1000	450	385[8]	200	1.5	1450	385	175	85	1.4	1000	1450	1.4
20F-V2	SP/SP			450	385[8]	200	1.6	1450	385	175	85	1.4	1050	1550	1.4
20F-V3	SP/SP			385[8,9]	385[8]	200	1.4	1600	385	175	85	1.4	1000	1500	1.4

Table B-5. (Continued)

COMBINATION SYMBOL[20]	SPECIES OUTER LAMINATIONS/CORE LAMINATIONS[5]	BENDING ABOUT X-X AXIS — Loaded Perpendicular to Wide Faces of Laminations						BENDING ABOUT Y-Y AXIS — Loaded Parallel to Wide Faces of Laminations					AXIALLY LOADED		
		Extreme Fiber in Bending F_{bxx}		Compression Perpendicular to Grain $F_{c\perp xx}$		Horizontal Shear F_{vxx} psi	Modulus of Elasticity E_{xx}^{22} x10⁶ psi	Extreme Fiber in Bending[6,18] F_{byy} psi	Compression Perpendicular to Grain $F_{c\perp yy}$ psi	Horizontal Shear F_{vyy} psi	Horizontal Shear F_{vyy} psi (For members with multiple-piece laminations which are not edge glued)[21]	Modulus of Elasticity $E_{yy}^{22} \times 10^6$ psi	Tension Parallel to Grain F_t psi	Compression Parallel to Grain F_c psi	Modulus of Elasticity E^{22} x10⁶ psi
		Tension Zone Stressed in Tension[6] psi	Compression Zone Stressed in Tension[7] psi	Tension Face psi	Compression Face psi										
1	2	3	4	5	6	7	8	9	10	11	12	13	14	15	16
20F-V4	SP/SP	2000	1000	450	385[8]	90[17]	1.5	1100	270	150	75	1.3	725	950	1.3
The following combination is balanced and intended for members continuous or cantilevered over supports and provides equal capacity in both positive and negative bending.															
20F-V5	SP/SP	2000	2000	450	450	200	1.6	1450	385	175	85	1.4	1050	1550	1.4
22F-V1	SP/SP	2200.	1100	450	450	200	1.6	1600	385	175	85	1.5	1050	1650	1.5
22F-V2	SP/SP			385[8,9]	385[8]	200	1.4	1600	385	175	85	1.4	1000	1500	1.4
22F-V3	SP/SP			450	385[8]	200	1.6	1500	385	175	85	1.4	1050	1500	1.4
The following combination is intended for straight or slightly cambered members for dry use and industrial appearance.[11]															
22F-V4	SP/SP	2200	1100	450	385[8]	90[17]	1.6	1250	270	155	75	1.4	825	1000	1.4
The following combination is balanced and intended for members continuous or cantilevered over supports and provides equal capacity in both positive and negative bending.															
22F-V5	SP/SP	2200	2200	450	450	200	1.6	1600	385	175	85	1.5	1050	1600	1.5
24F-V1	SP/SP	2400	1200	450	385[8]	200	1.7	1500	385	175	85	1.5	1100	1350	1.5
24F-V2	SP/SP			450	450	200	1.7	1600	385	175	85	1.5	1100	1600	1.5
24F-V3	SP/SP			450	450	200	1.8	1600	385	175	85	1.6	1150	1700	1.6
The following combination is intended for straight or slightly cambered members for dry use and industrial appearance.[11]															
24F-V4	SP/SP	2400	1200	450	385[8]	90[17]	1.7	1250	270	155	75	1.4	850	1050	1.4

Table B-5. (Continued)

COMBINATION SYMBOL[20]	SPECIES OUTER LAMINATIONS/CORE LAMINATIONS[5]	BENDING ABOUT X-X AXIS — Loaded Perpendicular to Wide Faces of Laminations — Extreme Fiber in Bending F_{bxx} — Tension Zone Stressed in Tension[6] psi	Compression Zone Stressed in Tension[7] psi	Compression Perpendicular to Grain $F_{c\perp xx}$ — Tension Face psi	Compression Face psi	Horizontal Shear F_{vxx} psi	Modulus of Elasticity E_{xx}[22] x10⁶ psi	BENDING ABOUT Y-Y AXIS — Loaded Parallel to Wide Faces of Laminations — Extreme Fiber in Bending[6,18] F_{byy} psi	Compression Perpendicular to Grain $F_{c\perp yy}$ psi	Horizontal Shear F_{vyy} psi	Horizontal Shear F_{vyy} psi (For members with multiple-piece laminations which are not edge glued)[21]	Modulus of Elasticity E_{yy}[22] x10⁶ psi	AXIALLY LOADED — Tension Parallel to Grain F_t psi	Compression Parallel to Grain F_c psi	Modulus of Elasticity E[22] x10⁶ psi
1	2	3	4	5	6	7	8	9	10	11	12	13	14	15	16
24F-V5	SP/SP	2400	2400	450	450	200	1.7	1600	335	175	90	1.5	1150	1700	1.5
Wet-use factors[2]		0.8	0.8	0.667	0.667	0.875	0.833	0.8	0.667	0.875	0.875	0.833	0.8	0.73	0.833
The following combination is balanced and intended for members continuous or cantilevered over supports and provides equal capacity in both positive and negative bending.															
						E-Rated Southern Pine									
16F-E1	SP/SP	1600	800	385[15]	385[15]	200	1.6	1550	385	175	90	1.5	1050	1600	1.5
The following combination is intended for straight or slightly cambered members for dry use and industrial appearance.[11]															
16F-E2	SP/SP	1600	800	385[15]	385[15]	90[17]	1.6	950	270	145	75	1.3	700	1050	1.3
The following combination is balanced and intended for members continuous or cantilevered over supports and provides equal capacity in both positive and negative bending.															
16F-E3	SP/SP	1600	1600	385[15]	385[15]	200	1.6	1700	385	175	90	1.5	1100	1650	1.5
20F-E1	SP/SP	2000	1000	385[15]	385[15]	200	1.7	1600	385	175	90	1.5	1050	1600	1.5
The following combination is balanced and intended for members continuous or cantilevered over supports and provides equal capacity in both positive and negative bending.															
20F-E2	SP/SP	2000	1000	450	385[15]	90[17]	1.6	1100	270	150	75	1.4	750	1000	1.4
The following combination is balanced and intended for members continuous or cantilevered over supports and provides equal capacity in both positive and negative bending.															
20F-E3	SP/SP	2000	2000	385[15]	385[15]	200	1.7	1800	385	175	90	1.5	1150	1700	1.5
22F-E1	SP/SP	2200	1100	450	385[15]	200	1.7	1600	385	175	90	1.5	1050	1650	1.5

Table B-5.   (Continued)

COMBINATION SYMBOL[20]	SPECIES OUTER LAMINATIONS/ CORE LAMINATIONS[5]	BENDING ABOUT X-X AXIS (Loaded Perpendicular to Wide Faces of Laminations)						BENDING ABOUT Y-Y AXIS (Loaded Parallel to Wide Faces of Laminations)					AXIALLY LOADED		
		Extreme Fiber in Bending F_{bxx}		Compression Perpendicular to Grain $F_{c\perp xx}$		Horizontal Shear F_{vxx} psi	Modulus of Elasticity E_{xx}^{22} x10⁶ psi	Extreme Fiber in Bending[6,18] F_{byy} psi	Compression Perpendicular to Grain $F_{c\perp yy}$ psi	Horizontal Shear F_{vyy} psi	Horizontal Shear F_{vyy} psi (For members with multiple-piece laminations which are not edge glued)[21]	Modulus of Elasticity $E_{yy}^{22} \times 10^6$ psi	Tension Parallel to Grain F_t psi	Compression Parallel to Grain F_c psi	Modulus of Elasticity E^{22} x10⁶ psi
		Tension Zone Stressed in Tension[6] psi	Compression Zone Stressed in Tension[7] psi	Tension Face psi	Compression Face psi										
1	2	3	4	5	6	7	8	9	10	11	12	13	14	15	16
The following combination is intended for straight or slightly cambered members for dry use and industrial appearance.[11]															
22F-E2	SP/SP	2200	1100	450	385[15]	90[17]	1.6	1250	270	155	80	1.4	850	1050	1.4
The following combination is balanced and intended for members continuous or cantilevered over supports and provides equal capacity in both positive and negative bending.															
22F-E3	SP/SP	2200	2200	450	450	200	1.7	1750	385	175	90	1.5	1150	1650	1.5
24F-E1	SP/SP	2400	1200	450	385[15]	200	1.8	1600	385	175	90	1.6	1100	1750	1.6
24F-E2	SP/SP	2400	1200	450	450	200	1.9	1700	385	175	90	1.6	1150	1700	1.6
The following combination is intended for straight or slightly cambered memebers for dry use and industrial appearance.[11]															
24F-E3	SP/SP	2400	1200	450	450	200[17]	1.8	1300	270	155	80	1.5	950	1100	1.5
The following combination is balanced and intended for members continuous or cantilevered over supports and provides equal capacity in both positive and negative bending.															
24F-E4	SP/SP	2400	2400	450	450	200	1.8	2000	385	175	90	1.6	1250	1750	1.6
Wet-use factors[2]		0.8	0.8	0.667	0.667	0.875	0.833	0.8	0.667	0.875	0.875	0.833	0.8	0.73	0.833

Footnotes for Table B-5. (This is Table 25-C-1, Part A, of the Uniform Building Code. Section numbers referred to below are those of UBC.)

[1] The combinations in this table are applicable to members consisting of four or more laminations and are intended primarily for members stressed in bending due to loads applied perpendicular to the wide faces of the laminations. Design values are tabulated, however, for loading both perpendicular and parallel to the wide faces of the laminations. For combinations and design values applicable to members loaded primarily axially or parallel to the wide faces of the laminations, see Table No. 25-C-1, Part B. For members of two or three laminations, see Table No. 25-C-1, Part B.

[2] The tabulated design values are for dry conditions of use. To obtain wet-use design values, multiply the tabulated values by the factors shown at the end of the table.

[3] The tabulated design values are for normal duration of loading. For other durations of loading, see Section 2504 (c) 4.

[4] The 22F and 24F combinations for members 15 inches and less in depth may not be readily available and the designer should check on availability prior to specifying. The 16F and 20F combinations are generally available for members 15 inches and less in depth.

Table B-5. *(Continued)*

[5]The symbols used for species are DF = Douglas fir-larch, DFS = Douglas fir south, HF = hem-fir, WW = western woods or Canadian softwood species, and SP = southern pine. (N3 refers to No. 3 structural joists and planks or structural light framing grade.)

[6]The tabulated design values in bending are applicable to members 12 inches or less in depth. For members greater than 12 inches in depth, the requirements of Section 2511 (d) 5 apply.

[7]Design values in this column are for extreme fiber stress in bending when the member is loaded such that the compression zone laminations are subjected to tensile stresses. The values in this column may be increased 200 psi where end joint spacing restrictions are applied to the compression zone when stressed in tension.

[8]Where specified, this value may be increased to 650 psi by providing in the bearing area at least one dense 2-inch nominal thickness lamination of Douglas fir-larch for western species combinations, or southern pine for southern pine combinations. These dense laminations must be backed by a medium-grain lamination of the same species.

[9]For bending members greater than 15 inches in depth, the design value for compression stress perpendicular-to grain is 650 psi on the tension face.

[10]Where specified, this value may be increased by providing at least two 2-inch nominal thickness Douglas fir-larch laminations in the bearing area. The compression-perpendicular-to-grain design values for Douglas fir-larch are 560 psi for medium grain and 650 psi for dense.

[11]These combinations are for dry conditions of use only because they may contain wane. They are recommended for industrial appearance grade and for straight or slightly cambered members only. If wane is omitted these restrictions do not apply.

[12]This value may be increased to 140 psi for western woods and to 155 psi for hem-fir when the member does not contain wane on both sides; to 115 psi for western woods and to 130 psi for hem-fir when the member does not contain wane on one side.

[13]This value may be increased to 110 psi when the member does not contain coarse-grain material; to 140 psi when the member does not contain wane on both sides or the member does not contain coarse-grain material and wane on one side; to 165 psi when the member does not contain coarse-grain material and wane on both sides.

[14]The compression-perpendicular-to-grain design value of 255 psi is based on the lowest strength species of the western woods group. If at least one 2-inch nominal thickness lamination of E-rated hem-fir with the same E value, or E-rated Douglas fir-larch 200,000 psi higher in modulus of elasticity (E) than that specified is used on the face of the member subjected to the compression-perpendicular-to-grain stress, $F_{c\perp}$ may be increased to 375 psi. If at least two 2-inch nominal thickness laminations of E-rated hem-fir with the same E value, or E-rated Douglas fir-larch 200,000 psi higher in modulus of elasticity than that specified are used in the bearing area on the face of the member subjected to the compression-perpendicular-to-grain stress, $F_{c\perp}$ may be increased to 500 psi.

[15]Where specified, this value may be increased to 650 psi by providing in the bearing area at least one 2-inch nominal thickness lamination of Douglas fir-larch for western species combinations, or one 2-inch nominal thickness lamination of southern pine for southern pine combinations having a modulus of elasticity (E) value 200,000 psi higher than the E value specified.

[16]E-rated Douglas fir-larch 200,000 psi higher in modulus of elasticity may be substituted for the specified E-rated hem-fir.

[17]This value may be increased to 140 psi when the member does not contain coarse-grain material or when the member does not contain wane on both sides; to 165 psi when members do not contain coarse-grain material or wane on one side; or to 200 psi when the member does not contain both coarse-grain material and wane on both sides of the member.

[18]Footnote 6 to Table No. 25-C-1, Part B, also applies.

[19]When Douglas fir south is used in place of all of the western wood laminations required in western species combinations 16F-V1, 16F-V4, 20F-V1, 20F-V5, 22F-V1, 22F-V5, 24F-V1, 24F-V6, 16F-E1, 16F-E4, 20F-E1, 20F-E4, 22F-E3, 24F-E6 and 24F-E7, the design value for horizontal shear (F_{vxx} = 165 psi and F_{vyy} = 145 psi for L3; and F_{vxx} = 90 psi and F_{vyy} = 135 psi for N3).

[20]The combination symbols relate to a specific combination of grades and species in U.B.C. Standard No. 25-11 that will provide the design values shown for the combinations. The first two numbers in the combination symbol correspond to the design value in bending shown in column 3. The letter in the combination symbol (either a ''V'' or an ''E'') indicates whether the combination is made from visually graded (V) or E-rated (E) lumber in the outer zones.

[21]These values for horizontal shear, F_{vyy} apply to members manufactured using multiple-piece laminations with unbonded edge joints. For members manufactured using single-piece laminations or using multiple-piece laminations with bonded edge joints, the horizontal shear values in column 11 apply.

[22]The duration of load modification factors given in Section 2504 (c) 4 shall not apply.

Table B-5 and footnotes are reproduced from the 1985 edition of the Uniform Building Code (copyright © 1985) with permission of the publishers, the International Conference of Building Officials.

Table B-6. Design Allowable Stresses and Modulus of Elasticity for Structural Glued Laminated Timber Under Normal-Duration Loading and Stressed Primarily in Axial Tension or Compression or in Bending with the Load Applied Parallel to the Wide Face of the Laminations.

					AXIALLY LOADED			BENDING ABOUT Y-Y AXIS (Loaded Parallel to Wide Faces of Laminations)							BENDING ABOUT X-X AXIS		
			MODULUS OF ELASTICITY[14] $E \times 10^6$ psi	COMPRESSION PERPENDICULAR TO GRAIN $F_c\perp$ psi	Tension Parallel to Grain F_t	Compression Parallel to Grain F_c		Extreme Fiber in Bending[6] F_{byy}			Horizontal Shear[7] F_{ryy}				Extreme Fiber in Bending[8] F_{bxx}		Horizontal Shear[9] F_{vxx}
COMBINATION SYMBOL	SPECIES[4]	GRADES[5]			2 or More Lams psi	4 or More Lams psi	2 or 3 Lams psi	4 or More Lams psi	3 Lams psi	2 Lams psi	4 or More Lams (For members with multiple-piece laminations.)[13]	4 or More Lams psi	3 Lams psi	2 Lams psi	2 Lams to 15 in. Deep[9] psi	4 or More Lams[10] psi	2 or More Lams psi
1	2	3	4	5	6	7	8	9	10	11	12	13	14	15	16	17	18
Visually Graded Western Species																	
1		L3	1.5	385[11]	900	1550	1200	1450	1250	1000	75	145	135	125	1250	1500	165
2		L2	1.7	385[11]	1250	1900	1600	1800	1600	1300	75	145	135	125	1700	2000	165
3		L2D	1.8	450	1450	2300	1850	2100	1850	1550	75	145	135	125	2000	2300	165
4	DF	L1CL	1.9	410[11]	1400	2100	1900	2200	2000	1650	75	145	135	125	1900	2200	165
5		L1	2.0	450	1600	2400	2100	2400	2100	1800	75	145	135	125	2200	2400	165
6		N3C	1.4	270	350	875	550	550	550	550	60	120	115	105	450	—	140
7		N3M	1.5	385	900	1550	700	1450	1250	1000	75	145	135	125	1000	—	165
8		N2	1.6	385[11]	1000	1550	1150	1600	1550	1300	75	145	135	125	1350	1600	165
9		N2D	1.8	450	1150	1800	1350	1850	1800	1500	75	145	135	125	1600	1850	165
10		N1	1.8	385[11]	1300	1950	1450	1950	1750	1500	75	145	135	125	1750	2100	165
11	DF	N1D	2.0	450	1500	2300	1700	2300	2100	1750	75	145	135	125	2100	2400	165
12		SS	1.8	385[11]	1400	1950	1650	2100	1950	1650	75	145	135	125	1900	2200	165
13		SSD	2.0	450	1600	2300	1950	2400	2300	1950	75	145	135	125	2200	2400	165

Table B-6. (Continued)

COMBINATION SYMBOL	SPECIES[4]	GRADES[5]	MODULUS OF ELASTICITY[14] $E \times 10^6$ psi	COMPRESSION PERPENDICULAR TO GRAIN $F_c\perp$ psi	AXIALLY LOADED — Tension Parallel to Grain F_t — 2 or More Lams psi	AXIALLY LOADED — Compression Parallel to Grain F_c — 4 or More Lams psi	AXIALLY LOADED — Compression Parallel to Grain F_c — 2 or 3 Lams psi	BENDING ABOUT Y-Y AXIS — Extreme Fiber in Bending[6] F_{byy} — 4 or More Lams psi	BENDING ABOUT Y-Y AXIS — Extreme Fiber in Bending[6] F_{byy} — 3 Lams psi	BENDING ABOUT Y-Y AXIS — Extreme Fiber in Bending[6] F_{byy} — 2 Lams psi	BENDING ABOUT Y-Y AXIS — Horizontal Shear[7] F_{vyy} — 4 or More Lams (For members with multiple-piece laminations)[13]	BENDING ABOUT Y-Y AXIS — Horizontal Shear[7] F_{vyy} — 4 or More Lams psi	BENDING ABOUT Y-Y AXIS — Horizontal Shear[7] F_{vyy} — 3 Lams psi	BENDING ABOUT Y-Y AXIS — Horizontal Shear[7] F_{vyy} — 2 Lams psi	BENDING ABOUT X-X AXIS — Extreme Fiber in Bending[8] F_{bxx} — 2 Lams to 15 in. Deep[9] psi	BENDING ABOUT X-X AXIS — Extreme Fiber in Bending[8] F_{bxx} — 4 or More Lams[10] psi	BENDING ABOUT X-X AXIS — Horizontal Shear F_{vxx} — 2 or More Lams psi
1	2	3	4	5	6	7	8	9	10	11	12	13	14	15	16	17	18
14	HF	L3	1.3	245[11]	800	1100	975	1200	1050	850	70	135	130	115	1100	1300	155
15		L2	1.4	245[11]	1050	1350	1300	1500	1350	1100	70	135	130	115	1450	1700	155
16		L1	1.6	245[11]	1200	1500	1450	1750	1550	1300	70	135	130	115	1600	1900	155
17		L1D	1.7	385	1400	1750	1700	2000	1850	1550	70	135	130	115	1900	2200	155
18		N3	1.3	245	425	900	575	700	700	700	70	135	130	115	575	—	155
19		N2	1.4	245[11]	850	1300	975	1350	1300	1100	70	135	130	115	1150	1350	155
20		N1	1.6	245[11]	975	1450	1250	1550	1500	1250	70	135	130	115	1350	1550	155
21		SS	1.6	245[11]	1100	1450	1350	1750	1650	1400	70	135	130	115	1500	1750	155
22	WW	L3	1.0	190	525	850	675	800	700	550	60	120	115	105	725	850	140
23		N3	1.0	190	275	625	450	450	450	450	60	120	115	105	400	—	140
24		N2	1.1	190	550	900	700	900	875	725	60	120	115	105	775	900	140
25		N1	1.2	190	650	1000	875	1050	1000	850	60	120	115	105	875	1050	140
26		SS	1.2	190	750	1000	1000	1150	1100	925	60	120	115	105	1000	1150	140
59	DFS	L3	1.1	335	800	1400	1050	1200	1050	850	75	145	135	125	1050	1250	165
60		L2	1.3	335	1050	1750	1400	1750	1550	1150	75	145	135	125	1450	1700	165
61		L1	1.5	395	1350	2200	1850	2000	1800	1500	75	145	135	125	1850	2200	165

Table B-6. (*Continued*)

COMBINATION SYMBOL	SPECIES[4]	GRADES[5]	MODULUS OF ELASTICITY[14] $E \times 10^6$ psi	COMPRESSION PERPENDICULAR TO GRAIN $F_c \perp$ psi	AXIALLY LOADED Tension Parallel to Grain F_t 2 or More Lams psi	AXIALLY LOADED Compression Parallel to Grain F_c 4 or More Lams psi	AXIALLY LOADED Compression Parallel to Grain F_c 2 or 3 Lams psi	BENDING ABOUT Y-Y AXIS Extreme Fiber in Bending[6] F_{byy} 4 or More Lams psi	BENDING ABOUT Y-Y AXIS Extreme Fiber in Bending[6] F_{byy} 3 Lams psi	BENDING ABOUT Y-Y AXIS Extreme Fiber in Bending[6] F_{byy} 2 Lams psi	Horizontal Shear[7] F_{vyy} 4 or More Lams (For members with multiple-piece laminations.)[13]	Horizontal Shear[7] F_{vyy} 4 or More Lams psi	Horizontal Shear[7] F_{vyy} 3 Lams psi	Horizontal Shear[7] F_{vyy} 2 Lams psi	BENDING ABOUT X-X AXIS Extreme Fiber in Bending[8] F_{bxx} 2 Lams to 15 in. Deep[9] psi	BENDING ABOUT X-X AXIS Extreme Fiber in Bending[8] F_{bxx} 4 or More Lams[10] psi	Horizontal Shear F_{vxx} 2 or More Lams psi
1	2	3	4	5	6	7	8	9	10	11	12	13	14	15	16	17	18
								E-Rated Western Species									
27		½-1.8E	1.8	385	900	1750	1200	1450	1250	1000	75	145	135	125	1250	1500	165
28		½-2.0E	2.0	450	1100	2000	1400	1450	1250	1000	75	145	135	125	1500	1750	165
29		½-2.2E	2.2	450	1250	2300	1550	1650	1400	1150	75	145	135	125	1700	2000	165
30	DF	⅙-1.8E	1.8	385	1550	2100	1700	2400	2400	2100	75	145	135	125	1800	2100	165
31		⅙-2.0E	2.0	450	1800	2400	1900	2400	2400	2400	75	145	135	125	2100	2400	165
32		⅙-2.2E	2.2	450	1800	2400	2100	2400	2400	2400	75	145	135	125	2300	2400	165
62		½-2.1E	2.1	450	1150	2200	1500	1550	1350	1100	75	145	135	125	1600	1900	165
63		⅙-2.1E	2.1	450	1800	2400	2000	2400	2400	2400	75	145	135	125	2200	2400	165
33		½-1.5E	1.5	245	800	1050	950	1200	1050	850	70	135	130	115	1100	1300	155
34		½-1.8E	1.8	385	900	1300	1200	1450	1250	1000	70	135	130	115	1250	1500	155
35	HF	½-2.0E	2.0	385	1100	1550	1400	1450	1250	1000	70	135	130	115	1500	1750	155
36		¼-1.5E	1.5	245	1200	1450	1300	2100	1900	1700	70	135	130	115	1400	1650	155
37		⅙-1.8E	1.8	385	1550	1950	1700	2400	2400	2100	70	135	130	115	1800	2100	155
38		⅙-2.0E	2.0	385	1800	2400	1900	2400	2400	2400	70	135	130	115	2100	2400	155

Table B-6. (Continued)

COMBINATION SYMBOL	SPECIES[4]	GRADES[5]	MODULUS OF ELASTICITY[14] $E \times 10^6$ psi	COMPRESSION PERPENDICULAR TO GRAIN $F_c\perp$ psi	AXIALLY LOADED Tension Parallel to Grain F_t 2 or More Lams psi	Compression Parallel to Grain F_c 4 or More Lams psi	Compression Parallel to Grain F_c 2 or 3 Lams psi	BENDING ABOUT Y-Y AXIS — Extreme Fiber in Bending[6] F_{byy} 4 or More Lams psi	3 Lams psi	2 Lams psi	Horizontal Shear[7] F_{ryy} 4 or More Lams (For members with multiple-piece laminations.)[13]	4 or More Lams psi	3 Lams psi	2 Lams psi	BENDING ABOUT X-X AXIS — Extreme Fiber in Bending[8] F_{bxx} 2 Lams to 15 in. Deep[9] psi	4 or More Lams[10] psi	Horizontal Shear F_{uxx} 2 or More Lams psi
1	2	3	4	5	6	7	8	9	10	11	12	13	14	15	16	17	18
39	WW	½-1.5E	1.5	190	800	1200	950	1200	1050	850	60	120	115	105	1100	1300	140
40		½-1.8E	1.8	190	900	1500	1200	1450	1250	1000	60	120	115	105	1250	1500	140
41		½-20E	2.0	190	1100	1750	1400	1450	1250	1000	60	120	115	105	1500	1750	140
42		¼-1.5E	1.5	190	1200	1550	1300	2100	1900	1700	60	120	115	105	1400	1650	140
43		⅙-1.8E	1.8	190	1550	1950	1700	2400	2400	2100	60	120	115	105	1800	2100	140
44		⅙-2.0E	2.0	190	1800	2200	1900	2400	2400	2400	60	120	115	105	2100	2400	140
Visually Graded Southern Pine																	
45	SP	N3C	1.1	270	325	850	550	550	550	550	60	120	115	105	450	—	140
46		N3M	1.3	385	900	1500	675	1450	1250	1000	90	175	165	150	1000	—	200
47		N2M[12]	1.4	385[11]	1200	1900	1150	1750	1550	1300	90	175	165	150	1400	1600	200
48		N2D[12]	1.7	450	1350	2200	1350	2000	1800	1500	90	175	165	150	1600	1900	200
49		N1M[12]	1.7	385[11]	1350	2100	1450	1950	1750	1500	90	175	165	150	1800	2100	200
50		N1D[12]	1.9	450	1550	2300	1700	2300	2100	1750	90	175	165	150	2100	2400	200
51		SSM	1.7	385[11]	1300	1900	1600	2100	1950	1650	90	175	165	150	1750	2100	200
52		SSD	1.9	450[11]	1500	2200	1850	2400	2300	1950	90	175	165	150	2100	2400	200

Table B-6.   (Continued)

					AXIALLY LOADED			BENDING ABOUT Y-Y AXIS							BENDING ABOUT X-X AXIS		
								Loaded Parallel to Wide Faces of Laminations							Loaded Perpendicular to Wide Faces of Laminations		
					Tension Parallel to Grain F_t	Compression Parallel to Grain F_c		Extreme Fiber in Bending F_{byy}			Horizontal Shear F_{ryy}				Extreme Fiber in Bending F_{bxx}		Horizontal Shear F_{vxx}
COMBINATION SYMBOL	SPECIES[4]	GRADES[5]	MODULUS OF ELASTICITY[14] E x 10^6 psi	COMPRESSION PERPENDICULAR TO GRAIN $F_c \perp$ psi	2 or More Lams psi	4 or More Lams psi	2 or 3 Lams psi	4 or More Lams psi	3 Lams psi	2 Lams psi	4 or More Lams (For members with multiple-piece laminations.)[13]	4 or More Lams psi	3 Lams psi	2 Lams psi	2 Lams to 15 in. Deep[9] psi	4 or More Lams[10] psi	2 or More Lams psi
1	2	3	4	5	6	7	8	9	10	11	12	13	14	15	16	17	18
E-Rated Southern Pine																	
53	SP	$\frac{1}{-}$1.8E	1.8	385	900	1900	1200	1450	1250	1000	90	175	165	150	1250	1500	200
54		$\frac{1}{-}$2.0E	2.0	450	1100	2300	1400	1450	1250	1000	90	175	165	150	1600	1750	200
55		$\frac{1}{-}$2.2E	2.2	450	1250	2400	1550	1650	1400	1150	90	175	165	150	1700	2000	200
56		$\frac{1}{-6}$1.8E	1.8	385	1550	1850	1700	2400	2400	2100	90	175	165	150	1800	2100	200
57		$\frac{1}{-6}$2.0E	2.0	450	1800	2400	1900	2400	2400	2400	90	175	165	150	2100	2400	200
58		$\frac{1}{-6}$2.2E	2.2	450	1800	2400	2100	2400	2400	2400	90	175	165	150	2300	2400	200
Wet-use factors[2]	—		0.833	0.667	0.8	0.73	0.73	0.8	0.8	0.8	0.875	0.875	0.875	0.875	0.8	0.875	0.875

(This is Table 25-C-1 Part B in the Uniform Building Code. Section numbers referred to below are those of UBC. Where Table 25-C-1 Part A is referred to, see Table B-5 of this book.)

[1]The combinations in this table are intended primarily for members loaded either axially or in bending with the loads acting parallel to the wide faces of the laminations. Design values for bending due to loading applied perpendicular to the wide faces of the laminations are also included: however, the combinations in Table No. 25-C-1, Part A, are usually better suited for this condition of loading. The design values for bending about the X-X axis (F_{bxx}) shown in Column 15 are for members from two laminations to 15 inches deep without tension laminations. Design values approximately 15 percent higher for members with four or more laminations are shown in Column 16. These higher design values, however, require special tension laminations which may not be readily available.

[2]The tabulated design values are for dry conditions of use. To obtain wet-use design values, multiply tabulated values by the factors shown at the end of the table.

[3]The tabulated design values are for normal duration of loading. For other durations of loading, see Section 2504 (c) 4.

[4]The symbols used for species are DF = Douglas fir-larch, DFS = Douglas fir south, HF = hem-fir, WW = western woods and Canadian softwood species, and SP = southern pine.

Table B-6.   (Continued)

[5] Grade designations are as follows:

Visually Graded Western Species

L1 is L1 laminating grade (dense for Douglas fir-larch and Douglas fir south).
L1D is L1 dense laminating grade for hem-fir.
L1CL is L1 close grain laminating grade.
L2D is L2 laminating grade (dense).
L2 is L2 laminating grade (medium grain).
L3 is L3 laminating grade (medium grain for Douglas fir-larch, Douglas fir south and hem-fir).
SSD is dense select structural, structural joists and planks, or structural light framing grade (dense).
SS is select structural, structural joists and planks, or structural light framing grade (medium grain for Douglas fir-larch).
N1D is dense No. 1 structural joists and planks, or structural light framing grade (dense).
N1 is No. 1 structural joists and planks, or structural light framing grade (medium grain for Douglas fir-larch).
N2D is dense No. 2 structural joists and planks, or structural light framing grade (medium grain for Douglas fir-larch).
N3M is No. 3 structural joists and planks, or structural light framing grade (medium grain).
N3C is No. 3 structural joists and planks, or structural light framing grade (coarse grain).
N3 is No. 3 structural joists and planks, or structural light framing grade.

Visually Graded Southern Pine

SSD is dense select structural, structural joists and planks, or structural light framing grade (dense).
SSM is select structural, structural joists and planks, or structural light framing grade (medium grain).
N1D is No. 1 dense structural joists and planks, or structural light framing grade or No. 1 boards graded as dense.
N1M is No. 1 structural joists and planks, or structural light framing grade or No. 1 boards all with a medium grain rate of growth.
N2D is No. 2 dense structural joists and planks, or structural light framing grade or No. 2 boards graded as dense.
N2M is No. 2 structural joists and planks, or structural light framing grade or No. 2 boards all with a medium grain rate of growth.
N3M is No. 3 structural joists and planks, or structural light framing grade or No. 3 boards all with a medium grain rate of growth.
N3C is No. 3 structural joists and planks, or structural light framing grade or No. 3 boards all with coarse grain rate of growth.

E-Rated Grades—All Species

$\frac{1}{6}$-2.2E has $\frac{1}{6}$ edge characteristic with 2.2E
$\frac{1}{6}$-2.1E has $\frac{1}{6}$ edge characteristic with 2.1E.
$\frac{1}{6}$-2.0E has $\frac{1}{6}$ edge characteristic with 2.0E.
$\frac{1}{4}$-1.8E has $\frac{1}{6}$ edge characteristic with 1.8E.
$\frac{1}{4}$-1.5E has $\frac{1}{6}$ edge characteristic with 1.5E.
$\frac{1}{2}$-2.2E, $\frac{1}{2}$-2.1E, $\frac{1}{2}$-2.0E, $\frac{1}{2}$-1.8E are E-rated grades with edge characteristics occupying up to one half of cross section.

[6] The values of F_{bxx} were calculated based on members 12 inches in depth (bending about Y-Y axis). When the depth is less than 12 inches, the values of F_{bxy} can be increased by multiplying by the following factors:

Depth, Inches	Multiplying Factor
10.75	1.01
8.75	1.04
6.75	1.07
5.125	1.10
3.125	1.14

[7] The design values in horizontal shear contained in this table are based on members without wane.

[8] The tabulated design values in bending are applicable to members 12 inches or less in depth. For members greater than 12 inches in depth, the requirements of Section 2511 (d) 5 apply.

[9] The design values in Column 15 are for members of from two laminations to 15 inches in depth without tension laminations.

[10] The design values in Column 16 are for members of four or more laminations in depth and require special tension laminations. When these values are used in design and the member is specified by combination symbol, the designer should also specify the required design value in bending.

Table B-6. (Continued)

[11] When tension laminations are used to obtain the design value for F_{bxx} shown in Column 16, the compression perpendicular to grain value, $F_c\perp$, for the tension face may be increased to 650 psi for Douglas fir-larch and southern pine, and to 500 psi for hem-fir because the tension laminations are required to be dense.

[12] Combinations 47, 48, 49 and 50 have more restrictive slope of grain requirements than the basic slope of grain of the grades of lumber used in order to obtain higher tension-parallel-to-grain values and design values in bending when loaded perpendicular to the wide faces of the laminations. The slopes of grain used to calculate the design values in Table No. 25-C-1, Part B, were: Combination 47, 1:14; Combination 48, 1:14; Combination 49, 1:16; and Combination 50, 1:14. When design stresses are lower than the design values shown, or when a less restrictive slope of grain provides the same design value, a less restrictive slope of grain may be used. The following table gives the design values of these combinations for various slopes of grain: Values of F_{bxx} in column 5 are for members of two laminations to 15 inches in depth without tension laminations, and values in Column 6 are for members of four or more laminations with tension laminations.

[13] These values for horizontal shear, F_{vxx}, apply to members manufactured using multiple-piece laminations with unbonded edge joints. For members using single-piece laminations or using multiple-piece laminations with bonded edge joints, the horizontal shear values tabulated in renumbered Columns 13, 14 and 15 apply.

[14] The duration of load modification factors given in Section 2504 (c) 4 shall not apply.

Slope of Grain	Comb. No.	Tension Parallel to Grain (F_t) 2 or More Lams psi 2	Comp. Parallel to Grain (F_c) 2 or 3 Lams psi 3	Comp. Parallel to Grain (F_c) 4 or More Lams psi 4	Bending about the X-X Axis F_{bxx}		Bending about the Y-Y Axis F_{byy}		
					2 Lams to 15 in. psi 5	4 or More Lams psi 6	2 Lams psi 7	3 Lams psi 8	4 or More Lams psi 9
1									
1:12	47	1200	1150	1900	1400	1600	1300	1550	1750
	48	1400	1350	2200	1600	1900	1500	1800	2000
	49	1300	1450	1900	1750	2100	1500	1750	1950
	50	1550	1700	2200	2100	2400	1750	2100	2300
1:10	47	1150	1150	1700	1400	1600	1300	1550	1750
	48	1350	1350	2000	1600	1900	1500	1800	2000
	49	1150	1450	1700	1550	1850	1500	1750	1850
	50	1350	1700	2000	1800	2100	1750	2100	2100
1:8	47	1000	1150	1500	1350	1600	1300	1550	1600
	48	1150	1350	1750	1600	1850	1500	1800	1850
	49	—	—	—	—	—	—	—	—
	50	—	—	—	—	—	—	—	—

Table B-6 and footnotes are reproduced from the 1985 edition of the Uniform Building Code (copyright © 1985) with permission of the publishers, the International Conference of Building Officials.

Table B-7.   Allowable Parallel-to-Grain Compressive Stress (psi) in End Bearing, F_g.

Species	Wet service	Dry service		
		Sawn lumber @ MC ≤ 19%		
		≤ 4″ thick	>4″ thick	Glulams
Calif. redwood	1560	2270	1720	2620
Calif. redwood, open grain	1150	1670	1270	1940
D. fir-larch	1340	2020	1480	2350
D. fir-larch, north	1340	2020	1480	2350
D. fir-larch, dense	1570	2360	1730	2750
D. fir, south	1220	1820	1340	2130
Hem-fir	1110	1670	1220	1940
Southern pine	1320	1970	1450	2300
Southern pine, dense	1540	2310	1690	2690

Reprinted with permission of the National Forest Products Association, Washington, D.C.

Appendix C

Table C-1.   Species Grouping for Design of Connections.

Species	Avg. Spec. Grav. *	Groups for timber connectors				Groups for nails, spikes, lag screws, wood screws, metal plate connectors			
		A	B	C	D	I	II	III	IV
Alpine fir	0.36				x				x
Ash, commercial white	0.62	x				x			
Balsam fir	0.36				x				x
Beech	0.68	x				x			
Birch, sweet, yellow	0.66	x				x			
Calif. redwood	0.42			x				x	
Calif. redwood (open grain)	0.37				x				x
Coast species	0.39								x
Douglas fir-larch	0.51		x				x		
Douglas fir-larch (dense)		x							
D. fir, south	0.48			x				x	
Eastern spruce	0.41				x				x
Engelmann spruce	0.36				x				x
Hem-fir	0.42			x				x	
Lodgepole pine	0.44			x				x	
Maple, black or sugar	0.66	x				x			
Mountain hemlock	0.47			x				x	
Oak, red or white	0.67	x				x			
Ponderosa pine	0.49			x				x	
Sitka spruce	0.43			x				x	
Southern cypress	0.48			x				x	
Southern pine	0.55		x				x		
Southern pine (dense)		x							
Virginia pine	0.54						x		
Western cedars	0.35				x				x
Western hemlock	0.48			x				x	
Yellow poplar	0.46			x				x	

*Average specific gravities, based on oven-dry weight and volume.

Reprinted with permission of the National Forest Products Association, Washington, D.C.

Table C-2.   Fastener Load Modification for Moisture Content.

Fastener type	Moisture condition		Modification factor
	When installed	In service	
Wire nails and spikes	Under withdrawal load		
	Dry	Dry	1.0
	Dry	Subject to wetting & drying	0.25
	Partially seasoned or wet	Dry	0.25
	Partially seasoned or wet	Remaining wet	1.0
	Under lateral load		
	Dry	Dry	1.0
	All other conditions		0.75
Bolts and lag screws	Dry	Dry	1.0
	Dry or wet	Exposed to weather	0.75
	Dry or wet	Wet	0.67
	(See also Table C-3.)		
Wood screws	Dry or wet	Dry	1.0
	Dry or wet	Exposed to weather	0.75
	Dry or wet	Wet	0.67
Timber connectors	Dry	Dry	1.0
	Wet	Dry	0.8
	Dry or wet	Partially seasoned or wet	0.67
Metal plate connectors	Dry	Dry	1.0
	Partially seasoned or wet	Dry or wet	0.8

Meanings of terms used above are:
Dry—MC 19% or less.
Wet—MC at or above the fiber saturation point (approximately 30%).
Partially seasoned—MC is above 19% but less than FSP.
For timber connectors, the moisture conditions listed apply to a depth of 3/4″ on the side in which the connector is installed.
Factors for timber connectors, bolts, and laterally loaded lag screws may be obtained by interpolation of the above when they are installed in partially seasoned wood that will become dry before full load is applied.

Reprinted with permission of the National Forest Products Association, Washington, D.C.

Table C-3.   Modification of Allowable Load for Laterally Loaded Bolts and Lag Screws Installed in Timber that is Seasoned in Place.

The multiplying factors below apply when the wood is above the fiber saturation point when the connection is made, but dries to 19% MC or less before full design load is applied. Factors apply to connections using either wood or metal splice plates.

Pattern of connection	Modification factor
One fastener only	
Two or more fasteners in a single line parallel to the grain	1.0
Two or more lines of fasteners parallel to the grain, but with separate splice plates for each line	
All other patterns	0.4

Reprinted with permission of the National Forest Products Association, Washington, D.C.

Table C-4.   Design Allowable Lateral Loads for Nails and Spikes (Lb).

Length	2″	2½″	3″	3½″	4″	4½″	5″	5½″	6″
Penny-wt	6d	8d	10d	16d	20d	30d	40d	50d	60d
				COMMON WIRE NAILS					
Dia. (in.)	0.113	0.131	0.148	0.162	0.192	0.207	0.225	0.244	0.263
Std. Penetration:*									
Group I	1.13″	1.31″	1.48″	1.62″	1.92″	2.07″	2.25″	2.44″	2.63″
Group II	1.24″	1.44″	1.63″	1.78″	2.11″	2.28″	2.48″	2.68″	2.89″
Group III	1.47″	1.70″	1.92″	2.11″	2.50″	2.69″	2.93″	3.17″	3.42″
Group IV	1.58″	1.83″	2.07″	2.27″	2.69″	2.90″	3.15″	3.42″	3.68″
Allowable Lateral Load									
Group I	77	97	116	133	172	192	218	246	275
Group II	63	78	94	108	139	155	176	199	223
Group III	51	64	77	88	114	127	144	163	182
Group IV	41	51	61	70	91	102	115	130	146
				BOX NAILS					
Dia. (in.)	0.099	0.113	0.128	0.135	0.148	0.148	0.162		
Std. Penetration:*									
Group I	0.99″	1.13″	1.28″	1.35″	1.48″	1.48″	1.62″		
Group II	1.09″	1.24″	1.41″	1.49″	1.63″	1.63″	1.78″		
Group III	1.29″	1.47″	1.66″	1.76″	1.92″	1.92″	2.11″		
Group IV	1.39″	1.58″	1.79″	1.89″	2.07″	2.07″	2.27″		
Allowable Lateral Load									
Group I	64	77	93	101	116	116	113		
Group II	51	63	76	82	94	94	108		
Group III	42	51	62	67	77	77	88		
Group IV	34	41	49	54	61	61	70		
				COMMON WIRE SPIKES					
Dia. (in.)			0.192	0.207	0.225	0.244	0.263	0.283	0.283
Std. Penetration:*									
Group I			1.92″	2.07″	2.25″	2.44″	2.63″	2.83″	2.83″
Group II			2.11″	2.28″	2.48″	2.68″	2.89″	3.11″	3.11″
Group III			2.50″	2.69″	2.93″	3.17″	3.42″	3.68″	3.68″
Group IV			2.69″	2.90″	3.15″	3.42″	3.68″	3.96″	3.96″
Allowable Lateral Load									
Group I			172	192	218	246	275	307	307
Group II			139	155	176	199	223	248	248
Group III			114	127	144	163	182	203	203
Group IV			91	102	115	130	146	163	163

*Values shown are for standard penetration or more, under normal-duration load. Penetration is the length of nail in the piece that contains the point. Standard penetrations are: Group I species, 10 diameters; Group II species, 11 diameters; Group III species, 13 diameters; and Group IV species, 14 diameters.

Spikes longer than 6″ are also available.

Diameter, standard penetrations, and allowable lateral loads for 12d nails are the same as for 10d. Length of 12d nails is 3¼″.

Reprinted with permission of the National Forest Products Association, Washington, D.C.

Table C-5.   Design Allowable Withdrawal Loads for Nails and Spikes.

Loads tabulated are pounds per inch of penetration in side grain of the member that contains the nail point, and are for normal-duration loading.

Length	2″	2½″	3″	3½″	4″	4½″	5″	5½″	6″
Penny-wt	6d	8d	10d	16d	20d	30d	40d	50d	60d
			COMMON WIRE NAILS						
Group I	47	55	62	68	80	86	94	102	110
Group II	29	34	38	42	49	53	58	63	67
Group III	18	21	23	26	30	33	35	38	41
Group IV	8	10	11	12	14	15	17	18	19
			BOX NAILS						
Group I	41	47	53	56	62	62	68		
Group II	25	29	33	35	38	38	42		
Group III	16	18	20	21	23	23	26		
Group IV	7	8	9	10	11	11	12		
			COMMON WIRE SPIKES						
Group I			80	86	94	102	110	118	118
Group II			49	53	58	63	67	73	73
Group III			30	33	35	38	41	45	45
Group IV			14	15	17	18	19	21	21

Spikes longer than 6″ are available.
Values for 12d nails ($3\frac{1}{4}$″ long) are the same as for 10d. Allowable loads shown are for specific gravities as follows:
Group I, 0.62; Group II, 0.51; Group III, 0.42; and Group IV, 0.31.
For values in wood of higher specific gravities, see NDS.
The allowable withdrawal load for nails in end grain should be taken as zero.
Reprinted with permission of the National Forest Products Association, Washington, D.C.

Table C-6.   Dimensions of Standard Lag Screws.

[All dimensions in inches]

D = Nominal diameter.
D_s = D = Diameter of shank.
D_R = Diameter at root of thread.
W = Width of bolt head across flats.
H = Height of bolt head.
L = Nominal length of bolt.
S = Length of shank.
T = Length of thread for effective embedment.
E = Length of tapered tip.
N = Number of threads per inch.

Dimensions of lag screw with nominal diameter (D) of—

Item	$\frac{1}{4}$	$\frac{5}{16}$	$\frac{3}{8}$	$\frac{7}{16}$	$\frac{1}{2}$	$\frac{9}{16}$	$\frac{5}{8}$	$\frac{3}{4}$	$\frac{7}{8}$	1	$1\frac{1}{8}$	$1\frac{1}{4}$
D_s = D	0.250	0.3125	0.375	0.4375	0.500	0.5625	0.625	0.750	0.875	1.000	1.125	1.250
D_R	.173	.227	.265	.328	.371	.435	.471	.579	.683	.780	.887	1.012
E	$\frac{3}{16}$	$\frac{1}{4}$	$\frac{1}{4}$	$\frac{9}{32}$	$\frac{5}{16}$	$\frac{3}{8}$	$\frac{3}{8}$	$\frac{7}{16}$	$\frac{1}{2}$	$\frac{9}{16}$	$\frac{5}{8}$	$\frac{3}{4}$
H	$\frac{11}{64}$	$\frac{13}{64}$	$\frac{1}{4}$	$\frac{19}{64}$	$\frac{21}{64}$	$\frac{3}{8}$	$\frac{27}{64}$	$\frac{1}{2}$	$\frac{19}{32}$	$\frac{21}{32}$	$\frac{3}{4}$	$\frac{27}{32}$
W	$\frac{3}{8}$	$\frac{1}{2}$	$\frac{9}{16}$	$\frac{5}{8}$	$\frac{3}{4}$	$\frac{7}{8}$	$\frac{15}{16}$	$1\frac{1}{8}$	$1\frac{5}{16}$	$1\frac{1}{2}$	$1\frac{11}{16}$	$1\frac{7}{8}$
N	10	9	7	7	6	6	5	$4\frac{1}{2}$	4	$3\frac{1}{2}$	$3\frac{1}{4}$	$3\frac{1}{4}$
D^2	.0625	.0976	.1406	.1914	.2500	.3164	.3906	.5625	.7656	1.0000	1.2656	1.5625
D^3	.354	.418	.479	.538	.594	.649	.703	.806	.905	1.000	1.0925	1.182

Nominal length of screw (L) in inches*

L	Item	$\frac{1}{4}$	$\frac{5}{16}$	$\frac{3}{8}$	$\frac{7}{16}$	$\frac{1}{2}$	$\frac{9}{16}$	$\frac{5}{8}$	$\frac{3}{4}$	$\frac{7}{8}$	1	$1\frac{1}{8}$	$1\frac{1}{4}$
1	S	$\frac{1}{4}$	$\frac{1}{4}$	$\frac{1}{4}$	$\frac{1}{4}$	$\frac{1}{4}$	—	—	—	—	—	—	—
	T	$\frac{9}{16}$	$\frac{1}{2}$	$\frac{1}{2}$	$\frac{15}{32}$	$\frac{7}{16}$	—	—	—	—	—	—	—
$1\frac{1}{2}$	S	$\frac{3}{8}$	$\frac{3}{8}$	$\frac{3}{8}$	$\frac{3}{8}$	$\frac{3}{8}$	—	—	—	—	—	—	—
	T	$\frac{15}{16}$	$\frac{7}{8}$	$\frac{7}{8}$	$\frac{27}{32}$	$\frac{13}{16}$	—	—	—	—	—	—	—
2	S	$\frac{1}{2}$	$\frac{1}{2}$	$\frac{1}{2}$	$\frac{1}{2}$	$\frac{1}{2}$	$\frac{1}{2}$	$\frac{1}{2}$	—	—	—	—	—
	T	$1\frac{5}{16}$	$1\frac{1}{4}$	$1\frac{1}{4}$	$1\frac{17}{32}$	$1\frac{3}{16}$	$1\frac{1}{8}$	$1\frac{1}{8}$	—	—	—	—	—
$2\frac{1}{2}$	S	1	1	$\frac{7}{8}$	$\frac{3}{4}$	$\frac{3}{4}$	1	$3\frac{1}{4}$	—	—	—	—	—
	T	$1\frac{5}{16}$	$1\frac{3}{8}$	$1\frac{3}{8}$	$1\frac{5}{32}$	$1\frac{7}{16}$	$3\frac{3}{8}$	$3\frac{3}{8}$	—	—	—	—	—
3	S	1	1	1	$1\frac{23}{32}$	$1\frac{11}{16}$	1	1	$1\frac{9}{16}$	1	$1\frac{7}{16}$	—	—
	T	$1\frac{13}{16}$	$1\frac{3}{4}$	$1\frac{3}{4}$	$1\frac{23}{32}$	$1\frac{11}{16}$	$1\frac{5}{8}$	$1\frac{5}{8}$	$1\frac{1}{2}$	$1\frac{1}{2}$	—	—	—
4	S	$1\frac{1}{2}$	$1\frac{1}{2}$	$1\frac{1}{2}$	$1\frac{1}{2}$	$1\frac{1}{2}$	$1\frac{1}{2}$	$1\frac{1}{2}$	$1\frac{1}{2}$	$1\frac{1}{2}$	$1\frac{1}{2}$	$1\frac{1}{2}$	$1\frac{1}{2}$
	T	$2\frac{5}{16}$	$2\frac{1}{4}$	$2\frac{1}{4}$	$2\frac{7}{32}$	$2\frac{3}{16}$	$2\frac{1}{8}$	$2\frac{1}{8}$	$2\frac{1}{16}$	2	$1\frac{15}{16}$	$1\frac{7}{8}$	$1\frac{3}{4}$

5	S	2	2	2	2	2	2	2	2	2	2	2
	T	$2\frac{13}{16}$	$2\frac{3}{4}$	$2\frac{23}{32}$	$2\frac{11}{16}$	$2\frac{5}{8}$	$2\frac{5}{8}$	$2\frac{9}{16}$	$2\frac{1}{2}$	$2\frac{7}{16}$	$2\frac{3}{8}$	$2\frac{1}{4}$
6	S	$2\frac{1}{2}$	$2\frac{1}{2}$	$2\frac{1}{2}$	$2\frac{1}{2}$	$2\frac{1}{2}$	$2\frac{1}{2}$	$2\frac{1}{2}$	$2\frac{1}{2}$	$2\frac{1}{2}$	$2\frac{1}{2}$	$2\frac{1}{2}$
	T	$3\frac{5}{16}$	$3\frac{1}{4}$	$3\frac{7}{32}$	$3\frac{3}{16}$	$3\frac{1}{8}$	$3\frac{1}{8}$	$3\frac{1}{16}$	3	$2\frac{15}{16}$	$2\frac{7}{8}$	$2\frac{3}{4}$
7	S	3	3	3	3	3	3	3	3	3	3	3
	T	$3\frac{13}{16}$	$3\frac{3}{4}$	$3\frac{23}{32}$	$3\frac{11}{16}$	$3\frac{5}{8}$	$3\frac{5}{8}$	$3\frac{9}{16}$	$3\frac{1}{2}$	$3\frac{7}{16}$	$3\frac{3}{8}$	$3\frac{1}{4}$
8	S	$3\frac{1}{2}$	$3\frac{1}{2}$	$3\frac{1}{2}$	$3\frac{1}{2}$	$3\frac{1}{2}$	$3\frac{1}{2}$	$3\frac{1}{2}$	$3\frac{1}{2}$	$3\frac{1}{2}$	$3\frac{1}{2}$	$3\frac{1}{2}$
	T	$4\frac{5}{16}$	$4\frac{1}{4}$	$4\frac{7}{32}$	$4\frac{3}{16}$	$4\frac{1}{8}$	$4\frac{1}{8}$	$4\frac{1}{16}$	4	$3\frac{15}{16}$	$3\frac{7}{8}$	$3\frac{3}{4}$
9	S	4	4	4	4	4	4	4	4	4	4	4
	T	$4\frac{13}{16}$	$4\frac{3}{4}$	$4\frac{23}{32}$	$4\frac{11}{16}$	$4\frac{5}{8}$	$4\frac{5}{8}$	$4\frac{9}{16}$	$4\frac{1}{2}$	$4\frac{7}{16}$	$4\frac{3}{8}$	$4\frac{1}{4}$
10	S	$4\frac{3}{4}$	$4\frac{3}{4}$	$4\frac{3}{4}$	$4\frac{3}{4}$	$4\frac{3}{4}$	$4\frac{3}{4}$	$4\frac{3}{4}$	$4\frac{3}{4}$	$4\frac{3}{4}$	$4\frac{3}{4}$	$4\frac{3}{4}$
	T	$5\frac{5}{16}$	5	$4\frac{31}{32}$	$4\frac{15}{16}$	$4\frac{7}{8}$	$4\frac{7}{8}$	$4\frac{13}{16}$	$4\frac{3}{4}$	$4\frac{11}{16}$	$4\frac{5}{8}$	$4\frac{1}{2}$
11	S	$5\frac{1}{2}$	$5\frac{1}{2}$	$5\frac{1}{2}$	$5\frac{1}{2}$	$5\frac{1}{2}$	$5\frac{1}{2}$	$5\frac{1}{2}$	$5\frac{1}{2}$	$5\frac{1}{2}$	$5\frac{1}{2}$	$5\frac{1}{2}$
	T	$5\frac{9}{32}$	$5\frac{1}{4}$	$5\frac{7}{32}$	$5\frac{3}{16}$	$5\frac{1}{8}$	$5\frac{1}{8}$	$5\frac{1}{16}$	5	$4\frac{15}{16}$	$4\frac{7}{8}$	$4\frac{3}{4}$
12	S	6	6	6	6	6	6	6	6	6	6	6
	T	$5\frac{13}{16}$	$5\frac{3}{4}$	$5\frac{23}{32}$	$5\frac{11}{16}$	$5\frac{5}{8}$	$5\frac{5}{8}$	$5\frac{9}{16}$	$5\frac{1}{2}$	$5\frac{7}{16}$	$5\frac{3}{8}$	$5\frac{1}{4}$

*Length of thread on intervening bolt lengths is the same as that of the next shorter bolt length listed. The length of the thread on standard bolt lengths in excess of 12 inches is equal to 1/2 the bolt length.

Source: Courtesy of Western Wood Products Association.

Table C-7. Design Allowable Lateral Loads for Lag Screws in Side Grain and with Wood Side Members.

Side member thickness (in.)	Lag screw length (in.)	Lag screw shank diam. (in.)	Allowable lateral load per lag screw (1b)							
			Group I		Group II		Group III		Group IV	
			P	Q	P	Q	P	Q	P	Q
$1\frac{1}{2}$	4	$\frac{1}{4}$	200	190	170	170	130	120	100	100
		$\frac{5}{16}$	280	230	210	180	150	130	120	100
		$\frac{3}{8}$	320	240	240	180	170	130	140	100
		$\frac{7}{16}$	350	250	270	190	190	140	150	110
		$\frac{1}{2}$	390	250	290	190	210	140	170	110
		$\frac{5}{8}$	470	280	360	210	260	150	200	120
	5	$\frac{1}{4}$	230	220	200	190	180	170	160	150
		$\frac{5}{16}$	330	280	290	250	230	200	190	160
		$\frac{3}{8}$	430	330	370	280	260	200	210	160
		$\frac{7}{16}$	540	380	400	290	290	210	230	160
		$\frac{1}{2}$	580	380	440	280	310	200	250	160
		$\frac{5}{8}$	710	420	530	320	380	230	310	180
	6	$\frac{1}{4}$	270	260	230	220	200	200	180	180
		$\frac{5}{16}$	380	320	330	280	290	250	260	220
		$\frac{3}{8}$	480	370	420	320	360	280	290	220
		$\frac{7}{16}$	590	420	510	360	400	280	320	230
		$\frac{1}{2}$	700	460	600	390	430	280	340	220
		$\frac{5}{8}$	850	510	710	430	510	310	410	250
	7	$\frac{1}{4}$	280	270	240	230	210	210	190	180
		$\frac{5}{16}$	400	340	350	300	310	270	280	230
		$\frac{3}{8}$	520	400	450	340	410	310	360	270
		$\frac{7}{16}$	640	450	560	390	500	350	420	290
		$\frac{1}{2}$	760	490	660	430	550	360	440	290
		$\frac{5}{8}$	910	540	780	470	640	380	510	310
$2\frac{1}{2}$	6	$\frac{3}{8}$	450	340	370	280	270	200	210	160
		$\frac{7}{16}$	570	400	430	310	310	220	250	180
		$\frac{1}{2}$	620	410	470	310	340	220	270	180
		$\frac{5}{8}$	730	440	550	330	390	240	320	190
		$\frac{3}{4}$	820	450	620	340	440	240	360	200
		$\frac{7}{8}$	930	490	710	370	500	260	400	210
		1	1040	520	790	390	560	280	450	230
	7	$\frac{3}{8}$	500	380	430	330	370	280	300	220
		$\frac{7}{16}$	660	470	570	410	420	300	340	240
		$\frac{1}{2}$	830	540	650	420	460	300	370	240
		$\frac{5}{8}$	1000	600	750	450	540	320	430	260
		$\frac{3}{4}$	1110	610	840	460	600	330	480	270
		$\frac{7}{8}$	1260	660	950	500	680	360	550	290
		1	1420	710	1070	540	770	380	620	310
	8	$\frac{3}{8}$	550	420	480	360	430	320	380	290
		$\frac{7}{16}$	720	510	630	440	550	390	440	310
		$\frac{1}{2}$	890	580	770	500	600	390	480	310
		$\frac{5}{8}$	1230	740	970	580	700	420	560	340
		$\frac{3}{4}$	1430	790	1080	600	780	430	620	340
		$\frac{7}{8}$	1590	830	1200	630	860	450	690	360
		1	1800	900	1360	680	970	490	780	390
	9	$\frac{3}{8}$	600	460	520	390	460	350	410	310
		$\frac{7}{16}$	790	550	680	480	610	430	540	380
		$\frac{1}{2}$	960	630	830	540	740	480	600	390
		$\frac{5}{8}$	1310	790	1130	680	860	520	690	420
		$\frac{3}{4}$	1670	920	1340	740	960	530	770	420
		$\frac{7}{8}$	1920	1000	1450	760	1040	540	830	430
		1	2170	1090	1640	820	1180	590	940	470

Allowables are for normal-duration loads.
   P is allowable load parallel to grain.
   Q is allowable load perpendicular to grain.
Where different authorities quote different allowables, the lower value is shown above.

Reprinted with permission of the National Forest Products Association, Washington, D.C.

Table C-8.   Design Allowable Lateral Loads for Lag Screws in Side Grain and with $\frac{1}{2}$" Thick Metal Side Member.

Lag screw length (in.)	Lag screw shank diam. (in.)	Allowable lateral load per lag screw (lb)							
		Group I		Group II		Group III		Group IV	
		P	Q	P	Q	P	Q	P	Q
3	$\frac{1}{4}$	240	185	210	160	160	120	130	100
	$\frac{5}{16}$	350	240	265	180	190	130	150	100
	$\frac{3}{8}$	420	250	315	190	225	140	180	110
	$\frac{7}{16}$	480	270	360	200	260	140	210	115
	$\frac{1}{2}$	535	280	400	210	290	150	230	120
	$\frac{5}{8}$	645	310	480	230	350	170	280	130
4	$\frac{1}{4}$	270	210	235	180	210	160	185	145
	$\frac{5}{16}$	410	280	350	240	290	200	230	160
	$\frac{3}{8}$	570	350	480	290	340	210	275	170
	$\frac{7}{16}$	725	405	550	310	390	220	310	175
	$\frac{1}{2}$	810	420	610	320	440	230	350	180
	$\frac{5}{8}$	980	470	740	355	530	250	425	200
5	$\frac{1}{4}$	285	220	245	190	220	170	195	150
	$\frac{5}{16}$	435	295	375	255	340	230	300	200
	$\frac{3}{8}$	615	375	530	320	470	285	375	230
	$\frac{7}{16}$	810	460	700	390	530	300	430	240
	$\frac{1}{2}$	1040	540	840	435	600	310	480	250
	$\frac{5}{8}$	1310	630	990	475	710	340	570	275
6	$\frac{1}{4}$	285	220	250	190	220	170	195	150
	$\frac{5}{16}$	445	300	385	260	340	230	300	205
	$\frac{3}{8}$	630	385	550	330	490	300	430	260
	$\frac{7}{16}$	850	480	730	410	660	370	540	300
	$\frac{1}{2}$	1100	570	950	490	760	395	600	320
	$\frac{5}{8}$	1640	790	1290	620	920	440	740	350
7	$\frac{3}{8}$	640	390	555	340	495	300	440	265
	$\frac{7}{16}$	870	485	750	420	670	375	590	330
	$\frac{1}{2}$	1120	580	970	500	870	450	740	380
	$\frac{5}{8}$	1710	820	1480	710	1130	540	900	430
	$\frac{3}{4}$	2380	1050	1840	810	1310	580	1050	460
8	$\frac{7}{16}$	880	490	760	420	680	380	600	330
	$\frac{1}{2}$	1140	590	980	510	880	455	775	400
	$\frac{5}{8}$	1750	840	1510	720	1320	630	1060	510
	$\frac{3}{4}$	2470	1090	2130	940	1560	685	1250	550
	$\frac{7}{8}$	3260	1360	2480	1030	1770	740	1420	590
	1	3680	1470	2780	1110	1990	795	1600	640
9	$\frac{1}{2}$	1150	600	990	510	890	460	780	410
	$\frac{5}{8}$	1770	850	1530	730	1370	655	1210	580
	$\frac{3}{4}$	2510	1110	2160	950	1790	780	1440	630
	$\frac{7}{8}$	3360	1400	2880	1200	2060	855	1650	685
	1	4280	1710	3230	1290	2310	925	1860	740

Allowables are for normal-duration loads.

  P is allowable load parallel to grain.

  Q is allowable load perpendicular to grain.

Where different authorities quote different allowables, the lower value is shown above.

Reprinted with permission of the National Forest Products Association, Washington, D.C.

Table C-9.   Design Allowable Withdrawal Loads for Lag Screws in Side Grain.

Lag Screw Shank Diameter (in.)	Allowable load (pounds) per inch of penetration of threaded portion only into the wood member containing the lag screw point.			
	Group I	Group II	Group III	Group IV
$\frac{1}{4}$	311	232	173	110
$\frac{5}{16}$	367	274	205	130
$\frac{3}{8}$	421	314	235	149
$\frac{7}{16}$	473	353	264	167
$\frac{1}{2}$	522	390	291	185
$\frac{5}{8}$	618	461	344	218
$\frac{3}{4}$	708	528	395	250
$\frac{7}{8}$	795	593	443	281
1	878	656	490	311

Above values are for the species of lowest average specific gravity of each Group; i.e., 0.62 for Group I, 0.51 for Group II, 0.42 for Group III, and 0.31 for Group IV. For species of known higher specific gravity, the above values may be increased. See National Design Specification.

The above values are for withdrawal from side grain only. Withdrawal from end grain should be avoided.

Reprinted with permission of the National Forest Products Association, Washington, D.C.

Table C-10.   Modification for Number of Fasteners in a Row; for Wood Side Plates and Timber Connectors, Bolts, or Lag Screws.

$\dfrac{N_A}{N} = \dfrac{\text{Actual load per row of fasteners (pounds)}}{\text{Allowable load per fastener (pounds)}}$

A_1/A_2	A_1(inches^2)	2	3	4	5	6	7	8	9	10	11	12
0.5	<12	2.00	2.76	3.36	3.80	4.08	4.27	4.40	4.41	—	—	—
	12 – <19	2.00	2.85	3.52	4.10	4.50	4.76	4.96	5.13	5.20	5.28	—
	19 – <28	2.00	2.91	3.72	4.40	4.92	5.39	5.68	6.03	6.30	6.49	6.60
	28 – <40	2.00	2.94	3.84	4.60	5.22	5.81	6.32	6.75	7.10	7.59	7.92
	40 – <64	2.00	3.00	3.88	4.70	5.40	6.02	6.64	7.11	7.60	8.14	8.64
	≥64	2.00	3.00	3.92	4.75	5.46	6.16	6.80	7.38	8.00	8.58	9.12
1.0	<12	2.00	2.91	3.68	4.25	4.68	4.97	5.20	5.31	5.40	6.16	—
	12 – <19	2.00	2.94	3.76	4.45	5.04	5.46	5.76	5.94	6.10	6.16	—
	19 – <28	2.00	3.00	3.88	4.65	5.34	5.95	6.40	6.84	7.20	7.48	7.68
	28 – <40	2.00	3.00	3.96	4.80	5.52	6.23	6.88	7.47	8.00	8.58	9.00
	40 – <64	2.00	3.00	4.00	4.85	5.64	6.37	7.04	7.65	8.40	9.02	9.60
	≥64	2.00	3.00	4.00	4.95	5.76	6.51	7.28	7.92	8.70	9.46	10.20
Number of fasteners required per row		2	3	4	5	6	7	8	9	10	11	12

A_1 = Cross-sectional area of main member(s) before boring or grooving.

A_2 = Sum of the cross-sectional areas of side members before boring or grooving

Note: For A_1/A_2 between 0 and 1.0, interpolate or extrapolate from the tabulated values. When A_1/A_2 exceeds 1.0, use A_2 instead of A_1, and A_2/A_1 instead of A_1/A_2.

Source: Courtesy of Western Wood Products Association.

The tabulated value above, divided by the number of fasteners required per row corresponds to the modification factor given by a similar table in the NDS. The NDS factor, when multiplied by the allowable load for a single fastener, gives a modified allowable for one fastener in the multiple-fastener row.

Table C-11. Modification for Number of Fasteners in a Row; for Metal Side Plates and Timber Connectors, Bolts, or Lag Screws.

A_1/A_2	A_1(inches²)	$\dfrac{N_A}{N} = \dfrac{\text{Actual load per row of fasteners (pounds)}}{\text{Allowable load per fastener (pounds)}}$										
	5 – <8	2.00	2.34	2.56	2.70	2.76	2.80	2.80	2.70	2.50	2.20	1.80
	8 – <16	2.00	2.55	2.92	3.15	3.24	3.36	3.36	3.42	3.40	3.30	3.12
	16 – <24	2.00	2.73	3.32	3.70	3.96	4.13	4.24	4.32	4.30	4.18	3.96
2–12	24 – <39	2.00	2.82	3.48	4.00	4.38	4.69	4.88	5.04	5.10	5.06	5.04
	39 – <64	2.00	2.88	3.68	4.35	4.86	5.25	5.60	5.94	6.20	6.38	6.60
	64 – <119	2.00	2.94	3.80	4.55	5.22	5.74	6.24	6.75	7.20	7.59	7.92
	119 – <199	2.00	2.97	3.88	4.75	5.52	6.23	6.88	7.56	8.10	8.69	9.36
	17 – <24	2.00	2.82	3.52	4.05	4.44	4.69	4.88	4.95	4.90	4.73	4.44
	24 – <39	2.00	2.88	3.64	4.30	4.80	5.18	5.44	5.58	5.60	5.50	5.28
12–18	39 – <64	2.00	2.94	3.76	4.50	5.10	5.60	6.00	6.30	6.70	6.82	6.96
	64 – <119	2.00	2.97	3.84	4.65	5.40	6.02	6.56	7.11	7.50	7.92	8.28
	119 – <199	2.00	3.00	3.92	4.80	5.64	6.44	7.12	7.74	8.30	8.80	9.36
	>199	2.00	3.00	4.00	4.90	5.82	6.65	7.44	8.19	9.00	9.68	10.44
	40 – <64	2.00	3.00	3.84	4.65	5.34	5.88	6.32	6.66	6.90	7.04	7.08
	64 – <119	2.00	3.00	3.88	4.70	5.52	6.23	6.88	7.47	8.00	8.36	8.76
18–24	119 – <199	2.00	3.00	3.96	4.90	5.76	6.58	7.36	8.10	8.80	9.46	10.20
	>199	2.00	3.00	4.00	5.00	5.88	6.72	7.60	8.37	9.20	10.12	10.92
	40 – <64	2.00	2.94	3.76	4.50	5.10	5.60	5.92	6.21	6.50	6.71	6.96
24–30	64 – <119	2.00	2.97	3.88	4.65	5.40	6.02	6.56	7.11	7.60	8.03	8.52
	119 – <199	2.00	3.00	3.92	4.80	5.64	6.44	7.12	7.83	8:50	9.13	9.72
	>199	2.00	3.00	3.96	4.90	5.82	6.65	7.44	8.28	9.00	9.79	10.68
	40 – <64	2.00	2.88	3.68	4.30	4.80	5.18	5.44	5.76	6.00	6.27	6.60
30–35	64 – <119	2.00	2.94	3.80	4.50	5.16	5.67	6.08	6.48	6.80	7.15	7.44
	119 – <199	2.00	2.97	3.88	4.75	5.52	6.16	6.80	7.38	8.00	8.58	9.24
	>199	2.00	3.00	3.92	4.85	5.70	6.51	7.20	8.01	8.70	9.46	10.20
	40 – <64	2.00	2.85	3.56	4.10	4.50	4.83	5.04	5.22	5.30	5.39	5.52
35–42	64 – <119	2.00	2.91	3.72	4.40	4.92	5.39	5.68	6.03	6.30	6.49	6.72
	119 – <199	2.00	2.94	3.84	4.65	5.34	5.95	6.48	7.02	7.60	8.03	8.52
	>199	2.00	2.97	3.92	4.80	5.58	6.30	6.96	7.56	8.20	8.80	9.36
Number of fasteners required per row		2	3	4	5	6	7	8	9	10	11	12

A_1 = Cross-sectional area of main member(s) before boring or grooving.
A_2 = Sum of cross-sectional areas of metal side plates before drilling.
Courtesy of Western Wood Products Association.
The tabulated value above, divided by the number of fasteners required per row corresponds to the modification factor given by a similar table in the NDS. The NDS factor, when multiplied by the allowable load for a single fastener, gives a modified allowable for one fastener in the multiple-fastener row.

Table C-12.   Design Allowable Loads for Bolts.

Allowable load (pounds) on one bolt of a standard double-shear connection.

Standard double-shear connection has wood side members one-half as thick as main member, as shown at right.

Main member thickness (in.) t	Bolt diam. (in.)	Species							
		* D.fir- larch		D. Fir- south		Hem- fir		Lodgepole pine	
		P	Q	P	Q	P	Q	P	Q
1½″	½	940	430	870	370	800	270	680	280
	5/8	1180	490	1090	420	1000	310	850	320
	3/4	1420	540	1310	470	1200	350	1020	350
	7/8	1660	600	1530	520	1400	380	1190	390
	1	1890	650	1750	570	1600	420	1360	420
2½″	½	1260	720	1290	620	1180	460	1080	470
	5/8	1820	810	1780	710	1620	520	1410	530
	3/4	2310	900	2180	790	1990	580	1700	590
	7/8	2740	990	2550	860	2330	630	1980	650
	1	3150	1080	2920	940	2670	690	2260	700
3½″	½	1270	980	1330	850	1210	640	1160	650
	5/8	1980	1130	2060	990	1890	720	1790	740
	3/4	2800	1260	2820	1100	2570	810	2310	820
	7/8	3580	1390	3480	1210	3180	890	2760	900
	1	4270	1520	4050	1320	3710	970	3170	990
4½″	5/8	1990	1400	2070	1220	1900	930	1820	950
	3/4	2860	1620	2980	1410	2730	1040	2610	1060
	7/8	3880	1790	3980	1560	3630	1140	3360	1160
	1	4930	1950	4920	1700	4500	1250	3990	1270
	1¼	6810	2280	6490	1990	5930	1460	5080	1490
5½″	5/8	1990	1410	2080	1220	1900	1010	1820	1030
	3/4	2860	1880	2990	1630	2730	1260	2620	1280
	7/8	3900	2180	4070	1900	3720	1400	3560	1420
	1	5070	2380	5270	2080	4820	1520	4550	1550
	1¼	7630	2790	7580	2430	6930	1780	6110	1820
7½″	5/8	1990	1260	2070	1100	1890	950	1820	960
	3/4	2860	1820	2990	1580	2730	1320	2620	1340
	7/8	3900	2420	4060	2110	3720	1730	3560	1760
	1	5080	3030	5300	2640	4850	2060	4650	2100
	1¼	7950	3800	8290	3310	7580	2430	7260	2480
9½″	3/4	2860	1640	2990	1430	2730	1250	2620	1270
	7/8	3890	2270	4070	1980	3720	1660	3560	1690
	1	5080	2960	5310	2580	4850	2130	4650	2170
	1¼	7950	4450	8290	3870	7580	3030	7270	3090
	1½	11470	5520	11960	4810	10930	3540	10470	3610

Allowable loads above are for normal load duration.
P is allowable load parallel to grain.
Q is allowable load perpendicular to grain.
For other angles to grain, use Hankinson's formula.
Tables of allowable bolt loads differ from one agency to another. The designer should use the values permitted by the applicable building code.
*Allowable bolt loads for Calif. redwood (close grain) and for Southern pine are the same as for D. fir-larch.

Reprinted with permission of the National Forest Products Association, Washington, D.C.

Table C-13. Withdrawal Resistance of Wood Screws in Side Grain.

Size or Gauge of Wood Screw	Diameter, inches	Allowable Withdrawal Load, lbs. per in. of Penetration			
		Group I	Group II	Group III	Group IV
6	.138	151	102	69	37
7	.151	165	112	76	41
8	.164	179	121	82	46
9	.177	193	131	89	48
10	.190	208	141	93	51
12	.216	236	160	108	60
14	.242	264	179	121	65
16	.268	293	198	134	74
18	.294	321	217	147	80
20	.320	350	237	161	88
24	.372	407	275	187	102

Courtesy of Western Wood Products Association.
Allowable loads shown above are for specific gravities as follows: Group I, 0.62; Group II, 0.51; Group III, 0.42; and Group IV, 0.31. For woods of higher specific gravities, see NDS.

Table C-14. Allowable Lateral Loads for Wood Screws in Side Grain.

Size or Gauge of Wood Screw	Diameter, inches	Allowable Lateral Load in pounds per Screw							
		Group I Side Plate		Group II Side plate		Group III Side Plate		Group IV Side Plate	
		Wood	Steel	Wood	Steel	Wood	Steal	Wood	Steel
6	.138	91	114	75	94	62	77	48	60
7	.151	109	137	90	113	74	92	57	72
8	.164	129	161	107	133	87	109	68	85
9	.177	150	188	124	155	102	127	79	99
10	.190	173	217	143	179	117	146	91	114
12	.216	224	280	185	231	151	189	118	147
14	.242	281	351	232	290	190	237	148	184
16	.268	345	431	284	356	233	291	181	226
18	.294	415	519	342	428	280	350	218	272
20	.320	492	614	406	507	332	415	258	323
24	.372	664	830	548	685	448	560	349	436

Courtesy of Western Wood Products Association.

Table C-15.   Design Allowable Loads for Shear Plates.

Wood piece min. actual thickness (in.)	Faces with plate on same bolt		Allowable load (lb) Species Grouping				Edge dist. (min.) Load direction Parallel to grain	Normal to grain
			A	B	C	D		
			Per $2\frac{5}{8}''$ shear unit (see note 2 below) with $\frac{3}{4}''$ bolt					
$1\frac{1}{2}$	1	P	(3110)	2670	2220	2010	$1\frac{1}{4}$	
		Q	1810	1550	1290	1110		$1\frac{1}{4}$ min.
		Q	2170	1860	1550	1330		$2\frac{3}{4}$*
$1\frac{1}{2}$	2	P	2420	2080	1730	1500	$1\frac{1}{4}$	
		Q	1410	1210	1010	870		$1\frac{1}{4}$ min.
		Q	1690	1450	1210	1040		$2\frac{3}{4}$ *
2	2	P	(3190)	2730	2270	1960	$1\frac{1}{4}$	
		Q	1850	1590	1320	1140		$1\frac{1}{4}$ min.
		Q	2220	1910	1580	1370		$2\frac{3}{4}$ *
$2\frac{1}{2}$	2	P	(3330)	2860	2380	2060	$1\frac{1}{4}$	
		Q	1940	1660	1380	1200		$1\frac{1}{4}$ min.
		Q	2320	1990	1650	1440		$2\frac{3}{4}$ *
			Per $4''$ shear unit (see note 2, below) with $\frac{3}{4}''$ or $\frac{7}{8}''$ bolt					
$1\frac{1}{2}$	1	P	4370	3750	3130	2700	$2\frac{3}{4}$	
		Q	2540	2180	1810	1550		$2\frac{3}{4}$ min.
		Q	3040	2620	2170	1860		$3\frac{3}{4}$ *
$1\frac{3}{4}$	1	P	(5090)	4360	3640	3140	$2\frac{3}{4}$	
		Q	2950	2530	2110	1810		$2\frac{3}{4}$ min.
		Q	3540	3040	2530	2200		$3\frac{3}{4}$ *
$1\frac{3}{4}$	2	P	3390	2910	2420	2090	$2\frac{3}{4}$	
		Q	1970	1680	1400	1250		$2\frac{3}{4}$ min.
		Q	2360	2020	1680	1410		$3\frac{3}{4}$ *
2	2	P	3790	3240	2700	2330	$2\frac{3}{4}$	
		Q	2200	1880	1570	1360		$2\frac{3}{4}$ min.
		Q	2640	2260	1880	1630		$3\frac{3}{4}$ *
$2\frac{1}{2}$	2	P	4310	3690	3080	2660	$2\frac{3}{4}$	
		Q	2500	2140	1780	1540		$2\frac{3}{4}$ min.
		Q	3000	2550	2140	1850		$3\frac{3}{4}$ *
3	2	P	4830	4140	3450	2980	$2\frac{3}{4}$	
		Q	2800	2400	2000	1720		$2\frac{3}{4}$ min.
		Q	3360	2880	2400	2060		$3\frac{3}{4}$ *
$3\frac{1}{2}$	2	P	(5030)	4320	3600	3110	$2\frac{3}{4}$	
		Q	2920	2500	2090	1800		$2\frac{3}{4}$ min.
		Q	3500	3000	2510	2160		$3\frac{3}{4}$ *

1. P is the allowable load parallel to the grain, and Q is the allowable normal to the grain, both unmodified. Loads are for one shear unit, installed in seasoned wood that will remain dry in service, and are for normal load duration.
2. A shear unit consists of either of the following:
   (a) two shear plates, back-to-back, with a single bolt, connecting wood members; or
   (b) a single shear plate and bolt, connecting a wood member to a metal member.
3. For load normal to the grain, edge distance required for the *unloaded* edge is that shown as "min."
4. For load normal to the grain, required edge distance for the *loaded* edge depends on the level of load, Q. Edge distances for two levels of Q are shown. For levels of Q between those shown, the required edge distance for the loaded edge is obtained by straight-line interpolation between the min. value and that marked with *.
5. Loads in parentheses ( ) are included only for use in Hankinson's formula. The allowable P value is limited to the value based on strength of the metal, as in note 6, below.
6. Regardless of the value in the above table, the allowable load per shear unit (in any direction) shall not exceed
   2900 lb for a $2\frac{5}{8}''$ shear plate
   4400 lb for a $4''$ shear plate with $\frac{3}{4}''$ bolt
   6000 lb for a $4''$ shear plate with $\frac{7}{8}''$ bolt
7. For load directions between 0 and 90 degrees to the grain, use Hankinson's formula with the tabulated values of P and Q, above.

Reprinted with permission of the National Forest Products Association, Washington, D.C.

Table C-16.   Required Connector
Spacing and End Distance.

Required Spacing —
For $2\frac{1}{2}''$ split ring or $2\frac{5}{8}''$ shear plate

Load Parallel to Grain	100% load	50% load
Spacing parallel to grain	$6\frac{3}{4}''$	$3\frac{1}{2}''$(min.)
Spacing perpendicular to grain	$3\frac{1}{2}''$	$3\frac{1}{2}''$(min.)

Load Perpendicular to Grain		
Spacing parallel to grain	$3\frac{1}{2}''$	$3\frac{1}{2}''$(min.)
Spacing perpendicular to grain	$4\frac{1}{4}''$	$3\frac{1}{2}''$(min.)

Required Spacing —
For 4" split ring or 4" shear plate

Load Parallel to Grain	100% load	50% load
Spacing parallel to grain	9"	5"(min.)
Spacing perpendicular to grain	5" (min.)	5"(min.)

Load Perpendicular to Grain		
Spacing parallel to grain	5"	5"(min.)
Spacing perpendicular to grain	6"	5"(min.)

Required End Distance—
For $2\frac{1}{2}''$ split ring or $2\frac{5}{8}''$ shear plate

Load Parallel to Grain	100% load	62.5% load
Tension member	$5\frac{1}{2}''$	$2\frac{3}{4}''$(min.)
Compression member	4"	$2\frac{1}{2}''$(min.)

Load Perpendicular to Grain		
Tension or compr. member	$5\frac{1}{2}''$	$2\frac{3}{4}''$(min.)

Required End Distance —
For 4" split ring or 4" shear plate

Load Parallel to Grain		
Tension member	7"	$3\frac{1}{2}''$(min.)
Compression member	$5\frac{1}{2}''$	$3\frac{1}{4}''$(min.)

Load Perpendicular to Grain		
Tension or compr. member	7"	$3\frac{1}{2}''$(min.)

Table C-17.   Design Allowable Loads for Split Rings.

Minimum Wood Thickness	Rings in		Allowable load, lb (unmodified) Species grouping				Edge distance (min.) Load direction	
			A	B	C	D	Parallel to Grain	Normal to Grain (Loaded edge)
			Per $2\frac{1}{2}''$ diameter split ring with $\frac{1}{2}''$ dia bolt					
1″	One	P	2630	2270	1900	1640	$1\frac{3}{4}''$	
	Face	Q	1580	1350	1130	970		$1\frac{3}{4}''$ min.
			to	to	to	to		
		Q	1900	1620	1350	1160		$2\frac{3}{4}''$ or more
$1\frac{1}{2}''$	One	P	3160	2730	2290	1960	$1\frac{3}{4}''$	
	Face	Q	1900	1620	1350	1160		$1\frac{3}{4}''$ min.
			to	to	to	to		
		Q	2280	1940	1620	1390		$2\frac{3}{4}''$ or more
$1\frac{1}{2}''$	Both	P	2430	2100	1760	1510	$1\frac{3}{4}''$	
	faces	Q	1460	1250	1040	890		$1\frac{3}{4}''$ min.
			to	to	to	to		
		Q	1750	1500	1250	1070		$2\frac{3}{4}''$ or more
2″	Both	P	3160	2730	2290	1960	$1\frac{3}{4}''$	
	faces	Q	1900	1620	1350	1160		$1\frac{3}{4}''$ min.
			to	to	to	to		
		Q	2280	1940	1620	1390		$2\frac{3}{4}''$ or more
			Per 4″ diameter split ring with $\frac{3}{4}''$ dia bolt					
1″	One	P	4090	3510	2920	2520	$2\frac{3}{4}''$	
	Face	Q	2370	2030	1700	1470		$2\frac{3}{4}''$ min.
			to	to	to	to		
		Q	2840	2440	2040	1760		$3\frac{3}{4}''$ or more
$1\frac{1}{2}''$	One	P	6020	5160	4280	3710	$2\frac{3}{4}''$	
	Face	Q	3490	2990	2490	2150		$2\frac{3}{4}''$ min.
			to	to	to	to		
		Q	4180	3590	2990	2580		$3\frac{3}{4}''$ or more
$1\frac{1}{2}''$	Both	P	4110	3520	2940	2540	$2\frac{3}{4}''$	
	faces	Q	2480	2040	1700	1470		$2\frac{3}{4}''$ min.
			to	to	to	to		
		Q	2980	2450	2040	1760		$3\frac{3}{4}''$ or more
$1\frac{5}{8}''$	One	P	6140	5260	4380	3790	$2\frac{3}{4}''$	
	face	Q	3560	3050	2540	2190		$2\frac{3}{4}''$ min.
			to	to	to	to		
		Q	4270	3660	3050	2630		$3\frac{3}{4}''$ or more
2″	Both	P	4950	4250	3540	3050	$2\frac{3}{4}''$	
	faces	Q	2870	2470	2050	1770		$2\frac{3}{4}''$ min.
			to	to	to	to		
		Q	3440	2960	2460	2120		$3\frac{3}{4}''$ or more
$2\frac{1}{2}''$	Both	P	5830	5000	4160	3600	$2\frac{3}{4}''$	
	faces	Q	3380	2900	2410	2080		$2\frac{3}{4}''$ min.
			to	to	to	to		
		Q	4050	3480	2890	2500		$3\frac{3}{4}''$ or more
3″	Both	P	6140	5260	4380	3790	$2\frac{3}{4}''$	
	faces	Q	3560	3050	2540	2190		$2\frac{3}{4}''$ min.
			to	to	to	to		
		Q	4270	3660	3050	2630		$3\frac{3}{4}''$ or more

P is the allowable load parallel to the grain.
Q is the allowable load normal to the grain.
For intermediate loaded edge distances, interpolate to obtain allowable Q.
For load-to-grain angles between 0 and 90 degrees, use Hankinson's formula.
All values above are for one ring, installed in seasoned wood that will remain dry in service. Allowables are for normal load duration.

Reprinted with permission of the National Forest Products Association, Washington, D.C.

Table C-18.   Design and Load Data for TECO Spike-Grid Connectors.

TECO SPIKE-GRID	Flat	Single curve	Circular
Type	Flat	Single curve	Circular
Size, square	$4\frac{1}{8}''$	$4\frac{1}{8}''$	$3\frac{1}{4}''$
Total depth of grids, maximum	$1''$	$1.38''$	$1.20''$
Diameter of bolt hole	$1.06''$	$1.06''$	$1.33''$
Weight, per 100 grids, lbs	50	75	26
LUMBER DIMENSIONS, minimum recommended for installation of flat grids			
Face width	$5\frac{1}{2}''$	$5\frac{1}{2}''$	$5\frac{1}{2}''$
Thickness			
Grids one face only	$1\frac{1}{2}''$	$1\frac{1}{2}''$	$1\frac{1}{2}''$
Grids opposite in both faces	$2\frac{1}{2}''$	—	$2\frac{1}{2}''$
Minimum diameter of pile for curved grids	—	$10''$	—
BOLT, diameter	$\frac{3}{4}''$ or $1''$	$\frac{3}{4}''$ or $1''$	$\frac{3}{4}''$ or $1''$
BOLT HOLE, diameter in timber	$\frac{13}{16}''$ or $1\frac{1}{16}''$	$\frac{13}{16}''$ or $1\frac{1}{16}''$	$\frac{13}{16}''$ or $1\frac{1}{16}''$
WASHERS			
Round, cast or malleable iron		Standard Size for Bolt Diameter Used.	
Square plate		$3'' \times 3'' \times \frac{3}{8}''$ Punched for Bolt Diameter Used.	
SPACING OF GRIDS, minimum, center to center			
0°–30° angle of load to grain			
Spacing parallel to grain	$7''$	$7''$	$7''$
Spacing perpendicular to grain	$5\frac{1}{2}''$	$5\frac{1}{2}''$	$5\frac{1}{2}''$
30°–90° angle of load to grain			
Spacing parallel or perpendicular to grain	$5\frac{1}{2}''$	$5\frac{1}{2}''$	$5\frac{1}{2}''$
END DISTANCES, center of grid to end of piece (tension or compression members)			
Standard	$7''$	$7''$	$7''$
Minimum, reduce loads 15%	$5''$	$5''$	$5''$
EDGE DISTANCES, center of grid to edge of piece			
Load applied at any angle to grain			
Standard	$3\frac{5}{8}''$	$3\frac{5}{8}''$	$3\frac{5}{8}''$
Minimum, reduce loads 15%	$2\frac{3}{4}''$	$2\frac{3}{4}''$	$2\frac{3}{4}''$
PROJECTED AREA for portion of one grid within member, square inches	2.06	2.06	1.95

Connector Load Grouping of Species When Structurally Graded

Connector Load Grouping

Group A Species	Group B Species	Group C Species
Douglas fir (dense)	Douglas fir (coast region)	Cypress, southern and tidewater red
Oak, red and white	Larch, western	Hemlock, west coast
Pine, southern (dense)	Pine, southern	Pine, Norway
		Redwood

Design Loads[1] for One Spike-Grid and Bolt in Single Shear

Group A			Group B			Group C		
Type of Grid	Bolt Diameter	Allowable Load	Type of Grid	Bolt Diameter	Allowable Load	Type of Grid	Bolt Diameter	Allowable Load
Flat	$\frac{3}{4}''$	3900#	Flat	$\frac{3}{4}''$	3500#	Flat	$\frac{3}{4}''$	3000#
	$1''$	4200#		$1''$	3800#		$1''$	3300#
Single Curve	$\frac{3}{4}''$	4200#	Single Curve	$\frac{3}{4}''$	3800#	Single Curve	$\frac{3}{4}''$	3200#
	$1''$	4500#		$1''$	4100#		$1''$	3500#
Circular	$\frac{3}{4}''$	3500#	Circular	$\frac{3}{4}''$	3100#	Circular	$\frac{3}{4}''$	2600#
	$1''$	3800#		$1''$	3400#		$1''$	2900#

[1] Allowable loads on spike-grids same for all angles of load to grain.
Source: Courtesy of TECO Products & Testing Corporation.

Index

Index